Rank-Based Methods for Shrinkage and Selection

Rank-Based Methods for Shrinkage and Selection

With Application to Machine Learning

A. K. Md. Ehsanes Saleh
Carleton University, Ottawa, Canada

Mohammad Arashi
Ferdowsi University of Mashhad, Mashhad, Iran

Resve A. Saleh
University of British Columbia, Vancouver, Canada

Mina Norouzirad
Center for Mathematics and Application of NOVA University Lisbon, Lisbon, Portugal

Registered Office
John Wiley & Sons, Inc., 111 River Street, Hoboken, NJ 07030, USA

Editorial Office
111 River Street, Hoboken, NJ 07030, USA

For details of our global editorial offices, customer services, and more information about Wiley products visit us at www.wiley.com.

Wiley also publishes its books in a variety of electronic formats and by print-on-demand. Some content that appears in standard print versions of this book may not be available in other formats.

Library of Congress Cataloging-in-Publication Data
Names: Saleh, A. K. Md. Ehsanes, author. | Arashi, M. (Mohammad), 1981- author. | Saleh, Resve A., 1957- author. | Norouzirad, Mina, author.
Title: Rank-based methods for shrinkage and selection : with application to machine learning / A.K. Md. Ehsanes Saleh, Mohammad Arashi, Resve A. Saleh, Mina Norouzirad.
Description: Hoboken, NJ : John Wiley & Sons, Inc., 2022. | Includes bibliographical references and index.
Identifiers: LCCN 2021042428 (print) | LCCN 2021042429 (ebook) | ISBN 9781119625391 (hardback) | ISBN 9781119625414 (pdf) | ISBN 9781119625421 (epub) | ISBN 9781119625438 (ebook)
Subjects: LCSH: Regression analysis. | Big data. | Machine learning.
Classification: LCC QA278.2 .R253 2022 (print) | LCC QA278.2 (ebook) | DDC 519.5/36–dc23/eng/20211105
LC record available at https://lccn.loc.gov/2021042428
LC ebook record available at https://lccn.loc.gov/2021042429

Cover image: Courtesy of Resve A. Saleh; s_maria/Shutterstock
Cover design by Wiley

Set in 9.5/12.5pt STIXTwoText by Integra Software Services Pvt. Ltd, Pondicherry, India

10 9 8 7 6 5 4 3 2 1

We dedicate this book to

Shahidara Saleh

Reihaneh Soleimani, Elena Arashi

Lynn Hilchie Saleh

Abbas Ali Norouzirad, Fereshteh Arefian

Contents in Brief

List of Figures *xvii*
List of Tables *xxi*
Foreword *xxv*
Preface *xxvii*

1 Introduction to Rank-based Regression *1*

2 Characteristics of Rank-based Penalty Estimators *47*

3 Location and Simple Linear Models *101*

4 Analysis of Variance (ANOVA) *149*

5 Seemingly Unrelated Simple Linear Models *191*

6 Multiple Linear Regression Models *215*

7 Partially Linear Multiple Regression Model *241*

8 Liu Regression Models *263*

9 Autoregressive Models *291*

10 High-Dimensional Models *307*

11 Rank-based Logistic Regression *329*

12 Rank-based Neural Networks *377*

Bibliography *433*
Author Index *443*
Subject Index *445*

Contents

List of Figures *xvii*
List of Tables *xxi*
Foreword *xxv*
Preface *xxvii*

1 Introduction to Rank-based Regression *1*
1.1 Introduction *1*
1.2 Robustness of the Median *1*
1.2.1 Mean vs. Median *1*
1.2.2 Breakdown Point *4*
1.2.3 Order and Rank Statistics *5*
1.3 Simple Linear Regression *6*
1.3.1 Least Squares Estimator (LSE) *6*
1.3.2 Theil's Estimator *7*
1.3.3 Belgium Telephone Data Set *7*
1.3.4 Estimation and Standard Error Comparison *9*
1.4 Outliers and their Detection *11*
1.4.1 Outlier Detection *12*
1.5 Motivation for Rank-based Methods *13*
1.5.1 Effect of a Single Outlier *13*
1.5.2 Using Rank for the Location Model *16*
1.5.3 Using Rank for the Slope *19*
1.6 The Rank Dispersion Function *20*
1.6.1 Ranking and Scoring Details *23*
1.6.2 Detailed Procedure for R-estimation *25*
1.7 Shrinkage Estimation and Subset Selection *30*
1.7.1 Multiple Linear Regression using Rank *30*
1.7.2 Penalty Functions *32*
1.7.3 Shrinkage Estimation *34*

1.7.4 Subset Selection *36*
1.7.5 Blended Approaches *39*
1.8 Summary *39*
1.9 Problems *41*

2 Characteristics of Rank-based Penalty Estimators *47*
2.1 Introduction *47*
2.2 Motivation for Penalty Estimators *47*
2.3 Multivariate Linear Regression *49*
2.3.1 Multivariate Least Squares Estimation *49*
2.3.2 Multivariate R-estimation *51*
2.3.3 Multicollinearity *51*
2.4 Ridge Regression *53*
2.4.1 Ridge Applied to Least Squares Estimation *53*
2.4.2 Ridge Applied to Rank Estimation *55*
2.5 Example: Swiss Fertility Data Set *56*
2.5.1 Estimation and Standard Errors *59*
2.5.2 Parameter Variance using Bootstrap *60*
2.5.3 Reducing Variance using Ridge *61*
2.5.4 Ridge Traces *62*
2.6 Selection of Ridge Parameter λ_2 *65*
2.6.1 Quadratic Risk *65*
2.6.2 K-fold Cross-validation Scheme *68*
2.7 LASSO and aLASSO *71*
2.7.1 Subset Selection *71*
2.7.2 Least Squares with LASSO *71*
2.7.3 The Adaptive LASSO and its Geometric Interpretation *73*
2.7.4 R-estimation with LASSO and aLASSO *77*
2.7.5 Oracle Properties *78*
2.8 Elastic Net (Enet) *82*
2.8.1 Naive Enet *82*
2.8.2 Standard Enet *83*
2.8.3 Enet in Machine Learning *84*
2.9 Example: Diabetes Data Set *85*
2.9.1 Model Building with R-aEnet *85*
2.9.2 MSE vs. MAE *88*
2.9.3 Model Building with LS-Enet *91*
2.10 Summary *94*
2.11 Problems *95*

3 Location and Simple Linear Models *101*
3.1 Introduction *101*

3.2 Location Estimators and Testing *104*
3.2.1 Unrestricted R-estimator of θ *104*
3.2.2 Restricted R-estimator of θ *107*
3.3 Shrinkage R-estimators of Location *108*
3.3.1 Overview of Shrinkage R-estimators of θ *108*
3.3.2 Derivation of the Ridge-type R-estimator *113*
3.3.3 Derivation of the LASSO-type R-estimator *114*
3.3.4 General Shrinkage R-estimators of θ *114*
3.4 Ridge-type R-estimator of θ *117*
3.5 Preliminary Test R-estimator of θ *118*
3.5.1 Optimum Level of Significance of PTRE *121*
3.6 Saleh-type R-estimators *122*
3.6.1 Hard-Threshold R-estimator of θ *122*
3.6.2 Saleh-type R-estimator of θ *123*
3.6.3 Positive-rule Saleh-type (LASSO-type) R-estimator of θ *125*
3.6.4 Elastic Net-type R-estimator of θ *127*
3.7 Comparative Study of the R-estimators of Location *129*
3.8 Simple Linear Model *132*
3.8.1 Restricted R-estimator of Slope *134*
3.8.2 Shrinkage R-estimator of Slope *135*
3.8.3 Ridge-type R-estimation of Slope *135*
3.8.4 Hard-Threshold R-estimator of Slope *136*
3.8.5 Saleh-type R-estimator of Slope *137*
3.8.6 Positive-rule Saleh-type (LASSO-type) R-estimator of Slope *138*
3.8.7 The Adaptive LASSO (aLASSO-type) R-estimator *138*
3.8.8 nEnet-type R-estimator of Slope *139*
3.8.9 Comparative Study of R-estimators of Slope *140*
3.9 Summary *141*
3.10 Problems *142*

4 Analysis of Variance (ANOVA) *149*
4.1 Introduction *149*
4.2 Model, Estimation and Tests *149*
4.3 Overview of Multiple Location Models *150*
4.3.1 Example: Corn Fertilizers *151*
4.3.2 One-way ANOVA *151*
4.3.3 Effect of Variance on Shrinkage Estimators *153*
4.3.4 Shrinkage Estimators for Multiple Location *156*
4.4 Unrestricted R-estimator *158*
4.5 Test of Significance *161*
4.6 Restricted R-estimator *162*

4.7 Shrinkage Estimators *163*
4.7.1 Preliminary Test R-estimator *163*
4.7.2 The Stein–Saleh-type R-estimator *164*
4.7.3 The Positive-rule Stein–Saleh-type R-estimator *165*
4.7.4 The Ridge-type R-estimator *167*
4.8 Subset Selection Penalty R-estimators *169*
4.8.1 Preliminary Test Subset Selector R-estimator *169*
4.8.2 Saleh-type R-estimator *170*
4.8.3 Positive-rule Saleh Subset Selector (PRSS) *171*
4.8.4 The Adaptive LASSO (aLASSO) *173*
4.8.5 Elastic-net-type R-estimator *177*
4.9 Comparison of the R-estimators *178*
4.9.1 Comparison of URE and RRE *179*
4.9.2 Comparison of URE and Stein–Saleh-type R-estimators *179*
4.9.3 Comparison of URE and Ridge-type R-estimators *179*
4.9.4 Comparison of URE and PTSSRE *180*
4.9.5 Comparison of LASSO-type and Ridge-type R-estimators *180*
4.9.6 Comparison of URE, RRE and LASSO *181*
4.9.7 Comparison of LASSO with PTRE *181*
4.9.8 Comparison of LASSO with SSRE *182*
4.9.9 Comparison of LASSO with PRSSRE *182*
4.9.10 Comparison of nEnetRE with URE *183*
4.9.11 Comparison of nEnetRE with RRE *183*
4.9.12 Comparison of nEnetRE with HTRE *183*
4.9.13 Comparison of nEnetRE with SSRE *184*
4.9.14 Comparison of Ridge-type vs. nEnetRE *184*
4.10 Summary *185*
4.11 Problems *185*

5 Seemingly Unrelated Simple Linear Models *191*
5.1 Introduction *191*
5.1.1 Problem Formulation *193*
5.2 Signed and Signed Rank Estimators of Parameters *194*
5.2.1 General Shrinkage R-estimator of β *198*
5.2.2 Ridge-type R-estimator of β *199*
5.2.3 Preliminary Test R-estimator of β *201*
5.3 Stein–Saleh-type R-estimator of β *202*
5.3.1 Positive-rule Stein–Saleh R-estimators of β *202*
5.4 Saleh-type R-estimator of β *203*
5.4.1 LASSO-type R-estimator of the β *205*
5.5 Elastic-net-type R-estimators *206*

5.6 R-estimator of Intercept When Slope Has Sparse Subset *207*
5.6.1 General Shrinkage R-estimator of Intercept *207*
5.6.2 Ridge-type R-estimator of θ *209*
5.6.3 Preliminary Test R-estimators of θ *209*
5.7 Stein–Saleh-type R-estimator of θ *210*
5.7.1 Positive-rule Stein–Saleh-type R-estimator of θ *211*
5.7.2 LASSO-type R-estimator of θ *213*
5.8 Summary *213*
5.8.1 Problems *214*

6 Multiple Linear Regression Models *215*
6.1 Introduction *215*
6.2 Multiple Linear Model and R-estimation *215*
6.3 Model Sparsity and Detection *218*
6.4 General Shrinkage R-estimator of β *221*
6.4.1 Preliminary Test R-estimator *222*
6.4.2 Stein–Saleh-type R-estimator *224*
6.4.3 Positive-rule Stein–Saleh-type R-estimator *225*
6.5 Subset Selectors *226*
6.5.1 Preliminary Test Subset Selector R-estimator *226*
6.5.2 Stein–Saleh-type R-estimator *228*
6.5.3 Positive-rule Stein–Saleh-type R-estimator (LASSO-type) *229*
6.5.4 Ridge-type Subset Selector *231*
6.5.5 Elastic Net-type R-estimator *231*
6.6 Adaptive LASSO *232*
6.6.1 Introduction *232*
6.6.2 Asymptotics for LASSO-type R-estimator *233*
6.6.3 Oracle Property of aLASSO *235*
6.7 Summary *238*
6.8 Problems *239*

7 Partially Linear Multiple Regression Model *241*
7.1 Introduction *241*
7.2 Rank Estimation in the PLM *242*
7.2.1 Penalty R-estimators *246*
7.2.2 Preliminary Test and Stein–Saleh-type R-estimator *248*
7.3 ADB and ADL$_2$-risk *249*
7.4 ADL$_2$-risk Comparisons *253*
7.5 Summary: L_2-risk Efficiencies *260*
7.6 Problems *262*

8 Liu Regression Models *263*
8.1 Introduction *263*
8.2 Linear Unified (Liu) Estimator *263*
8.2.1 Liu-type R-estimator *266*
8.3 Shrinkage Liu-type R-estimators *268*
8.4 Asymptotic Distributional Risk *269*
8.5 Asymptotic Distributional Risk Comparisons *271*
8.5.1 Comparison of SSLRE and PTLRE *272*
8.5.2 Comparison of PRSLRE and PTLRE *274*
8.5.3 Comparison of PRLRE and SSLRE *276*
8.5.4 Comparison of Liu-Type Rank Estimators With Counterparts *277*
8.6 Estimation of d *279*
8.7 Diabetes Data Analysis *280*
8.7.1 Penalty Estimators *281*
8.7.2 Performance Analysis *284*
8.8 Summary *288*
8.9 Problems *288*

9 Autoregressive Models *291*
9.1 Introduction *291*
9.2 R-estimation of ρ for the AR(p)-Model *292*
9.3 LASSO, Ridge, Preliminary Test and Stein–Saleh-type R-estimators *294*
9.4 Asymptotic Distributional L_2-risk *296*
9.5 Asymptotic Distributional L_2-risk Analysis *299*
9.5.1 Comparison of Unrestricted vs. Restricted R-estimators *300*
9.5.2 Comparison of Unrestricted vs. Preliminary Test R-estimator *300*
9.5.3 Comparison of Unrestricted vs. Stein–Saleh-type R-estimators *300*
9.5.4 Comparison of the Preliminary Test vs. Stein–Saleh-type R-estimators *302*
9.6 Summary *303*
9.7 Problems *304*

10 High-Dimensional Models *307*
10.1 Introduction *307*
10.2 Identifiability of β^* and Projection *309*
10.3 Parsimonious Model Selection *309*
10.4 Some Notation and Separation *311*
10.4.1 Special Matrices *311*
10.4.2 Steps Towards Estimators *312*
10.4.3 Post-selection Ridge Estimation of $\beta^*_{S_1}$ and $\boldsymbol{\beta}^*_{S_2}$ *312*

10.4.4 Post-selection Ridge R-estimators for $\beta^*_{S_1}$ and $\beta^*_{S_2}$ 313
10.5 Post-selection Shrinkage R-estimators 315
10.6 Asymptotic Properties of the Ridge R-estimators 316
10.7 Asymptotic Distributional L_2-Risk Properties 321
10.8 Asymptotic Distributional Risk Efficiency 324
10.9 Summary 326
10.10 Problems 327

11 Rank-based Logistic Regression 329
11.1 Introduction 329
11.2 Data Science and Machine Learning 329
11.2.1 What is Robust Data Science? 329
11.2.2 What is Robust Machine Learning? 332
11.3 Logistic Regression 333
11.3.1 Log-likelihood Setup 334
11.3.2 Motivation for Rank-based Logistic Methods 338
11.3.3 Nonlinear Dispersion Function 341
11.4 Application to Machine Learning 342
11.4.1 Example: Motor Trend Cars 344
11.5 Penalized Logistic Regression 347
11.5.1 Log-likelihood Expressions 347
11.5.2 Rank-based Expressions 348
11.5.3 Support Vector Machines 349
11.5.4 Example: Circular Data 353
11.6 Example: Titanic Data Set 359
11.6.1 Exploratory Data Analysis 359
11.6.2 RLR vs. LLR vs. SVM 365
11.6.3 Shrinkage and Selection 367
11.7 Summary 370
11.8 Problems 371

12 Rank-based Neural Networks 377
12.1 Introduction 377
12.2 Set-up for Neural Networks 379
12.3 Implementing Neural Networks 381
12.3.1 Basic Computational Unit 382
12.3.2 Activation Functions 382
12.3.3 Four-layer Neural Network 384
12.4 Gradient Descent with Momentum 386
12.4.1 Gradient Descent 386
12.4.2 Momentum 388

12.5 Back Propagation Example *389*
12.5.1 Forward Propagation *390*
12.5.2 Back Propagation *392*
12.5.3 Dispersion Function Gradients *394*
12.5.4 RNN Algorithm *395*
12.6 Accuracy Metrics *396*
12.7 Example: Circular Data Set *400*
12.8 Image Recognition: Cats vs. Dogs *405*
12.8.1 Binary Image Classification *406*
12.8.2 Image Preparation *406*
12.8.3 Over-fitting and Under-fitting *409*
12.8.4 Comparison of LNN vs. RNN *410*
12.9 Image Recognition: MNIST Data Set *414*
12.10 Summary *421*
12.11 Problems *421*

Bibliography *433*
Author Index *443*
Subject Index *445*

List of Figures

1.1 Four plots using different versions of the telephone data set with fitted lines. 8
1.2 Histograms and ordered residual plots of LS and Theil estimators. 12
1.3 Effect of a single outlier on LS and rank estimators. 14
1.4 Gradients of absolute value ($B_n'(\theta)$) and dispersion ($D_n'(\theta)$) functions. 18
1.5 Scoring functions $\phi(u) = \sqrt{12}(u - 0.5)$ and $\phi^+(u) = \sqrt{3}u$. 25
1.6 Dispersion functions and derivative plots for Figure 1.1(d). 28
1.7 Key shrinkage characteristics of LASSO and ridge. 34
1.8 Geometric interpretation of ridge. 35
1.9 Geometric interpretation of LASSO. 37
2.1 The first-order nature of shrinkage due to ridge. 56
2.2 Two outliers found in the Q–Q plot for the Swiss data set. 61
2.3 Sampling distributions of rank estimates. 62
2.4 Shrinkage of β_5 due to increase in ridge tuning parameter, λ_2. 63
2.5 Ridge traces for orthonormal, diagonal, LS, and rank estimators ($m = 40$). 64
2.6 MSE Derivative plot to find optimal λ_2 for the diagonal case. 68
2.7 Bias, variance and MSE for the Swiss data set (optimal $\lambda_2 = 70.8$). 69
2.8 MSE for training, CV and test sets, and coefficients from the ridge trace. 70
2.9 The first-order nature of shrinkage due to LASSO. 74
2.10 Diamond-warping effect of weights in the aLASSO estimator for $p = 2$. 75
2.11 Comparison of LASSO and aLASSO traces for the Swiss data set. 78
2.12 Variable ordering from R-LASSO and R-aLASSO traces for the Swiss data set. 79

2.13 Ranked residuals of the diabetes data set. (Source: Rfit() package in R.) 86
2.14 Rank-aLASSO trace of the diabetes data set showing variable importance. 87
2.15 Diabetes data set showing variable ordering and adjusted R^2 plot. 87
2.16 Rank-aLASSO cleaning followed by rank-ridge estimation. 89
2.17 R-ridge traces and CV scheme with optimal λ_2. 91
2.18 MSE and MAE plots for five-fold CV scheme producing similar optimal λ_2. 92
2.19 LS-Enet traces for $\alpha = 0.0, 0.2, 0.4, 0.8, 1.0$. 93
2.20 LS-Enet traces and five-fold CV results for $\alpha = 0.6$ from glmnet(). 93
3.1 Key shrinkage R-estimators to be considered. 109
3.2 The ADRE of the shrinkage R-estimator using the optimal c and URE. 116
3.3 The ADRE of the preliminary test (or hard threshold) R-estimator for different Δ^2 based on $\lambda^* = \sqrt{2\ln(2)}$. 124
3.4 The ADRE of nEnet R-estimators. 128
3.5 Figure of the ADRE of all R-estimators for different Δ^2. 129
4.1 Boxplot and Q–Q plot using ANOVA table data. 153
4.2 LS-ridge and ridge R traces for fertilizer problem from ANOVA table data. 154
4.3 LS-LASSO and LASSOR traces for the fertilizer problem from the ANOVA table data. 155
4.4 Effect of variance on shrinkage using ridge and LASSO traces. 156
4.5 Hard threshold and positive-rule Stein–Saleh traces for ANOVA table data. 157
8.1 Left: the Q-Q plot for the diabetes data sets; Right: the distribution of the residuals. 281
11.1 Sigmoid function. 335
11.2 Outlier in the context of logistic regression. 339
11.3 LLR vs. RLR with one outlier. 345
11.4 LLR vs. RLR with no outliers. 346
11.5 LLR vs. RLR with two outliers. 347
11.6 Binary classification – nonlinear decision boundary. 353
11.7 Binary classification comparison – nonlinear boundary. 354
11.8 Ridge comparison of number of correct solutions with $n = 337$. 356
11.9 LLR-ridge regularization showing the shrinking decision boundary. 356
11.10 LLR, RLR and SVM on the circular data set with mixed outliers. 358
11.11 Histogram of passengers: (a) age and (b) fare. 361

11.12 Histogram of residuals associated with the null, LLR, RLR, and SVM cases for the Titanic data set. SVM probabilities were extracted from the sklearn.svm package. 366
11.13 RLR-ridge trace for Titanic data set. 369
11.14 RLR-LASSO trace for the Titanic data set. 369
11.15 RLR-aLASSO trace for the Titanic data set. 370
12.1 Computational unit (neuron) for neural networks. 382
12.2 Sigmoid and relu activation functions. 383
12.3 Four-layer neural network. 385
12.4 Neural network example of back propagation. 391
12.5 Forward propagation matrix and vector operations. 392
12.6 ROC curve and random guess classifier line based on the RLR classifier on the Titanic data set of Chapter 11. 400
12.7 Neural network architecture for the circular data set. 402
12.8 LNN and RNN on the circular data set ($n = 337$) with nonlinear decision boundaries. 403
12.9 Convergence plots for LNN and RNN for the circular data set. 403
12.10 ROC plots for LNN and RNN for the circular data set. 405
12.11 Typical setup for supervised learning methods. The training set is used to build the model. 405
12.12 Examples from test data set with cat = 1, dog = 0. 407
12.13 Unrolling of an RGB image into a single vector. 407
12.14 Effect of over-fitting, under-fitting and regularization. 410
12.15 Convergence plots for LLN and RNN (test size = 35). 411
12.16 ROC plots for LLN and RNN (test size = 35). 412
12.17 Ten representative images from the MNIST data set. 415
12.18 LNN and RNN convergence traces – loss vs. iterations (×100). 417
12.19 Residue histograms for LNN (0 outliers) and RNN (50 outliers). 417
12.20 These are 49 potential outlier images reported by RNN. 418
12.21 LNN (0 outliers) and RNN (144 outliers) residue histograms. 419
12.22 LNN and RNN confusion matrices and MCC scores. 420

List of Tables

1.1 Comparison of mean and median on three data sets. 3
1.2 Examples comparing order and rank statistics. 5
1.3 Belgium telephone data set. 7
1.4 Comparison of LS and Theil estimations of Figures 1.1(a) and (d). 11
1.5 Walsh averages for the set {0.1, 1.2, 2.3, 3.4, 4.5, 5.0, 6.6, 7.7, 8.8, 9.9, 10.5}. 19
1.6 The individual terms that are summed in $D_n(\beta)$ and $L_n(\beta)$ for the telephone data set. 27
1.7 The terms that are summed in $D_n(\theta)$ and $L_n(\theta)$ for the telephone data set. 29
1.8 The LS and R estimations of slope and intercept for Figure 1.1 cases. 29
1.9 Interpretation of L_1/L_2 loss and penalty functions. 40
2.1 Swiss fertility data set. 57
2.2 Swiss fertility data set definitions. 59
2.3 Swiss fertility estimates and standard errors for least squares (LS) and rank (R). 59
2.4 Swiss data subset ordering using | t.value |. 79
2.5 Swiss data models with adjusted R^2 values. 80
2.6 Estimates with outliers from diabetes data before standardization. 88
2.7 Estimates. MSE and MAE for the diabetes data. 92
2.8 Enet estimates, training MSE and test MSE as a function of α for the diabetes data. 94
3.1 The ADRE values of ridge for different values of Δ^2. 117
3.2 Maximum and minimum guaranteed ADRE of the preliminary test R-estimator for different values of α. 122

3.3 The ADRE values of the Saleh-type R-estimator for $\lambda^*_{\max} = \sqrt{\frac{2}{\pi}}$ and different Δ^2. *125*
3.4 The ADRE values of the positive-rule Saleh R-estimator for $\lambda^*_{\max} = \sqrt{\frac{2}{\pi}}$ and different Δ^2. *127*
3.5 The ADRE of all R-estimators for different Δ^2. *130*
4.1 Table of (hypothetical) corn crop yield from six different fertilizers. *152*
4.2 Table of p-values from pairwise comparisons of fertilizers. *153*
8.1 The VIF values of the diabetes data set. *281*
8.2 Estimations for the diabetes data*. (The numbers in parentheses are the corresponding standard deviations). *286*
11.1 LLR algorithm. *343*
11.2 RLR algorithm. *343*
11.3 Car data set. *345*
11.4 Ridge accuracy vs. λ_2 with $n = 337$ (six outliers). *355*
11.5 RLR-LASSO estimates vs. λ_1 with number of correct predictions. *357*
11.6 Sample of Titanic training data. *360*
11.7 Specifications for the Titanic data set. *361*
11.8 Number of actual data entries in each column. *362*
11.9 Cross-tabulation of survivors based on sex. *363*
11.10 Cross-tabulation using Embarked for the Titanic data set. *363*
11.11 Sample of Titanic numerical training data. *364*
11.12 Number of correct predictions for Titanic training and test sets. *365*
11.13 Train/test set accuracy for LLR-ridge. Optimal value at (*). *367*
11.14 Train/test set accuracy for RLR-ridge. Optimal value at (*). *367*
11.15 Train/Test set accuracy for LLR-LASSO. Optimal value at (*). *368*
11.16 Train/test set accuracy for RLR-LASSO. Optimal value at (*). *368*
12.1 RNN-ridge algorithm. *397*
12.2 Interpretation of the confusion matrix. *398*
12.3 Confusion matrix for Titanic data sets using RLR (see Chapter 11). *398*
12.4 Number of correct predictions (percentages) and AUROC of LNN-ridge. *404*
12.5 Input (x_{ij}), output (y_i) and predicted values $\tilde{p}(\boldsymbol{x}_i)$ for the image classification problem. *408*
12.6 Confusion matrices for RNN and LNN (test size = 35). *412*
12.7 Accuracy metrics for RNN vs. LNN (test size = 35). *412*
12.8 Train/test set accuracy for LNN. F_1 score is associated with the test set. *413*
12.9 Train/test set accuracy for RNN. F_1 score is associated with the test set. *413*
12.10 Confusion matrices for RNN and LNN (test size = 700). *414*

12.11 Accuracy metrics for RNN vs. LNN (test size = 700). 414
12.12 MNIST training with 0 outliers. 417
12.13 MNIST training with 90 outliers. 420
12.14 MNIST training with 180 outliers. 420
12.15 MNIST training with 270 outliers. 421
12.16 Table of responses and probability outputs. 422

Foreword

It is my pleasure to write this foreword for Professor Saleh's latest book, "Rank-based Methods for Shrinkage and Selection with Application to Machine Learning". I have known Professor Saleh for many decades as a leader in Canadian statistics and looked forward to meeting him regularly at the Annual Meeting of the Statistical Society of Canada.

We are well into the golden age of probability and statistics with the emergence of data science and machine learning. Many decades ago, we could attract many bright students to this field of endeavor but today the interest is overwhelming. The connection between theoretical statistics and applied statistics is an important part of machine learning and data science. In order to engage fully in data science, one needs a solid understanding of both the theoretical and the practical aspects of probability and statistics.

The book is unique in presenting a comprehensive approach to inference in regression models based on ranks. It starts with the basics, which enables rapid understanding of the innovative ideas in the rest of the book. In addition to the more familiar aspects of rank-based methods such as comparisons among groups, linear regression, and time series, the authors show how many machine-learning tools can be made more robust via rank-based methods. Modern approaches to model selection, logistic regression, neural networks, elastic net, and penalized regression are studied through this lens. The work is presented clearly and concisely, and highlights many areas for further investigation.

Professor Saleh's wealth of experience with ridge regression, Stein's method, and preliminary test estimation informs the arc of the book, and he and his co-authors have built on his expertise to expand the application of rank-based approaches to modern big-data and high-dimensional settings. Careful attention is paid throughout to both theoretical rigor and engaging applications. The applications are made accessible through detailed discussion of computational methods and their implementation in the R and Python computing environments. Its broad coverage of topics, and careful attention to both theory and methods, ensures that this book will be an invaluable resource for students and researchers in statistics.

The authors have identified many areas of useful future research that could be pursued by graduate students and practitioners alike. In this regard, this book is an important contribution in the ongoing research towards robust data science.

Professor N. Reid
University of Toronto
June 2021

Preface

The objective of this book is to introduce the audience to the theory and application of robust statistical methodologies using rank-based methods. We present a number of new ideas and research directions in machine learning and statistical analysis that the reader can and should pursue in the future. We begin by noting that the well-known least squares and likelihood principles are traditional methods of estimation in machine learning and data science. One of the most widely read books is the *Introduction to Statistical Learning* (James et al., 2013) which describes these and other methods. However, it also properly identifies many of their shortcomings, especially in terms of robustness in the presence of outliers. Our book describes a number of novel ideas and concepts to resolve these problems, many of which are worthy of further investigation. Our goal is to motivate the interest of more researchers to pursue further activities in this field. We build on this motivation to carry out a rigorous mathematical analysis of rank-based penalty estimators.

From our point of view, outliers are present in almost all real-world data sets. They may be the result of human error, transmission error, measurement error or simply due to the nature of the data being collected. Whatever be the reason, we must first recognize that all data sets have some form of outliers and then build solutions based on this fact. Outliers may greatly affect the estimates and lead to poor prediction accuracy. As a result, operations such as data cleaning, outlier detection and robust regression methods are extremely important in building models that provide suitably accurate prediction capability. Here, we describe rank-based methods to address many such problems. Indeed, many researchers are now involved in these and other methods towards *robust data science.* Most of the methods and results presented in this book were derived from our implementations in *R* and *Python* which are languages used routinely by statisticians and by practitioners in machine learning and data science. Some of the problems at the end of each chapter involve the use of R. The reader will be well-served to follow the descriptions in the book while implementing the ideas wherever possible in R or Python. This is the best way to get the most out of this book.

Rank regression is based on the linear rank dispersion function described by Jaeckel (1972). The dispersion function replaces the least squares loss function to enable estimates based on the median rather than the mean. This book is intended to guide the reader in this direction starting with basic principles such as the importance of the median vs. the mean, comparisons of rank vs. least squares methods on simple linear problems, and the role of penalty functions in improving the accuracy of prediction. We present new practical methods of data cleaning, subset selection and shrinkage estimation in the context of rank-based methods. We then begin our theoretical journey starting with basic rank statistics for location and simple linear models, and then move on to multiple regression, ANOVA and problems in a high-dimensional setting. We conclude with new ideas not published elsewhere in the literature in the area of rank-based logistic regression and neural networks to address classification problems in machine learning and data science.

We believe that most practitioners today are still employing least squares and log-likelihood methods that are not robust in the presence of outliers. This is due to the long history of these estimation methods in statistics and their natural adoption in the machine learning community over the past two decades. However, the history of estimation theory actually changed its course radically many decades prior when Stein (1956) and James and Stein (1961) proved that the sample mean based on a sample from a p-dimensional multivariate normal distribution is inadmissible under a quadratic loss function for $p \geq 3$. This result gave birth to a class of shrinkage estimators in various forms and set-ups. Due to the immense impact of Stein's theory, scores of technical papers appeared in the literature covering many areas of application. Beginning in the 1970s, the pioneering work of Saleh and Sen (1978, 1983, 1984b,a, 1985a,a,b,c,d,e, 1986, 1987) expanded the scope of this class of shrinkage estimators using the "quasi-empirical Bayes" method to obtain robust (such as R-, L-, and M-estimation) Stein-type estimators. Details are provided in Saleh (2006).

Of particular interest here is the use of penalty estimators in the context of robust R-estimation. Next generation "shrinkage estimators" known as "ridge regression estimators" for the multiple linear regression model were developed by Hoerl and Kennard (1970) based on "Tikhonov's regularization" (Tikhonov, 1963). The ridge regression (RR) estimator is the result of minimizing the penalized least squares criterion using an L_2-penalty function. Ridge regression laid the foundation of penalty estimation. Later, Tibshirani (1996) proposed the "least absolute shrinkage and selection operator" (LASSO) by minimizing the penalized least squares criterion using an L_1-penalty function which went viral in the area of model selection.

Unlike the RR estimator, LASSO simultaneously selects and estimates variables. It is the reminiscent of "subset selection". The subset selection rule is extremely variable due to its inherent discreteness (Breiman, 1996; Fan and Li, 2001). It is also highly variable and often trapped into a locally optimal solution rather than the globally optimal solution. LASSO is a continuous process and stable; however, it is not suggested to be used in multicollinear situations. Zou and Hastie (2005) proposed a compromised penalty function which is a combination of L_1 and L_2 penalty giving rise to the "elastic net" estimator. It can select groups of correlated variables. It is metaphorically like a stretchable fishing net retaining all potentially big fish.

Although LASSO simultaneously estimates and selects variables, it does not possess "oracle properties" in general. To overcome this problem Fan and Li (2001) proposed the "smoothly clipped absolute deviation" (SCAD) penalty function. Following Fan and Li (2001), Zou (2006) modified LASSO using a weighted L_1-penalty function. Zou (2006) called this estimator an adaptive LASSO (aLASSO). Later, Zhang (2010) suggested a minimax concave penalty (MCP) estimator. All results found in the above literature are based on penalized least squares criterion.

This book contains a thorough study of rank-based estimation with three basic penalty estimators, namely, ridge regression, LASSO and "elastic net". It also includes preliminary test and Stein-type R-estimators for completeness. Efforts are made to present a clear and balanced introduction of rank-based estimators with mathematical comparisons of the properties of various estimators considered. The book is directed towards graduate students, researchers of statistics, economics, bio-statistical biologists and for all applied statisticians, economists and computer scientists and data scientists, among others. The literature is very limited in the area of robust penalty and other shrinkage estimators in the context of rank-based estimation. Here, we provide both theoretical and practical aspects of the subject matter.

The book is spread over twelve chapters. Chapter 1 begins with an introductory examination of the median, outliers and robust rank-based methods, along with a brief look at penalty estimators. Chapter 2 continues with the characteristics of rank-based penalty estimators and demonstrates their enormous value in machine learning. Chapter 3 provides the preliminaries of rank-based theory and various aspects of it, along with a description of penalty estimators, which are then applied to location and simple linear models. Chapters 4 deals with ANOVA and Chapter 5 with seemingly unrelated simple linear models. Chapter 6 considers the multiple linear model and Chapter 7 expands on the "partially linear regression model" (PLM). The Liu regression estimator is discussed in Chapter 8. Chapter 9 introduces the AR(p) model. Chapter 10 covers selection and shrinkage of variables in high-dimensional data analysis. Chapter 11 deals with multivariate rank-based logistic regression models. Finally, Chapter 12 concludes with applications of rank-based neural networks.

To our knowledge, this is one of the first books to combine advanced statistical analysis with advanced machine learning. Each chapter is self-contained but those interested in machine learning may consider Chapters 1 and 2, and 11 and 12, while those interested in statistics may consider Chapters 3–10. A good mix of the two would be derived from Chapters 1–4 and 11 and 12. It is our hope that readers in both fields will find something of value, and that it will lead to many areas of future research.

The authors wish to thank the developers of **Rfit** (Kloke and McKean, 2012) and **glmnet** (Stanford University) which are extremely useful packages for R-estimation and penalized maximum likelihood estimation, respectively. We also thank Professor Brent Johnson (University of Rochester) for the rank-based LASSO and aLASSO code (Johnson and Peng, 2008) provided on his website.

Professor A.K. Md. E. Saleh is grateful to NSERC for supporting his research for more than four decades and is appreciative of Professors P.K. Sen (U. of North Carolina), J. Jurečková (Charles U., Prague), M. Ghosh (Gainsville, Florida), H. L. Koul (Michigan State U.), T. Kubokawa (Tokyo, Japan), T. Shiraishi (Tsukuba, Japan) and C. Robert (Université Paris-Dauphine) for their active participation during the time of these grants. He is grateful to his loving wife, Shahidara Saleh, for her support over 70 years of marriage. He also appreciates his grandchildren Jasmine Alam, Sarah Alam, Migel Saleh and Malique Saleh for their loving care.

Professor M. Arashi wishes to thank his family in Iran, specifically his wife, Reihaneh Arashi (maiden name: Soleimani) for her everlasting love and support and his daughter, Elena Arashi. This research is supported by Ferdowsi University of Mashhad, Iran and in part by the National Research Foundation (NRF) SARChI Research Chair UID: 71199, and DST-NRF Centre of Excellence in Mathematical and Statistical Sciences (CoE-MaSS) of South Africa.

Professor R. Saleh wishes to thank his parents, Dr. Ehsanes Saleh and Mrs. Shadidara Saleh, for their love and support. He also wishes to acknowledge valuable discussions with Victor Aken'Ova and Sohaib Majzoub regarding logistic regression and neural networks, and the encouragement and support from Lynn Saleh, Isme Alam, Raihan Saleh and Jody Fast during the writing of this book.

Dr. M. Norouzirad deeply thanks her parents, Abbas Ali Norouzirad and Fereshteh Arefian for their unconditional trust, timely encouragement, and endless patience. She also thanks her sister, Mehrnoosh Norouzirad, for heartwarming kindness. She also wishes to acknowledge funding provided by National Funds through the FCT - Fundação para a Ciência e a Tecnologia, I.P., under the scope of the project UIDB/00297/2020 (Center for Mathematics and Applications).

A.K. Md. Ehsanes Saleh
Mohammad Arashi
Resve A. Saleh
Mina Norouzirad

1

Introduction to Rank-based Regression

1.1 Introduction

The purpose of this book is to lay the groundwork for robust data science using rank-based methods. The field of machine learning has not yet fully embraced a class of robust estimators that addresses issues that limit the value of least squares estimation. For example, outliers in data sets may produce misleading results that are not suitable for inference. They can also affect results obtained from penalty estimators. We believe that robust estimators for regression problems are well-suited to data science. This book is intended to provide both practical and mathematical foundations in the study of rank-based methods. It will introduce a number of new ideas and approaches to the practice and theory of robust estimation and encourage readers to pursue further investigation in this field. While the main goal of this book is to provide a rigorous treatment of the subject matter, we begin with some introductory material to build insight and intuition about rank-based regression and penalty estimators, especially for those who are new to the topic and those looking to understand key concepts. To motivate the need for such methods, we will start with a discussion of the median as it is the key to rank-based methods and then build on that concept towards the notion of robust data science.

1.2 Robustness of the Median

1.2.1 Mean vs. Median

Our starting point is a brief review of two useful statistics: the mean and the median. We want to assess their suitability and usefulness in the context of simple linear regression. We state up front that the mean is the basis for the most

Rank-Based Methods for Shrinkage and Selection: With Application to Machine Learning.
First Edition. A. K. Md. Ehsanes Saleh, Mohammad Arashi, Resve A. Saleh, and
Mina Norouzirad.

popular regression methods but it is worth revisiting this approach relative to other methods.

Given a set of n values, $\boldsymbol{x} = \{x_1, \dots, x_n\}$, the arithmetic mean (mean for short in our study) is the average of the values while the median is the middle value when placed in sorted order. The mean is given by

$$\text{mean}(\boldsymbol{x}) = \bar{x} = \frac{1}{n}\sum_{i=1}^{n} x_i, \tag{1.2.1}$$

which is the sum of the values divided by n.

The median of the set depends on whether n is even or odd. First, we order the values from smallest to largest. After ordering, the new set is denoted by $\{x_{(1)}, \dots, x_{(n)}\}$, where $x_{(1)} \le \cdots \le x_{(n)}$. These are referred to as the order statistics of $\boldsymbol{x}$ with $x_{(1)} = \min\{x_i,\ i = 1, \dots, n\}$ and $x_{(n)} = \max\{x_i,\ i = 1, \dots, n\}$, and $x_{(i)}$ is the ordered observations. If n is odd, we take the middle value in the ordered set, i.e. $\text{median}(x) = x_{(\frac{n+1}{2})}$. If n is even, we take the average of the two values surrounding the middle. This case can be expressed compactly as follows. Assuming the above ordering, then

$$\text{median}(\boldsymbol{x}) = \frac{1}{2}(x^{(l)} + x^{(u)}), \tag{1.2.2}$$

where $x^{(l)} = x_{(\frac{n}{2})}$ and $x^{(u)} = x_{(\frac{n}{2}+1)}$. The lower and upper values are averaged to obtain the median. We can already see a problem in that the mean is easy to understand using one equation, but the median requires a number of steps and a somewhat complicated description. However, the median has significant value when it comes to finding the essence of a data set, as we will see shortly.

Table 1.1 compares and contrasts these two statistics. Three data sets are provided to illustrate the characteristics of each statistic. The first set $\{3, 1, 4, 2, 5, 3\}$ results in the same value of 3 for the mean and median. This is consistent with a visual examination of the values in the set. However, if we change one of the values from 3 to 100 to produce the second set, $\{100, 1, 4, 2, 5, 3\}$, the mean increases significantly to 19.2, while the median produces a value of 3.5 close to the previous value. In this case, the value of 100 is considered an *outlier* and we can observe its dramatic effect on the two statistics. The mean is greatly affected by one outlier which makes it unstable. On the other hand, the median seems rather stable in the presence of this outlier. Mean instability is more apparent with two outliers as in set 3.

If we now let $x_{(n)}$ represent the outlier value in set 2, hence $x_{(n)} = 100$ initially, we note that as $x_{(n)} \to \infty$, the mean tends to ∞ while the median remains at 3.5. For example, $x_{(n)} = 1000$ changes the mean to 169.2 while the median stays at 3.5. This is problematic enough, but the standard deviation that uses the mean is also greatly affected. Let μ be the population mean, and let $\bar{x}$ of Eq. (1.2.1) be the estimate of μ. Further, let the population standard deviation, σ, be defined by

$$\sigma = \sqrt{\frac{1}{n}\sum_{i=1}^{n}(x_i - \mu)^2},$$

and the estimator of σ, which is the standard deviation (sd) of the sample set, be given by[1]:

$$\text{sd}(\boldsymbol{x}) = \hat{\sigma} = \sqrt{\frac{1}{n-1}\sum_{i=1}^{n}(x_i - \bar{x})^2}. \tag{1.2.3}$$

Then, as the value of the outlier increases to infinity, the mean goes to infinity and, as a consequence, the standard deviation goes to infinity. Hence, one outlier can have a domino effect on various statistics based on the mean.

A robust statistic for a set containing outliers is the median absolute deviation (MAD) given by

$$\text{MAD}(\boldsymbol{x}) = \underset{1 \le i \le n}{\text{median}}\left(|x_i - \text{median}(\boldsymbol{x})|\right). \tag{1.2.4}$$

A consistent estimator of the population standard deviation based on MAD can be computed as follows[2]

$$\hat{\eta} = \text{mad}(\boldsymbol{x}) = k \times \text{MAD}(\boldsymbol{x}), \tag{1.2.5}$$

where $k = 1.4826$ (Johnson and Peng, 2008, cf.)[3]. If we take set 1 in Table 1.1, the original set without outliers, then $\text{sd}(\boldsymbol{x}) = 1.41$ and $\text{mad}(\boldsymbol{x}) = 1.48$. For the

Table 1.1 Comparison of mean and median on three data sets.

Measure	Set 1 {3,1,4,2,5,3}	Set 2 {100,1,4,2,5,3}	Set 3 {100,100,1,4,2,5,3}
Mean	3.0	19.2	30.7
Median	3.0	3.5	4.0

1 We follow the naming convention of R where $\text{sd}(x)$ is the standard deviation.
2 We follow the naming convention of the R command where $\text{mad}(x) = k \times \text{MAD}(\boldsymbol{x})$.
3 In this book, $\hat{\eta}$ is used to denote the robust estimator of this quantity. We use a multiplying factor of 1.4826 to make it an unbiased summary statistic.

case of set 2 with $x_{(n)} = 100$, $\text{sd}(\boldsymbol{x}) = 39.6$ and $\text{mad}(\boldsymbol{x}) = 2.22$. If we now set $x_{(n)} = 1000$, then $\text{sd}(\boldsymbol{x}) = 407.0$ and $\text{mad}(\boldsymbol{x}) = 2.22$; that is, it remains the same. Clearly, a single large outlier can disrupt the integrity of the mean and standard deviation while the median and $\text{mad}(\boldsymbol{x})$ remain stable even as $x_{(n)} \to \infty$.

Next, if we examine set 3 listed as $\{100, 100, 1, 4, 2, 5, 3\}$ in the final column of Table 1.1, we now have two outliers and the mean shifts further beyond the range of the original elements of set 1. On the other hand, the median maintains relative stability. We refer to the median as having the *robustness* property for this reason.

1.2.2 Breakdown Point

To illustrate further, we could continue to add more outliers to the data set that lie between 100 and 120, up to the *breakdown point* when 50% of the data are outliers, and the median would still provide a good representation of the original data set. Of course, if we exceed the 50% point, the values in the original data become outliers and the outliers become the actual data. Hence, the breakdown point of an estimator lies between 0 and 0.5, with 0.5 being the best and 0 being the worst.

The breakdown point (Wilcox, 2012) is a measure of the tolerance level of a statistic or estimator to outliers. The breakdown point of an estimator is the proportion of incorrect (i.e. arbitrarily large) observations an estimator can handle before giving an incorrect (i.e. arbitrarily large) result. The higher the breakdown point of an estimator, the more robust it is. The mean statistic is the least tolerant (in fact, the breakdown point is $\frac{1}{n}$ for a sample of size n observations and nearly zero for large sample sizes) whereas the median may tolerate up to 50% of the data as outliers (the statistic remains in a compact set). This is rather remarkable and provides a strong basis to pursue rank-based methods that are based on the median.

Intuitively, we can understand that a breakdown point cannot exceed 50% because if more than half the observations are contaminated, it is not possible to distinguish between the underlying distribution and the contaminating distribution (Rousseeuw and Leroy, 1987). Therefore, the maximum breakdown point is 0.5 and there are estimators that achieve such a breakdown point. However, there is often a trade-off in such estimators where we may sacrifice certain regularity conditions, such as consistency and asymptotic normality, to be discussed later. It is better to retain these regularity properties even if the breakdown point is reduced below 0.5. In particular, the rank-based methods have a breakdown point around 0.3 but possess the desired regularity conditions and are very efficient estimators.

1.2.3 Order and Rank Statistics

It is important to clarify the difference between "ordered" and "rank" statistics. Let $X_1, ..., X_n$ be a random sample of variables. Now, $X_{(i)}$ is the ith smallest random variable and $X_{(1)}, ..., X_{(n)}$ are referred to as *ordered* statistics of $X_1, ..., X_n$. By definition, $X_{(1)} \leq ... \leq X_{(n)}$. We say that X_i has *rank* R_i among $X_1, ..., X_n$ if $X_i = X_{(R_i)}$ assuming the R_ith order statistic is uniquely defined. By "uniquely defined", we are assuming that ties do not occur. That is $X_{(i)} \neq X_{(j)}$ for all $i \neq j$. Table 1.2 provides examples to clearly illustrate these notions.

Based on the definition of rank, the median is the observation for which the rank equals $\frac{n+1}{2}$ in case of odd n, and the mean of two observations that are ranked $\frac{n}{2}$ and $\frac{n}{2} + 1$, when n is even. In Table 1.2, for the first three sets, $n = 3$ (odd case), so the value of rank 2 is the median; hence, this is the second value in sets X and Y, and the first value in the third set, Z. For the fourth set, n is even, so we select the two values with ranks 2 and 3; thus, the median is average of these two values, i.e. $(4+5)/2 = 4.5$. Outliers, if any, will be at the extremes of order statistics. The fourth set is said to have a potential outlier at the extreme order statistic, $W_{(4)} = 99$.

The above series of examples provides some intuition as to why *rank-based* methods that use the median (and some other measures that will be identified later) may be of significant value in regression problems relative to least squares estimation (LSE) which is based on the mean. This not only applies to LSE, it also applies to any mean regression, such as generalized linear models (GLM). Many realistic data sets have a non-zero percentage of outliers and we may be interested in finding them and/or utilizing estimators that are robust when outliers are present. This is the value proposition of rank-based methods.

Table 1.2 Examples comparing order and rank statistics.

Sets	Ordered statistics	Rank statistics
$X = \{1, 2, 3\}$	$X_{(1)} = 1, X_{(2)} = 2, X_{(3)} = 3$	$R_1 = 1, R_2 = 2, R_3 = 3$
$Y = \{3, 2, 1\}$	$Y_{(1)} = 1, Y_{(2)} = 2, Y_{(3)} = 3$	$R_1 = 3, R_2 = 2, R_3 = 1$
$Z = \{5, 6, 4\}$	$Z_{(1)} = 4, Z_{(2)} = 5, Z_{(3)} = 6$	$R_1 = 2, R_2 = 3, R_3 = 1$
$W = \{5, 3, 99, 4\}$	$W_{(1)} = 3, W_{(2)} = 4,$	$R_1 = 3, R_2 = 1,$
	$W_{(3)} = 5, W_{(4)} = 99$	$R_3 = 4, R_4 = 2$

1.3 Simple Linear Regression

We have studied the characteristics of the mean and median for simple data sets to understand the differences between the two. We now explore the use of the mean and median for simple linear regression. A one-dimensional linear regression problem will be sufficient for our purposes here. We refer to the dimensionality of this problem as $p = 1$; for higher dimensions, $p \geq 2$. Suppose we are given a data set comprised of $\boldsymbol{x} = (x_1, \dots, x_n)^\top \in \mathbb{R}^n$ and $\boldsymbol{y} = (y_1, \dots, y_n)^\top \in \mathbb{R}^n$. Our objective is to fit a line to the data such that the line has the smallest error, $\boldsymbol{\varepsilon} = (\varepsilon_1, \dots, \varepsilon_n)^\top \in \mathbb{R}^n$, and provides a good representation of the data. More concretely, we describe the simple linear model with the following equation

$$\boldsymbol{y} = \theta \mathbf{1}_n + \beta \boldsymbol{x} + \boldsymbol{\varepsilon}, \tag{1.3.1}$$

where the two parameters of interest are the slope of the line, β, and the y-intercept, θ, and $\mathbf{1}_n = (1, \dots, 1)^\top$ is an n-vector of 1. We seek to estimate θ and β using the least squares (LS) method. Note that $\boldsymbol{x}$, $\boldsymbol{y}$ and $\boldsymbol{\varepsilon}$ are all vectors of length n, assuming we are given n data points from which to fashion a line.

1.3.1 Least Squares Estimator (LSE)

The classical LSE to determine β and θ is based on minimization of the residual sum of squares (RSS). We denote the estimated values from LSE as $\hat{\beta}_n^{\text{LS}}$ and $\hat{\theta}_n^{\text{LS}}$, respectively. For arbitrary values of $\hat{\theta}_n$ and $\hat{\beta}_n$ and a specific data point (x_i, y_i), the residual is defined as $\hat{\varepsilon}_i = e_i = y_i - (\hat{\theta}_n - \hat{\beta}_n x_i)$. Then the residual sum of the squares (RSS) is simply

$$\text{RSS} = e_1^2 + e_2^2 + \cdots + e_n^2. \tag{1.3.2}$$

If we minimize the RSS for a given data set $\boldsymbol{x}$ and $\boldsymbol{y}$, then the estimates are

$$\hat{\beta}_n^{\text{LS}} = \frac{\sum_{i=1}^{n} (x_i - \bar{x})(y_i - \bar{y})}{\sum_{i=1}^{n} (x_i - \bar{x})^2}, \tag{1.3.3}$$

and

$$\hat{\theta}_n^{\text{LS}} = \bar{y} - \hat{\beta}_n^{\text{LS}} \bar{x}. \tag{1.3.4}$$

Note the use of the mean values, $\bar{x}$ and $\bar{y}$, to estimate the two parameters. This is an important point to make since LSE provides the *optimal unbiased estimation* of the slope and intercept for a given set of data points. More generally, it is the maximum likelihood estimate under the normality assumptions for the error.

1.3.2 Theil's Estimator

We can also estimate the parameters using the median by way of Theil's estimator (Theil, 1950). For a given set of points, $\{(x_i, y_i)\}_{i=1}^n$, Theil's estimators of β and θ can be obtained by finding the median of the pairwise slopes of all points in the data set, denoted by $\hat{\beta}_n^{\text{Theil}}$, and the median of partial-residuals, $(y_i - \hat{\beta}_n^{\text{Theil}} x_i)$, which we denote by $\hat{\theta}_n^{\text{Theil}}$. The actual residuals are $(y_i - (\hat{\theta}_n^{\text{Theil}} + \hat{\beta}_n^{\text{Theil}} x_i))$ which are obtained after estimation. See Hallander and Wolfe (1999) for more details. Hence, we have that

$$\hat{\beta}_n^{\text{Theil}} = \underset{1 \le i < j \le n, i \ne j}{\text{median}} \left(\frac{y_j - y_i}{x_j - x_i} \right), \tag{1.3.5}$$

and

$$\hat{\theta}_n^{\text{Theil}} = \underset{1 \le i \le n}{\text{median}} \left(y_i - \hat{\beta}_n^{\text{Theil}} x_i \right). \tag{1.3.6}$$

Theil's method is an unbiased estimator. Its breakdown point is known to be 29.3%. While this is not as high as 50%, it can tolerate outliers to a higher degree than LSE making it a more robust estimator.

1.3.3 Belgium Telephone Data Set

It is instructive to qualitatively compare the mean versus median approaches using a very simple data set. For this purpose, we will use the Belgium telephone data set (Kloke and McKean, 2012). This data set is provided in Table 1.3 and contains the number of calls (in units of tens of millions) made in Belgium in the years between 1950 and 1973. It has 24 data points, with 6 outliers. The outliers are due to a change in measurement technique without re-calibration for 6 years, as is often cited.

Table 1.3 Belgium telephone data set.

x	1950	1951	1952	1953	1954	1955	1956	1957	1958	1959	1960	1961
y	0.44	0.47	0.47	0.59	0.66	9.73	0.81	0.88	1.06	1.2	1.35	1.49

x	1962	1963	1964	1965	1966	1967	1968	1969	1970	1971	1972	1973
y	1.61	2.12	11.9	12.4	14.2	15.9	18.2	21.2	4.3	2.4	2.7	2.9

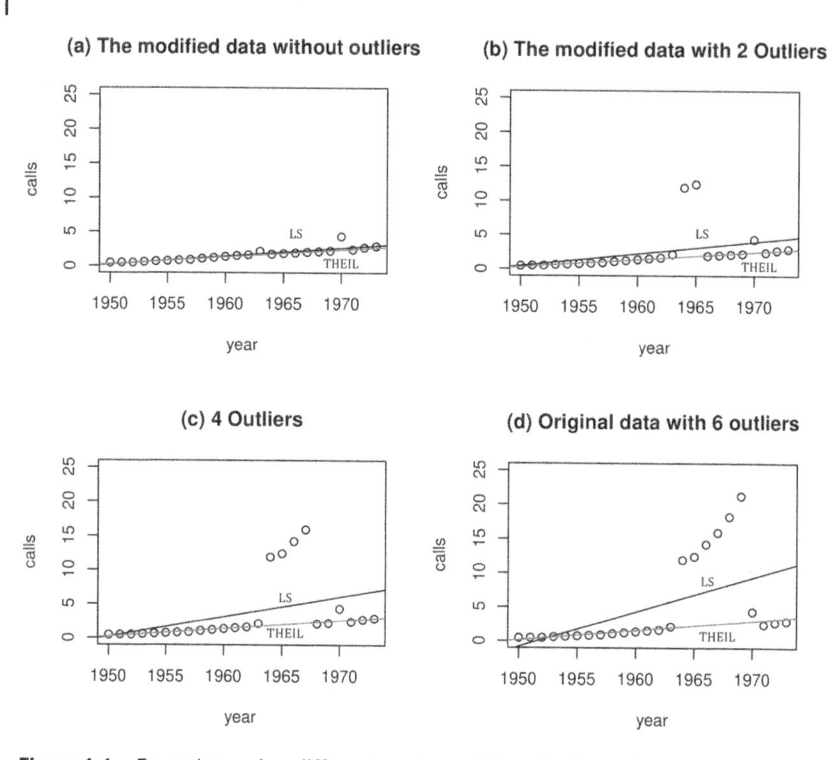

Figure 1.1 Four plots using different versions of the telephone data set with fitted lines. The median-based approach (Theil estimation) is robust and stable while the LS line varies depending on the number of outliers.

We will mainly use this data set throughout this chapter. However, we will also adjust the data values to control the number of outliers in order to accentuate the differences between the two competing approaches.

In Figure 1.1, we show four different cases. The first three data sets, Figures 1.1(a), (b) and (c) are modified to have zero, two, and four outliers, respectively. Figure 1.1(d) is the original data set with six outliers, as given in Table 1.3.

The results of linear regression using LSE and Theil are shown as fitted lines on the scatter plots of Figure 1.1. Note that, as we add more outliers in the data set, the lines associated with LS change and move toward the outliers, while the lines based on the median (Theil) remain relatively stable. This figure gives us an indication of how easily LSE values can shift due to a few outliers. The process of parameter estimation, the calculation of the standard error statistic and hypothesis testing for each case is discussed in the next section.

1.3.4 Estimation and Standard Error Comparison

We will now carry out a quantitative comparison. The equations for each estimator were provided earlier. We begin this section with a description of the standard error for each estimator and end with hypothesis testing. The standard error (s.e.) is a measure of the variability or accuracy of parameter estimation.

The LSE standard errors for the respective estimates are given by

$$\text{s.e.}(\hat{\theta}_n^{\text{LS}}) = \sqrt{\hat{\sigma}^2\left[\frac{1}{n} + \frac{\bar{x}^2}{\sum_{i=1}^{n}(x_i - \bar{x})^2}\right]}, \tag{1.3.7}$$

and

$$\text{s.e}(\hat{\beta}_n^{\text{LS}}) = \sqrt{\frac{\hat{\sigma}^2}{\sum_{i=1}^{n}(x_i - \bar{x})^2}}, \tag{1.3.8}$$

where σ can be estimated using the residual standard error (RSE) as follows:

$$\hat{\sigma} \approx \text{RSE} = \sqrt{\text{RSS}/(n-2)}. \tag{1.3.9}$$

For Theil's estimator, we employ MAD to calculate the standard errors. Using following definition of partial-residuals, $r_i = (y_i - \hat{\beta}_n^{\text{Theil}} x_i)$, we obtain

$$\text{s.e.}(\hat{\theta}_n^{\text{Theil}}) = k \times \underset{1 \le i \le n}{\text{median}} \left(|r_i - \hat{\theta}_n^{\text{Theil}}|\right), \tag{1.3.10}$$

and we use the slopes, $s_{ij} = (y_j - y_i)/(x_j - x_i)$, to obtain

$$\text{s.e.}(\hat{\beta}_n^{\text{Theil}}) = k \times \underset{1 \le i < j \le n, i \ne j}{\text{median}} \left(|s_{ij} - \hat{\beta}_n^{\text{Theil}}|\right). \tag{1.3.11}$$

Our quantitative comparison involves only Figures 1.1(a) (no outliers) and 1.1(d) (six outliers). We use Eqs. (1.3.3) and (1.3.4) to obtain the LS estimates. First, we compute $\hat{\beta}_n^{\text{LS}}$ from Eq. (1.3.3) and use that result in Eq. (1.3.4) to compute $\hat{\theta}_n^{\text{LS}}$. For Figure 1.1(a), we obtain $\hat{\beta}_n^{\text{LS}} = 0.12$ and $\hat{\theta}_n^{\text{LS}} = -234.2$; however, for Figure 1.1(d) we obtain $\hat{\beta}_n^{\text{LS}} = 0.50$ and $\hat{\theta}_n^{\text{LS}} = -983.9$. Clearly, there is a significant change in the LS estimates when outliers are added.

To obtain the estimates using the median due to Theil, we apply Eqs. (1.3.5) and (1.3.6). First, we compute $\hat{\beta}_n^{\text{Theil}}$ from Eq. (1.3.5) and use that result in Eq. (1.3.6) to compute $\hat{\theta}_n^{\text{Theil}}$. For Figure 1.1(a), we obtain $\hat{\beta}_n^{\text{Theil}} = 0.10$ and $\hat{\theta}_n^{\text{Theil}} = -201.8$; and for Figure 1.1(d) we obtain $\hat{\beta}_n^{\text{Theil}} = 0.14$ and $\hat{\theta}_n^{\text{Theil}} = -270.4$. Both sets of results are summarized in Table 1.4. Also provided in the table are the standard error values for each case. Notably, the values do not change by much but the s.e. of the intercept improves greatly over LSE.

For hypothesis testing in least squares estimation, we use the t-statistic as follows:

$$t\text{-value}(\hat{\beta}_n^{\text{LS}}) = \frac{\hat{\beta}_n^{\text{LS}} - 0}{\text{s.e}(\hat{\beta}_n^{\text{LS}})}. \tag{1.3.12}$$

Once the t-statistic is computed, it is used to determine the p-value for hypothesis testing; in particular, $\mathcal{H}_o : \beta = 0$ vs. $\mathcal{H}_A : \beta \neq 0$. The LSE p-values provided in the table are very small indicating that the null hypothesis, $\mathcal{H}_o$, can be rejected in favor of the alternate hypothesis, $\mathcal{H}_A$.

For the Theil's estimator, we can not use the same method as LSE for hypothesis testing since we do not have any information about the distribution of the median. Theil's estimator falls into a special category known as nonparametric statistics (Corder and Foreman, 2014). This is a branch of statistics in which the underlying distribution is not known so alternative methods are used.

Instead of the approach taken above, the Wilcoxon test can be utilized here. The test essentially calculates the difference between pairs of data and analyzes their differences to establish if they are significantly different from one another in a statistical sense. The step-by-step procedure for comparing two related samples using *Wilcoxon signed-rank test statistic*, T, based on Corder and Foreman (2014) is as follows:

(1) For each item in a sample of n items, obtain a difference score D_i between two measurements (i.e. $D_i = y_i - (\hat{\theta}_n^{\text{Theil}} + \hat{\beta}_n^{\text{Theil}} x_i)$ $i = 1, \ldots, n$).
(2) Take the absolute value of the differences, $|D_i|$.
(3) Omit difference scores of zero, giving you a set of n' non-zero absolute difference scores, where $n' \leq n$. Thus, n' becomes the actual sample size.
(4) Then, assign ranks R_i from 1 to n' to each of the $|D_i|$ such that the smallest gets rank 1 and the largest gets rank n'. If two or more $|D_i|$ are equal, they are all assigned same average rank (i.e. midrank) of the ranks they would have been assigned individually had ties not occurred.
(5) Next, apply the sign + or − to each of the n' ranks, R_i, depending on whether D_i was originally positive or negative, producing $+R_i$ or $-R_i$, respectively.
(6) The Wilcoxon test T-statistic is taken as the absolute sum of either the positive ranks, $T = |\sum_i(+R_i)|$, or negative ranks, $T = |\sum_i(-R_i)|$, whichever is *smaller*. Hence, T is always positive.

Next, the T-statistic can be examined for significance. As n' increases, the sampling distribution converges to a normal distribution. Thus, for $n' \geq 20$, a z-score can be calculated. We compute the z-scores for large samples using

Table 1.4 Comparison of LS and Theil estimations of Figures 1.1(a) and (d).

Case (LSE)	Estimators	Coefficient	s.e	*t*-value	*p*-value
Figure 1.1(a)	$\hat{\theta}_n^{\mathrm{LS}}$	−234.2	23.4	−10.0	<0.001
	$\hat{\beta}_n^{\mathrm{LS}}$	0.120	0.01	10.1	<0.001
Figure 1.1(d)	$\hat{\theta}_n^{\mathrm{LS}}$	−983.9	325.2	−3.0	0.006
	$\hat{\beta}_n^{\mathrm{LS}}$	0.504	0.166	3.0	0.006
Case (Theil)	**Estimators**	**Coefficient**	**s.e**	***T*-statistic**	z^*
Figure 1.1(a)	$\hat{\theta}_n^{\mathrm{Theil}}$	−201.8	0.06	–	–
	$\hat{\beta}_n^{\mathrm{Theil}}$	0.104	0.02	107.0	−1.23
Figure 1.1(d)	$\hat{\theta}_n^{\mathrm{Theil}}$	−270.4	0.32	–	–
	$\hat{\beta}_n^{\mathrm{Theil}}$	0.139	0.18	116.5	−0.96

$$\bar{x}_T = \frac{n'(n'+1)}{4}, \quad s_T = \sqrt{\frac{n'(n'+1)(2n'+1)}{24}}, \text{ and } z^* = \frac{T - \bar{x}_T}{s_T},$$

where $\bar{x}_T$ is the mean and s_T is the standard deviation, and z^* is the z-score for an approximation of the data to the normal distribution.

To perform a two-sided test, we reject $\mathcal{H}_o$ if $z_{\text{critical}} < |z^*|$. Specifically, for $\alpha = 0.05$, $z_{\text{critical}} = 1.96$. In the case of Figure 1.1(d), we have $n' = 24$ and obtain $T = 116.5$ (the negative sum is smaller), so $z^* = (116.5 - 150)/35 = -0.96$. Therefore, we cannot reject $\mathcal{H}_o$. This suggests there is no difference in the two samples. Similarly, for Figure 1.1(a), we find that $T = 107.0$ so that $z^* = (107.0 - 150)/35 = -1.23$. Again, we cannot reject $\mathcal{H}_o$ suggesting no difference in the two samples. Table 1.4 shows the results of this approach[4] on Theil's estimator for the telephone data set in the two cases of Figures 1.1(a) and 1.1(d).

1.4 Outliers and their Detection

Typically, outliers are created by errors in transcribing the data, or noise in the data collection process, or simply anomalies in data values. It is relatively easy to observe an outlier on a graph in a one-variable setting, but much more difficult to identify in the case of multiple linear regression. Hence, a clear

4 The results are based on the Python library scipy.stats.wilcoxon() that provides the T-statistic.

definition of an outlier and reliable methods for their detection are difficult to provide. One approach is to examine the residuals after analysis for both LSE and median-based methods to determine if outliers can be separated from the main data. Order statistics can also be used to identify outliers.

1.4.1 Outlier Detection

The residuals for simple linear regression can be computed after estimation using a general equation as follows

$$\hat{e}_i = y_i - (\hat{\theta}_n + \hat{\beta}_n x_i), \ i = 1, \dots, n, \tag{1.4.1}$$

where $\hat{\theta}_n$ and $\hat{\beta}_n$ are arbitrary estimators of θ and β, respectively. We note that if there is an outlier present in the residuals, $\{\hat{e}_1, \dots, \hat{e}_n\}$, at least one of the extreme order statistics $\hat{e}_{(1)}$ or $\hat{e}_{(n)}$ must be an outlier. Therefore, it is possible to identify outliers by their position in the order statistics. Regardless of its size, the outlier will always be positioned at one extreme or the other. In reality, multiple outliers may exist in the data set and the residuals associated with them would form a set of order statistics at one extreme or the other, or both. This residual ordering can be done for LSE and the median-based method and then compared.

Consider Figure 1.2 showing the computed residuals for both LSE and median-based cases on the original telephone data set of Table 1.3 and depicted in Figure 1.1(d). In the upper half of the figure, we present the histograms of

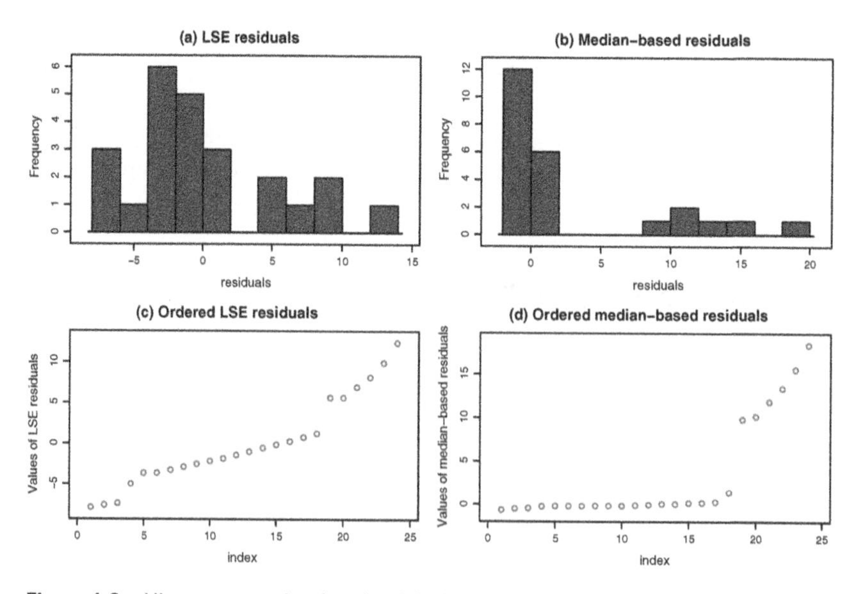

Figure 1.2 Histograms and ordered residual plots of LS and Theil estimators.

the computed residuals for the two cases. In the lower half, we plot the order statistics of each set of residuals from 1 to 24.

The LSE residuals form a distribution that makes it difficult to separate outliers from the rest of the data points. It is not easy to make a definitive statement about which bars on the histogram are associated with the six outliers. There appear to be outliers on both sides of 0 but one cannot be sure. On the other hand, the residuals for the median-based approach show a clear separation between the outliers and the rest of the data. A cutoff value of 5.0 provides a suitable dividing line to sift out the six outliers. Note that ordering does not help to separate the outliers from the data in the LSE case as shown in the ordered residuals below the histogram. In stark contrast, we see a clear distinction between the outliers and the rest of the data in the median case. Their respective order statistics are $\hat{e}_{(19)}, \ldots, \hat{e}_{(24)}$ since there are 24 points and the six outliers are large and positive.

These six data points may now be inspected to determine if they should be retained, modified or removed from the data set. Some points may be easily corrected if it was a manual entry mistake, or a measurement calibration error that can be fixed (as is the case here). The decision whether or not a data point is a true outlier is based on well-established rules or metrics for a given application, either based on, for example, 4 or 5 standard deviations away from the mean, or say, several quartiles away from the inner first and third quartiles (25% to 75% in terms of percentiles of the residuals). These approaches are prone to error and must be done carefully.

While this method of using order statistics is not foolproof, it shows promise and will be workable for higher dimension problems without modification. In general, it is difficult, error-prone and somewhat tedious to identify and remove outliers. A better approach is to use robust statistical procedures that are relatively insensitive to outliers so that they do not have to be detected or removed. A method using rank estimation is such an approach and is introduced in the next section.

1.5 Motivation for Rank-based Methods

1.5.1 Effect of a Single Outlier

As a preview of the robustness of rank-based methods, we show four cases in Figure 1.3 using the telephone data set to demonstrate that a single outlier can have deleterious effects on LSE but little or no effect on rank estimates, depending on its position relative to other observations in the data set. Outliers are generally viewed as nuisance observations in a data set and their effects are considered to be relatively benign so robust methods have not been widely embraced as yet. To provide a clear need for robust regression, we

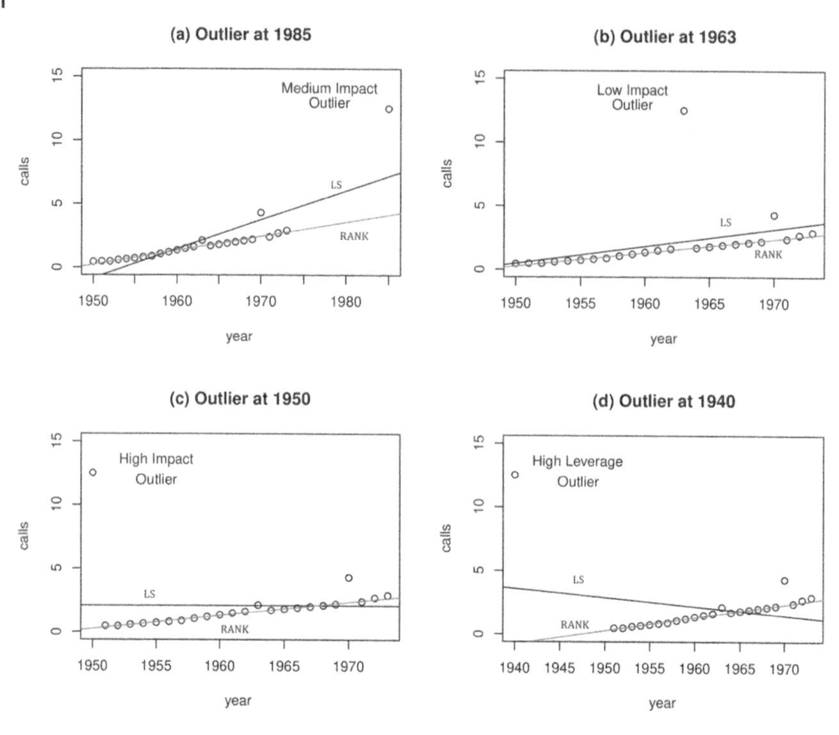

Figure 1.3 Effect of a single outlier on LS and rank estimators. The line due to rank is relatively unchanged compared to LS line.

provide 4 interesting cases but note that the results are specific to this data set. These examples should leave no doubt that robust methods should be used in mainstream machine learning tools, and that research in this area should be increased to resolve any issues that may limit their use.

Starting with Figure 1.3(a), we see that the insertion of one strategically placed outlier on the far right of the data set at (year = 1985, calls = 12.5), has a medium impact on the LSE result, relative to the other cases shown, while the rank result remains unaffected. However, when the outlier is moved to (year = 1963, calls = 12.5) as in Figure 1.3(b), the LSE and rank results are similar; hence, this is a low-impact outlier in this data set. It is only meant to show that in some cases, an obvious outlier may not influence the LSE result by much whereas the rank result remains the same as before.

Now consider the case of the outlier at (year = 1950, calls = 12.5) in Figure 1.3(c). This position constitutes a high-impact outlier for this data set. In fact, LSE tells us that there is no relationship between calls and year, which we know is incorrect. We will say more about this case shortly. In the last case, Figure 1.3(d), the added outlier at (year = 1940, calls = 12.5) is called a *high-leverage* point. This type of outlier can have serious consequences on the results of LSE.

In fact, it shows a negative correlation between year and calls which is clearly wrong. The point of all these examples is to illustrate the unpredictability of even a single outlier in determining the LSE, let alone a number of outliers and their effects. They show that not all outliers are created equal but the results of rank are quite robust in all four cases.

To emphasize the potential hazard of Figure 1.3(c), assume we had to select a small subset from 10 explanatory variables and year was just one of them. A result such as the high-impact case with a slope of 0 would suggest that we remove year as a variable of interest. Any method that relies on the LSE result to select a subset of explanatory variables would fail in this case. However, the rank estimate virtually ignores both the high-impact and high-leverage points and therefore would be a better basis for subset selection.

In essence, LSE makes an inherent assumption that if outliers exist, they are distributed uniformly in all directions. Of course, rank methods do not assume this but it would be a big limitation if rank methods were not accurate when outliers are balanced in all directions. Fortunately, the rank estimates are in line with LSE in the balanced case but provide robustness in the y-space for unbalanced cases. This is an important point to note if one were to abandon LSE in favor of rank methods.

Thus far, we have seen that using the median has certain barriers that have limited its use in the past, not least of which are more complicated expressions, the need for nonparametric analysis, and the cumbersome nature of the solution presented to this point. But given that Theil's method performed much better than LSE in the presence of outliers, other methods of obtaining the slope and intercept using the median deserves some attention. Before describing such methods, there are a number of drawbacks in Theil's method to review.

First, all pairwise slopes must be computed before the median is taken. Assuming n data points, this results in $n(n-1)/2$ slope calculations of the form $(y_j - y_i)/(x_j - x_i)$ before using Eq. (1.3.5). For our telephone example where $n = 24$, there are 276 slope calculations. Clearly this is a problem as n gets large since the computational complexity[5] grows according to $O(n^2)$. The number of partial-residuals of the form $(y_i - \hat{\beta}_n^{\text{Theil}} x_i)$ to be computed using Eq, (1.3.6) is $O(n)$, which is $n = 24$ in this case, so there is no issue here. Second, all of the points were spaced evenly on the x-axis which produced good accuracy using Theil; however, points are not generally evenly spaced. Therefore, accuracy could be reduced in uneven cases. And third, the process becomes unwieldy as the dimensionality of the problem increases. So far we have considered a simple

5 The computational complexity of an algorithm can be characterized using the $O(f(n))$ notation. For example, $O(n)$ refers to an algorithm that runs in linear time w.r.t. n, whereas $O(n^2)$ has a quadratic run time. Likewise, $O(2^n)$ has an exponential run time. A logarithmic run time would be represented as $O(\log(n))$, etc.

linear case ($p = 1$). However, if $p \gg 1$, this method is much more complicated to implement and the breakdown point decreases. All of these concerns must be addressed if LSE is to be replaced by a median-based method. Fortunately, rank-based methods will alleviate these issues.

1.5.2 Using Rank for the Location Model

To introduce rank-based methods, consider the location model

$$y_i = \theta + \varepsilon_i, \quad i = 1, \ldots, n,$$

where $\varepsilon_1, \ldots, \varepsilon_n$ are i.i.d. random variables. A location model attempts to estimate the center of a given sample, $y_1, \ldots, y_n$, of a random variable Y. This could be the sample mean or median or any other similar type of estimate depending on the underlying distribution. We will describe how to address this simple case to compare two similar estimators that seek the median.

In theory, the median is obtained using an L_1 loss function. To estimate the location parameter, we could use the absolute value loss function given by

$$B_n(\theta) = \sum_{i=1}^{n} |y_i - \theta|. \tag{1.5.1}$$

The median estimate, $\hat{\theta}_n^{\text{median}}$, is obtained as

$$\hat{\theta}_n^{\text{median}} = \underset{\theta \in \mathbb{R}}{\text{argmin}} \{B_n(\theta)\}. \tag{1.5.2}$$

The gradient function is

$$B_n'(\theta) = -\sum_{i=1}^{n} \text{sgn}(y_i - \theta). \tag{1.5.3}$$

The absolute value loss function has a discontinuous first derivative that creates some theoretical problems since it does not satisfy certain regularity conditions.

Now consider multiplying the terms in the L_1 loss function by their respective ranks to form a rank-based loss function as follows

$$D_n(\theta) = \sum_{i=1}^{n} |y_i - \theta| \, R_{n_i}^{+}(\theta), \tag{1.5.4}$$

where $R_{n_i}^{+}(\theta)$ are the ranks of $|y_i - \theta|$ among $|y_1 - \theta|, \ldots, |y_n - \theta|$. The "+" sign indicates that we are taking the absolute value of the quantities before ranking them. Note that, in the new loss function, we are simply multiplying the absolute value of $y_i - \theta$ by its rank, $R_{n_i}^{+}(\theta)$. By doing so, we have a new loss function that produces the median of the pairwise averages of y_i. The pairwise averages of a set of numbers, including the original numbers themselves, are referred to as the Walsh averages. In effect, we obtain the median of the Walsh averages using this rank-based loss function. The result is equivalent to the Hodges–Lehmann (HL) estimator (cf. Hettmansperger and McKean, 2011) given by:

$$\hat{\theta}_n^{\mathrm{HL}} = \underset{1\le i\le j\le n}{\mathrm{median}}\left\{\frac{y_i+y_j}{2}\right\}.$$

We will find it convenient to normalize the ranks by defining $u = R^+_{n_i}(\theta)/(n+1)$ such that $0 < u < 1$, and then use it to define a scoring function $\phi^+(u) = u$. This is a specific scoring function that we have chosen here. Since a variety of other scoring functions are possible, we use the notation, $a_n^+(R^+_{n_i}(\theta))$, for a generic score function and embed the actual function $\phi^+(u)$ inside this generic function. Combining the above, we define the rank dispersion function as follows,

$$D_n(\theta) = \sum_{i=1}^{n} |y_i - \theta|\, a_n^+(R^+_{n_i}(\theta)), \tag{1.5.5}$$

with the corresponding gradient given by

$$D_n'(\theta) = -\sum_{i=1}^{n} \mathrm{sgn}(y_i - \theta)\, a_n^+(R^+_{n_i}(\theta)). \tag{1.5.6}$$

The rank estimate is

$$\hat{\theta}_n^{\mathrm{R}} = \underset{\theta\in\mathbb{R}}{\mathrm{argmin}}\,\{D_n(\theta)\}. \tag{1.5.7}$$

All we have done thus far is to create a new loss function by applying a weight consisting of a normalized rank to the L_1 loss function. Instead of obtaining the median, it produces the median of the Walsh averages. Although the median is a robust estimator of the location, we will find that the median of the Walsh averages is also a robust estimator but with a lower breakdown point. Note that the median is the 50th percentile of the data, but any estimator that provides a result that is in the vicinity of the median is also robust.

To illustrate the difference using an example, consider the set,

$$\boldsymbol{y} = (0.1, 1.2, 2.3, 3.4, 4.5, 5.0, 6.6, 7.7, 8.8, 9.9, 10.5)^{\mathsf{T}}.$$

We seek to estimate the location parameter, θ. In doing so, we can minimize Eq. (1.5.1) or Eq. (1.5.4). Which one should we choose? To answer this question, let us first examine their derivatives as a function of θ. As shown in Figure 1.4,

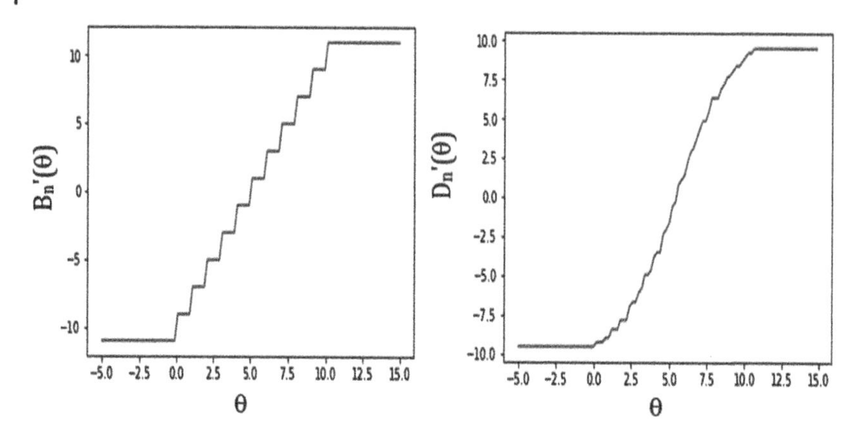

Figure 1.4 Gradients of absolute value ($B_n'(\theta)$) and dispersion ($D_n'(\theta)$) functions.

the gradient of the absolute value loss function, $B_n'(\theta)$, is a series of steps. Each step occurs at a different value of y_i. Hence, the number of steps is n which in this case is 11. However, the gradient of dispersion function, $D_n'(\theta)$, is similar in nature but smoothed out by the scoring function $a_n^+(\cdot)$ with more steps generated in the same interval. Each step occurs at values associated with each one of the Walsh averages. As a result, the number of steps is $\frac{n(n+1)}{2}$ which is 66 in this case, as shown in the graph on the right. Under the assumptions listed earlier, it is possible to show that the estimator of Eq. (1.5.7) is consistent and asymptotically normal (Hettmansperger and McKean, 2011) whereas the estimator of Eq. (1.5.2) does not have these desirable properties. This is why we should choose to minimize Eq. (1.5.4).

Of course, when we minimize $D_n(\theta)$ and $B_n(\theta)$, we do not obtain the same results. From inspection of the numbers, it is clear that the median is 5.0 which is obtained by minimizing $B_n(\theta)$. However, when we minimize $D_n(\theta)$ we obtain 5.55, which is not the median of the data set. As mentioned above, $D_n(\theta)$ produces the median of the Walsh averages, i.e. the pairwise averages of the given numbers. The Walsh averages for this data set are provided in Table 1.5 in sorted order. We can count a total of 66 averages, some of which are duplicated. As a result, there are a maximum of 66 steps in the gradient function, but if there are duplicates in the Walsh averages, the number of steps can be below 66. If we select the median of the Walsh averages from the table, we would select the average of the 33rd and 34th values which is 5.55 (shown in bold).

Although we do not obtain the actual median, the rank-based approach still provides both robustness and asymptotic normality. From Figure 1.4, we can intuitively understand why the dispersion function leads to the regularity

Table 1.5 Walsh averages for the set {0.1, 1.2, 2.3, 3.4, 4.5, 5.0, 6.6, 7.7, 8.8, 9.9, 10.5}.

0.10	0.65	1.20	1.20	1.75	1.75	2.30	2.30	2.30	2.55	2.85	2.85
3.10	3.35	3.40	3.40	3.65	3.90	3.90	3.95	4.20	4.45	4.45	4.45
4.50	4.75	5.00	5.00	5.00	5.00	5.00	5.30	**5.55**	**5.55**	5.55	5.55
5.80	5.85	6.10	6.10	6.10	6.35	6.40	6.60	6.65	6.65	6.90	6.95
7.15	7.20	7.45	7.50	7.70	7.70	7.75	8.25	8.25	8.55	8.80	8.80
9.10	9.35	9.65	9.90	10.20	10.50	—	—	—	—	—	—

conditions that we seek in an estimator and why it inherits the robustness property similar to the median, while not actually computing the true median. The asymptotic properties of the linear rank-based estimators and their relative efficiencies can be found in Puri and Sen (1985) for the interested reader.

1.5.3 Using Rank for the Slope

If we now wish to define a linear rank estimate of the slope for the simple linear model, we can use another formulation of the dispersion function:

$$D_n(\beta) = \sum_{i=1}^{n} (y_i - \beta x_i) a_n(R_{n_i}(\beta))$$

where $R_{n_i}(\beta) = R_{n_i}(y_i - \beta x_i)$ is the rank of $(y_i - \beta x_i)$ among $(y_1 - \beta x_1), \ldots, (y_n - \beta x_n)$, and $a_n(R_{n_i}(\beta))$ is a different scoring function which uses $\phi(u) = u - 1/2$, with $u = R_{n_i}(\beta)/(n+1)$ as before. Note that we do not use the "+" notation here since absolute values of the quantities are not taken. The minimization of this dispersion function does not typically produce the median of the slopes as was the case in Theil's estimator. Rather, it produces an estimate in the vicinity of the median. The actual quantile associated with the estimate depends on the data. The median is, of course, at the 50th percentile but we can retain some of the robustness properties (albeit at a lower breakdown point) by selecting a slope near the 50th percentile. This is best illustrated using another example.

Consider the following data set:

x	0.1	0.2	0.3	0.4	0.5	0.6	0.7	0.8	0.9	1.2
y	3.2	4.0	4.2	4.7	6.5	5.5	6.7	20.2	22.0	8.0

If we use a simple linear model, then we seek an estimate of the slope, β. Since $n = 10$, there are a total of $n(n-1)/2 = 45$ slopes to compute. These ordered slopes are shown in the table below. The median is the 23rd entry which is 6.25.

However, the rank estimate, obtained by minimizing $D_n(\beta)$, is 5.40 which is the 21st entry (in bold). It is at the 46th percentile.

1	−46.67	−30.50	−10.00	1.00	2.00	2.14	2.60	3.50	3.75	4.00	4.00	4.13
13	4.17	4.22	4.33	4.36	4.60	5.00	5.00	5.00	**5.40**	5.83	6.25	6.67
25	8.00	8.25	8.33	11.50	12.00	18.00	18.00	23.5	24.29	25.71	27.00	29.67
37	32.00	34.60	38.75	38.75	45.67	55.00	73.50	76.50	135.00	—	—	—

Therefore, while not computing the median of the slopes, the rank-based method produces an estimate that is near the median. The benefits of using a median-based approach such as rank should be clear at this point. We will show how we can perform linear regression using rank that parallels the methodology of the least squares estimator but provides robustness. Although Theil's method is effective, we will see that rank-based methods provide a more efficient solution and they are the basis for the theory and methods to be described in the rest of this book.

1.6 The Rank Dispersion Function

In this section, we pursue a formal approach inspired by the robustness of the median in linear regression. We refer to this approach as a rank-based method in which values in the vicinity of the median are used for estimation. It would be unwieldy and cumbersome to use Theil's method to perform linear regression due to its somewhat ad hoc nature. Fortunately, a procedure was developed in the early 1970s in the work of Jurečková (1971) and Jaeckel (1972) that produces similar results, does not require the calculation of $O(n^2)$ slopes, provides for outlier detection, and extends easily to higher dimensions, i.e. multiple linear regression.

Recall that in LS estimation, we minimize a quantity we call the *quadratic* or L_2 loss function,

$$\text{SSE} = \sum_{i=1}^{n}(y_i - (\theta + \beta x_i))^2, \tag{1.6.1}$$

which is the sum of the squares of the error (SSE). Jaeckel (1972) developed a loss function which allows for rank-based linear regression. The rank estimator requires simply replacing the quadratic loss function with Jaeckel's dispersion function and then minimizing it for R-estimation (i.e. rank-based estimation). The dispersion function can be viewed either as a pseudo-norm, as noted in Kloke and McKean (2012), or a weighted sum of the residuals. The pseudo-norm designation is due to the fact that it behaves somewhat like the well-known L_1-norm. As before, we will first compute the slope and use it to compute the intercept. For the slope, the dispersion function is given by

$$D_n(\beta) = \sum_{i=1}^{n}(y_i - (\theta + \beta x_i))\, a_n(R_{n_i}(y_i - (\theta + \beta x_i)). \tag{1.6.2}$$

This function requires some explanation. The first term is the residual itself. Note that it is a function of θ and β but we seek only β at this stage. The term $R_{n_i}(y_i - (\theta + \beta x_i))$ is the rank of the ith residual. However, the intercept θ is a constant and, as such, it will not change the rankings. That is, a constant added to a set of values will not change their relative ranks. Therefore, we can remove θ from the ranking function, i.e. we simply rank the partial-residuals

$$R_{n_i}(\beta) = R_{n_i}(y_i - \beta x_i).$$

Furthermore, we will see shortly that

$$\sum_{i=1}^{n} a_n(R_{n_i}(\beta)) = 0. \tag{1.6.3}$$

The term $a_n(\cdot)$ is a weighting function for the residuals, and contains a scoring function. Since this function sums to zero, the dispersion function is invariant w.r.t. θ. In particular, we can show that

$$\begin{aligned} D_n(\theta,\beta) &= \sum_{i=1}^{n}(y_i - (\theta + \beta x_i))a_n(R_{n_i}(y_i - \beta x_i)) \\ &= \sum_{i=1}^{n}(y_i - \beta x_i)a_n(R_{n_i}(y_i - \beta x_i)) - \theta \sum_{i=1}^{n} a_n(R_{n_i}(y_i - \beta x_i)) \\ &= \sum_{i=1}^{n}(y_i - \beta x_i)a_n(R_{n_i}(y_i - \beta x_i)) = D_n(\beta), \end{aligned} \tag{1.6.4}$$

which we can write compactly as

$$D_n(\beta) = \sum_{i=1}^{n}(y_i - \beta x_i)a_n(R_{n_i}(\beta)). \tag{1.6.5}$$

It can be shown that $D_n(\beta)$ is a non-negative, piece-wise linear, continuous, convex function of β and therefore is quite suitable to quickly obtain an estimate using numerical optimization routines. This was the essential breakthrough that made rank-based methods workable. Furthermore, the solution to Eq. (1.6.5) is consistent and satisfies certain asymptotic normality conditions

(Hettmansperger and McKean, 2011). In addition, the rank estimate is invariant to whether the data is centered or uncentered. Therefore, it is important to consider the intercept in rank methods.

The rank estimate $\hat{\beta}_n^{\mathrm{R}}$ can be found by minimizing the function as follows

$$\hat{\beta}_n^{\mathrm{R}} = \underset{\beta \in \mathbb{R}}{\operatorname{argmin}}\; D_n(\beta). \tag{1.6.6}$$

Another alternative to obtaining the optimal value for β is to take the derivative,

$$\frac{\partial D_n(\beta)}{\partial \beta} = -\sum_{i=1}^{n} x_i a_n(R_{n_i}(\beta)), \tag{1.6.7}$$

and set it to zero. Therefore, an estimate can be obtained for β from $\partial D_n(\beta)/\partial\beta = 0$ and produces the same $\hat{\beta}_n^{\mathrm{R}}$ as above. In this case, a root-finding method is needed to solve the problem.

Let us now define

$$L_n(\beta) = -\frac{1}{\sqrt{n}}\frac{\partial D_n(\beta)}{\partial \beta} = \frac{1}{\sqrt{n}}\sum_{i=1}^{n} x_i a_n(R_{n_i}(\beta)). \tag{1.6.8}$$

Then, a gradient test for $\mathcal{H}_o : \beta = 0$ vs. $\mathcal{H}_A : \beta \neq 0$ can be developed by defining

$$L_n(0) = -\frac{1}{\sqrt{n}}\frac{\partial D_n(\beta)}{\partial \beta}\bigg|_{\beta=0}.$$

The next step is to estimate the intercept θ given that we know $\hat{\beta}_n^{\mathrm{R}}$. This requires another dispersion function that uses the absolute values of the residuals as follows:

$$D_n(\theta) = \sum_{i=1}^{n} |y_i - (\theta + \hat{\beta}_n^{\mathrm{R}} x_i)|\, a_n^+(R_{n_i}(|y_i - (\theta + \hat{\beta}_n^{\mathrm{R}} x_i)|)), \tag{1.6.9}$$

which we can write compactly as

$$D_n(\theta) = \sum_{i=1}^{n} |y_i - (\theta + \hat{\beta}^{\mathrm{R}} x_i)|\, a_n^+(R_{n_i}^+(\theta)), \tag{1.6.10}$$

where

$$R_{n_i}^+(\theta) = R_{n_i}(|y_i - (\theta + \hat{\beta}_n^{\mathrm{R}} x_i)|).$$

Note that all terms are included here since the invariance property no longer holds. In addition, $R_{n_i}^+(\cdot)$ is the rank of the absolute values of the residuals, and $a_n^+(\cdot)$ uses a different scoring function. From this function, the intercept is obtained as follows

$$\hat{\theta}_n^{\text{R}} = \underset{\theta \in \mathbb{R}}{\operatorname{argmin}}\, D_n(\theta). \tag{1.6.11}$$

As before, an alternative to obtaining $\hat{\theta}_n^{\text{R}}$ is to take the derivative and set it to zero, i.e. $\partial D_n(\theta)/\partial\theta = 0$, where

$$\frac{\partial D_n(\theta)}{\partial\theta} = -\sum_{i=1}^{n} \operatorname{sgn}(y_i - (\theta + \hat{\beta}_n^{\text{R}} x_i))a_n^+(R_{n_i}^+(\theta)). \tag{1.6.12}$$

We can define the function $T_n(\theta)$ as

$$T_n(\theta) = -\frac{1}{\sqrt{n}}\frac{\partial D_n(\theta)}{\partial\theta} = \frac{1}{\sqrt{n}}\sum_{i=1}^{n} \operatorname{sgn}(y_i - (\theta + \hat{\beta}_n^{\text{R}} x_i))a_n^+(R_{n_i}^+(\theta)). \tag{1.6.13}$$

A gradient test for $\mathcal{H}_o : \theta = 0$ vs. $\mathcal{H}_A : \theta \neq 0$ can be developed by defining

$$T_n(0) = -\frac{1}{\sqrt{n}}\left.\frac{\partial D_n(\theta)}{\partial\theta}\right|_{\theta=0}.$$

1.6.1 Ranking and Scoring Details

The ranking and scoring functions will now be elaborated. We first want to solve the linear regression problem by minimizing the dispersion function, $D_n(\beta)$, w.r.t. β. The ranking function, $R_i = R(y_i - \beta x_i)$ ranks each of the partial-residuals $(y_i - \beta x_i)$ among $(y_1 - \beta x_1, \dots, y_n - \beta x_n)$. For convenience, we use $R_{n_i}(\beta)$ to refer to the rank of the ith partial-residual. The rank itself will lie in the set $\{1, \cdots, n\}$ which means that its range is dependent on the number of observations, n. To avoid this, the rank is normalized by a factor of $(n + 1)$ so that $0 < R_{n_i}(\beta)/(n+1) < 1$. The weighting component, $a_n(\cdot)$ in $D_n(\beta)$, involves a scoring function which is a function of these normalized ranks. The normalized rank values applied to a scoring function can be defined in either of the following ways

$$a_n(i) = \mathbb{E}\left[\phi\left(U_{n_i}\right)\right], \quad i = 1, \dots, n \tag{1.6.14}$$

where $U_{n_1} \le U_{n_2} \le \cdots \le U_{n_n}$ are the ordered statistics corresponding to the sample size of n from the $U(0,1)$ distribution; or

$$a_n(R_{n_i}(\beta)) = \phi\left(\frac{R_{n_i}(\beta)}{n+1}\right). \tag{1.6.15}$$

Here, $\phi(u) = 2u - 1, 0 < u < 1$ is a non-decreasing, square-integrable scoring function. Without loss of generality, standardized scores can be defined such that

(1) $\int_0^1 \phi(u)du = 0$, i.e. the integral in [0,1] is 0

(2) $\int_0^1 \phi^2(u)\mathrm{d}u = 1$, i.e. the square integral in [0,1] is 1
(3) $\phi(1-u) = -\phi(u)$, i.e. it is skew-symmetric about $\frac{1}{2}$.

The Wilcoxon score function, given by

$$\phi(u) = \sqrt{12}\left(u - \frac{1}{2}\right), \tag{1.6.16}$$

satisfies the above conditions, since clearly

$$\int_0^1 \sqrt{12}\left(u - \frac{1}{2}\right)\mathrm{d}u = \sqrt{12}\left(\frac{u^2}{2} - \frac{u}{2}\right|_0^1 = 0,$$

and

$$\int_0^1 12\left(u - \frac{1}{2}\right)^2 \mathrm{d}u = \int_0^1 12\left(u^2 - u + \frac{1}{4}\right)\mathrm{d}u = 12\left(\frac{u^3}{3} - \frac{u^2}{2} + \frac{u}{4}\right|_0^1 = 1,$$

and

$$\phi(1-u) = \sqrt{12}\left(1 - u - \frac{1}{2}\right) = -\sqrt{12}\left(u - \frac{1}{2}\right) = -\phi(u).$$

In order to compute $\hat{\theta}_n^{\mathrm{R}}$, we minimize $D_n(\theta)$ of Eq. (1.6.10) which involves $R^+(\cdot)$ and $a_n^+(\cdot)$. The function $R_{n_i}^+(\cdot)$ ranks $|y_i - (\theta + \beta x_i)|$ among $|y_1 - (\theta + \beta x_1)|, \dots, |y_n - (\theta + \beta x_n)|$. The corresponding signed-rank scores are generated using one of the following two ways

$$a_n^+(i) = \mathbb{E}\left[\phi^+\left(U_{n_i}\right)\right], \quad i = 1, \dots, n \tag{1.6.17}$$

where the ordered $U_{n_i} \sim U(0,1)$, for $i = 1, \dots, n$; or

$$a_n^+(R_{n_i}^+(\beta)) = \phi^+\left(\frac{R_{n_i}^+(\beta)}{n+1}\right), \tag{1.6.18}$$

where the function,

$$\phi^+(u) = \phi\left(\frac{u+1}{2}\right), \tag{1.6.19}$$

can be found from considering Wilcoxon score function of Eq. (1.6.16) to be

$$\phi^+(u) = \sqrt{12}\left(\frac{u+1}{2} - \frac{1}{2}\right) = \sqrt{3}u. \tag{1.6.20}$$

The plots for the standardized $\phi(u)$ and $\phi^+(u)$ are provided in Figure 1.5. We see that both functions are linear with respect to u, with the range $\frac{1}{n+1} < u < \frac{n}{n+1}$. From these plots, we can understand that $\sum_{i=1}^{n} a_n(\cdot) = 0$ but $\sum_{i=1}^{n} a_n^+(\cdot) \neq 0$. The residuals are weighted by these two scoring functions to form the loss functions.

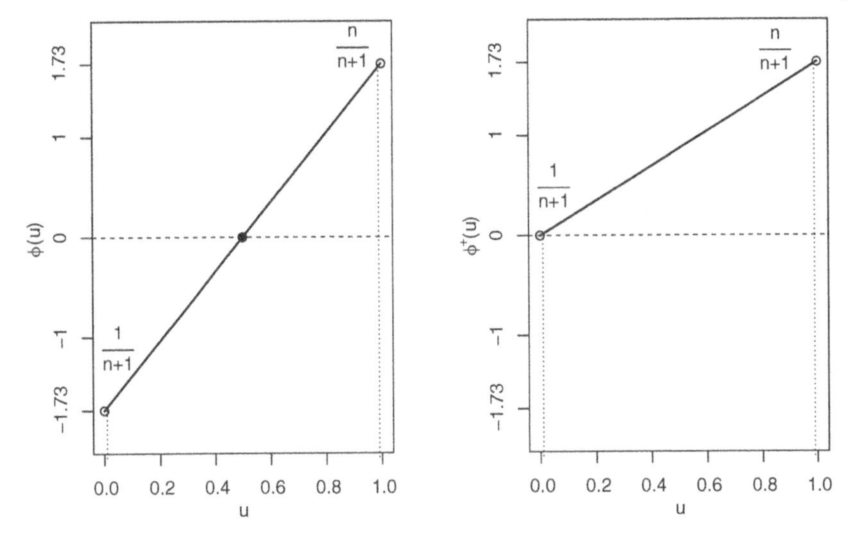

Figure 1.5 Scoring functions $\phi(u) = \sqrt{12}(u - 0.5)$ and $\phi^{+}(u) = \sqrt{3}u$.

The loss functions are then minimized to obtain the estimates. Both of these functions will be used for R-estimation in the next section.

1.6.2 Detailed Procedure for R-estimation

For demonstration purposes, we will now use the dispersion function to numerically obtain the R-estimations of θ and β for the original Belgium telephone data set (see Table 1.3 which is plotted in Figure 1.1(d)). We will detail the steps for numerical calculations and then show the results graphically to provide the complete picture. We first seek to estimate β using Eq. (1.6.5). The estimator $\hat{\beta}_n^{\mathrm{R}}$ is given by

$$\hat{\beta}_n^{\mathrm{R}} = \underset{\beta}{\operatorname{argmin}} \left\{ \sum_{i=1}^{n} (y_i - \beta x_i) \sqrt{12} \left(\frac{R_{n_i}(\beta)}{n+1} - \frac{1}{2} \right) \right\},$$

where $R_{n_i}(\beta) = R_{n_i}(y_i - \beta x_i)$. One can use a statistical software, e.g., the function `optim` of the package `CVXR` in R, to numerically solve this nonlinear optimization problem. Here, we will show how to do this by hand so that a deeper understanding can be obtained.

Clearly, we need to begin by picking a number of values for β to proceed with the solution to this problem until we find the minimum. For this purpose, we consider the three somewhat arbitrary values for β: $\beta^{(0)} = 0$, $\beta^{(1)} = 0.145$, and $\beta^{(2)} = 0.2$, knowing the minimum will likely be in this range.

The calculations to obtain $D_n(\beta)$ are reported in Table 1.6. We show column values for x, y, R, a_n and D_n as defined in the caption. There is one row in this table for each data point in the telephone data set. The D_n column is to be summed for each of the $\beta^{(0)}$, $\beta^{(1)}$, and $\beta^{(2)}$ cases, as required by Eq. (1.6.5). Thus, among these three values, the solution, i.e. $\hat{\beta}_n^{\mathrm{R}}$, is the one that gives the minimum value of "sum" in the last row of Table 1.6. In this example, $\hat{\beta}_n^{\mathrm{R}} = \beta^{(1)} = 0.145$ which will be confirmed later. But, more generally, determining the value of β by hand to find a global solution to the optimization problem is not an easy task and, as mentioned before, we need to use software to numerically find the minimum value. The detailed step-by-step procedure given here is meant to solidify our understanding of Eq. (1.6.5).

Equivalently, instead of the dispersion function, $D_n(\beta)$, one can use the root of its derivative, $L_n(\beta)$, Eq. (1.6.8), as

$$\hat{\beta}_n^{\mathrm{R}} = \left\{\beta : L_n(\beta) = -\frac{1}{\sqrt{n}}\sum_{i=1}^{n} x_i\sqrt{12}\left(\frac{R_{n_i}(\beta)}{n+1} - \frac{1}{2}\right) = 0\right\}.$$

The function `uniroot.all` from package *rootSolve* in R can be used here. We provide the required values in Table 1.6 in the column labeled L_n. Here we seek the "sum" that is closest to zero. Again, the zero-crossing occurs at $\hat{\beta}^{\mathrm{R}} = \beta^{(1)} = 0.145$.

We have shown two methods for R-estimation of β. Typically, we use $D_n(\beta)$ to compute the estimate, and $L_n(\beta)$ for hypothesis testing. It is instructive to examine a graphical equivalent of the procedure outlined above. This is shown in Figure 1.6. Consider the two panels on the left. The top panel shows the graph of $D_n(\beta)$ which is clearly a convex function of β. We note that the minimum is shown to be at 0.145. Similarly, the bottom panel is the function $L_n(\beta)$. We see the zero crossing of this function at 0.145 also, with 276 steps, one for each slope. The values of these two graphs can be matched with the data for $\beta^{(0)}$, $\beta^{(1)}$, and $\beta^{(2)}$ in Table 1.6.

Now we turn to the rank estimation of θ using Eq. (1.6.10). In this regard, based on the related Wilcoxon score in Eq. (1.6.20), it can be estimated as

$$\hat{\theta}_n^{\mathrm{R}} = \operatorname{argmin}\left\{\sum_{i=1}^{n} \left|y_i - (\theta + \hat{\beta}_n^{\mathrm{R}} x_i)\right| \sqrt{3}\left(\frac{R_{n_i}^{+}(\theta)}{n+1}\right)\right\},$$

where

$$R_{n_i}^{+}(\theta) = R_{n_i}(|y_i - (\theta + \hat{\beta}_n^{\mathrm{R}} x_i)|).$$

Here, we are given that $\hat{\beta}_n^{\mathrm{R}} = 0.145$ for the telephone data set so the only unknown is θ.

Table 1.6 The individual terms that are summed in $D_n(\beta)$ and $L_n(\beta)$ for the telephone data set using initial values $\beta^{(0)} = 0$, $\beta^{(1)} = 0.145$, and $\beta^{(2)} = 0.2$. Here, $R = R_{n_i}(\beta)$, $a_n = a_{n_i}(R)$, $D_n = (y - \beta\, x)a_n$, $L_n = xa_n/\sqrt{n}$.

		$\beta^{(0)} = 0$				$\beta^{(1)} = 0.145$				$\beta^{(2)} = 0.2$			
x	y	R	a_n	D_n	L_n	R	a_n	D_n	L_n	R	a_n	D_n	L_n
1950	0.44	1	−1.59	−0.70	−634.27	17	0.62	−177.13	248.19	18	0.76	−296.88	303.35
1951	0.47	2.50	−1.39	−0.65	−551.83	16	0.48	−137.82	193.14	16	0.48	−189.01	193.14
1952	0.47	2.50	−1.39	−0.65	−552.11	14	0.21	−59.10	82.82	15	0.35	−135.08	138.03
1953	0.59	4	−1.18	−0.69	−469.53	13	0.07	−19.70	27.62	14	0.21	−81.06	82.86
1954	0.66	5	−1.04	−0.69	−414.51	12	−0.07	19.71	−27.63	13	0.07	−27.03	27.63
1955	0.73	6	−0.90	−0.66	−359.42	11	−0.21	59.13	−82.94	12	−0.07	27.04	−27.65
1956	0.81	7	−0.76	−0.62	−304.28	10	−0.35	98.58	−138.31	11	−0.21	81.14	−82.99
1957	0.88	8	−0.62	−0.55	−249.09	4	−1.18	335.25	−470.49	9	−0.48	189.39	−193.73
1958	1.06	9	−0.48	−0.51	−193.83	9	−0.48	138.03	−193.83	8	−0.62	243.52	−249.21
1959	1.20	10	−0.35	−0.42	−138.52	7	−0.76	216.91	−304.75	7	−0.76	297.68	−304.75
1960	1.35	11	−0.21	−0.28	−83.16	8	−0.62	177.47	−249.47	6	−0.90	351.85	−360.34
1961	1.49	12	−0.07	−0.10	−27.73	6	−0.90	256.35	−360.53	5	−1.04	406.04	−415.99
1962	1.61	13	0.07	0.11	27.75	5	−1.04	295.81	−416.20	4	−1.18	460.27	−471.70
1963	2.12	14	0.21	0.44	83.28	15	0.35	−98.48	138.81	10	−0.35	135.27	−138.81
1964	11.90	19	0.90	10.72	361.08	19	0.90	−247.37	361.08	19	0.90	−343.06	361.08
1965	12.40	20	1.04	12.89	416.84	20	1.04	−285.05	416.84	20	1.04	−395.53	416.84
1966	14.20	21	1.18	16.72	472.66	21	1.18	−321.11	472.66	21	1.18	−446.38	472.66
1967	15.90	22	1.32	20.93	528.53	22	1.32	−356.85	528.53	22	1.32	−496.93	528.53
1968	18.20	23	1.45	26.48	584.47	23	1.45	−391.27	584.47	23	1.45	−546.18	584.47
1969	21.20	24	1.59	33.78	640.45	24	1.59	−423.99	640.45	24	1.59	−593.73	640.45
1970	4.30	18	0.76	3.28	306.46	18	0.76	−215.77	306.46	17	0.62	−242.99	250.74
1971	2.40	15	0.35	0.83	139.37	1	−1.59	454.41	−641.11	1	−1.59	624.33	−641.11
1972	2.70	16	0.48	1.31	195.22	2	−1.45	414.67	−585.65	2.50	−1.39	542.76	−557.77
1973	2.90	17	0.62	1.81	251.12	3	−1.32	375.11	−530.15	2.50	−1.39	542.76	−558.05
Sum				122.78	28.95			107.78	0.0			108.19	−2.32

As before, consider three selected values for θ as $\theta^{(0)} = -285$, $\theta^{(1)} = -283$, $\theta^{(2)} = -280$. We already know that this is the proper range for the solution so these are suitable choices. Table 1.7 provides the calculations for these values. Again, we show column values for x, y, R^+, a_n^+ and D_n as defined in the caption. If we sum the D_n columns for each case, we arrive at the totals at the bottom of the table. We note that 124 is the smallest value and therefore we choose $\hat{\theta}^{\mathrm{R}} = \theta^{(1)} = -283.0$ as the solution based only on the three choices given.

It is possible to find the root of $T_n(\theta)$ in Eq. (1.6.13) rewritten as

$$\hat{\theta}_n^{\mathrm{R}} = \left\{\theta : T_n(\theta) = \frac{1}{n}\sum_{i=1}^{n} \operatorname{sgn}(y_i - (\theta + \hat{\beta}_n^{\mathrm{R}} x_i))\sqrt{3}\left(\frac{R_{n_i}^+(\theta)}{n+1}\right) = 0\right\}.$$

In this regard, we provide the columns labeled T_n in Table 1.7 for each θ. This time we are interested in the column sum that is closest to zero. Of course, it is the same column as before and we find that $\hat{\theta}^{\mathrm{R}} = -283.0$. For a graphical

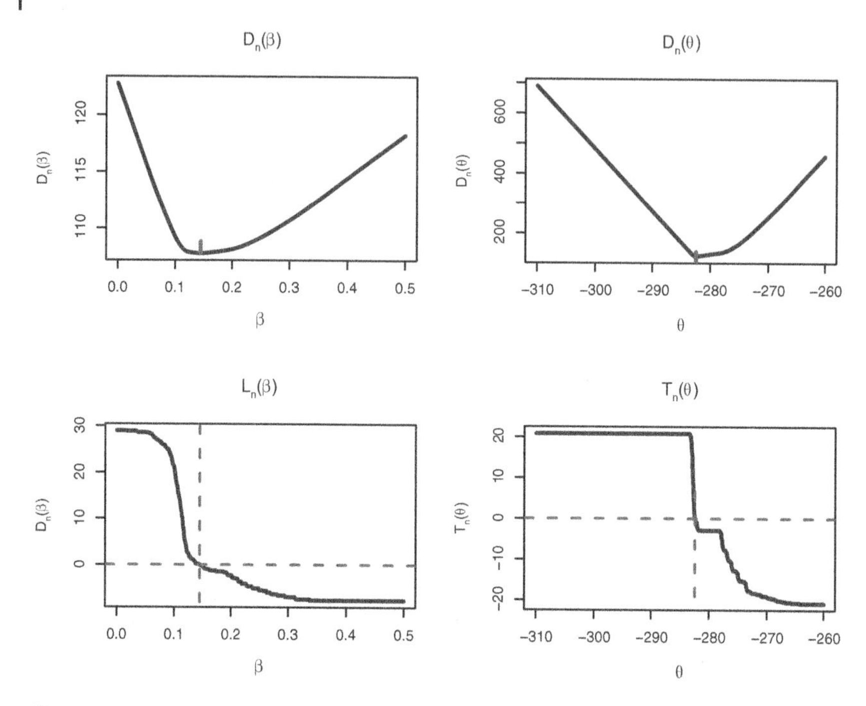

Figure 1.6 Dispersion functions and derivative plots for Figure 1.1(d).

equivalent of this process, we can return to Figure 1.6 and examine the panels on the right. We see the plot for $D_n(\theta)$ at the top and the minimum of the function is at −283.0 as expected. The "sum" values in Table 1.7 can be validated using this graph. Similarly, the bottom panel for $T_n(\theta)$ is the derivative plot with 24 steps and has a zero crossing at −283.0. This completes the detailed calculations for the original telephone data set.

Using the procedure outlined above, the R-estimates for all four cases shown in Figure 1.1 can be computed. This was done in Python using built-in optimization routines and is provided in Table 1.8 along with the LS results. We validated the results in R using Rfit (Kloke and McKean, 2012) which performs rank-based linear regression using the techniques described here. The slope and intercept R-estimates are only marginally affected by the outliers. On the other hand, large changes occur in the LS estimates in Table 1.8 for the corresponding cases. In addition, the rank estimates are similar to Table 1.4 which was based on Theil's method.

If we return to in Figure 1.6 which graphically illustrates the procedure to obtain estimates, we observe that $D_n(\beta)$ and $D_n(\theta)$ are both convex functions. This is a key feature of the dispersion function that allows it to be used to find the estimates using optimization routines. And we now have a formal

Table 1.7 The terms that are summed in $D_n(\theta)$ and $L_n(\theta)$ for the telephone data set using initial values $\theta^{(0)} = -285$, $\theta^{(1)} = -283$, and $\theta^{(2)} = -280$. Here, $R^+ = R^+_{n_i}(\theta)$, $a^+_n = a^+_{n_i}(R^+)$, $D_n = |y - \theta - \hat{\beta}^{\mathrm{R}}\, x|\, a^+_n$, $L_n = \mathrm{sgn}(y - \theta - \hat{\beta}^{\mathrm{R}}\, x)a^+_n/\sqrt{n}$.

		$\theta^{(0)} = -285$				$\theta^{(1)} = -283$				$\theta^{(2)} = -280$			
x	y	R^+	a^+_n	D_n	L_n	R^+	a^+_n	D_n	L_n	R^+	a^+_n	D_n	L_n
1950	0.44	17	1.18	7.46	0.24	3	0.21	0.02	0.04	2	0.14	0.51	−0.03
1951	0.47	16	1.11	6.89	0.23	1	0.07	0	−0.01	3	0.21	0.80	−0.04
1952	0.47	14	0.97	5.88	0.20	4	0.28	0.04	−0.06	5	0.35	1.38	−0.07
1953	0.59	13	0.90	5.43	0.18	5	0.35	0.06	−0.07	6	0.42	1.66	−0.09
1954	0.66	12	0.83	4.95	0.17	6	0.42	0.10	−0.09	7	0.48	1.94	−0.10
1955	0.73	11	0.76	4.47	0.16	7	0.48	0.16	−0.10	8	0.55	2.26	−0.11
1956	0.81	10	0.69	4.02	0.14	8	0.55	0.21	−0.11	9	0.62	2.59	−0.13
1957	0.88	4	0.28	1.61	0.06	14	0.97	0.45	−0.20	15	1.04	4.43	−0.21
1958	1.06	9	0.62	3.58	0.13	9	0.62	0.27	−0.13	10	0.69	2.91	−0.14
1959	1.20	7	0.48	2.77	0.10	11	0.76	0.33	−0.16	12	0.83	3.51	−0.17
1960	1.35	8	0.55	3.18	0.11	10	0.69	0.30	−0.14	11	0.76	3.21	−0.16
1961	1.49	6	0.42	2.42	0.09	12	0.83	0.36	−0.17	13	0.90	3.80	−0.18
1962	1.61	5	0.35	2.01	0.07	13	0.90	0.42	−0.18	14	0.97	4.13	−0.20
1963	2.12	15	1.04	6.36	0.21	2	0.14	0.01	−0.03	4	0.28	1.09	−0.06
1964	11.90	19	1.32	20.78	0.27	19	1.32	12.59	0.27	19	1.32	7.58	0.27
1965	12.40	20	1.39	22.38	0.28	20	1.39	13.75	0.28	20	1.39	8.48	0.28
1966	14.20	21	1.45	25.74	0.30	21	1.45	16.74	0.30	21	1.45	11.24	0.30
1967	15.90	22	1.52	29.35	0.31	22	1.52	19.91	0.31	22	1.52	14.15	0.31
1968	18.20	23	1.59	34.13	0.32	23	1.59	24.25	0.32	23	1.59	18.23	0.32
1969	21.20	24	1.66	40.37	0.34	24	1.66	30.06	0.34	24	1.66	23.77	0.34
1970	4.30	18	1.25	9.09	0.26	18	1.25	1.33	0.26	1	0.07	0.19	−0.01
1971	2.40	1	0.07	0.37	0.01	17	1.18	1.16	−0.24	18	1.25	5.97	−0.26
1972	2.70	2	0.14	0.75	0.03	16	1.11	0.92	−0.23	17	1.18	5.45	−0.24
1973	2.90	3	0.21	1.14	0.04	15	1.04	0.81	−0.21	16	1.11	5.07	−0.23
Sum				245	4.24			124	0.00			136	−0.60

Table 1.8 The LS and R estimations of slope and intercept for Figure 1.1 cases.

Estimators	Case (a)	Case (b)	Case (c)	Case (d)	Estimators	Case (a)	Case (b)	Case (c)	Case (d)
$\hat{\theta}^{\mathrm{LS}}_n$	−234.2	−340.1	−563.8	−983.9	$\hat{\theta}^{\mathrm{R}}_n$	−205.3	−210.7	−224.8	−283.7
$\hat{\beta}^{\mathrm{LS}}_n$	0.120	0.175	0.289	0.504	$\hat{\beta}^{\mathrm{R}}_n$	0.105	0.108	0.115	0.145

procedure to perform rank-based linear regression through the dispersion functions, $D_n(\beta)$ and $D_n(\theta)$. The derivative functions, $L_n(0)$ and $T_n(0)$ also provide useful gradient test statistics, although this is not generally used in machine learning applications. The theoretical underpinnings of the approach may be found in Hettmansperger and McKean (2011).

1.7 Shrinkage Estimation and Subset Selection

We have discussed the use of rank-based methods for simple linear regression with one explanatory variable ($p = 1$) and shown that it offers enormous value compared to LSE. This section extends the previous study to multiple linear regression, i.e. when $p \geq 2$, where p is the number of explanatory variables. Real-world data sets typically have a large number of parameters associated with them. We will see that equations associated with $p \geq 2$ are natural extensions of the $p = 1$ case.

The goal of any regression method used in machine learning is to produce a compact model with high prediction accuracy using a minimum number of parameters. However, new problems arise in multiple linear regression that are not encountered in simple linear regression. In particular, problems may arise due to multicollinearity, over-fitting, under-fitting, identifying the subset of important variables, and shrinking the size of parameter values to improve model accuracy. A penalty function applied to the loss function allows us to mitigate the problems listed above while achieving the desired goals of regression. This subject will be detailed in Chapter 2.

The purpose of this section is to provide background and intuition surrounding the use of penalty functions for multiple linear regression. It gives an overview of penalty functions and offers a number of geometric interpretations of their effects. We will present the basic equations and why they are so useful for machine learning. In addition, the reader will become familiar with some of the terminology used in this field. This will also set the stage for the rest of this book as it focuses on penalty estimation as it pertains to rank-based methods in general.

1.7.1 Multiple Linear Regression using Rank

Most realistic problems fall into the category of multiple linear regression given by the matrix expression

$$\boldsymbol{y} = \theta \mathbf{1}_n + \boldsymbol{X}\boldsymbol{\beta} + \varepsilon, \tag{1.7.1}$$

where $\boldsymbol{y}$ is an $n \times 1$ vector of responses, θ is the intercept (which is also referred to as β_0), $\mathbf{1}_n = (1, \dots, 1)^{\top}$ is an n-vector, $\boldsymbol{X}$ is an $n \times p$ design matrix, $\boldsymbol{\beta}$ is a $p \times 1$

vector of parameters, and ε is an $n \times 1$ error term. Each row of $\boldsymbol{X}$ represents one observation out of a total of n, while each column is a vector of values, one for each of the p variables. For the moment, and without loss of generality, if we assume that $\theta = 0$ for a centered problem (i.e. subtracting their respective means from all data columns), then we have

$$\boldsymbol{y} = \boldsymbol{X}\boldsymbol{\beta} + \boldsymbol{\varepsilon}. \tag{1.7.2}$$

The LSE of this equation has an elegant closed-form solution as follows:

$$\hat{\boldsymbol{\beta}}_n^{\mathrm{LS}} = (\boldsymbol{X}^\top\boldsymbol{X})^{-1}\boldsymbol{X}^\top\boldsymbol{y}. \tag{1.7.3}$$

Two potential problems may arise from this solution. First, if $p > n$, then the matrix $\boldsymbol{X}^\top\boldsymbol{X}$ is not invertible and the parameters are indeterminate. This is easy to check and can be resolved by either adding more data points or reducing the number of parameters (Tibshirani, 1996; Zou and Hastie, 2005). Second, if any of the variables are linearly related, then $\boldsymbol{X}^\top\boldsymbol{X}$ may not be invertible and the parameters are again indeterminate. The solution here is not so straightforward since it implies that there may exist an infinite number of possible solutions. Detecting this multicollinearity condition requires checking the eigenvalues of the correlation matrix, or its condition number, or the variance inflation factor (VIF) (Saleh et al., 2019).

As a simple illustration, let $p = 20$ such that we have variables $\boldsymbol{x}_1, \dots, \boldsymbol{x}_{20}$. Further, let

$$\boldsymbol{x}_7 = 2\boldsymbol{x}_2,$$
$$\boldsymbol{x}_1 = 0.6\boldsymbol{x}_{11} + 0.4\boldsymbol{x}_4.$$

Then, we will encounter multicollinearity in the $\boldsymbol{X}^\top\boldsymbol{X}$ matrix since $\boldsymbol{x}_2$ is linearly related to $\boldsymbol{x}_7$ due to the first equation, and $\boldsymbol{x}_1$ is linearly related to $\boldsymbol{x}_4$ and $\boldsymbol{x}_{11}$ in the second equation. Therefore, the matrix will not be invertible for reasons cited above.

Outliers could also play a role in this context. We note that high-leverage observations may inadvertently contribute to multicollinearity when in fact there is none between certain variables. The use of rank-based methods would assist in producing robust estimates, or perhaps the identification, removal or correction of these influential observations. However, there may still exist inherent multicollinearity even in the clean data set if the design matrix is not of full rank[6]. Interestingly, the search for methods to address multicollinearity eventually led to the widespread use of penalty functions in machine learning, often for reasons unrelated to this problem.

---

6 Here we refer to the rank of a matrix.

Let us examine the solution to this problem in the context of multiple regression using penalty functions. This topic will be elaborated greatly in Chapter 2 but we provide the basic formulations here. For multiple linear regression, the unpenalized loss function for the LS method to obtain unbiased estimates is given by

$$J_n(\boldsymbol{\beta}) = \frac{1}{n}\sum_{i=1}^{n}(y_i - \boldsymbol{x}_i^{\top}\boldsymbol{\beta})^2 \tag{1.7.4}$$

where $\boldsymbol{x}_i = (x_{i1}, \ldots, x_{ip})^{\top}$. This is the sum of squares of the errors (SSE) for the multivariate case with an extra factor of $1/n$ to average the result. The term $\boldsymbol{x}_i$ is the ith row of the $\boldsymbol{X}$ matrix. In machine learning, this convex loss function is solved numerically as an optimization problem to obtain the unbiased estimates.

The rank-based loss function is the dispersion function for multiple linear regression as follows

$$D_n(\boldsymbol{\beta}) = \frac{1}{n}\sum_{i=1}^{n}(y_i - \boldsymbol{x}_i^{\top}\boldsymbol{\beta})a_n(R_{n_i}(\boldsymbol{\beta})). \tag{1.7.5}$$

This convex loss function is also minimized using an optimization loop to obtain the unbiased estimates.

Each of these loss functions can be modified to include a penalty function and solved iteratively in the same way to produce biased estimates. We note that the penalty is applied only to the p variables and not to the intercept. Therefore, it is best to center the data, that is, subtract the column means from each column of the design matrix and response, before applying a penalty function. In effect, we are setting the intercept to 0. This is not needed in rank-based methods since the explanatory variables are estimated separately from the intercept parameter. However, we will center the data for both to maintain consistency.

1.7.2 Penalty Functions

The importance of the penalty function cannot be overstated. The basic idea is to look into the parameter space to find an alternate set of estimates that provide a more compact, general and accurate model for prediction under the guidance of a penalty function. This process is commonly referred to in machine learning circles as *regularization*. Finding the right penalty function makes machine learning more of an art form rather than a science. In fact, the proper selection and tuning of a penalty function controls prediction accuracy and compactness of a model, as will be seen shortly.

To use a penalty in estimation, we start with a least squares loss function and simply add the penalty function to it. The penalized loss function forms a new objective function that is minimized to produce the estimates. A penalty estimator has the form

$$J_n^{\text{LS-Pen}}(\boldsymbol{\beta}) = \frac{1}{n}\sum_{i=1}^{n}(y_i - \boldsymbol{x}_i^{\top}\boldsymbol{\beta})^2 + p_\lambda(\boldsymbol{\beta}), \tag{1.7.6}$$

where $p_\lambda(\boldsymbol{\beta})$ is the penalty function and λ is the tuning parameter that controls the amount of shrinkage or model complexity (i.e. subset selection).

Likewise the general form of penalty estimators for rank-based methods using the dispersion function is as follows,

$$D_n^{\text{R-Pen}}(\boldsymbol{\beta}) = \frac{1}{n}\sum_{i=1}^{n}(y_i - \boldsymbol{x}_i^{\top}\boldsymbol{\beta})a_n(R_{n_i}(\boldsymbol{\beta})) + p_\lambda(\boldsymbol{\beta}). \tag{1.7.7}$$

Penalty functions will not only mitigate the effects of multicollinearity, but they are effective at addressing other aforementioned issues that arise in multiple linear regression. In particular, if $p = 50$, for example, and we wanted to build a more compact model with say $p = 20$, using the most important explanatory variables, how should we go about doing this? We can choose a suitable penalty function for this purpose. The process is generally referred to as *subset selection*.

Another issue is that if $p = 50$ and we find that we are over-fitting the data, but with $p = 10$ we are under-fitting the data, how do we choose the proper value of p to get the "best" fit? Some penalty functions provide this facility. More generally, we use a penalty to shrink the magnitudes of the parameters to improve model prediction. Also, in the areas of logistic regression and neural networks, the iterations of an optimization loop may lead to large values which may cause overflow. In this case, we are interested in penalty estimators that tend to shrink parameter values.

There are two basic ways to shrink a parameter value: subtraction and division (or equivalently, multiply by a number less than 1). For example, assume we have estimate, $\hat{\beta}_n$, and we want to reduce its value. We can simply let $\hat{\beta}_n^{\text{shrinkage}} = \hat{\beta}_n - \lambda$ and increase λ until we achieve the desired reduction in value. If $\lambda = \hat{\beta}_n$, then $\hat{\beta}_n^{\text{shrinkage}} = 0$ and no further reduction is allowed. A second approach is to set $\hat{\beta}_n^{\text{shrinkage}} = \hat{\beta}_n/(1 + \lambda)$ and increase λ until the required reduction is obtained. However, in this case, $\hat{\beta}_n^{\text{shrinkage}}$ does not reach 0, except at $\lambda = \infty$. As simple as these two cases may appear, they are the basis of the two most popular penalty estimators in use today: LASSO and ridge. The key characteristics of these two important penalty functions are illustrated in Figure 1.7.

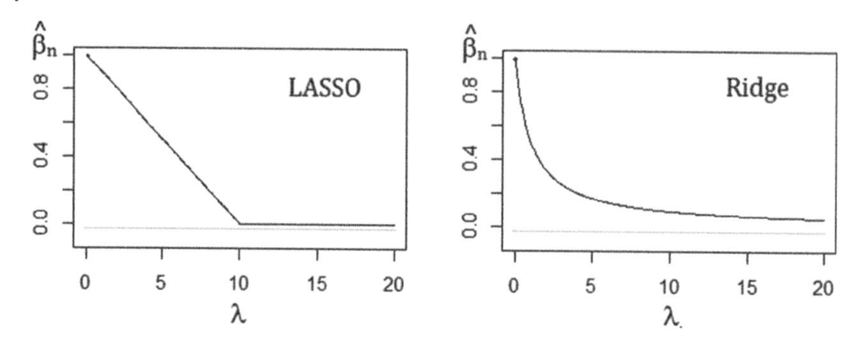

Figure 1.7 Key shrinkage characteristics of LASSO and ridge.

1.7.3 Shrinkage Estimation

The most widely used form of penalty estimation is referred to as a *ridge regression* (Hoerl and Kennard, 1970). In this case,

$$p_\lambda(\boldsymbol{\beta}) = \lambda \sum_{j=1}^{p} |\beta_j|^2, \tag{1.7.8}$$

where λ is a tuning parameter. It is often referred to as the L_2 penalty since it is based on the L_2-norm. Its primary purpose is to shrink the parameters thus moving the estimates away from the unbiased values and towards the null hypothesis, $\boldsymbol{\beta} = 0$. By properly tuning λ, it is possible to apply enough shrinkage to the parameters such that the resulting biased parameters provide an accurate model for prediction.

The objective function using the least squares (LS) loss and ridge penalty functions is given by

$$J_n^{\text{LS-Ridge}}(\boldsymbol{\beta}) = \frac{1}{n}\sum_{i=1}^{n}(y_i - \boldsymbol{x}_i^{\top}\boldsymbol{\beta})^2 + \lambda \sum_{j=1}^{p} |\beta_j|^2. \tag{1.7.9}$$

The objective function using the rank loss and ridge penalty functions is given by

$$D_n^{\text{R-Ridge}}(\boldsymbol{\beta}) = \frac{1}{n}\sum_{i=1}^{n}(y_i - \boldsymbol{x}_i^{\top}\boldsymbol{\beta})a_n(R_{n_i}(\boldsymbol{\beta})) + \lambda \sum_{j=1}^{p} |\beta_j|^2. \tag{1.7.10}$$

One can show that the penalty function effectively places a constraint on the sum of the squares of the parameters. If we had only two parameters, β_1 and β_2, the constraint would be as follows,

$$\beta_1^2 + \beta_2^2 \leq c, \tag{1.7.11}$$

where c is a constant. Note that this constraint would form a circular region about the origin, as shown in Figure 1.8. The figure represents the two-dimensional plane of possible β_1 and β_2 values. The circles represent the constraint $\beta_1^2 + \beta_2^2 = c$ due to the ridge penalty function for different values of c, as set by λ in the penalty function. The point marked "X" represents the unbiased rank or LS estimates. The contours around the estimates represent lines of equal value of the unpenalized loss function. The largest circle (labeled 0) is associated with $\lambda = 0$, while the point the origin (where $\beta_1 = \beta_2 = 0$) is due to $\lambda = \infty$.

Starting at the unbiased solution on the circle labeled 0, with $\lambda_0 = 0$, as we apply the penalty function with tuning factor λ, the estimates shrink and follow some hypothetical trajectory depicted in the figure until the edge of the circle 1 is reached at $\lambda = \lambda_1 > 0$. As λ increases to $\lambda = \lambda_2 > \lambda_1$ (circle 2), we shrink the estimates by a greater degree. At $\lambda = \lambda_3 > \lambda_2$, we reach the circle 3. This continues until all estimates reach 0 at $\lambda = \infty$ at the origin. Essentially, we are shrinking the estimates towards the null hypothesis, $\boldsymbol{\beta} = 0$. In practice, we seek

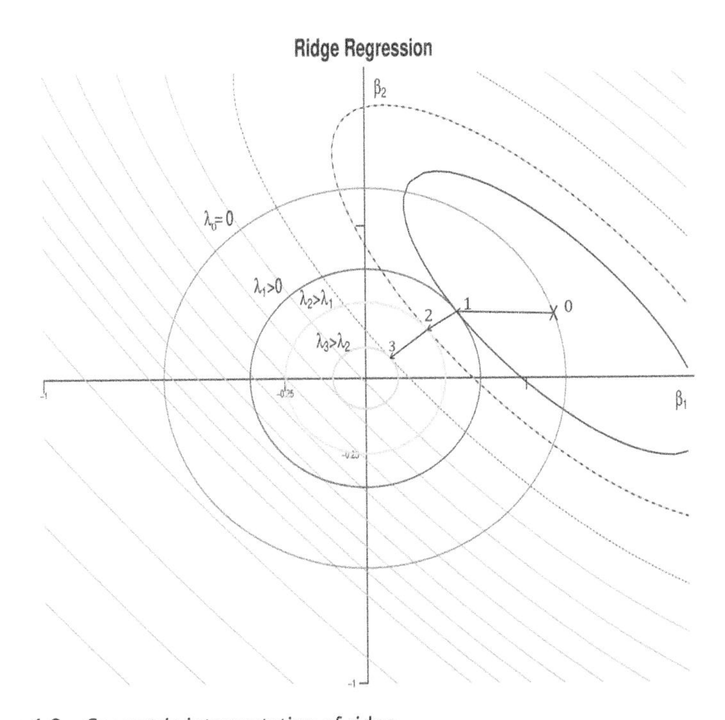

Figure 1.8 Geometric interpretation of ridge.

some intermediate point between the unbiased estimate and the origin. If we have over-fitting or under-fitting problems[7], we can try a number of different values of λ until a proper balance is struck between the two. The process of finding the optimal value, λ_{opt}, to balance this *bias-variance trade-off* is called regularization and is used to obtain high model prediction accuracy.

1.7.4 Subset Selection

The generic term "subset selection" refers to the process of identifying the most important explanatory variables from a large set of given variables. In particular, if there are p variables, the goal is to find a smaller subset, $p_0 < p$, with which to build an accurate predictive model. Unfortunately, this requires that 2^p models be built and tested to determine which one offers the best subset. This would be the "oracle" subset meaning the subset we would ideally want in our compact model. Of course, this process is computationally prohibitive so we seek other methods that have the potential to find this oracle subset with much less effort. Any approach that is able to find the oracle subset for any given p_0 is said to have the oracle property.

A first approach in this direction combining shrinkage and subset selection is referred to as *LASSO* (least absolute shrinkage and selection operator) (Tibshirani, 1996). In this case,

$$p_\lambda(\boldsymbol{\beta}) = \lambda \sum_{j=1}^{p} |\beta_j|, \tag{1.7.12}$$

where λ is the tuning parameter. This is also referred to as the L_1 penalty because it is reminiscent of the L_1-norm.

The objective function with least squares loss and LASSO penalty functions is given by

$$J_n^{\text{LS-LASSO}}(\boldsymbol{\beta}) = \frac{1}{n}\sum_{i=1}^{n}(y_i - \boldsymbol{x}_i^{\mathsf{T}}\boldsymbol{\beta})^2 + \lambda \sum_{j=1}^{p} |\beta_j|. \tag{1.7.13}$$

The objective function with rank loss and LASSO penalty functions is given by

$$D_n^{\text{R-LASSO}}(\boldsymbol{\beta}) = \frac{1}{n}\sum_{i=1}^{n}(y_i - \boldsymbol{x}_i^{\mathsf{T}}\boldsymbol{\beta})a_n(R_{n_i}(\boldsymbol{\beta})) + \lambda \sum_{j=1}^{p} |\beta_j|. \tag{1.7.14}$$

---

7 The definitions of over-fitting and under-fitting will be elaborated in the next chapter. Briefly, over-fitting refers to the tendency to try to fit every point in the data set as closely as possible, whereas under-fitting is inadvertently not fitting enough points in an attempt to create a generalized model.

Using this formulation, a constraint, given by c, is imposed by the penalty function as follows,

$$|\beta_1| + |\beta_2| \leq c. \tag{1.7.15}$$

The resulting constraint geometry is a diamond shape in two dimensions rather than the circle, as shown in Figure 1.9. The adjustment of $\lambda > 0$ has the effect of decreasing the size of the diamond. If $\lambda_0 = 0$, we obtain the unbiased estimates. As λ is increased, the diamond shrinks in size, as do the resulting parameters so it is a shrinkage estimator. Eventually, when $\lambda = \infty$, the origin is reached, as was the case in ridge regression.

The hypothetical trajectory for LASSO starts at the unbiased LSE or rank estimates ("X") on diamond 0 and terminates on diamond 1 with $\lambda = \lambda_1 > 0$. Both estimates shrink but neither one is 0 yet. However, when $\lambda = \lambda_2 > \lambda_1$ and the path lands on one corner of diamond 2, where $\beta_2 = 0$, and we are left with a model with only $\beta_1 \neq 0$. We can now remove β_2 and keep β_1. As such, LASSO is a subset selection estimator. Further increases in λ eventually forces $\beta_1 = 0$ at the origin.

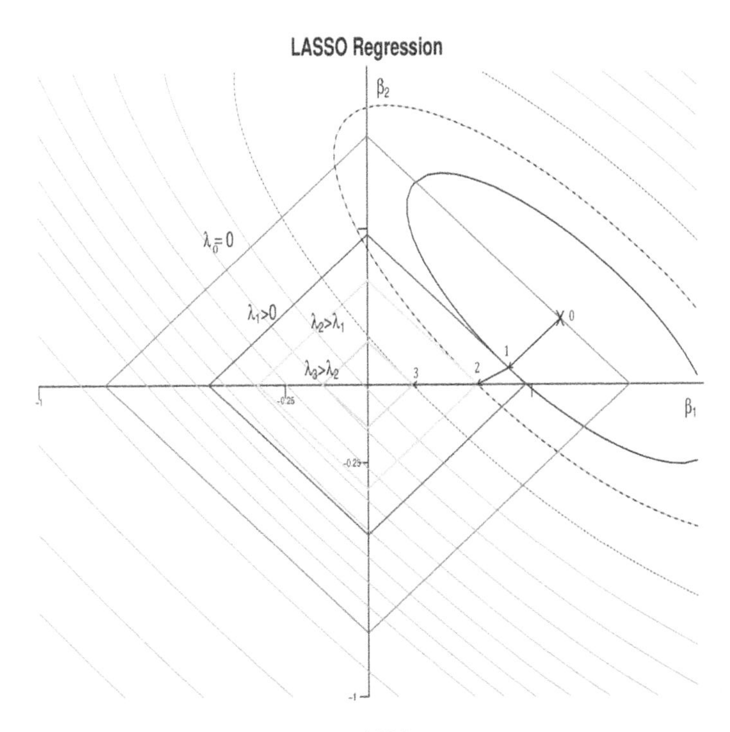

Figure 1.9 Geometric interpretation of LASSO.

Subset selection is very important when the number of parameters is very large. One can imagine a data set with $p = 50$ being reduced to $p_0 = 30$ by this method, and ideally only the important explanatory variables would be retained in the compact model. Unfortunately, LASSO does not have the oracle property; that is, it is possible that the "wrong corner" of the diamond may be reached during shrinkage.

For example, let $p = 7$ and assume we know a priori that $\beta_4 = 0$ and $\beta_6 = 0$ while all other parameters are non-zero. If a penalty estimator such as LASSO reports that initially we should set $\beta_3 = 0$ and $\beta_7 = 0$, then it does not have the oracle property. However, it may still produce a compact model but the parameter values may not be as meaningful. This is important because of model interpretability. We want to be able to eliminate parameters from the model that are not explanatory in terms of the response variable, leaving only those that are important. Whenever model interpretability is important, we require the oracle property be present in the penalty function during subset selection.

Improved methods have been developed that do indeed possess the oracle properties. One in particular is relevant to the discussion here: adaptive LASSO (aLASSO) (Zou, 2006). In the original LASSO, the λ value was applied uniformly to all parameters. In aLASSO, an additional weight is applied to each parameter as follows,

$$p_\lambda(\boldsymbol{\beta}) = \lambda \sum_{j=1}^{p} w_j |\beta_j|. \tag{1.7.16}$$

For example, the weight for each parameter could be set to $w_j = 1/|\hat{\beta}_j^*|$ to produce the new penalty function

$$p_\lambda(\boldsymbol{\beta}) = \lambda \sum_{j=1}^{p} \frac{|\beta_j|}{|\hat{\beta}_j^*|}, \tag{1.7.17}$$

where $\hat{\beta}_j^*$, a known quantity, is the an unpenalized arbitrary estimate for β_j, i.e. the unbiased least squares or rank estimate. We usually consider a $\sqrt{n}$-consistent estimator as $\hat{\beta}_j^*$ to retain good properties for the penalized estimator.

We can understand the adaptive LASSO approach by realizing that small $\hat{\beta}_j^*$ values will lead to large weights, which is equivalent to a high λ with the diamond close to the origin (see Figure 1.9), while large $\hat{\beta}_j^*$ values would be equivalent to a small λ and accordingly a large diamond constraint. Hence, each parameter has its own effective λ based on the unpenalized estimates. This makes sense intuitively since the unpenalized solution has valuable information about the magnitude of the estimates, which in turn can be used as a weighting factor to guide the aLASSO estimation in the right direction for the removal of unimportant variables.

1.7.5 Blended Approaches

Another possible approach is to combine the ridge and LASSO penalties into a single penalty referred to as the *elastic net* (Zou and Hastie, 2005) as follows

$$p_\lambda(\boldsymbol{\beta}) = \lambda_1 \sum_{j=1}^{p} |\beta_j| + \lambda_2 \sum_{j=1}^{p} |\beta_j|^2. \tag{1.7.18}$$

We now have two tuning parameters, λ_1 and λ_2. We note that if $\lambda_1 = 0$ we obtain ridge regression, whereas if $\lambda_2 = 0$ we obtain LASSO. A slight modification of elastic net involves replacing LASSO with aLASSO in order to include the oracle property, as follows

$$p_\lambda(\boldsymbol{\beta}) = \lambda_1 \sum_{j=1}^{p} \frac{|\beta_j|}{|\hat{\beta}_j^*|} + \lambda_2 \sum_{j=1}^{p} |\beta_j|^2. \tag{1.7.19}$$

The two parameters, λ_1 and λ_2, are adjusted until the desired bias/variance trade-off is achieved (i.e. balancing over-fitting and under-fitting). If we let $\lambda_1 = \lambda\alpha$ and $\lambda_2 = \lambda(1-\alpha)/2$, then it is clear that if $\alpha = 0$ we are using ridge while if $\alpha = 1$ we are using aLASSO. In practice, we seek values of λ and α that provide the desired model accuracy. One can understand that the oracle property will be enforced only when $\alpha = 1$, but not necessarily as α approaches 0. However, it provides a better combination of the two penalty functions since it offers a solution to multicollinearity, shrinkage, and oracle subset selection.

1.8 Summary

To summarize this chapter, we have made the case for rank-based estimation as a robust alternative to least squares estimation (LSE) when dealing with outliers, which are contained in most realistic data sets. It was shown that LSE may produce misleading results in the presence of outliers. A number of different cases for the Belgium telephone data set were presented to compare and contrast rank and LSE. The rank estimates were demonstrably better in terms of robustness, especially in high-impact and high-leverage cases of a single outlier.

Further, to improve prediction accuracy, we have described the use of penalty functions which can be viewed as additional constraints on parameter values. Ridge regression provides parameter shrinkage and can be used to reduce the magnitude of parameter values. LASSO is another penalty function that features shrinkage and subset selection, although it does not hold the oracle property. aLASSO, a variant of LASSO, has the oracle property but typically requires unbiased estimates as weighting terms. We also presented the elastic net which is a combination of the two. Much more detail about these penalty functions will be presented in Chapter 2.

We described geometric interpretations of the effect of penalty functions on the estimates. They effectively introduce constraints on the parameter values to lie on a circle in the case of ridge, or on a diamond shape in the case of LASSO. Constraints on LSE and R-estimates provide effective solutions to the problems of multicollinearity, over-fitting/under-fitting, identifying important explanatory variables and shrinking parameter values.

A conceptual view of LSE, rank, ridge and LASSO is provided in Table 1.9. It offers another context of where rank-based methods fit into the overall picture. In particular, the quadratic loss function is based on the L_2-norm, as is the ridge penalty function. However, rank methods are (loosely) based on the L_1-norm, as is the LASSO penalty. Given that rank-based methods produce robust estimates, we say that it behaves like an L_1 loss function. We note the symmetry between the loss and penalty functions in this table. We also specify the key attribute of each method.

We have focused on rank-based methods for linear regression in this chapter. However, the rank-based approach can also be extended to logistic regression and neural networks as we shall demonstrate in the last two chapters of this book on machine learning. Furthermore, penalty functions can also be applied to logistic regression and neural networks in a similar manner. In the next few

Table 1.9 Interpretation of L_1/L_2 loss and penalty functions

Norm	Loss function	Penalty function	Interpretation of entries
L_2	Least squares (mean-based)	Ridge (shrinkage)	L_2 loss is LSE L_2 penalty is ridge
L_1	**Rank (median-based)**	LASSO (selection)	L_1 loss is R-estimation L_1 penalty is LASSO

chapters, we provide a more rigorous treatment of these and other subjects as they pertain to the theory and practice of rank-based methods.

1.9 Problems

1.1. We want to build a simple linear model using Theil's estimator for

$$y_i = \theta + \beta x_i, \quad i = 1, \dots, 100.$$

The Theil estimator of a set of points (x_i, y_i) requires finding the median of pairwise slopes of all points,$(\frac{y_j - y_i}{x_j - x_i})$, and the median of the partial-residuals, $(y_i - \hat{\beta}_n^{\text{Theil}} x_i)$. The estimates are given by

$$\hat{\beta}_n^{\text{Theil}} = \underset{1 \le i < j \le n, i \ne j}{\text{median}} \left(\frac{y_j - y_i}{x_j - x_i} \right), \text{ and } \hat{\theta}_n^{\text{Theil}} = \underset{1 \le i \le n}{\text{median}} \left(y_i - \hat{\beta}_n^{\text{Theil}} x_i \right).$$

If $n = 100$, how many slope calculations do we need to compute $\hat{\beta}_n^{\text{Theil}}$? How many for $\hat{\theta}_n^{\text{Theil}}$? Which is $O(n^2)$ or $O(n)$ in computational complexity?

1.2. Consider the following small data set.

x_i	0.10	0.20	0.30	0.40	0.50	0.60
y_i	3.2	4.0	4.2	4.7	6.5	5.5

Let the data be modeled as

$$\boldsymbol{y} = \theta + \beta \boldsymbol{x} + \varepsilon.$$

Perform hand calculations, or use R or Python, to do the following.

(a) Find the least squares estimates of θ and β using

$$\hat{\beta}_n^{\text{LS}} = \frac{\sum_{i=1}^{n} (x_i - \bar{x})(y_i - \bar{y})}{\sum_{i=1}^{n} (x_i - \bar{x})^2}, \quad \text{and} \quad \hat{\theta}_n^{\text{LS}} = \bar{y} - \hat{\beta}_n^{\text{LS}} \bar{x}.$$

(b) Based on (a), find the outlier(s) using a residual histogram plot.

(c) Find median-based estimates for the intercept and slope parameters using Theil's method.

(d) Based on (c), find the outlier(s) from Theil's estimator using another residual histogram plot.

(e) Remove the outlier(s) from the data set found in part (d) and repeat the process. Comment on the results obtained with and without outliers.

1.3. As an exercise, show that $\phi(u) = -\sqrt{2}\cos(\pi u)$ satisfies the constraints of $\phi(u), 0 < u < 1$, such that $\int_0^1 \phi(u)du = 0$ and $\int_0^1 \phi^2(u)du = 1$. Show that it is also skew-symmetric about 1/2. What is $\phi^+(u)$ in this case? Plot the new scoring functions as in Figure 1.5 along with the Wilcoxon score functions from $0 < u < 1$.

1.4. For the data set in Problem 1.2, obtain the estimators for θ and β using a rank-based method as follows. You can calculate the solution by hand, or use R or Python.

(a) The estimator $\hat{\beta}_n^{\text{R}}$ is given by

$$\hat{\beta}_n^{\text{R}} = \underset{\beta}{\text{argmin}} \left\{ \sum_{i=1}^{n} (y_i - \beta x_i)\sqrt{12} \left(\frac{R_{n_i}(\beta)}{n+1} - \frac{1}{2} \right) \right\},$$

where $R_{n_i}(\beta) = R_{n_i}(y_i - \beta x_i)$. Carry out the minimization step by computing the sum

$$D_n(\beta) = \sum_{i=1}^{n} (y_i - \beta x_i) a_n(R_{n_i}(\beta)).$$

for $\beta = 4.5, 4.6, 4.7, 4.8, 4.9, 5.0$.

(b) The estimator $\hat{\theta}_n^{\text{R}}$ is given by

$$\hat{\theta}_n^{\text{R}} = \text{argmin} \left\{ \sum_{i=1}^{n} \left| y_i - (\theta + \hat{\beta}^{\text{R}} x_i) \right| \sqrt{3} \left(\frac{R_{n_i}^{+}(\theta)}{n+1} \right) \right\},$$

where

$$R_{n_i}^{+}(\theta) = R_{n_i}(|y_i - (\theta + \hat{\beta}^{\text{R}} x_i)|).$$

Carry out the minimization step by computing the sum

$$D_n(\theta) = \sum_{i=1}^{n} |y_i - (\theta + \hat{\beta}^{\text{R}} x_i)| \, a_n^{+}(R_{n_i}^{+}(\theta)),$$

for $\theta = 2.5, 2.6, 2.7, 2.8, 2.9, 3.0$.

(c) Plot the linear regression results from Problem 1.2 and the above results against one another superimposed on the original data set.

(d) Compare and contrast the three estimators used on this data set.

1.5. For the following data set, it is best to use R to solve the problems below. In R, the least square estimates can be obtained using a built-in function `lm()`. However, a package called Rfit (Kloke and McKean, 2012) should be installed in order to quickly obtain the rank-based results. (If you are doing hand calculations only, use the last six of the eight (x, y) pairs of data in the second row.)

x	5551	794	1619	2079	918	1231	3641	4314
y	40	29	28	23	21	17	27	39
x	2786	3989	376	5428	2628	4497	3545	−2108
y	38	46	24	48	34	43	38	94

(a) Draw a scatter plot for this data set. What type of outlier(s) exist in the data, based on the cases given in Figure 1.3?
(b) Fit a line using a least squares method. Plot the associated line on the scatter plot.
(c) Fit a line using a rank-based method. Plot the associated line on the same scatter plot.
(d) Discuss the reasons why the two plots are so different. Which of the two estimates (rank or LS) is more representative of the data?

1.6. Given that the rank-based slope parameter, β, can be estimated using

$$D_n(\beta) = \sum_{i=1}^{n}(y_i - \beta x_i)a_n(R_{n_i}(\beta)),$$

verify this equation by writing a program in R or Python to reproduce the results of Table 1.6 using the telephone data set of Table 1.3. Identify the three points that were computed in Figure 1.6.

1.7. Given that the rank-based intercept, θ, of a simple linear regression problem can be estimated using

$$D_n(\theta) = \sum_{i=1}^{n} |y_i - (\theta + \hat{\beta}^{R} x_i)| a_n^{+}(R_{n_i}^{+}(\theta)),$$

where

$$R_{n_i}^{+}(\theta) = R_{n_i}(|y_i - (\theta + \hat{\beta}^{R} x_i)|).$$

verify this equation by writing a program in R or Python that reproduces the results of Table 1.7 using the telephone data set of Table 1.3. Identify the three points that were computed in Figure 1.6.

1.8. A bridge rank-based penalty estimator would take the form,

$$D_n^{\text{R-bridge}}(\boldsymbol{\beta}) = \frac{1}{n}\sum_{i=1}^{n}(y_i - \boldsymbol{x}_i\boldsymbol{\beta})a_n(R_{n_i}(\boldsymbol{\beta})) + \lambda \sum_{j=1}^{p} |\beta_j|^{\gamma},$$

where the exponent γ has been introduced in the last term.

(a) What type of penalty function is produced if $\gamma = 1$?
(b) What type of penalty function is produced if $\gamma = 2$?
(c) What if $\gamma = 1.5$ or any value between 1 and 2? How would you characterize this type of penalty function?

1.9. For a centered problem, the least squares estimator has the form:

$$\hat{\beta}_n^{\text{LS}} = (X_c^\top X_c)^{-1} X_c^\top \mathbf{y}.$$

One can show that the LS-ridge estimator has the form:

$$\hat{\beta}_n^{\text{LS-Ridge}} = (X_c^\top X_c + \lambda)^{-1} X_c^\top \mathbf{y}.$$

Using the following data from a previous problem, center the data and the solve the following problems.

x_i	0.10	0.20	0.30	0.40	0.50	0.60
y_i	3.2	4.0	4.2	4.7	5.0	5.5

(a) Compute $\hat{\beta}_n^{\text{LS}}$ and $\hat{\theta}_n^{\text{LS}}$ after centering.
(b) Compute $\hat{\beta}_n^{\text{R}}$ and $\hat{\theta}_n^{\text{R}}$.
(c) Find λ such that $\hat{\beta}_n^{\text{LS-Ridge}}$ has the same value as $\hat{\beta}_n^{\text{R}}$.
(d) Plot the results of (b) and (c) and comment on the results and suggest which is the better solution. In particular, compare the intercept of the two methods. Also, is it possible to use LS-ridge to produce the results obtained by rank-based methods?

1.10. (PROJECT) In this problem, you will use R to compare rank with LS. If you have not installed R on your computer, this would be a good time to do so. You should also use `install.packages("Rfit")` to install the Rfit package.

(a) Load the Rfit library and examine the telephone data used in this chapter.

```
library("Rfit")
data(telephone)
telephone
```

(b) Run rfit() on the telephone data and compare the rank results with Table 1.8, case (d).

```
fit <- rfit(calls ~ year, data=telephone)
summary(fit)
```

(c) Run lm() on the telephone data and compare the LS results with Table 1.8, case (d).

```
lmfit <- lm(calls ~ year, data=telephone)
summary(lmfit)
```

(d) Plot the results of the two regression methods. Which is the rank line and which is the LS line?

```
plot(calls ~ year,data=telephone,
        main="Rank vs. LSE",ylim=c(0,25))
abline(coef(lmfit),lwd=2,col="black")
abline(coef(fit),lwd=2,col="red")
```

2

Characteristics of Rank-based Penalty Estimators

2.1 Introduction

In this chapter, we examine practical aspects of rank-based penalty estimators to understand their characteristics. We will demonstrate that, with the proper combination of rank-based linear regression and suitable penalty functions, it is possible to create a compact and robust model that provides improved prediction accuracy and model interpretability. We have already seen in the previous chapter that rank-based regression effectively filters outliers from the data set, acting as an implicit data cleaning agent. It is called a robust estimator for this reason, i.e. it is tolerant of a variety of different types of outliers. Once we identify the outliers, we may choose to reduce the number of observations used in regression from n to n_o, where $n_o \leq n$, but R-estimation inherently does this for us. We now turn our attention to penalty functions. Our goal here is to reduce the number of variables from p to a proper subset p_o using a penalty function, where $p_o \leq p$. This subset can, in turn, be used to create a compact and interpretable model by applying another penalty function to shrink the estimates. By using rank-based penalty estimators, we will in effect be taking a large data set that is $n \times p$, reducing it to a smaller data set that is $n_o \times p_o$, and using that data set to create a model that provides the required prediction accuracy.

2.2 Motivation for Penalty Estimators

Penalty functions and their effective use are very important in machine learning, especially for multiple linear regression, logistic regression, and neural networks. The literature on the subject of penalized rank-based estimators is rather sparse, with the notable exceptions of the work of Johnson and Peng (2008), Turkmen and Ozturk (2014) and Saleh et al. (2018). We will

Rank-Based Methods for Shrinkage and Selection: With Application to Machine Learning.
First Edition. A. K. Md. Ehsanes Saleh, Mohammad Arashi, Resve A. Saleh, and
Mina Norouzirad.

cover rank-based linear regression in this chapter, and then rank-based logistic regression and neural networks in the last two chapters. The original motivation for the penalty function was to address multicollinearity in linear regression. However, the advantages of using a penalty function even in the absence of multicollinearity soon became apparent and it was re-purposed for regularization, a term that refers to adjusting the estimates to improve prediction accuracy of a model and make it more general. Other penalty functions were introduced, each one providing a new capability or improvement on an existing method. We will be describing a variety of such penalty estimators throughout this book.

In this chapter, we sequence through well-known and often-used penalty functions, and apply them to least squares (LS) and rank loss functions to compare and contrast their properties. We will also present a family of related orthogonal and diagonal penalty estimators based on the well-known penalty functions which are simple to use and inherit many of the characteristics of the original penalty estimators from which they are derived.

To begin the study, we will motivate the use of ridge regression (Hoerl and Kennard, 1970) and carry out simple analysis to understand its properties and characteristics. We will illustrate that this penalty function can improve prediction accuracy of a regression model by tuning the degree of regularization on a training set until the desired accuracy is achieved. Once the groundwork for ridge regression is laid, we compare penalized LS and rank estimation using ridge traces on an example data set. Ridge traces provide valuable insight into how ridge works, and this is true of the traces of other penalty functions. We will find that ridge shrinks the parameter values towards the null and, as such, it is called a shrinkage estimator. We refer the reader to Saleh et al. (2019) for a detailed treatment of the subject.

We then move to the LASSO (Tibshirani, 1996) and adaptive LASSO (aLASSO) (Zou, 2006) penalty functions which conduct feature subset selection alongside shrinkage. We compare and contrast these two approaches in the context of LS and rank estimation. We describe the oracle property of aLASSO that can provide, under certain regularity conditions, a set of estimates that asymptotically behave as if the true model were known in advance. Perhaps more importantly, it can provide a nearly optimal ordering of the variables (from most important to least important) for subset selection by examining the zero-crossings of aLASSO traces. LASSO is known not to have the oracle property but is effective in shrinking certain variables to zero to reduce the model size. The key oracle feature of aLASSO can be exploited for subset selection using an algorithm that runs in linear time with respect to p (i.e. the number of variables) using the zero-crossings of the parameters in an aLASSO trace, but requires an additional m models to be built for the trace[1].

Next, we describe blended approaches including "naive" elastic net (nEnet), standard elastic net (Zou and Hastie, 2005) and our improvement called adaptive Enet (aEnet). Of these, aEnet holds the most promise of the blended approaches as it provides shrinkage due to ridge and oracle variable selection due to aLASSO. It is suitable when collinearity is not present and in low-dimensional settings. In order to define the weights associated with the aLASSO penalty function, we may also need the matrix $\boldsymbol{C} = \boldsymbol{X}^{\top}\boldsymbol{X}$ to be invertible for reasons that will become clear later in this chapter. Many data sets satisfy these requirements so we believe this adaptive Enet penalty function may be important in data science.

Finally, we conclude the chapter with an example to bring all of the key points together. We show that while combining all the methods provides a seemingly elegant solution, it may be more practical to sequence the loss and penalty functions to maximize their effectiveness. We demonstrate how the combination of rank, aLASSO and ridge can produce a compact model possessing both prediction accuracy and model interpretability. In upcoming chapters, a rigorous mathematical treatment of rank-based penalty estimators will be presented but here we want to describe the practical usage of them as it pertains to machine learning.

2.3 Multivariate Linear Regression

2.3.1 Multivariate Least Squares Estimation

Consider the multiple regression model in matrix form as follows,

$$\boldsymbol{y} = \theta \mathbf{1}_n + \boldsymbol{X}\boldsymbol{\beta} + \boldsymbol{\varepsilon}, \tag{2.3.1}$$

where $\boldsymbol{y} = (y_1, \dots, y_n)^{\top}$ is a response vector, $\boldsymbol{X}$ is an $n \times p$ design matrix of constants, and $\boldsymbol{\varepsilon} = (\varepsilon_1, \dots, \varepsilon_n)^{\top}$ is the error term. Regression parameters are $(\theta, \boldsymbol{\beta})^{\top}$, where θ is the intercept and $\boldsymbol{\beta} = (\beta_1, \dots, \beta_p)^{\top}$ are the p coefficients of interest. Also, $\mathbf{1}_n = (1, \dots, 1)^{\top}$ is an n-vector of 1s. Here we assume $\boldsymbol{\varepsilon} \sim \mathcal{N}_n(\mathbf{0}, \sigma^2 \boldsymbol{I}_n)$, the normal distribution with mean $\mathbf{0} = (0, \dots, 0)^{\top}$, an $n \times 1$ vector, and variance-covariance matrix $\sigma^2 \boldsymbol{I}_n$. If the data is centered such that

$$\sum_{i=1}^{n} y_i = 0, \text{ and } \quad \sum_{i=1}^{n} x_{ij} = 0 \quad \text{for} \quad j = 1, \dots, p, \tag{2.3.2}$$

1 The approach here requires $p+m$ models. Most selection algorithms are typically quadratic, $O(p^2)$, while the best subset selection is exponential, $O(2^p)$, as described in James et al. (2013).

then we can eliminate the intercept from the analysis and we will find it convenient to do so. Notably, the penalty function is not applied to θ so it can be generally excluded from the discussion by centering. We can now use the formulation given by

$$\boldsymbol{y} = \boldsymbol{X}\boldsymbol{\beta} + \boldsymbol{\varepsilon}. \tag{2.3.3}$$

Our goal is to study the estimation of $\boldsymbol{\beta}$ in a variety of different ways depending on the requirements of the problem. In our first approach, the quadratic or L_2 loss function is as follows:

$$J(\boldsymbol{\beta}) = (\boldsymbol{y} - \boldsymbol{X}\boldsymbol{\beta})^\top(\boldsymbol{y} - \boldsymbol{X}\boldsymbol{\beta}). \tag{2.3.4}$$

To obtain the least squares estimate, $\hat{\boldsymbol{\beta}}_n^{\text{LS}}$, we take the derivative of $J(\boldsymbol{\beta})$ w.r.t. $\boldsymbol{\beta}$ and set it to 0, i.e.

$$\frac{\partial J(\boldsymbol{\beta})}{\partial \boldsymbol{\beta}} = -2\boldsymbol{X}^\top(\boldsymbol{y} - \boldsymbol{X}\boldsymbol{\beta}) = 0. \tag{2.3.5}$$

Then, it follows that

$$-\boldsymbol{X}^\top\boldsymbol{y} + \boldsymbol{X}^\top\boldsymbol{X}\boldsymbol{\beta} = 0, \tag{2.3.6}$$

which can be solved to obtain

$$\hat{\boldsymbol{\beta}}_n^{\text{LS}} = (\boldsymbol{X}^\top\boldsymbol{X})^{-1}\boldsymbol{X}^\top\boldsymbol{y} \tag{2.3.7}$$

provided $\boldsymbol{X}^\top\boldsymbol{X}$ is invertible. This is the direct matrix solution for least squares estimation (LSE) and it provides the best linear unbiased estimate. It is also referred to in the literature as the ordinary least squares (OLS) estimator.

The variance–covariance matrix of the estimates is given by

$$\text{Var}(\hat{\boldsymbol{\beta}}_n^{\text{LS}}) = \sigma^2(\boldsymbol{X}^\top\boldsymbol{X})^{-1} = \sigma^2\boldsymbol{C}^{-1}, \tag{2.3.8}$$

where $\boldsymbol{C} = \boldsymbol{X}^\top\boldsymbol{X}$. Then, the variance of the jth parameter is

$$\text{Var}(\hat{\beta}_j^{\text{LS}}) = \sigma^2 C^{jj}, \quad j = 1, \dots, p, \tag{2.3.9}$$

where C^{jj} is the jth diagonal entry of $\boldsymbol{C}^{-1}$. Hence, the standard error (s.e.) is

$$\text{s.e.}(\hat{\beta}_j^{\text{LS}}) = \hat{\sigma}\sqrt{C^{jj}}, \quad j = 1, \dots, p, \tag{2.3.10}$$

where $\hat{\sigma} = \sqrt{\text{SSE}/(n-p)}$ is an unbiased estimator of σ, and SSE is the sum of squares of the errors defined as

$$\text{SSE} = (\boldsymbol{y} - \boldsymbol{X}\hat{\beta}_n^{\text{LS}})^\top(\boldsymbol{y} - \boldsymbol{X}\hat{\beta}_n^{\text{LS}}). \tag{2.3.11}$$

This can be summarized as $\hat{\boldsymbol{\beta}}_n^{\text{LS}} \sim \mathcal{N}_p(\boldsymbol{\beta}, \sigma^2\boldsymbol{C}^{-1})$.

2.3.2 Multivariate R-estimation

Rank-based estimation was described in detail in Chapter 1. The dispersion function for the multiple regression model is as follows

$$D_n(\boldsymbol{\beta}) = \sum_{i=1}^{n}(y_i - \boldsymbol{x}_i^\top \boldsymbol{\beta})a_n(R_{n_i}(\boldsymbol{\beta})), \tag{2.3.12}$$

where $\boldsymbol{x}_i = (x_{i1}, \dots, x_{ip})^\top$ and $R_{n_i}(\boldsymbol{\beta}) = R_{n_i}(y_i - \boldsymbol{x}_i\boldsymbol{\beta})$ is the rank of the residuals. This is a piece-wise continuous, convex function that can be minimized using an optimization loop to find the estimates, $\hat{\boldsymbol{\beta}}_n^{\text{R}}$. It can be demonstrated that the solution is consistent and asymptotically normal (Hettmansperger and McKean, 2011) Similar to LS, the asymptotic characteristics can be summarized as $\hat{\boldsymbol{\beta}}_n^{\text{R}} \sim \mathcal{N}(\boldsymbol{\beta}, \eta^2\boldsymbol{C}^{-1})$, where η^2 is a scaling parameter that depends on the error probability density function and the score function $\phi(u)$, as described in chapter 3 of Hettmansperger and McKean (2011)[2]. Hence, in a manner similar to LS, we say that

$$\text{s.e.}(\hat{\beta}_j^{\text{R}}) = \eta\sqrt{C^{jj}}, \quad j = 1, \dots, p, \tag{2.3.13}$$

where C^{jj} is the jth diagonal entry of $\boldsymbol{C}^{-1}$, provided $\boldsymbol{C} = \boldsymbol{X}^\top\boldsymbol{X}$ is invertible.

2.3.3 Multicollinearity

The requirement mentioned before that matrix $\boldsymbol{X}^\top\boldsymbol{X}$ be invertible deserves further discussion. Specifically, the matrix must be of full rank[3]. In other words, the variables cannot be linearly dependent on one another. If two variables are linearly or nearly linearly related, we refer to this as collinearity, and when several variables are linearly dependent, and perhaps with multiple occurrences of such linear dependence, the term multicollinearity is invoked. In these cases, LSE and rank will produce estimates with unduly large sampling variances such that meaningful statistical inference becomes difficult.

The original motivation for the use penalty functions was due to multicollinearity. The implication of multicollinearity is that there exist an infinite number of solutions to the regression problem with no constraint to decide which one is most suitable. There are several ways to determine if multicollinearity exists. One way is to find out if the matrix $\boldsymbol{C} = \boldsymbol{X}^\top\boldsymbol{X}$ is invertible. For example, if $\det(\boldsymbol{C}) = 0$, then the matrix is singular and therefore not invertible. Near collinearity is also a problem. For this, we can compute the condition number and, if very large, we declare the matrix as ill-conditioned

2 Their notation uses τ_ϕ^2 in place of η^2. We will be using η^2 for rank-based methods throughout this book.
3 The term "rank" used in this context is referring to the rank of a matrix.

and potentially non-invertible. Other tests include the variance inflation factor (VIF), and a ratio test of the largest and smallest eigenvalues of the matrix. We will not describe these techniques in this chapter, but many resources are available for the interested reader to obtain further details (Montgomery et al., 2012; Saleh et al., 2019).

One way to address the problem of multicollinearity is to somehow constrain the problem to produce a unique solution. In fact, linear regression with a penalty function can be viewed as a constrained minimization problem. Consider a function $h(\boldsymbol{\beta})$ that has an infinite number of solutions when we try to minimize it w.r.t. $\boldsymbol{\beta}$. We now apply a constraint function $g(\boldsymbol{\beta})$ as follows

$$\hat{\boldsymbol{\beta}}_n^{\text{Pen}} = \underset{\boldsymbol{\beta}}{\text{argmin}}\ h(\boldsymbol{\beta}) \qquad \text{subject to} \quad g(\boldsymbol{\beta}) \leq c, \tag{2.3.14}$$

where $\hat{\boldsymbol{\beta}}_n^{\text{Pen}}$ is the constrained estimator. The constant c associated with the constraint function allows a specific solution to be selected. To apply the constraint in a single equation, we can reformulate the problem as a Lagrangian multiplier as follows

$$J^{\text{Pen}}(\boldsymbol{\beta}) = h(\boldsymbol{\beta}) + \lambda g(\boldsymbol{\beta}), \tag{2.3.15}$$

where λ is the multiplier or tuning parameter. In the above, $J^{\text{Pen}}(\boldsymbol{\beta})$ is the objective function, $h(\boldsymbol{\beta})$ plays the role of the loss function, $g(\boldsymbol{\beta})$ plays the role of the penalty function and λ plays the role of c. More concretely, what we call penalty estimation is, in general, minimization of a class of loss functions subject to a penalty function constraint, $p_\lambda(|\boldsymbol{\beta}|)$ with $|\boldsymbol{\beta}| = (|\beta_1|, \dots, |\beta_p|)^\top$. The solution

$$\hat{\boldsymbol{\beta}}_n^{\text{Pen}} = \underset{\boldsymbol{\beta}}{\text{argmin}}\ J^{\text{Pen}}(\boldsymbol{\beta}) \tag{2.3.16}$$

is called a penalized estimator (PE). Note, in most penalty function cases, we have $p_\lambda(|\boldsymbol{\beta}|) = \lambda \sum_{i=1}^{p} p(|\beta_i|)$.

For instance, for LS we will have

$$J^{\text{LS-Pen}}(\boldsymbol{\beta}) = (\boldsymbol{y} - \boldsymbol{X}\boldsymbol{\beta})^\top (\boldsymbol{y} - \boldsymbol{X}\boldsymbol{\beta}) + p_\lambda(|\boldsymbol{\beta}|), \tag{2.3.17}$$

whereas for rank, we will have

$$D^{\text{R-Pen}}(\boldsymbol{\beta}) = \sum_{i=1}^{n} (y_i - \boldsymbol{x}_i^\top \boldsymbol{\beta}) a_n(R_{n_i}(\boldsymbol{\beta})) + p_\lambda(|\boldsymbol{\beta}|). \tag{2.3.18}$$

This is the form we can use to solve regression problems with multicollinearity where the first term is the loss function (either least squares or rank) and the second term is the penalty function (either ridge, LASSO, elastic net or any other variant).

We should note here that penalty functions are widely used even in the absence of multicollinearity. For example, they can be used to improve model prediction using regularization or for subset selection to reduce model complexity and improve model interpretability. Referring back to Chapter 1, consider the geometric plots of Figures 1.8 and 1.9. The contour lines around the unbiased estimate, indicated by "x", are associated with the loss function (either LS or rank). The circles and diamonds represent the constraints of the penalty functions (ridge and LASSO, respectively). As the degree of regularization is increased through λ, different estimates are obtained. Typically, PEs shift unbiased estimates to a series of biased estimates from which we can choose the optimal one that provides accurate prediction. This ultimate goal will drive our discussions in the sections to follow.

2.4 Ridge Regression

Ridge regression (Hoerl and Kennard, 1970) was developed as a way to mitigate the effects of high variance due to multicollinearity by producing a set of biased estimates with lower variance using an L_2 penalty function applied to the quadratic loss function. The degree of bias is adjusted using a tuning parameter λ which can range from 0 to infinity, but usually an optimal value is selected as described later.

2.4.1 Ridge Applied to Least Squares Estimation

The penalty function of ridge applied to the quadratic loss function is given by

$$J^{\text{LS-Ridge}}(\boldsymbol{\beta}) = (\boldsymbol{y} - \boldsymbol{X\beta})^{\top}(\boldsymbol{y} - \boldsymbol{X\beta}) + \lambda\boldsymbol{\beta}^{\top}\boldsymbol{\beta}. \tag{2.4.1}$$

To obtain the ridge estimates, we take the derivative of $J^{\text{LS-Ridge}}(\boldsymbol{\beta})$ w.r.t. $\boldsymbol{\beta}$ and set it to 0, i.e.

$$\frac{\partial J^{\text{LS-Ridge}}(\boldsymbol{\beta})}{\partial \boldsymbol{\beta}} = -2\boldsymbol{X}^{\top}(\boldsymbol{y} - \boldsymbol{X\beta}) + 2\lambda\boldsymbol{\beta} = 0. \tag{2.4.2}$$

Then, it follows that

$$-\boldsymbol{X}^{\top}\boldsymbol{y} + \boldsymbol{X}^{\top}\boldsymbol{X\beta} + \lambda\boldsymbol{\beta} = 0, \tag{2.4.3}$$

and re-arranging gives the ridge regression estimator,

$$\hat{\boldsymbol{\beta}}_n^{\text{LS-ridge}} = (\boldsymbol{X}^{\top}\boldsymbol{X} + \lambda\boldsymbol{I}_p)^{-1}\boldsymbol{X}^{\top}\boldsymbol{y}, \tag{2.4.4}$$

where $\boldsymbol{I}_p$ is a $p \times p$ identity matrix. This is the standard form of the ridge estimate. We can see that problems arising from multicollinearity are circumvented by adding λ to the diagonal elements of $\boldsymbol{X}^\top\boldsymbol{X}$. For some $\lambda > 0$, the matrix $(\boldsymbol{X}^\top\boldsymbol{X} + \lambda\boldsymbol{I}_p)$ is invertible. Furthermore, by careful selection of λ, an accurate prediction model can be obtained in the presence (or absence) of multicollinearity.

To gain insight into the ridge mechanism, it is possible to re-write Eq. (2.4.4) in terms of $\hat{\boldsymbol{\beta}}^{\text{LS}}$ by multiplying all terms in Eq. (2.4.3) by $(\boldsymbol{X}^\top\boldsymbol{X})^{-1}$ and rearranging to obtain

$$-\hat{\boldsymbol{\beta}}_n^{\text{LS}} + \boldsymbol{\beta} + \lambda\boldsymbol{C}^{-1}\boldsymbol{\beta} = 0, \tag{2.4.5}$$

where $\boldsymbol{C}^{-1} = (\boldsymbol{X}^\top\boldsymbol{X})^{-1}$. In doing so, we are assuming $\boldsymbol{C}$ is invertible and hence, there is assumed to be no multicollinearity in $\boldsymbol{X}$; in general, it may not be the case. Then

$$\hat{\boldsymbol{\beta}}_n^{\text{LS-ridge}} = (\boldsymbol{I}_p + \lambda\boldsymbol{C}^{-1})^{-1}\hat{\boldsymbol{\beta}}_n^{\text{LS}}. \tag{2.4.6}$$

In this form of ridge, we can view $\boldsymbol{W}(\lambda) = (\boldsymbol{I}_p+\lambda\boldsymbol{C}^{-1})^{-1}$ as a weight matrix that acts as a shrinkage operator on $\hat{\boldsymbol{\beta}}_n^{\text{LS}}$. The higher the value of λ, the greater the shrinkage. As mentioned, many data sets do not have multicollinearity problems and, even in those cases, ridge is very effective in building a better model for prediction.

We have stated that ridge shrinks the unbiased estimates but have not described the manner in which this shrinkage occurs. To develop some insight, we can make a simplifying assumption that $\boldsymbol{C}$ is the identity matrix, $\boldsymbol{I}_p$. That is, $\boldsymbol{X}$ is an orthonormal matrix. If applied to Eq. (2.3.7), we obtain $\hat{\boldsymbol{\beta}}_n^{\text{O-LS}} = \boldsymbol{X}^\top\boldsymbol{y}$, the orthonormal-LS estimator. And if applied to Eq. (2.4.6), it reduces to

$$\hat{\boldsymbol{\beta}}_n^{\text{O-ridge}} = \frac{\hat{\boldsymbol{\beta}}_n^{\text{LS}}}{1+\lambda}, \tag{2.4.7}$$

where $\hat{\boldsymbol{\beta}}_n^{\text{O-ridge}}$ is the estimator for the orthonormal case. This case tells us something useful about the basic mechanism of ridge. It illustrates that ridge regression applies a factor of $1/(1+\lambda)$ to shrink the $\hat{\boldsymbol{\beta}}_n^{\text{LS}}$ values until they reach 0 at $\lambda = \infty$. This is called the orthogonal ridge-type estimator in this book.

However, ridge does not actually scale all parameters by the same amount as implied in the above equation. To get a better handle on the true nature of ridge, let us assume that $\boldsymbol{C}^{-1}$ is a diagonal matrix with non-zero values C^{jj}, $j = 1, \ldots, p$. All off-diagonal terms are set to zero implying no correlation between the parameters. Therefore, we can write that for the jth parameter that

$$\hat{\beta}_j^{\text{D-ridge}} = \frac{\hat{\beta}_j^{\text{LS}}}{1+\lambda C^{jj}}, \tag{2.4.8}$$

where $\hat{\beta}_j^{\text{D-ridge}}$ is the estimate for the diagonal $\boldsymbol{C}^{-1}$ case. This is the diagonal ridge-type estimator. We may recall that the variance of $\hat{\boldsymbol{\beta}}_n^{\text{LS}}$ is given by the covariance matrix

$$\text{Var}(\hat{\boldsymbol{\beta}}_n^{\text{LS}}) = \sigma^2(\boldsymbol{X}^\top \boldsymbol{X})^{-1} = \sigma^2 \boldsymbol{C}^{-1}, \tag{2.4.9}$$

where σ^2 is the variance of the residuals. Since $\boldsymbol{C}^{-1}$ is assumed to be diagonal, we can set $\lambda = \lambda_2 \sigma^2$ in Eq. (2.4.8) and re-write the above as

$$\hat{\beta}_j^{\text{D-ridge}} = \frac{\hat{\beta}_j^{\text{LS}}}{1 + \lambda_2 \sigma^2 C^{jj}} = \frac{\hat{\beta}_j^{\text{LS}}}{1 + \lambda_2 \,\text{Var}(\hat{\beta}_j^{\text{LS}})}. \tag{2.4.10}$$

As λ_2 increases, each $\hat{\beta}_j^{\text{LS}}$ shrinks at a rate determined by the variance of $\hat{\beta}_j^{\text{LS}}$ to produce $\hat{\beta}_j^{\text{D-ridge}}$. We can consider the following cases:

(a) If $\sigma^2 C^{jj}$ is very small, i.e. much smaller than 1, then $\hat{\beta}_j^{\text{D-ridge}}$ will not change rapidly as λ_2 increases.
(b) If $\sigma^2 C^{jj}$ is very large, i.e. much greater than 1, then $\hat{\beta}_j^{\text{D-ridge}}$ will be quickly reduced towards zero as λ_2 increases.
(c) Moderate values of $\sigma^2 C^{jj} \approx 1$ will lie between the other two cases.

This is illustrated in Figure 2.1.

The key intuition developed here is that the shrinkage rate to first-order is controlled by the variance of each estimate. That is, a higher variance forces a smaller estimate and a correspondingly smaller variance. Of course, in reality, the true ridge estimator considers the entire covariance matrix, including all C^{ij} terms where $i, j = 1, \dots, p$, to obtain the full effects of ridge shrinkage. But it is instructive to view ridge regression in this fashion to gain insight into its behavior. As we will see shortly, ridge traces capture the complex trajectories of each of β_j as λ_2 increases. We will typically use λ_2 as the ridge tuning parameter.

2.4.2 Ridge Applied to Rank Estimation

The rank loss function is the dispersion function for multiple linear regression as follows

$$D_n(\boldsymbol{\beta}) = \sum_{i=1}^{n} (y_i - \boldsymbol{x}_i^\top \boldsymbol{\beta}) a_n(R_{n_i}(\boldsymbol{\beta})). \tag{2.4.11}$$

Ridge regression is applied to rank-based methods in a manner similar to LS:

$$D^{\text{R-ridge}}(\boldsymbol{\beta}) = D_n(\boldsymbol{\beta}) + \lambda_2 \boldsymbol{\beta}^\top \boldsymbol{\beta}, \tag{2.4.12}$$

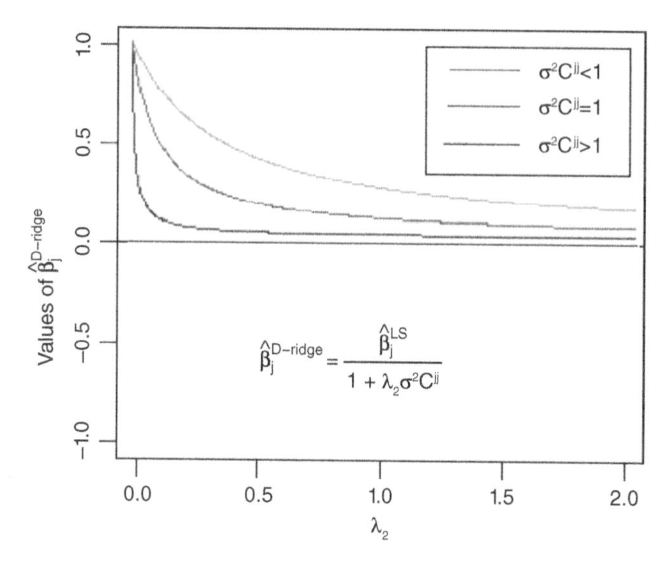

Figure 2.1 The first-order nature of shrinkage due to ridge.

where λ_2 is the ridge tuning parameter, and in expanded form, we have

$$D^{\text{R-ridge}}(\boldsymbol{\beta}) = \sum_{i=1}^{n}(\boldsymbol{y}_i - \boldsymbol{x}_i^\top \boldsymbol{\beta})a_n(R_{n_i}(\boldsymbol{\beta})) + \lambda_2 \boldsymbol{\beta}^\top \boldsymbol{\beta}. \tag{2.4.13}$$

The penalized rank estimates, $\hat{\boldsymbol{\beta}}_n^{\text{R-ridge}}$, are obtained by minimizing Eq. (2.4.13). All properties of LS-ridge regression described above hold equally for R-ridge regression.

2.5 Example: Swiss Fertility Data Set

We now use a well-known Swiss fertility[4] data set to demonstrate some of the features of ridge regression. The data itself is provided in Table 2.1 and the variable definitions and their abbreviations are provided in Table 2.2.

The data set indicates five features that may influence the fertility rates (F) in a 1888 study with 47 observations. The abbreviations of the five variables are I, Ex, Ed, A, and C and the objective is to find the coefficient values or estimates for these five parameters in order to predict F. Hence, $n = 47$ and $p = 5$.

4 This data set is part of the R environment in the *datasets* package.

Table 2.1 Swiss fertility data set (F = fertility, A = agriculture, Ex = examination, Ed = education, C = Catholic, I = infant mortality).

Region	*F*	*A*	*Ex*	*Ed*	*C*	*I*	Region	*F*	*A*	*Ex*	*Ed*	*C*	*I*
Courtelary	80.20	17.00	15	12	9.96	22.20	Delemont	83.10	45.10	6	9	84.84	22.20
Franches-Mnt	92.50	39.70	5	5	93.40	20.20	Moutier	85.80	36.50	12	7	33.77	20.30
Neuveville	76.90	43.50	17	15	5.16	20.60	Porrentruy	76.10	35.30	9	7	90.57	26.60
Broye	83.80	70.20	16	7	92.85	23.60	Glane	92.40	67.80	14	8	97.16	24.90
Gruyere	82.40	53.30	12	7	97.67	21.00	Sarine	82.90	45.20	16	13	91.38	24.40
Veveyse	87.10	64.50	14	6	98.61	24.50	Aigle	64.10	62.00	21	12	8.52	16.50
Aubonne	66.90	67.50	14	7	2.27	19.10	Avenches	68.90	60.70	19	12	4.43	22.70
Cossonay	61.70	69.30	22	5	2.82	18.70	Echallens	68.30	72.60	18	2	24.20	21.20
Grandson	71.70	34.00	17	8	3.30	20.00	Lausanne	55.70	19.40	26	28	12.11	20.20
La Vallee	54.30	15.20	31	20	2.15	10.80	Lavaux	65.10	73.00	19	9	2.84	20.00
Morges	65.50	59.80	22	10	5.23	18.00	Moudon	65.00	55.10	14	3	4.52	22.40
Nyone	56.60	50.90	22	12	15.14	16.70	Orbe	57.40	54.10	20	6	4.20	15.30
Oron	72.50	71.20	12	1	2.40	21.00	Payerne	74.20	58.10	14	8	5.23	23.80

(Continued)

Table 2.1 *(Continued)*

Region	*F*	*A*	*Ex*	*Ed*	*C*	*I*	Region	*F*	*A*	*Ex*	*Ed*	*C*	*I*
Paysd'enhaut	72.00	63.50	6	3	2.56	18.00	Rolle	60.50	60.80	16	10	7.72	16.30
Vevey	58.30	26.80	25	19	18.46	20.90	Yverdon	65.40	49.50	15	8	6.10	22.50
Conthey	75.50	85.90	3	2	99.71	15.10	Entremont	69.30	84.90	7	6	99.68	19.80
Herens	77.30	89.70	5	2	100.00	18.30	Martigwy	70.50	78.20	12	6	98.96	19.40
Monthey	79.40	64.90	7	3	98.22	20.20	St Maurice	65.00	75.90	9	9	99.06	17.80
Sierre	92.20	84.60	3	3	99.46	16.30	Sion	79.30	63.10	13	13	96.83	18.10
Boudry	70.40	38.40	26	12	5.62	20.30	La Chauxdfnd	65.70	7.70	29	11	13.79	20.50
Le Locle	72.70	16.70	22	13	11.22	18.90	Neuchatel	64.40	17.60	35	32	16.92	23.00
Val de Ruz	77.60	37.60	15	7	4.97	20.00	ValdeTravers	67.60	18.70	25	7	8.65	19.50
V. De Geneve	35.00	1.20	37	53	42.34	18.00	Rive Droite	44.70	46.60	16	29	50.43	18.20
Rive Gauche	42.80	27.70	22	29	58.33	19.30							

Table 2.2 Swiss fertility data set definitions.

Feature	Symbol	Meaning
Region	R	French-speaking provinces of Switzerland in 1888
Fertility	F	Ig, 'common standardized fertility measure'
Agriculture	A	% of males involved in agriculture as occupation
Examination	Ex	% draftees receiving highest mark on army examination
Education	Ed	% education beyond primary school for draftees
Catholic	C	% "Catholic" (as opposed to "Protestant")
Infant mortality	I	Live births who live less than 1 year

Table 2.3 Swiss fertility estimates and standard errors for least squares (LS) and rank (R).

Feature	$\hat{\beta}_n^{LS}$	s.e.($\hat{\beta}_n^{LS}$)	$\hat{\beta}_n^{R}$	s.e.($\hat{\beta}_n^{R}$)
Intercept (β_0)	66.92	10.706	66.43	10.518
Agriculture (β_1)	−0.17	0.070	−0.20	0.069
Examination (β_2)	−0.26	0.253	−0.26	0.249
Education (β_3)	−0.87	0.183	−0.89	0.179
Catholic (β_4)	0.10	0.035	0.11	0.034
Infant mortality (β_5)	1.08	0.381	1.20	0.375

The goal is to create a model suitable for prediction using ridge regression for both LS and rank. This book will make further use of this data set so readers are encouraged to familiarize themselves with it and to work through the examples provided in this chapter to reinforce the material being presented.

We first note that all the data in Table 2.1 lies in the range 0–100 so scaling is not essential and the estimates can be reasonably interpreted relative to one another. Typically, we would need to scale the data so that columns have the same standard deviation (usually set to 1), as described later. We skip that step here.

2.5.1 Estimation and Standard Errors

We have chosen a data set that has an unknown number of outliers prior to working with the data, but we intend to compute the LS and rank estimates, and then determine how many outliers actually exist.

To obtain the LSE, we first center the data, both the design matrix (comprised of columns A, Ex, Ed, C and I) and the response (column F). Then we use Eq. (2.3.7) to obtain the estimates. For the standard errors of LSE, we can first compute the value of $\hat{\sigma}^2 = \text{SSE}/(n-p) = 50.1$, noting that SSE is from Eq. (2.3.11). Then we apply Eqs. (2.3.8) and (2.3.10) to obtain the s.e. We can obtain the LS intercept, called β_0 or θ_n, using *uncentered data* of $\boldsymbol{y}$ and $\boldsymbol{X}$ from

$$\hat{\theta}_n^{\text{LS}} = \frac{1}{n}\sum_{i=1}^{n}(y_i - \boldsymbol{x}_i^\top \hat{\boldsymbol{\beta}}_n^{\text{LS}}). \tag{2.5.1}$$

For R-estimates, we do not need to center the data prior to estimation. We simply minimize (2.3.12) to obtain the parameters, and use (2.3.13) to obtain the s.e. with $\hat{\eta}^2 = 49.0$. The rank intercept can be obtained using the median of the residuals from the original uncentered data of $\boldsymbol{y}$ and $\boldsymbol{X}$

$$\hat{\theta}_n^{\text{R}} = \underset{1\le i\le n}{\text{median}}\left(y_i - \boldsymbol{x}_i^\top \hat{\boldsymbol{\beta}}_n^{\text{R}}\right). \tag{2.5.2}$$

The parameter values from LSE and rank methods are shown in Table 2.3. We find that the LSE and rank results[5] are similar, both for estimates and standard error (s.e.), implying that there are no outliers. However, a check of the Q–Q plot for the residuals indicates two outliers, one in each tail, as shown in Figure 2.2.

Although both LS and rank produce roughly similar results, this does not imply the absence of outliers in the data, as seen around (−2.5,−2.5) and (2.5,2.5) of the Q–Q plot. So one cannot simply assume no outliers when LS and rank estimates are similar. We should keep this in mind when interpreting the results of the two estimators. What we do know is that both estimators are equivalent in this balanced scenario. Of course, any imbalance in the number outliers in either tail may lead to different LS and rank estimates. In any case, we will first study this balanced outlier situation, where we expect similar performance, and then later examine the unbalanced case using a different data set.

2.5.2 Parameter Variance using Bootstrap

We draw the reader's attention to the standard error columns in Table 2.3. It is well known that these values quantify the uncertainty in the estimates. If we square the standard error, we obtain the estimated variance for each parameter. We can illustrate this more concretely by sampling the given data set multiple times and then computing estimates for each sample. That is, we can create sampling distributions of the estimates using the bootstrap method

5 These results are obtained from R using `lm()` for LSE and `rfit()` for R-estimates.

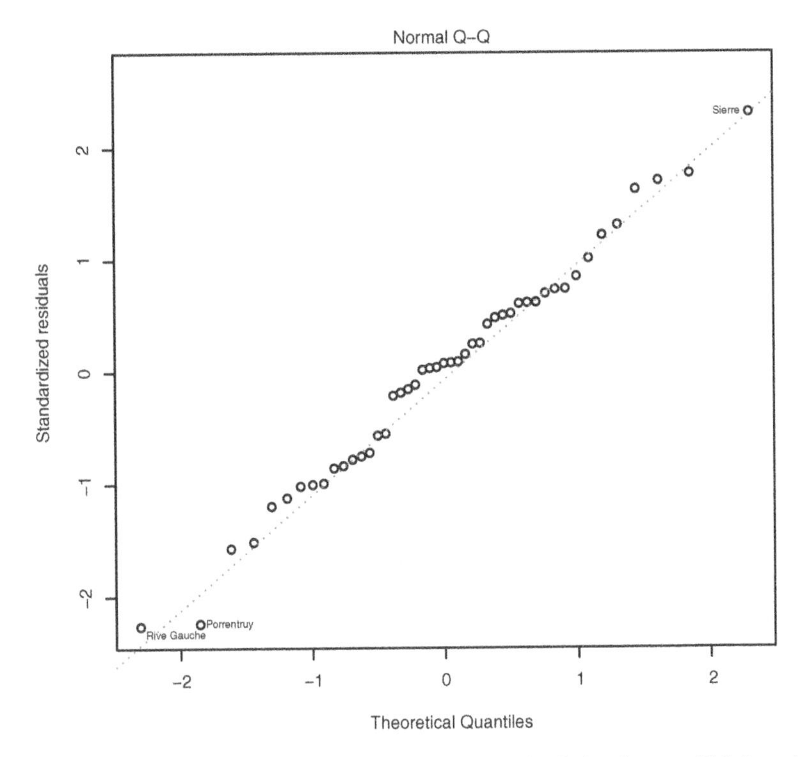

Figure 2.2 Two outliers found in the Q–Q plot for the Swiss data set. This is a plot of the percentiles of the residuals (dots) against the standard normal distribution (line).

(James et al., 2013). There are 47 observations in this data set. We randomly pick observations (with replacement) to create a new data set containing 47 observations, which implies that we may have repeated observations in our synthesized data set. We then compute the estimates and repeat this procedure multiple times to generate parameter distributions. Figure 2.3 shows the LS estimate distributions after sampling the data set in this manner. The mean values of each distribution are the unbiased estimates. The spread of the estimates around the mean captures the variance. We note that β_2, β_3 and β_5 have large variances, as confirmed by Table 2.3. In addition, β_0 has a large variance but we will not be applying the ridge penalty to it; only $\beta_1 \ldots \beta_5$ will be penalized because they are directly associated with the response. In fact, we will center the data so that the intercept is not even calculated when applying ridge.

2.5.3 Reducing Variance using Ridge

In order to reduce the variance, we can apply ridge regression. It is useful to visually examine the shrinkage in action of one parameter and its variance, as shown in Figure 2.4, with embedded boxplots. This figure was produced by

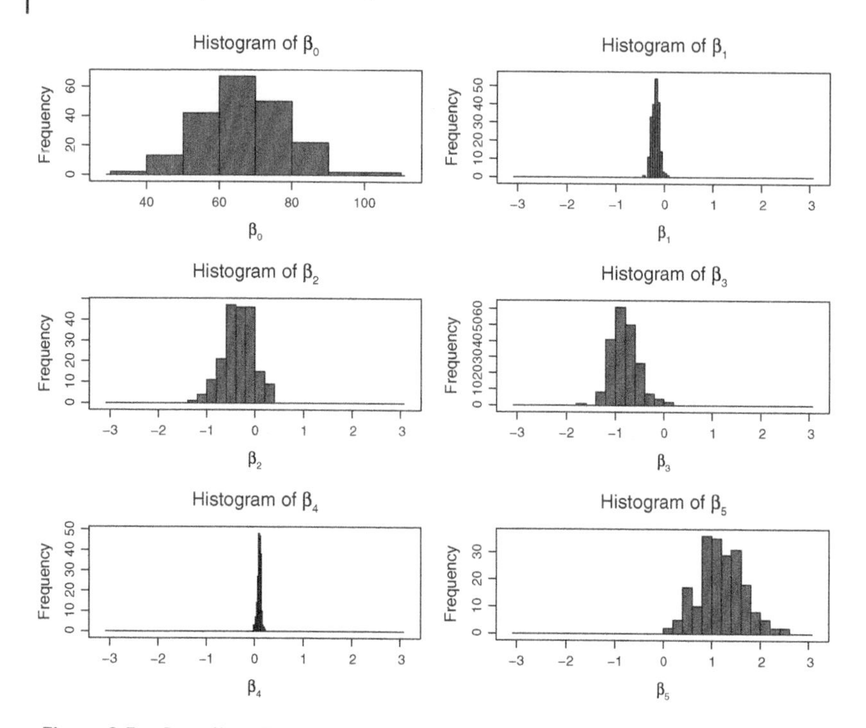

Figure 2.3 Sampling distributions of rank estimates.

taking the ridge weight matrix of Eq. (2.4.6) and progressively increasing the tuning parameter λ_2 by an order of magnitude. The values chosen were 1, 10, 100 and 1000. As we can observe, the effect of the LS-ridge on β_5 is that the sample mean value is reduced towards the null while the bootstrapped sample variance is likewise reduced. This occurs for all other parameters at different rates, and for both LS and rank methods.

2.5.4 Ridge Traces

It is instructive to know what happens to each parameter as we vary λ_2. As we increase λ_2, the estimates tend to decrease but exactly how they decrease is not as easy to predict. In this section, we examine the so-called ridge traces whereby the estimates are plotted as a function of $\log(\lambda_2)$. Typically, plotting the estimates as a function of λ_2 is not as informative as $\log(\lambda_2)$ because of the large scale over which shrinkage occurs. In this and other chapters, we will be using traces to illustrate many of the characteristics and properties of penalty estimators. We use ridge to introduce the traces and their meaning.

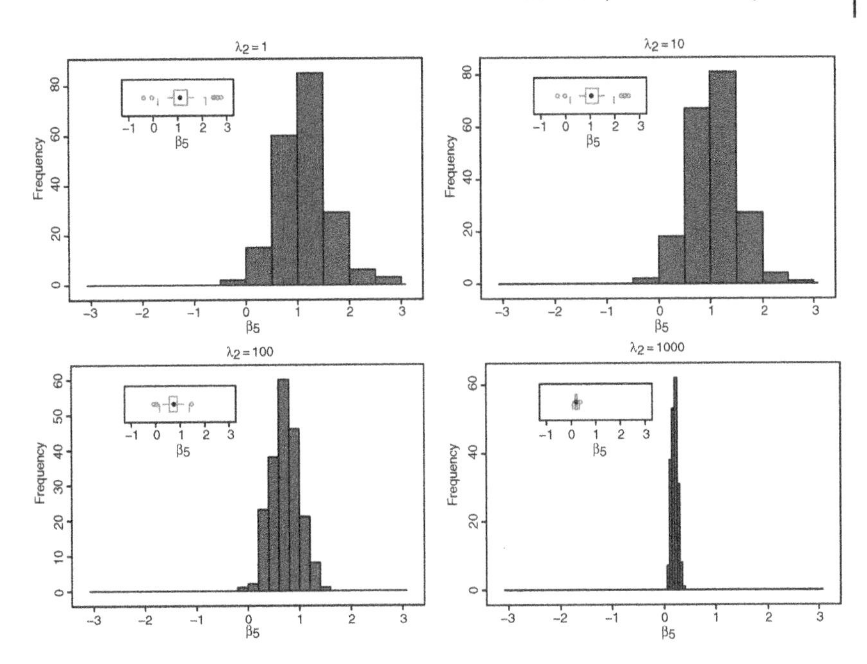

Figure 2.4 Shrinkage of β_5 due to increase in ridge tuning parameter, λ_2.

For a given trace, we have to build m models, one for each selected λ_2. Earlier, we considered three different equations with $\boldsymbol{C}^{-1}$ being an orthonormal matrix in Eq. (2.4.7), a diagonal matrix in Eq. (2.4.10) and a full matrix in Eq. (2.4.6). These cases are shown in Figure 2.5 for O-ridge, D-ridge and LS-ridge, respectively. Additionally, the R-ridge trace using Eq. (2.4.12) is also provided. They are all different in their shrinkage behavior which gives us some insight into their inherent properties. Two important points are worth mentioning here. First, if we pick any vertical line associated with a specific λ_2 value, the intersection of that line and the curves would give us the penalized estimates. We should note that there is nothing magical about a particular λ_2 value. It depends on the scale of values associated with the corresponding estimator. A low or high value of λ_2 does not have any specific meaning. Second, the traces for ridge do not theoretically reach 0 until $\lambda = \infty$, although many may appear to reach 0 at a much earlier value. This point will become more meaningful when we discuss the LASSO traces.

It is interesting to observe that the orthonormal case exhibits the basic shrinkage operation of ridge. We see all five variables of the top-left of Figure 2.5 converging towards zero at the same rate because their decay is controlled only by $1/(1+\lambda_2)$. The diagonal case is much more interesting since the trajectories

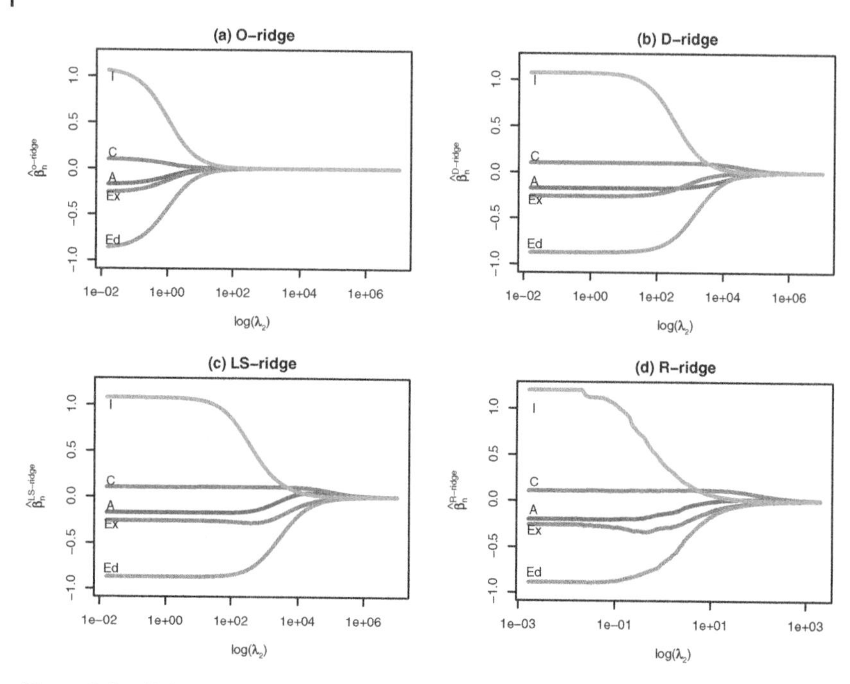

Figure 2.5 Ridge traces for orthonormal, diagonal, LS, and rank estimators ($m = 40$).

are controlled by $1/(1+\lambda_2 C^{jj})$. Note that the top-right of Figure 2.5 more closely resembles the accurate plots of LS-ridge and R-ridge shown in the two bottom panels, respectively. This validates the equations showing that C^{jj} controls most of the trace characteristics. The diagonal traces only capture the *uncorrelated trajectory* of each parameter so the traces monotonically converge to zero without ever crossing zero. The full ridges traces may not be monotonic in their behavior and may cross zero and change signs. Still, the diagonal case gives us insight into the nature of the traces. In fact, the orthogonal and diagonal cases serve as simple but effective penalty estimators in their own right (cf. Saleh et al., 2019).

The traces for LS-ridge and R-ridge illustrate the differences between these two cases. In the R-ridge case, variable Ex is clearly non-monotonic but this is not as noticeable in the LS-ridge case. The main difference between the two traces lies along the $\log(\lambda_2)$-axis due the difference in the unpenalized loss functions for each method. A simple forward or backward translation of one or the other would align the traces. In general, the shape of LS and rank traces for ridge may differ slightly with no outliers, but more noticeably when outliers are present.

2.6 Selection of Ridge Parameter λ_2

Naturally, questions arise as to how to set the proper value of λ_2 and the nature of the trade-off as we shrink the estimates and their variances. In short, we are trading off bias against variance (i.e. under-fitting vs. over-fitting), and the proper value of λ_2 is based on a desired prediction performance of our model. We now delve deeper into each question in the context of LS, which will be similar to rank. Note that an optimal λ_2 value is unique to an estimator and generally cannot be compared or transferred between estimators. Additionally, the optimal λ_2 usually lies in a range of values in practice, as opposed to being one unique value.

2.6.1 Quadratic Risk

We start by describing the infamous *bias-variance trade-off* which is at the heart of selecting the optimal λ_2. We know the meaning of variance – the degree of uncertainty associated with an estimate. A working definition of the bias is the difference between the expected value of a given estimator, $\mathbb{E}(\hat{\boldsymbol{\beta}}_n)$, and the true value, $\boldsymbol{\beta}$. An estimator is said to be unbiased if $\mathbb{E}(\hat{\boldsymbol{\beta}}_n) = \boldsymbol{\beta}$. When we decrease the variance, we tend to increase the bias. We seek to balance the two by choosing the appropriate λ_2.

It is well known that LSE provides the minimum variance of a set of linear unbiased estimates so it would appear that we cannot improve it. However, the variance may still be rather large if there is strong correlation between variables, i.e. a certain amount of collinearity exists among the columns of the design matrix. By applying a penalty function, we increase the bias and decrease the variance until the estimates minimize a given statistic.

Consider the use of the mean-square error (MSE) (Montgomery et al., 2012), also referred to as quadratic risk (and ADL$_2$-risk in Chapter 3), for the purposes of finding the optimal λ_2 to balance the bias and variance. The MSE of the estimates given by

$$\text{MSE}(\hat{\boldsymbol{\beta}}_n^{\text{LS-ridge}}) = \text{tr}(\text{Var}(\hat{\boldsymbol{\beta}}_n^{\text{LS-ridge}})) + \text{bias}^2(\hat{\boldsymbol{\beta}}_n^{\text{LS-ridge}}) \tag{2.6.1}$$

where $\text{tr}(\boldsymbol{M})$ is the trace of some matrix, $\boldsymbol{M}$, which involves adding up the diagonal terms of $\boldsymbol{M}$, and let $\text{bias}^2(\hat{\boldsymbol{\beta}}_n^{\text{LS-ridge}}) = \text{bias}^{\mathsf{T}}(\hat{\boldsymbol{\beta}}_n^{\text{LS-ridge}})\text{bias}(\hat{\boldsymbol{\beta}}_n^{\text{LS-ridge}})$. The formulas for bias and variance for LS-ridge are given by

$$\text{bias}^2(\hat{\boldsymbol{\beta}}_n^{\text{LS-ridge}}) = \lambda_2^2\, \boldsymbol{\beta}^{\mathsf{T}}(\boldsymbol{X}^{\mathsf{T}}\boldsymbol{X} + \lambda_2 \boldsymbol{I}_p)^{-2}\boldsymbol{\beta}, \tag{2.6.2}$$

which is a scalar quantity, and

$$\mathrm{Var}(\hat{\boldsymbol{\beta}}_n^{\text{LS-ridge}}) = \sigma^2(\boldsymbol{X}^\top\boldsymbol{X} + \lambda_2\boldsymbol{I}_p)^{-1}(\boldsymbol{X}^\top\boldsymbol{X})(\boldsymbol{X}^\top\boldsymbol{X} + \lambda_2\boldsymbol{I}_p)^{-1}, \tag{2.6.3}$$

which is a matrix quantity (for which trace is taken), respectively.

The above expressions are cumbersome and difficult to gain any understanding. To simplify the analysis and to develop intuition, let $\boldsymbol{X}^\top\boldsymbol{X} = \boldsymbol{I}_p$. This yields

$$\mathrm{bias}^2(\hat{\boldsymbol{\beta}}_n^{\text{LS-ridge}}) = \frac{\lambda_2^2}{(1+\lambda_2)^2}\boldsymbol{\beta}^\top\boldsymbol{\beta}, \tag{2.6.4}$$

and

$$\mathrm{tr}(\mathrm{Var}(\hat{\boldsymbol{\beta}}_n^{\text{LS-ridge}})) = \frac{p\sigma^2}{(1+\lambda_2)^2}. \tag{2.6.5}$$

Hence, the quadratic risk is given by

$$\mathrm{MSE}(\hat{\boldsymbol{\beta}}_n^{\text{LS-ridge}}) \approx \frac{p\sigma^2}{(1+\lambda_2)^2} + \frac{\lambda_2^2}{(1+\lambda_2)^2}\boldsymbol{\beta}^\top\boldsymbol{\beta}. \tag{2.6.6}$$

We see that if $\lambda_2 = 0$, then the bias is zero and the variance is given by $p\,\sigma^2$. Trying to enforce that the bias be zero may not lead to a desirable estimator due to the possibility of a large variance. In fact, as we increase λ_2, the bias increases while the variances decreases. As $\lambda_2 \to \infty$, the variance approaches zero while the bias converges to $\boldsymbol{\beta}^\top\boldsymbol{\beta}$, provided this quantity is bounded. This situation is also undesirable because the bias is now large. The optimal choice of λ_2 is then a balance between enough bias to avoid over-fitting and enough variance to avoid under-fitting. Our goal becomes finding the optimal value $\lambda_{2(\text{opt})}$ that minimizes the quadratic risk.

To obtain the optimal λ_2 for the orthonormal case, we take the partial derivative of Eq. (2.6.6) w.r.t. λ_2 and set it to zero. Solving for λ_2, we obtain

$$\lambda_{2(\text{opt})}^{\text{O}} = \frac{p\sigma^2}{\boldsymbol{\beta}^\top\boldsymbol{\beta}}. \tag{2.6.7}$$

For the Swiss data set, assume that $\hat{\boldsymbol{\beta}}_n^{\text{LS}}$ are the unbiased estimates, $p = 5$ and $\hat{\sigma}^2 = 50.1$. Then, the approximate value of the optimal λ_2 is

$$\hat{\lambda}_{2(\text{opt})}^{\text{O}} = \frac{p\hat{\sigma}^2}{(\hat{\boldsymbol{\beta}}^{\text{LS}})^\top\hat{\boldsymbol{\beta}}^{\text{LS}}} = 123.7.$$

If we examine Figure 2.5, we would place a vertical line at $\lambda_2 = 123.7$ in the bottom left panel labelled LS-ridge to obtain the ridge estimates. However, we should not put too much weight on this result because the assumptions were unrealistic; but it does illustrate the key points of how to use the optimal λ_2.

To obtain a slightly more accurate $\lambda_{2(\text{opt})}$, assume that $(\boldsymbol{X}^\top\boldsymbol{X})^{-1}$ is a diagonal matrix with diagonal terms C^{jj}, $j = 1, \dots, p$. In that case, the quadratic risk is

$$\text{MSE}(\hat{\boldsymbol{\beta}}_n^{\text{LS-ridge}}) = \sum_{j=1}^{p} \frac{\sigma^2 C^{jj}}{(1+\lambda_2 C^{jj})^2} + \sum_{j=1}^{p} \frac{\lambda_2^2 \beta_j^2 (C^{jj})^2}{(1+\lambda_2 C^{jj})^2}. \tag{2.6.8}$$

If we take the partial derivative of Eq. (2.6.8) w.r.t λ_2 and set it to zero, we can determine its optimal value.

$$\frac{\partial \text{MSE}(\hat{\boldsymbol{\beta}}_n^{\text{LS-ridge}})}{\partial \lambda_2} = 2\sum_{j=1}^{p} \frac{(C^{jj})^2}{(1+\lambda_2 C^{jj})^3} (\lambda_2 \beta_j^2 - \sigma^2) = 0. \tag{2.6.9}$$

Unfortunately, this equation does not have a closed-form solution. We can either solve it numerically or plot a graph to estimate $\lambda_{2(\text{opt})}$. Again, assume we have obtained $\hat{\boldsymbol{\beta}}_n^{\text{LS}}$, $\boldsymbol{C}^{-1}$ and $\hat{\sigma}^2$. Then we can apply these values to the above equations. We plot the MSE and its derivative in Figure 2.6. In the MSE plot, we observe the effects of the bias-variance tradeoff since the function reduces in value as we shrink the parameters (and reduce the variance) but then reaches a minimum and then increases (as the bias increases). The optimal value appears to be $\hat{\lambda}_{2(\text{opt})}^{\text{D}} = 77.8$. This is both the minimum of the MSE plot and the zero-crossing of the derivative plot. Note that the ridge estimator is not uniformly better than the LS. We indicate the range over which ridge will produce a lower MSE compared to the LS. In particular, if $\lambda_2 < 168.0$, the LS-ridge will be better than LSE. Beyond this point, it is better to use the unpenalized least squares estimates.

We now consider using the full set of Eqs. (2.6.1)–(2.6.3) on the Swiss data set to obtain the optimal λ_2. We have plotted the results for the square of the bias, the trace of the variance matrix and the resulting MSE of the estimates. This is shown in Figure 2.7. We see that the bias is initially zero when $\lambda_2 = 0$ and the variance is large. As we increase λ_2, the variance drops while the bias increases and eventually the bias determines the MSE. In between the extremes, the MSE reaches a minimum at $\hat{\lambda}_{2(\text{opt})} = 70.8$ which is in the vicinity of the value produced using the diagonal approximation for the MSE equation. We can now take this value back to Figure 2.5 and draw a vertical line in the LS-ridge plot at this value. The resulting LS-ridge estimates with $\hat{\lambda}_{2(\text{opt})} = 70.8$ obtained by minimizing the quadratic risk are

$$\hat{\boldsymbol{\beta}}_n^{\text{LS-ridge}} = (A = -0.17, Ex = -0.27, Ed = -0.86, C = 0.11, I = 0.91)^\top.$$

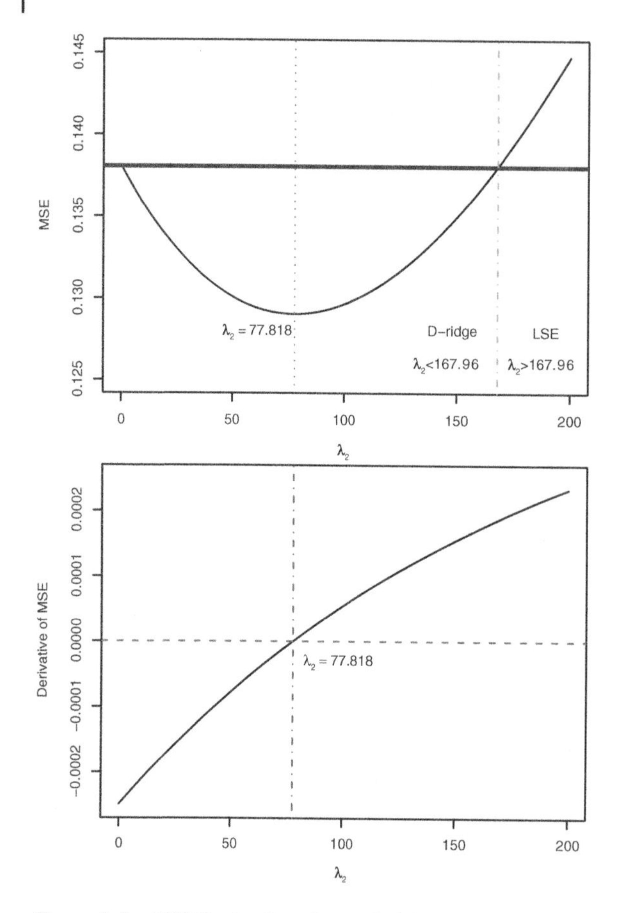

Figure 2.6 MSE Derivative plot to find optimal λ_2 for the diagonal case.

2.6.2 K-fold Cross-validation Scheme

In machine learning applications, we do not compute the optimal λ_2 in the manner described in the previous section. Instead, a cross-validation scheme is used to produce estimates that are suitable for prediction. If our goal is to produce a generalized model, then we should use the data set in a way that reflects this goal. Given a data set, we first divide it into a training set and a test set (which we view as an "unseen" or "hold-out" test set). The test set cannot be used to select λ_2 but can be used to assess our final prediction accuracy. We do not want to inadvertently tune λ_2 to this specific test set since the prediction accuracy may not generalize to other test sets. Instead, it is common to further subdivide the training set into K disjoint sets, called cross-validation sets or "folds", as discussed below. These sets will be used to select the optimal λ_2.

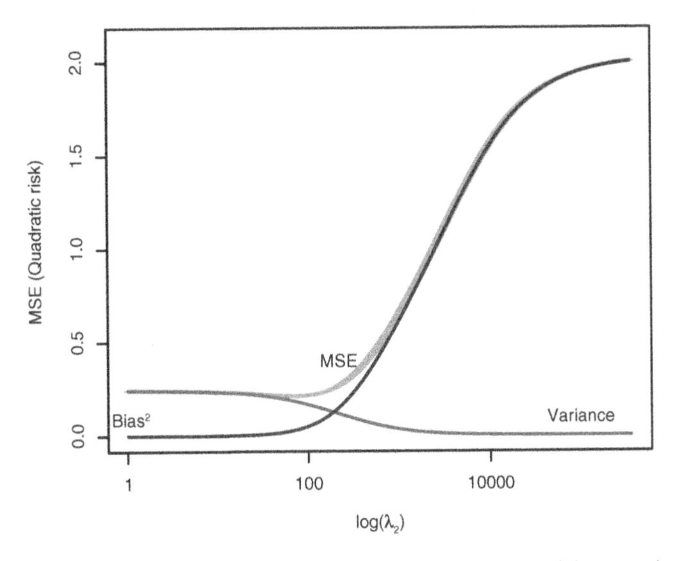

Figure 2.7 Bias, variance and MSE for the Swiss data set (optimal $\lambda_2 = 70.8$).

We now describe the cross-validation (CV) scheme that operates on a number of folds. We begin subdividing the training set into, say, five smaller training and test sets, hence five folds. For each value of λ_2, the estimates from the five smaller training sets are used to compute the mean-square prediction error (MSE)[6], or the average squared-residual error, on each of their associated five smaller test sets as follows,

$$\text{MSE}^{(j)} = \frac{1}{n}\sum_{i=1}^{n}(y_i - (\hat{\theta}_n^{(j)} + \boldsymbol{x}_i^\top \hat{\boldsymbol{\beta}}_n^{(j)}))^2 \quad j = 1, \ldots, 5 \tag{2.6.10}$$

where $\hat{\theta}_n^{(j)}$ and $\hat{\boldsymbol{\beta}}_n^{(j)}$ are the estimates of the jth fold. We average the $\text{MSE}^{(j)}$ over the five folds. Next, we sweep λ_2 and repeat the process for a number of different λ_2 values. Finally, we choose the λ_2 that minimizes the average MSE among all selected values of λ_2 used during cross-validation. The residual MSE is a useful statistic here since it can be readily computed for any data set once we obtain the average estimates during cross-validation.

The method above is called a K-fold cross-validation scheme where $K = 5$ in this case. Usually $K = 5$ or $K = 10$, depending of the size of the data set. In particular, let us illustrate the procedure to determine $\lambda_{2(\text{opt})}$, but this time for the R-ridge case. In our current Swiss data set, we have $n = 47$ rows in $\boldsymbol{X}$,

6 Note: MSE was used previously for the quadratic risk but, from here on, it refers to the prediction error.

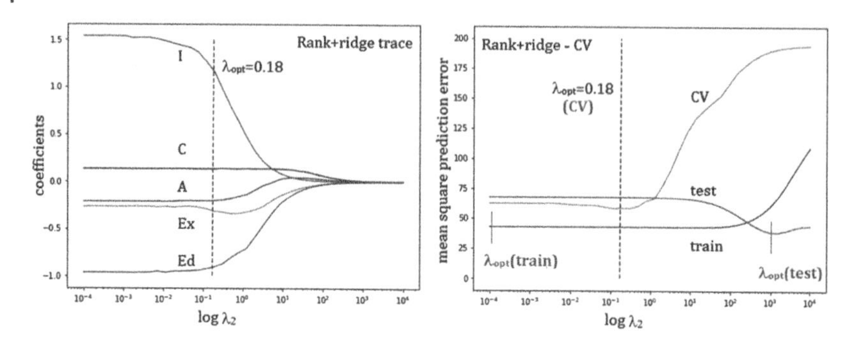

Figure 2.8 MSE for training, CV and test sets, and coefficients from the ridge trace.

so we will set aside 12 rows as our "unseen" test set and use 35 rows as our training set. Then, we randomly divide this 35 remaining rows of the training set into five subsets with 7 rows in each. Next, we "hold out" 7 rows as a cross-validation (CV) test set, while using the remaining 28 rows as the CV training set to determine the estimates. This can be done five times to produce five different estimates. These estimates are each used for prediction on the 7 row CV test set to compute the MSE. Repeating this five times on different CV sets is a five-fold cross-validation procedure, for which we would have five MSE values that we would average. This five-fold procedure is repeated for each λ_2 that we sweep from, say, 10^{-4} to 10^4. Finally, we can choose the λ_2 that delivers the lowest average MSE for the five-fold cross-validation sets.

This is the exact procedure used to generate Figure 2.8 where we juxtapose the R-ridge trace with the CV results. The R-ridge trace is the average of the estimates in each fold. That is why the estimates for $\lambda_2 = 0$ are different from the original R-estimates. They are a result of using the CV training subsets (folds) rather than the entire data set. For the panel on the right, the optimal value, $\hat{\lambda}_{2(\text{opt})} \approx 0.18$, was selected to minimize the CV error, not the training or test error, or quadratic risk. The training MSE always increases as λ_2 increases, and the test MSE is generally unavailable, so this is why we optimize based on the five-fold cross-validation sets. Had we optimized for the test set, we would pick a much larger $\lambda_{2(\text{opt})}(\text{test})$ value which would then not generalize to a broader set of test data.

The resulting estimates with $\hat{\lambda}_{2(\text{opt})} = 0.18$ from the panel on the left are

$$\hat{\boldsymbol{\beta}}_n^{\text{R-ridge}} = (A = -0.20, Ex = -0.29, Ed = -0.92, C = 0.12, I = 1.20)^{\top}.$$

We already know the unbiased estimates for the entire data set as they were previously reported in Table 2.3. This is close to the original rank estimates except that it was derived from a portion of the total data set (the training portion) rather than the entire set.

2.7 LASSO and aLASSO

2.7.1 Subset Selection

We now turn our attention to subset selection. It is important to produce a model that uses the fewest variables while still retaining high prediction accuracy. This is because the residual variance can also be large if there are a large number of variables in the model. To understand this point, consider the estimated residual variance for LS estimates which is given by

$$\hat{\sigma}^2 = \frac{1}{n-p}(\boldsymbol{y} - \boldsymbol{X}\hat{\boldsymbol{\beta}}_n^{\text{LS}})^{\top}(\boldsymbol{y} - \boldsymbol{X}\hat{\boldsymbol{\beta}}_n^{\text{LS}}). \tag{2.7.1}$$

As p approaches n, the variance goes to infinity. In effect, we have another bias-variance trade-off involving the number of variables in our model (sometimes called complexity of the model). If we have too many variables, we risk over-fitting the data set and that leads to a high variance. If we have too few variables, we risk under-fitting which creates a high bias situation. Our next goal is to seek the optimal number of variables, $p_o < p$, via subset selection.

We have described subset selection in the previous chapter as a way of reducing model size while maintaining the required prediction accuracy. LASSO provides both shrinkage and subset selection based on an L_1 penalty function. It sets certain parameter values to zero during regularization and those parameters may be removed from the initial set. Subset selection is performed by LASSO with a suitable choice of its tuning parameter. In order to understand the nature of LASSO, we first derive the equations and then investigate its properties under different assumptions.

2.7.2 Least Squares with LASSO

The LASSO penalty applied to the quadratic loss function is given by

$$J^{\text{LS-LASSO}}(\boldsymbol{\beta}) = (\boldsymbol{y} - \boldsymbol{X}\boldsymbol{\beta})^{\top}(\boldsymbol{y} - \boldsymbol{X}\boldsymbol{\beta}) + 2\lambda \sum_{j=1}^{p} |\beta_j|, \tag{2.7.2}$$

which we can re-write as

$$J^{\text{LS-LASSO}}(\boldsymbol{\beta}) = (\boldsymbol{y} - \boldsymbol{X}\boldsymbol{\beta})^{\top}(\boldsymbol{y} - \boldsymbol{X}\boldsymbol{\beta}) + 2\lambda \mathbf{1}_p^{\top}|\boldsymbol{\beta}|, \tag{2.7.3}$$

where $|\boldsymbol{\beta}| = (|\beta_1|, \dots, |\beta_p|)^{\top}$ and $\mathbf{1}_p = (1, \dots, 1)^{\top}$ is a p-tuple of 1s.

To obtain the LASSO estimates, we take the derivative of $J^{\text{LS-LASSO}}(\boldsymbol{\beta})$ w.r.t. $\boldsymbol{\beta}$ and set it to 0, i.e.

$$\frac{\partial J^{\text{LS-LASSO}}(\boldsymbol{\beta})}{\partial \boldsymbol{\beta}} = -2\boldsymbol{X}^{\top}(\boldsymbol{y} - \boldsymbol{X}\boldsymbol{\beta}) + 2\lambda \text{sgn}(\boldsymbol{\beta}) = 0. \tag{2.7.4}$$

Then, it follows that

$$-2\boldsymbol{X}^{\top}\boldsymbol{y} + 2\boldsymbol{X}^{\top}\boldsymbol{X}\boldsymbol{\beta} + 2\lambda \text{sgn}(\boldsymbol{\beta}) = 0. \tag{2.7.5}$$

Pre-multiplying by $(\boldsymbol{X}^{\top}\boldsymbol{X})^{-1}$ and rearranging, we obtain an equation in terms of $\hat{\boldsymbol{\beta}}_n^{\text{LS}}$ and $\hat{\boldsymbol{\beta}}_n^{\text{LS-LASSO}}$ which leads to

$$-\hat{\boldsymbol{\beta}}_n^{\text{LS}} + \hat{\boldsymbol{\beta}}_n^{\text{LS-LASSO}} + \lambda \boldsymbol{C}^{-1}\text{sgn}(\hat{\boldsymbol{\beta}}_n^{\text{LS-LASSO}}) = 0, \tag{2.7.6}$$

where $\boldsymbol{C}^{-1} = (\boldsymbol{X}^{\top}\boldsymbol{X})^{-1}$. The matrix $(\boldsymbol{C}^{-1}$ does not have to be of full rank to apply LASSO. In fact, LASSO addresses the multicollinearity problem in a manner similar to ridge. However, we assume it is invertible for our purposes here. Then, we obtain the following equation.

$$\hat{\boldsymbol{\beta}}_n^{\text{LS-LASSO}} = \hat{\boldsymbol{\beta}}_n^{\text{LS}} + \lambda \boldsymbol{C}^{-1}\text{sgn}(\hat{\boldsymbol{\beta}}_n^{\text{LS-LASSO}}). \tag{2.7.7}$$

A closed-form solution cannot be found for $\hat{\boldsymbol{\beta}}_n^{\text{LS-LASSO}}$ but making some simplifying assumptions can be instructive as to how LASSO operates.

If we assume $\boldsymbol{C}^{-1}$ is the identity matrix (orthonormal case), the equation becomes

$$\hat{\boldsymbol{\beta}}_n^{\text{O-LASSO}} = \hat{\boldsymbol{\beta}}_n^{\text{LS}} + \lambda \text{sgn}(\hat{\boldsymbol{\beta}}_n^{\text{O-LASSO}}), \tag{2.7.8}$$

where $\hat{\boldsymbol{\beta}}_n^{\text{O-LASSO}}$ are the LASSO estimates for the orthonormal case. Studying the different cases to consider in the above, we can represent the solution compactly for each β_j as

$$\hat{\beta}_j^{\text{O-LASSO}} = \text{sgn}(\hat{\beta}_j^{\text{LS}})(|\hat{\beta}_j^{\text{LS}}| - \lambda)^{+}. \tag{2.7.9}$$

The notation $(\cdot)^{+}$ implies that the quantity inside the parentheses is set to zero only if it goes negative. Essentially, we are subtracting λ from $|\hat{\beta}_j^{\text{LS}}|$ to obtain LASSO shrinkage in the orthonormal case. The shrinkage is linear with respect to λ until it reaches zero and all parameters shrink at the same rate given by λ.

If we now assume that $\boldsymbol{C}^{-1}$ is a diagonal matrix with diagonal elements given by C^{jj}, $j = 1, \ldots, p$, this implies no correlation between variables. For this case, we can rewrite the LASSO equation as follows (see Saleh et al., 2019, chapter 5),

$$\hat{\beta}_j^{\text{D-LASSO}} = \text{sgn}(\hat{\beta}_n^{\text{LS}})(|\hat{\beta}_j^{\text{LS}}| - \lambda\, C^{jj})^+, \tag{2.7.10}$$

where $\hat{\beta}_j^{\text{D-LASSO}}$ is the LASSO estimate for the diagonal $\boldsymbol{C}^{-1}$ case. Noting that λ is simply a tuning parameter, we can define $\lambda = \lambda_1\sigma^2$. Then we can write

$$\hat{\beta}_j^{\text{D-LASSO}} = \text{sgn}(\hat{\beta}_j^{\text{LS}})(|\hat{\beta}_j^{\text{LS}}| - \lambda_1\, \sigma^2 C^{jj})^+, \tag{2.7.11}$$

or equivalently,

$$\hat{\beta}_j^{\text{D-LASSO}} = \text{sgn}(\hat{\beta}_j^{\text{LS}})(|\hat{\beta}_j^{\text{LS}}| - \lambda_1\, \text{Var}(\hat{\beta}_j^{\text{LS}}))^+. \tag{2.7.12}$$

The slope of the line, hence how fast an estimate reaches 0, is controlled by the variance of each $\hat{\beta}_j^{\text{LS}}$. This first-order analysis is useful in building intuition about the nature of the shrinkage in LASSO. Removal of a variable, called *subset selection*, is determined by $\text{Var}(\hat{\beta}_j^{\text{LS}}) = \sigma^2 C^{jj}$ in the diagonal case. The point at which the estimate reaches 0, the so-called "zero-crossing" of each trace, occurs at $\lambda_{1,j}^0 = |\hat{\beta}_j^{\text{LS}}| / \text{Var}(\hat{\beta}_j^{\text{LS}})$. This is telling us that a large $\hat{\beta}_j^{\text{LS}}$ and a small variance will lead to a zero-crossing at a high value of λ_1. A small $\hat{\beta}_j^{\text{LS}}$ and a large variance leads to a zero-crossing at a low value of λ_1. The implications of this are important. It means that the ordering of the $\hat{\beta}_j^{\text{D-LASSO}}$ values in reaching 0 is related to the ratio of $|\hat{\beta}_j^{\text{LS}}|$ and its variance or, more precisely, the variance and covariances of each parameter as given by the $\boldsymbol{C}^{-1}$ matrix in Eq. (2.7.7).

Consider the case of a particular parameter as illustrated in Figure 2.9. As λ_1 increases, $\hat{\beta}_j^{\text{LS}} = 1.0$ shrinks at a linear rate determined by the variance of $\hat{\beta}_j^{\text{LS}}$ to produce $\hat{\beta}_j^{\text{D-LASSO}}$. We show the following cases of Eq. (2.7.12):

(a) If $\sigma^2 C^{jj}$ is very small, i.e. much smaller than 1, then $\hat{\beta}_j^{\text{D-LASSO}}$ will gradually decrease linearly as λ_1 increases.
(b) If $\sigma^2 C^{jj}$ is very large, i.e. much greater than 1, then $\hat{\beta}_j^{\text{D-LASSO}}$ will quickly reduce linearly towards zero as λ_1 increases.
(c) Moderate values of $\sigma^2 C^{jj} \approx 1$ will lie between the other two cases.

Note that all three lines reach 0 at different points due to the different variances.

2.7.3 The Adaptive LASSO and its Geometric Interpretation

While LASSO is capable of performing subset selection, it is missing a key property – the oracle property. That is, the zero-crossing order of LASSO is interesting but it does not tell us about the importance of each β_j. However, this issue was addressed by Zou (2006) in the adaptive LASSO (aLASSO) method. An excellent comparison of rank-based variable selection may be found in Johnson and Peng (2008).

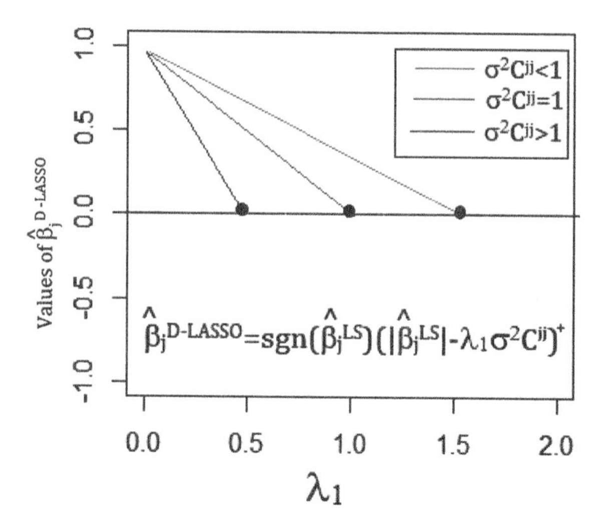

Figure 2.9 The first-order nature of shrinkage due to LASSO.

In order to incorporate the oracle property into LS regression, a weight is introduced into the penalty function, as follows

$$J^{\text{LS-aLASSO}}(\boldsymbol{\beta}) = (\boldsymbol{y} - \boldsymbol{X\beta})^{\top}(\boldsymbol{y} - \boldsymbol{X\beta}) + \lambda_1 \sum_{j=1}^{p} w_j |\beta_j|. \tag{2.7.13}$$

The effect of individual weights, w_j, on each parameter can be seen in the geometric interpretation of Figure 2.10. Weights effectively re-shape the diamond from a symmetrical one in LASSO (Figure 1.9) to an asymmetrical one in aLASSO (Figure 2.10). The asymmetry is due to the differences in weights. In the case of LASSO in Figure 1.9, the trajectory starting from "x" could reach either $\beta_1 = 0$ or $\beta_2 = 0$ since the diamond is symmetric. However, aLASSO would likely reach the $\beta_2 = 0$ corner first, which is the nearest corner, assuming the geometry in Figure 2.10. Then, the asymmetric diamond would continue to shrink in size as indicated by $\lambda_1 > 0$ and $\lambda_2 > \lambda_1$, etc., and eventually reach $\beta_1 = 0$ before or at $\lambda = \infty$. This sequence would imply that β_1 is more important than β_2 because aLASSO set $\beta_2 = 0$ first (i.e. because it is less important). The aLASSO trace captures the actual trajectory of each parameter. In particular, the *zero-crossing order* of the parameters in an aLASSO trace is a key feature of this oracle penalty function and we will explore this further in this chapter, and in more detail in subsequent chapters.

The method of incorporating the oracle property requires a suitable selection of the weights. Zou (2006) proposes the use of $w_j = 1/|\hat{\beta}_j^{\text{LS}}|$ and indeed this does provide the much desired oracle property. We stipulate here that the

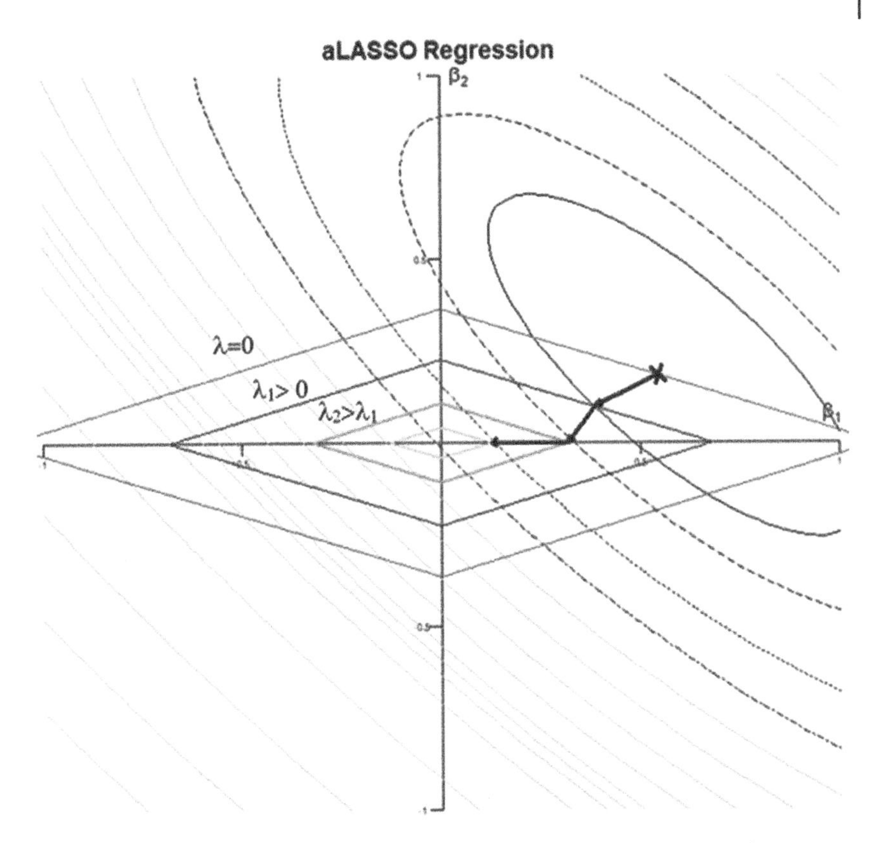

Figure 2.10 Diamond-warping effect of weights in the aLASSO estimator for $p = 2$.

LS (or rank) estimates must be known in advance but that is not an onerous requirement assuming that multicollinearity is not present. In that case, the loss function of aLASSO is given by

$$J^{\text{LS-aLASSO}}(\boldsymbol{\beta}) = (\boldsymbol{y} - \boldsymbol{X}\boldsymbol{\beta})^\top(\boldsymbol{y} - \boldsymbol{X}\boldsymbol{\beta}) + 2\lambda \sum_{j=1}^{p} \frac{|\beta_j|}{|\hat{\beta}_j^{\text{LS}}|}, \tag{2.7.14}$$

where each $|\beta_j|$ is divided by $|\hat{\beta}_j^{\text{LS}}|$. Again, we assume that $\hat{\boldsymbol{\beta}}^{\text{LS}}$ is available implying that $\boldsymbol{C} = \boldsymbol{X}^\top\boldsymbol{X}$ is invertible. Then, the estimates, assuming a diagonal $\boldsymbol{C}$ matrix, using aLASSO are given by

$$\hat{\beta}_j^{\text{D-aLASSO}} = \text{sgn}(\hat{\beta}_j^{\text{LS}}) \left(|\hat{\beta}_j^{\text{LS}}| - \lambda_1 \frac{\sigma^2 C^{jj}}{|\hat{\beta}_j^{\text{LS}}|} \right)^+ . \tag{2.7.15}$$

Since $\sigma^2 C^{jj}$ are the variances of each $\hat{\beta}_j^{\text{LS}}$, we can view the individual weights as simply the variance divided by the magnitude of each $\hat{\beta}_j^{\text{LS}}$ as follows

$$\hat{\beta}_j^{\text{D-aLASSO}} = \text{sgn}(\hat{\beta}_j^{\text{LS}}) \left(|\hat{\beta}_j^{\text{LS}}| - \lambda_1 \frac{\text{Var}(\hat{\beta}_j^{\text{LS}})}{|\hat{\beta}_j^{\text{LS}}|} \right)^+ . \tag{2.7.16}$$

The weight is effective at guiding the aLASSO solution to the corners of the asymmetric diamond shape in an order determined by the size of the variance as compared to the size of $\hat{\beta}_j^{\text{LS}}$. In fact, the oracle property will force the solution of aLASSO to visit the least important parameters first (in an attempt to set them to zero as in Figure 2.10) and eventually make its way to the most important parameters as λ_1 is increased from 0 to ∞.

We will need to identify the zero-crossings of aLASSO traces because it contains useful variable ordering information for subset selection. A first-order estimate for the zero-crossings can be performed quickly by setting the $(\cdot)^+$ term to zero in Eq. (2.7.16). Therefore, for a strictly diagonal design matrix, we can use

$$\lambda_j^0 = \frac{|\beta_j|^2}{\text{Var}(\hat{\beta}_j^{\text{LS}})} \approx \frac{|\beta_j|^2}{\text{s.e.}^2(\hat{\beta}_j^{\text{LS}})} = (t_j)^2. \tag{2.7.17}$$

The zero-crossing of β_j is simply its t statistic squared, i.e. $(t_j)^2$. If $(t_j)^2$, or equivalently $|t_j|$, are ordered, then a reasonable sequence of variables (although not necessarily oracle) is obtained[7]. This ordering assumes no correlation between variables, which is unlikely and hence not oracle, but it does provide a good first-order assessment of the importance of each variable.

If we compare the zero-crossings for LASSO and aLASSO for the orthonormal case, we find that $\lambda_{0,j}^{\text{LASSO}} = |\hat{\beta}_j^{\text{LS}}|$, and $\lambda_{0,j}^{\text{aLASSO}} = |\hat{\beta}_j^{\text{LS}}|^2$. Therefore the variable ordering would be the same for both and hence both have the oracle property for the orthonormal case. Once you introduce the diagonal case with $\text{Var}(\hat{\beta}_j^{\text{LS}})$, the two may no longer agree on the ordering since $\lambda_{0,j}^{\text{LASSO}} = |\hat{\beta}_j^{\text{LS}}| / \text{Var}(\hat{\beta}_j^{\text{LS}})$ and $\lambda_{0,j}^{\text{aLASSO}} = |\hat{\beta}_j^{\text{LS}}|^2 / \text{Var}(\hat{\beta}_j^{\text{LS}})$. And finally, with the full $\boldsymbol{C}^{-1}$ matrix, the correlations between variables may change the ordering significantly.

To summarize, LASSO has the oracle property only in the orthonormal case, whereas aLASSO retains it theoretically in all cases. Of course, in practice, the zero-crossings are data-dependent but generally have the capability of producing a near-optimal ordering in most cases, assuming enough observations.

---

7 The t statistic is used for null hypothesis testing as the ratio of the estimate to its standard error and is reported by `lm()` in R using the `summary()` command.

2.7.4 R-estimation with LASSO and aLASSO

For R-estimation, we formulate the objective function for the R-LASSO estimator as

$$D^{\text{R-LASSO}}(\boldsymbol{\beta}) = D_n(\boldsymbol{\beta}) + \lambda_1 \sum_{j=1}^{p} |\beta_j|. \tag{2.7.18}$$

If R-LASSO is applied to the Swiss data set using a five-fold cross-validation scheme, as described earlier, the following estimates are produced:

$$\hat{\boldsymbol{\beta}}_n^{\text{R-LASSO}} = (A = -0.16, Ex = -0.23, Ed = -0.80, C = 0.10, I = 1.07)^\top.$$

Note that none of the parameters have been set to 0. Hence, for this example, LASSO does not perform subset selection.

The R-aLASSO objective function is

$$D^{\text{R-aLASSO}}(\boldsymbol{\beta}) = D_n(\boldsymbol{\beta}) + \lambda_1 \sum_{j=1}^{p} \frac{|\beta_j|}{|\hat{\beta}_j^{\text{R}}|}. \tag{2.7.19}$$

Again, applying a five-fold cross-validation scheme using the Swiss data set produces the following estimates:

$$\hat{\boldsymbol{\beta}}_n^{\text{R-aLASSO}} = (A = -0.11, Ex = -0.003, Ed = -0.94, C = 0.11, I = 1.13)^\top.$$

This time, the variable Ex has been set to a value that is almost 0. This variable is effectively removed from the parameter set and we can observe the subset selection feature of aLASSO in this case. The important combination of *rank loss with aLASSO penalty* has robustness while ordering the variables so that optimal subset selection can be carried out. We will describe this in more detail shortly.

But first, to illustrate the oracle feature of aLASSO, we need to examine the traces associated with LASSO and aLASSO[8]. This is provided in Figure 2.11 for the Swiss data set. The number of models built is $m = 40$, one for each selected λ_1. The first thing to notice is that the traces all reach 0 well before $\lambda_1 = \infty$. These points are called the "zero-crossings" of the traces. This is a major feature difference between ridge and LASSO (or aLASSO). We again note that LS and rank are very similar, owing to the importance of $\boldsymbol{C}^{-1}$ in the trajectory of the traces in both sets. The differences in the values along the $\log(\lambda_1)$ axis are due to the magnitude of the LS vs. rank loss functions, as described earlier for ridge.

8 These results were generated using R code obtained from Brent Johnson, University of Rochester at the site `https://www.urmc.rochester.edu/labs/brent-johnson/software.aspx`.

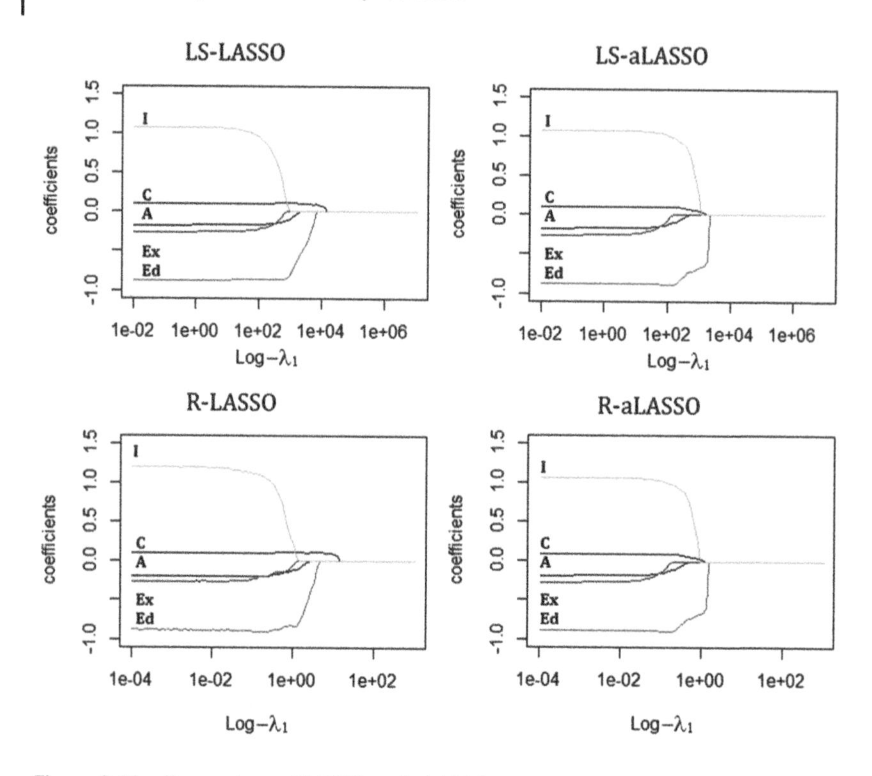

Figure 2.11 Comparison of LASSO and aLASSO traces for the Swiss data set.

Essentially, this figure shows that the properties of LS-LASSO and LS-aLASSO hold equally for R-LASSO and R-aLASSO.

2.7.5 Oracle Properties

Next, we want to focus on the nature of the differences between LASSO and aLASSO in their traces. Consider Figure 2.12 showing the traces from R-LASSO and R-aLASSO in more detail. While the two panels appear to be the same, the zero-crossings of the trajectories occur in a different order for LASSO and aLASSO. In particular, LASSO has the order, from right to left, of C, Ed, A, I, Ex whereas aLASSO has the order Ed, C, I, A, Ex. But which one is correct?

We know that aLASSO has the oracle property while LASSO does not. Therefore, we postulate that the aLASSO ordering is correct and that the LASSO ordering is incorrect. In fact, the order of the zero-crossings for aLASSO sequences from most important starting on the right to least important on the left. As a side check, we can examine the $|t|$ in Table 2.4. This table shows the

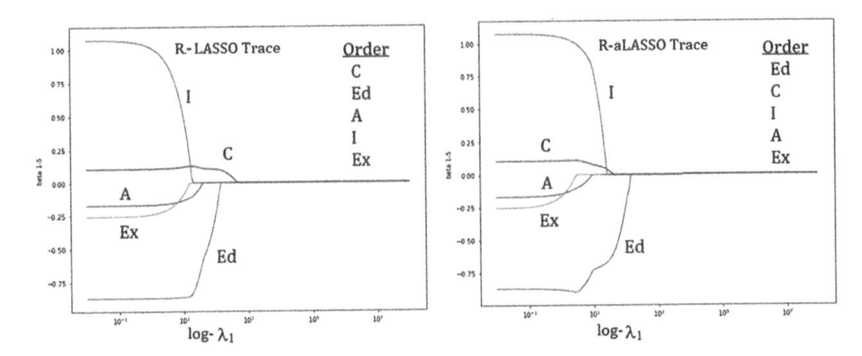

Figure 2.12 Variable ordering from R-LASSO and R-aLASSO traces for the Swiss data set.

Table 2.4 Swiss data subset ordering using | *t*.value |.

		Rank estimates from rfit()		
Feature	**Estimate**	**Std. error**	***t*.value**	**|*t*.value| order**
(Intercept)	66.43	10.51	6.3	0
Agriculture (*A*)	−0.20	0.06	−2.90	4
Examination (Ex)	−0.25	0.24	−1.04	5
Education (Ed)	−0.88	0.17	−4.92	1
Catholic (C)	0.10	0.03	3.10	3
Infant mortality (I)	1.19	0.37	3.19	2

rank estimates, standard errors and t statistics for the Swiss data set for $\lambda_1 = 0$. The $|t|$ statistic ordering is Ed, I, C, A, and Ex, which is similar to the aLASSO order. Strictly based on the magnitudes of the $|t|$ values in the table, we could state that Ed is the most important variable, Ex is the least important, and C, I, and A are roughly of the same importance. All three methods determined that Ex is the least important variable, but unfortunately LASSO did not set it to 0 while aLASSO did in the five-fold CV scheme for the Swiss data set, as shown earlier.

We can test these conjectures using a number of different variable selection schemes as described in James et al. (2013). The best subset selection builds all possible 2^p models and selects the best one for each set with the same number of variables. This process is shown in Table 2.5. We will use least squares analysis in R to obtain the needed information since the results for rank and least squares are similar. The rows labeled ($\star$) are the best subset, those labeled

Table 2.5 Swiss data models with adjusted R^2 values: best (★), R-aLASSO (△), $|t|$ (•), RLASSO (▽). Best subset ordering is obtained by trying all possible combinations of variables in each group and selecting the one with the highest adjusted R^2. R-aLASSO uses the results of the trace ordering in Figure 2.12. R-LASSO uses the results trace orderings in Figure 2.12. The $|t|$ uses the ordering from Table 2.4. In all cases the highest adjusted R^2 is selected.

Model	Adjusted-R^2	Subset selection
F ~ Ed	0.4282	★ △ •
F ~ C	0.1976	▽
F ~ A	0.1052	
F ~ Ex	0.4042	
F ~ I	0.1552	
F ~ Ed + C	0.5552	★ △ ▽
F ~ Ed + Ex	0.4830	
F ~ Ed + A	0.4242	
F ~ Ed + I	0.5450	•
F ~ C + Ex	0.4043	
F ~ C + I	0.3005	
F ~ C + A	0.2641	
F ~ Ex + I	0.5152	
F ~ Ex + A	0.4068	
F ~ I + A	0.2868	
F ~ C + Ed + I	0.6390	★ △ •
F ~ C + Ed + Ex	0.5452	
F ~ C + Ed + A	0.6173	▽
F ~ C + Ex + A	0.4074	
F ~ C + Ex + I	0.5083	
F ~ C + A + I	0.3040	
F ~ Ex + Ed + A	0.5259	
F ~ Ex + Ed + I	0.5935	
F ~ Ex + A + I	0.5077	
F ~ Ed + A + I	0.5368	
F ~ Ed + C + I + A	0.6707	★ △ • ▽
F ~ Ed + C + I + Ex	0.6319	
F ~ Ed + C + A + Ex	0.6164	
F ~ Ed + I + A + Ex	0.6105	
F ~ C + I + A + Ex	0.6319	
F ~ Ed + C + I + A + Ex	0.6710	★ △ • ▽

($\triangle$) are aLASSO, those labeled ($\bullet$) are based on $|t|$ and those labeled (∇) are LASSO. The criteria used for determining the best model is the adjusted R^2 statistic (James et al., 2013) given by

$$\text{adjusted } R^2 = 1 - \frac{\text{RSS}/(n-d-1)}{\text{TSS}/(n-1)}, \tag{2.7.20}$$

where

$$\text{TSS} = \sum_{i=1}^{n}(y_i - \bar{y})^2 \quad \text{and} \quad \text{RSS} = \sum_{i=1}^{n}(y_i - \hat{y})^2,$$

with $\bar{y}$ as the mean response and $\hat{y}_i = (\hat{\theta}_n^{\text{LS}} + \sum_{j=1}^{p} x_{ij}\hat{\beta}_j^{\text{LS}})$ as the predicted response, with $\hat{\theta}_n^{\text{LS}} = 0$ if the data is centered. Here, d is the number of parameters used in the model and can range from 1 to p. Typically, the adjusted R^2 value should increase as we add more parameters, then reach a maximum, and finally start to decrease as additional parameters begin to add noise to the model. As a practical matter, we will typically choose the number of parameters to maximize the adjusted R^2 statistic but there are situations where we may select fewer or greater than the optimal number if we have domain knowledge. In the absence of such knowledge, we attempt to balance under-fitting and over-fitting by maximizing the adjusted-R^2 statistic.

Table 2.5 shows that aLASSO is able to predict the best model in each category based on the zero-crossing order. Using R notation[9], the best one-variable model is ($F \sim Ed$). The best two-variable model is ($F \sim Ed+C$), the best three-variable model is ($F \sim Ed+C+I$), and so on. On the other hand, LASSO would incorrectly predict the best one-variable model is ($F \sim C$), correctly predict the best two-variable model is ($F \sim Ed + C$), incorrectly predict the best three-variable model is ($F \sim C + Ed + A$), and so on. They all correctly predict that the Ex variable could be removed with little or no impact since the adjusted R^2 values changes only by +0.0003 by including it. Therefore, a reasonable choice based on the adjusted R^2-values is the four-variable model ($F \sim Ed+C+I+A$), and aLASSO essentially found this model.

This demonstrates the oracle property of aLASSO can be very valuable for subset selection. Used in conjunction with the rank loss function, it provides a variable ordering that could be extracted from the R-aLASSO trace, assuming enough data is available, and the results could be used to perform optimal subset selection. The adjusted R^2 statistic is useful in this regard, as illustrated

9 The notation "$F \sim Ed$" means "F is a function of only Ed" in R notation. Specifically, $F \sim Ed$ is equivalent to the model $F = \beta_0 + \beta_1 Ed + \varepsilon$.

above. In practice, we often replace the adjusted R^2 with a five-fold CV MSE for model selection since we have easy access to this information.

2.8 Elastic Net (Enet)

We now consider blending the L_1 and L_2 penalty functions to determine if some additional benefit can be derived from their combination. Ridge provides a smooth shrinkage function that is strictly convex while LASSO provides the capability to zero out certain unimportant variables but is not a strictly convex function (Zou and Hastie, 2005). It is desirable and somewhat elegant to have the benefits of both in one penalty function called elastic net or Enet.

2.8.1 Naive Enet

The penalty function of Enet applied to a quadratic loss function is given by

$$J^{\text{LS-Enet}}(\boldsymbol{\beta}) = (\boldsymbol{y} - \boldsymbol{X}\boldsymbol{\beta})^{\top}(\boldsymbol{y} - \boldsymbol{X}\boldsymbol{\beta}) + \lambda_2 \sum_{j=1}^{p} |\beta_j|^2 + 2\lambda_1 \sum_{j=1}^{p} |\beta_j|, \tag{2.8.1}$$

which we can re-write as

$$J^{\text{LS-Enet}}(\boldsymbol{\beta}) = (\boldsymbol{y} - \boldsymbol{X}\boldsymbol{\beta})^{\top}(\boldsymbol{y} - \boldsymbol{X}\boldsymbol{\beta}) + \lambda_2 \boldsymbol{\beta}^{\top}\boldsymbol{\beta} + 2\lambda_1 \mathbf{1}^{\top}|\boldsymbol{\beta}|, \tag{2.8.2}$$

where $\lambda_1, \lambda_2 \in \mathbb{R}^+$ are the tuning parameters.

To obtain the Enet estimates, we take the derivative of $J(\boldsymbol{\beta})$ w.r.t. $\boldsymbol{\beta}$ and set it to 0, i.e.

$$\frac{\partial J^{\text{LS-Enet}}(\boldsymbol{\beta})}{\partial \boldsymbol{\beta}} = -2\boldsymbol{X}^{\top}(\boldsymbol{y} - \boldsymbol{X}\boldsymbol{\beta}) + 2\lambda_2\boldsymbol{\beta} + 2\lambda_1 \operatorname{sgn}(\boldsymbol{\beta}) = 0. \tag{2.8.3}$$

Then, it follows that

$$-2\boldsymbol{X}^{\top}\boldsymbol{y} + 2\boldsymbol{X}^{\top}\boldsymbol{X}\boldsymbol{\beta} + 2\lambda_2\boldsymbol{\beta} + 2\lambda_1 \operatorname{sgn}(\boldsymbol{\beta}) = 0. \tag{2.8.4}$$

Multiplying all terms by $\boldsymbol{C}^{-1}$ and rearranging, we obtain an equation in terms of $\hat{\boldsymbol{\beta}}_n^{\text{LS}}$ and a term referred to as the Naive elastic net (nEnet) estimate, $\hat{\boldsymbol{\beta}}_n^{\text{LS-nEnet}}$,

$$-\hat{\boldsymbol{\beta}}_n^{\text{LS}} + \hat{\boldsymbol{\beta}}_n^{\text{LS-nEnet}} + \lambda_2 \boldsymbol{C}^{-1}\hat{\boldsymbol{\beta}}_n^{\text{LS-nEnet}} + \lambda_1 \boldsymbol{C}^{-1}\operatorname{sgn}(\hat{\boldsymbol{\beta}}_n^{\text{LS-nEnet}}) = 0. \tag{2.8.5}$$

Finally,

$$(\boldsymbol{I}_p + \lambda_2 \boldsymbol{C}^{-1})\hat{\boldsymbol{\beta}}_n^{\text{LS-nEnet}} = \hat{\boldsymbol{\beta}}_n^{\text{LS}} + \lambda_1 \boldsymbol{C}^{-1}\operatorname{sgn}(\hat{\boldsymbol{\beta}}_n^{\text{LS-nEnet}}). \tag{2.8.6}$$

A closed-form solution cannot be obtained for $\hat{\boldsymbol{\beta}}_n^{\text{LS-nEnet}}$ but making some simplifying assumptions can be instructive as to how nEnet operates. As before, if we assume $\boldsymbol{C}^{-1}$ is the identity matrix (orthonormal case), the equation becomes

$$(1 + \lambda_2)\hat{\boldsymbol{\beta}}_n^{\text{O-nEnet}} = \hat{\boldsymbol{\beta}}_n^{\text{LS}} + \lambda_1 \text{sgn}(\hat{\boldsymbol{\beta}}_n^{\text{O-nEnet}}) \tag{2.8.7}$$

where $\hat{\boldsymbol{\beta}}^{\text{O-nEnet}}$ are the nEnet estimates for the orthogonal case. This simplifies to

$$\hat{\beta}_j^{\text{O-nEnet}} = \frac{\text{sgn}(\hat{\beta}_j^{\text{LS}})(|\hat{\beta}_j^{\text{LS}}| - \lambda_1)^+}{1 + \lambda_2}. \tag{2.8.8}$$

In other words, $\hat{\beta}_j^{\text{O-nEnet}}$ involves first computing $\hat{\beta}_j^{\text{LS}}$, then shrinking using LASSO, and then shrinking again due to ridge. We are actually performing double shrinkage on $\hat{\beta}_j^{\text{LS}}$, once due to LASSO by subtracting λ_1 and once due to ridge by dividing by $(1 + \lambda_2)$. This is not desirable so an adjustment is needed, as described in the next section.

If we now let $\boldsymbol{C}^{-1}$ be a diagonal matrix with non-zero values C^{jj}, $j = 1, \dots, p$, then we obtain

$$\hat{\beta}_j^{\text{D-nEnet}} = \frac{\text{sgn}(\hat{\beta}_j^{\text{LS}})(|\hat{\beta}_j^{\text{LS}}| - \lambda_1 C^{jj})^+}{1 + \lambda_2 C^{jj}}. \tag{2.8.9}$$

where $\hat{\boldsymbol{\beta}}^{\text{D-nEnet}}$ is the nEnet estimate for the diagonal case. As before, this informs us that C^{jj} is a controlling term for shrinkage in both ridge and LASSO in the diagonal nEnet penalty.

2.8.2 Standard Enet

By inspection of earlier equations, we find that nEnet performs a double shrinkage of the LS estimates. To compensate for the second shrinkage process, we can boost the result by an amount $(1 + \lambda_2)$ as follows:

$$\hat{\beta}_j^{\text{D-sEnet}} = (1 + \lambda_2)\frac{\text{sgn}(\hat{\beta}_j^{\text{LS}})(|\hat{\beta}_j^{\text{LS}}| - \lambda_1 C^{jj})^+}{1 + \lambda_2 C^{jj}}. \tag{2.8.10}$$

This is referred to as the standard elastic net (sEnet). As before, we can see that the variance term controls both ridge and LASSO as follows

$$\hat{\beta}_j^{\text{D-sEnet}} = (1 + \lambda_2)\frac{\text{sgn}(\hat{\beta}_j^{\text{LS}})(|\hat{\beta}_j^{\text{LS}}| - \lambda_1' \text{Var}(\hat{\beta}_j^{\text{LS}}))^+}{1 + \lambda_2' \text{Var}(\hat{\beta}_j^{\text{LS}})} \tag{2.8.11}$$

where $\lambda_1 = \lambda_1'\sigma^2$ and $\lambda_2 = \lambda_2'\sigma^2$.

We can go one step further and incorporate the oracle property by replacing LASSO with aLASSO. Then, the above equation becomes

$$\hat{\beta}_j^{\text{D-aEnet}} = (1+\lambda_2)\frac{\text{sgn}(\hat{\beta}_j^{\text{LS}})\left(|\hat{\beta}_j^{\text{LS}}| - \lambda_1' \frac{\text{Var}(\hat{\beta}_j^{\text{LS}})}{|\hat{\beta}_j^{\text{LSE}}|}\right)^+}{1+\lambda_2' \text{Var}(\hat{\beta}_j^{\text{LS}})}. \tag{2.8.12}$$

We call this the adaptive elastic net for the diagonal case (D-aEnet). Because of the reasons cited earlier regarding the oracle property, this aEnet formulation could potentially be better than nEnet or sEnet. Of course, the caveat is that the least squares estimates are needed for aLASSO and they may not be available.

2.8.3 Enet in Machine Learning

For machine learning, the form of the objective function for LS-Enet is usually expressed using λ and α, rather than λ_1 and λ_2, as follows

$$J^{\text{LS-Enet}}(\boldsymbol{\beta}) = (\boldsymbol{y}-\boldsymbol{X\beta})^\top(\boldsymbol{y}-\boldsymbol{X\beta}) + \lambda\left[\frac{1-\alpha}{2}\boldsymbol{\beta}^\top\boldsymbol{\beta} + \alpha\sum_{j=1}^{p}|\beta_j|\right], \tag{2.8.13}$$

where $\lambda > 0$ is the standard tuning parameter and $0 < \alpha < 1$ controls degree to which ridge or LASSO is applied.

Likewise, the form for R-Enet is

$$D^{\text{R-Enet}}(\boldsymbol{\beta}) = D_n(\boldsymbol{\beta}) + \lambda\left[\frac{(1-\alpha)}{2}\boldsymbol{\beta}^\top\boldsymbol{\beta} + \alpha\sum_{j=1}^{p}|\beta_j|\right]. \tag{2.8.14}$$

Finally, the objective function of the rank-based adaptive Enet (R-aEnet) is given by

$$D^{\text{R-aEnet}}(\boldsymbol{\beta}) = D_n(\boldsymbol{\beta}) + \lambda\left[\frac{(1-\alpha)}{2}\boldsymbol{\beta}^\top\boldsymbol{\beta} + \alpha\sum_{j=1}^{p}\frac{|\beta_j|}{|\hat{\beta}_j^{\text{R}}|}\right]. \tag{2.8.15}$$

The tuning parameter λ controls the impact of the penalty function while α controls the selection of ridge or aLASSO by setting it to 0 or 1. Other settings are possible but the most useful values are near 1. For example, the optimum value is $\alpha = 0.95$ for the Swiss data set using a five-fold CV scheme which produces the estimates:

$$\hat{\boldsymbol{\beta}}_n^{\text{R-aEnet}} = (A = -0.11, Ex = -0.0027, Ed = -0.94, C = 0.12, I = 1.14)^\top.$$

The estimate for variable Ex is small enough to remove, as was the case with R-aLASSO.

2.9 Example: Diabetes Data Set

We conclude the chapter with a sizable data set to review the procedures for model building using techniques of the previous sections with the overall goal to reduce the "noise" in the data set and be left with only the "signal". We will do this by effectively reducing the number of observations, if outliers are present, and reducing the number of variables, if certain explanatory variables are deemed to be unnecessary, and then applying a shrinkage estimator to build the model. The most suitable combination for data cleaning, variable selection and parameter shrinkage is the rank loss function combined with aLASSO and ridge penalty functions of Eq. (2.8.15).

2.9.1 Model Building with R-aEnet

We will use a well-known diabetes[10] data set (cf. Johnson and Peng, 2008), which is also used in Chapter 8, with $n = 442$ and $p = 10$, to illustrate the process of model building using R-aEnet. The first step is to set $\lambda = 0$ in Eq. (2.8.15) to obtain the rank estimates (given later in Table 2.6 without standarization of the data). These estimates can be used to identify outliers. Usually we use the LS estimates to find outliers but here we provide another method using rank. Generally, any outliers in the data set will appear at the extreme order statistics of the residuals. The results of the ranked absolute value of the residuals are shown in Figure 2.13 indicate three possible outliers in the data set. Since this is a simple demonstration, we could declare all three outliers as noise and remove them, resulting in $n_o = 439$ observations and $p = 10$ variables. However, the advantage of rank-based methods is that it will automatically do so without manual intervention.

The second step is to set $\alpha = 1$ in Eq. (2.8.15) and sweep λ so that we can obtain the aLASSO trace to identify and remove any unimportant variables. This is because model complexity introduces yet another bias-variance trade-off. Compact models with fewer variables tend to have lower variance, which is desirable, but it also leads to a higher bias. We need to strike a balance between having too many variables or too few variables to avoid over-fitting or under-fitting, respectively. Hence, we will select the minimum number of variables that provide prediction "signal" and remove the ones that add prediction "noise". This is accomplished using the ordering provided by aLASSO and the adjusted R^2 statistic to decide how many variables to keep. We could use

10 Obtained from the sklearn Python library of data sets and accessed using diab = datasets.load_diabetes().

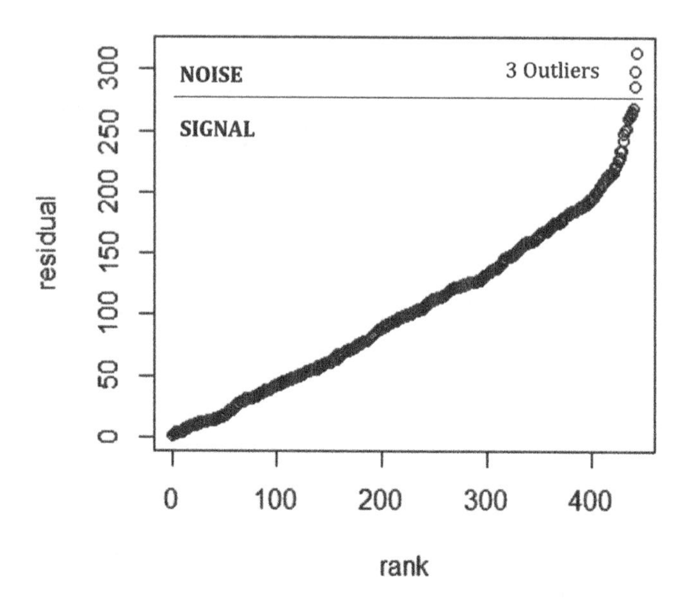

Figure 2.13 Ranked residuals of the diabetes data set. (Source: Rfit() package in R.)

aLASSO by itself as a shrinkage estimator, but here we take a different approach and use it to perform subset selection only.

The aLASSO trace (using data that has not been standardized) is shown in Figure 2.14. Note that some variables have zero-crossings in close proximity to one another, which we have clustered into most important, moderately important and least important variables. This indicates that the optimal order may be permutable for those variables. Local re-ordering within groups may be needed in some data sets to obtain the true optimal subset. This is due to the fact that correlations will exist between different parameters that could change the parameter combinations as the subset size increases. That is, the best subset of size 4 may not be a superset of the best subset of size 3. In our case, only p models will be built to find the optimal set rather than 2^p models, given the near-optimal ordering from aLASSO traces. Further, the selection criteria should be based on the three broad categories (most important, moderate importance, least important) rather than the exact sequential ordering offered by aLASSO.

Here, we use the exact aLASSO ordering of variables and plot the adjusted R^2 as shown in Figure 2.15. Then, we select the subset that maximizes the adjusted R^2 and declare the rest as noise. The curve based on aLASSO ordering (lower one) closely follows the best subset selection order (upper one), as shown in the plot. It tells us that six variables should be selected, namely, bmi, $s5$, bp, $s1$, sex, and $s2$, because the adjusted R^2 drops in value when the seventh variable is added to the model. It is interesting to note that the $|t.\text{value}|$ shown

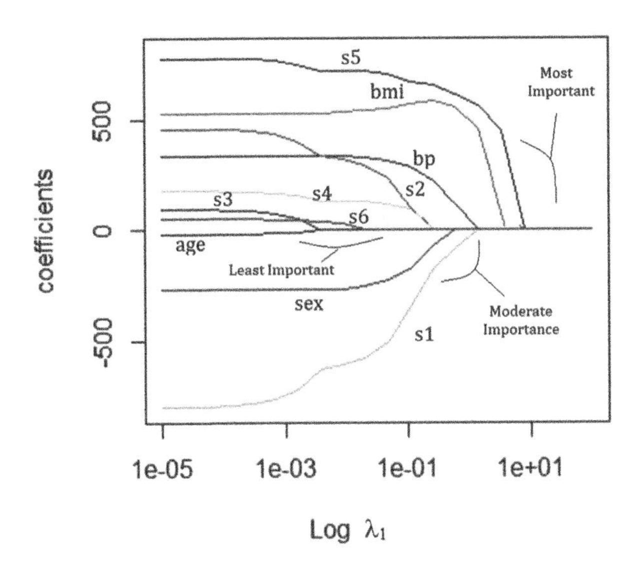

Figure 2.14 Rank-aLASSO trace of the diabetes data set showing variable importance.

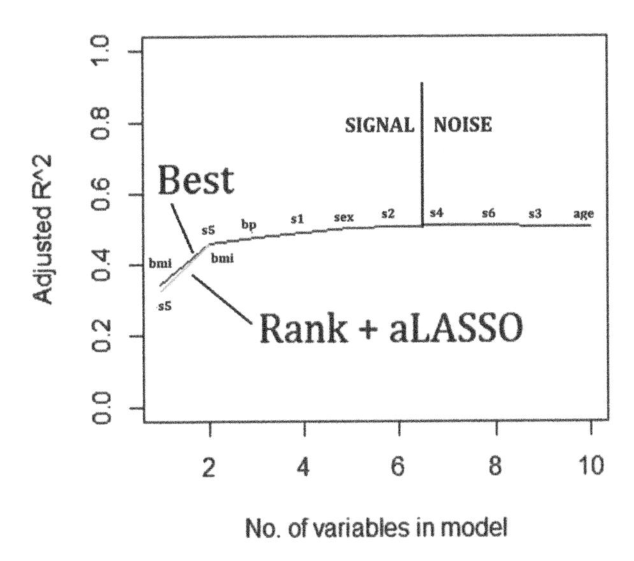

Figure 2.15 Diabetes data set showing variable ordering and adjusted R^2 plot.

Table 2.6 Estimates with outliers from diabetes data before standardization.

		Rank estimates from rfit()		
Explanatory variable	**Estimate (original)**	**Std. error**	***t*.value statistic**	**|*t*.value| ordering**
Intercept	151.9	3.83	39.6	0
bmi	526.8	70.3	7.4	1
bp	336.2	69.2	4.8	2
s5	775.1	181.8	4.2	3
sex	−268.4	64.7	−4.1	4
s1	−795.2	440.7	−1.8	5
s2	458.2	358.6	1.2	6 signal
s4	177.9	170.8	1.0	7 noise
s6	50.5	69.7	0.7	8
s3	93.1	224.8	0.4	9
age	−18.3	63.1	−0.3	10

in Table 2.6 selects the same six variables, assuming the objective is to build a six-variable model. The remaining four variables, namely, *age*, *s*3, *s*4, and *s*6, would add noise to our regression model so we can remove them. We may not know the meanings of all the variables, for example *s*1–*s*6, but we are able to ascertain their relative importance from the $|t.\text{value}|$, aLASSO traces ($m = 40$) and adjusted R^2 plot ($p = 10$). Hence, a total of $p + m$ models are needed for near-optimal subset selection.

Before going any further, we should recognize that the two steps completed thus far have effectively reduced the observations from $n = 442$ to $n_o = 439$, and the number of variables from $p = 10$ to $p_o = 6$, as illustrated conceptually in Figure 2.16. We can remove the columns and rows of data associated with the shaded area on the left of the full data set. The remaining data in white has now been effectively "cleaned" of outliers and unimportant variables. This is accomplished by using the rank (L1-type) loss with the aLASSO (L1-type) penalty.

2.9.2 MSE vs. MAE

We need to choose an appropriate evaluation statistic for our upcoming five-fold CV scheme. We may select the residual MSE statistic if the data set has

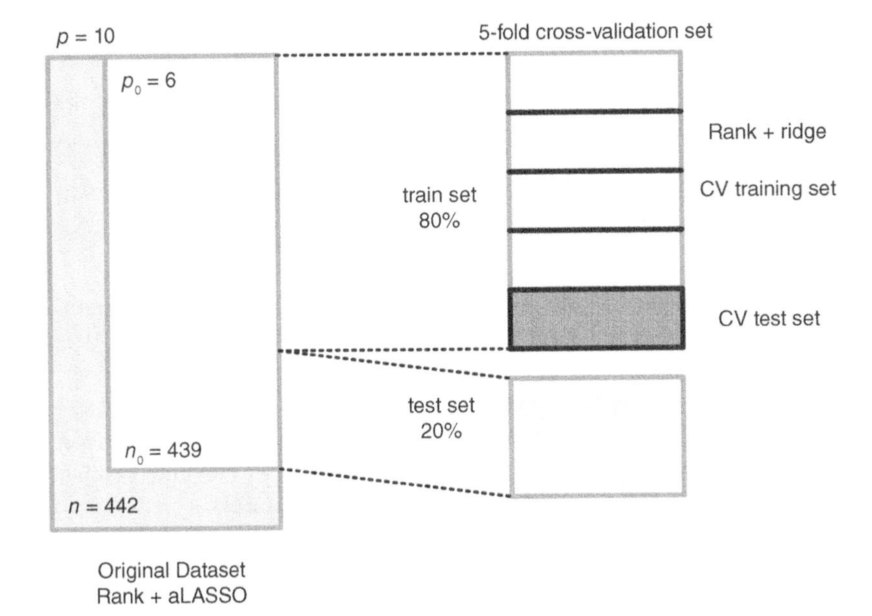

Figure 2.16 Rank-aLASSO cleaning followed by rank-ridge estimation.

no outliers or balanced outliers. However, in the general case of an unknown number of outliers and their locations, we need to use a robust statistic. One candidate is the residual median absolute error (MAE) given by

$$\text{MAE} = \text{median}(|y_1 - \hat{y}_1|, |y_2 - \hat{y}_2|, \dots, |y_n - \hat{y}_n|), \tag{2.9.1}$$

where $\hat{y}_i = \hat{\theta}_n^{\text{R}} + \sum_{j=1}^{p} x_{ij}\hat{\beta}_j^{\text{R}}$. We will first use MSE assuming no outliers are present. Later we will compare MSE with MAE and show that the results are similar with no outliers. This is important because we would like to use the MAE statistic for robust analysis even when there are no outliers.

Next, we divide the data set into two parts: a training set (80% of the data) and a hold-out test set (20% of the data), as in Figure 2.16. The rank estimates from the full training set can be used to determine the MSE of the training set and the test set as a baseline for our model accuracy. This requires that we set $\lambda = 0$ in Eq. (2.8.15). Of course, we must now standardize the $\boldsymbol{X}$ matrix before obtaining the estimates. This is done preemptively in order to avoid problems later on of having one coefficient value dominate over the others, especially in the case of ridge due to the fact that coefficients are squared in its penalty function (James and Stein, 1961). This could happen if the range of one column is, say, $0\text{–}10^2$ while another column is in the range of $0\text{–}10^6$. The standardization of $\boldsymbol{X}$ is performed to bring the column values into the same range. In addition, standardization improves interpretability of the coefficients which is perhaps a more important reason to use it.

Standardization is carried out by dividing each element in a column by the standard deviation of that column as follows:

$$\tilde{x}_{ij} = \frac{x_{ij}}{\sqrt{\frac{1}{n-1}\sum_{i=1}^{n}(x_{ij} - \bar{x}_j)}} \quad \text{for } i = 1, \ldots, n, \; j = 1, \ldots, p. \tag{2.9.2}$$

After standardization of the diabetes data set and computation of the rank estimates, the training set has an MSE = 2769.49, while the test set has a higher MSE = 3377.24. This is expected since the model has been fit to the training set, so its MSE would be smaller, and then used to predict on the "unseen" test set, which often has a higher MSE (although it could be lower in some cases). Given the fact that the MSE in the two cases are different, it is worthwhile trying to improve the results and to illustrate the model building process.

Our final step is to set $\alpha = 0$ in Eq. (2.8.15) to invoke ridge and then sweep λ to obtain a shrinkage model using R-ridge regression with a five-fold cross-validation scheme in the manner shown in Figure 2.16. The CV training set is used to obtain the estimates for each fold while the CV test set is used to obtain the MSE. This is repeated five times, once for each fold shown in the figure. The results are averaged to obtain the estimates and MSE for each λ value. The λ value producing the lowest MSE is chosen as the optimal value and the final estimates are based on this value.

The results are shown in Figure 2.17. On the left panel are the ridge traces for the six variables of interest using the standardized $\boldsymbol{X}$ matrix. On the right, the MSE vs. log λ_2 plots are shown for the training set, CV sets and test set. Note that the optimal λ_2 is based on the CV curve rather than the test curve as is common practice. Although it is tempting to use the minimum of the test MSE to pick the optimal λ_2, we do not have access to this test data typically, but it is shown to emphasize that we are optimizing based on the CV data. However, it can be used to compare different estimators and modeling approaches.

The comparison results described above are reported in Table 2.7 for the estimators described in this chapter. Included in the table are estimates and number of variables used, p_o, for LSE, LS-ridge, LS-LASSO and LS-Enet. The R package used to obtain these results will be described in the next section. We also show the R-estimates and R-aEnet estimates (which are essentially the R-ridge estimates with six variables). In all cases, the same five-fold CV scheme was used with 80% of the data for CV training MSE and 20% for final test MSE.

Since we are using robust methods, the use of mean-absolute error is warranted. We show the MAE and MSE results from a five-fold CV scheme on the

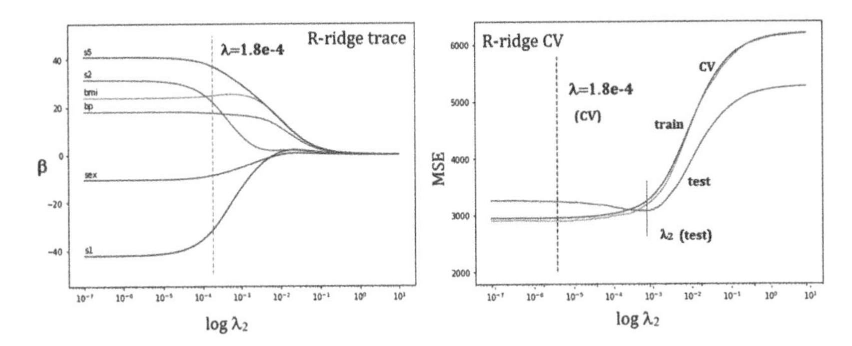

Figure 2.17 R-ridge traces and CV scheme with optimal λ_2.

training set, and the corresponding MAE and MSE on the test set in Table 2.7. The table indicates that the lowest test MSE and MAE is obtained in the final column for R-aEnet. In particular, the test MSE is 3156.56 while the test MAE is 32.53. By comparison, the nearest test MSE is 3338.88 from LSE while the nearest MAE is 39.80 due to LS-ridge. Hence, the approach of using the rank loss function combined with aLASSO for subset selection and ridge for shrinkage provides a compact and interpretable model with the best prediction accuracy. For R-ridge, the detailed CV plots are given in Figure 2.18, including the ranges of MSE and MAE values, respectively, for the five-fold CV scheme. We note that neither plot has a discernible minimum but in fact a minimum does exist for both at the same value as indicated in the figures.

2.9.3 Model Building with LS-Enet

The previous section presented multiple results from the LS-Enet estimator. The method of obtaining those results will now be described. Recall from Eq. (2.8.13) that we use two tuning parameters, α and λ. Essentially, it requires a grid search for the two parameters to find their optimal values. Since $\alpha \in [0, 1]$, it is useful to examine the evolution of the traces by picking six values of α over this range.

In Figure 2.19, we see traces for the different cases of the α tuning parameter. They all look about the same, but if we look carefully, we notice that as α moves from 0 to 1 in increments of 0.2, and the LS-ridge trace ($\alpha = 0.0$) begins to shrink and curl towards the λ-axis until it becomes the LS-LASSO trace ($\alpha = 1.0$). It is worthwhile experimenting with LS-Enet on this data set to show how a good model can be produced.

After standardizing the data as before, we can use `glmnet()` (R package available from Stanford University) to perform a sweep of α from 0 to 1 in increments

Table 2.7 Estimates. MSE and MAE for the diabetes data

	LS-ridge	LS-Enet	LS-LASSO	LSE	R-estimate	R-aEnet
AGE	0.29	0	0	−0.06	−1.05	0
SEX	−11.00	−11.60	−10.00	−12.25	−13.01	−12.21
BMI	25.30	25.74	25.70	25.67	25.81	23.02
BP	11.61	11.74	10.71	12.21	12.44	16.63
S1	−5.21	−11.14	−6.37	−30.97	−27.00	−30.31
S2	−4.65	0	0	15.23	10.93	22.0
S3	−10.73	−9.01	−12.34	−0.43	−1.47	0
S4	6.27	5.30	8.21	8.29	9.44	0
S5	21.91	25.88	25.07	32.93	32.32	38.32
S6	3.6	2.79	2.08	2.89	2.14	0
p_o	10	8	8	10	10	6
Train MSE	2781.48	2769.95	2784.40	2762.79	2769.49	2833.34
Train MAE	38.21	37.83	38.82	36.91	36.70	38.31
Test MSE	3373.13	3267.95	3363.78	3338.88	3377.24	3156.56
Test MAE	39.80	40.10	40.11	41.28	40.21	32.53

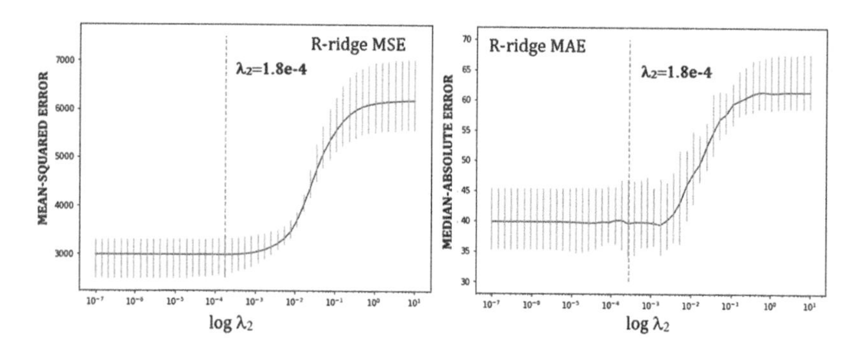

Figure 2.18 MSE and MAE plots for five-fold CV scheme producing similar optimal λ_2.

of 0.1, and a sweep of λ from 10^{-6} to 10^{6} using 100 steps. While the results fluctuate somewhat, the best values are around $\alpha = 0.6$. In Figure 2.20, we show the corresponding coefficient traces for $\alpha = 0.6$ and cross-validation results. The optimal λ lies in a bounded range indicated by the grey vertical lines in the figure on the right. The optimal value of $\lambda_{\text{opt}} = 0.43$ reported by `glmnet()` is the black vertical line which was transferred to the figure on the left thereby

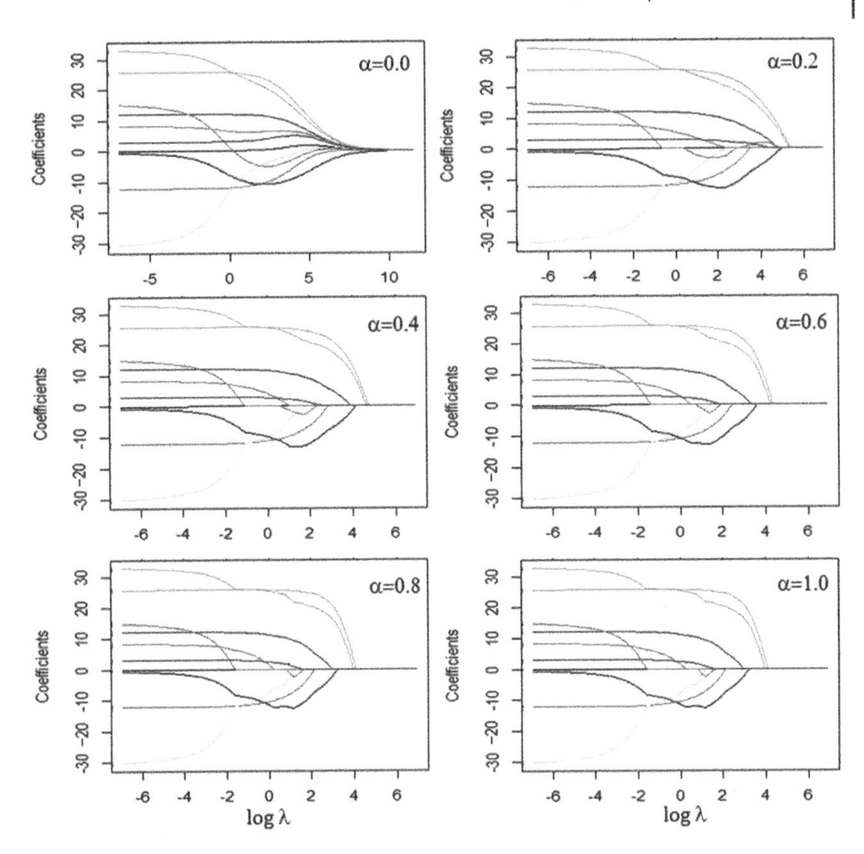

Figure 2.19 LS-Enet traces for $\alpha = 0.0, 0.2, 0.4, 0.8, 1.0$.

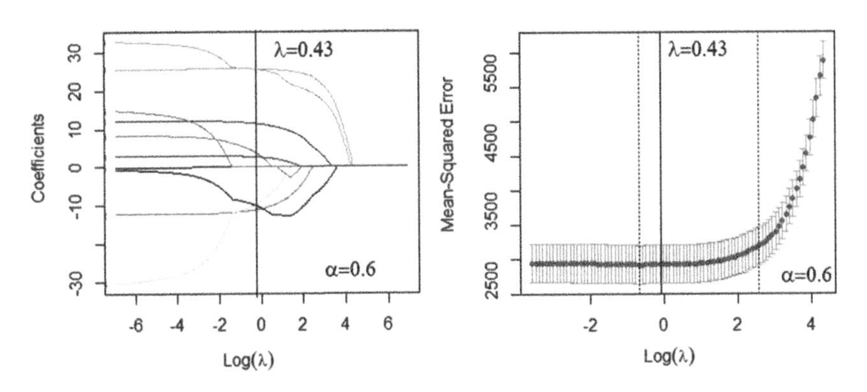

Figure 2.20 LS-Enet traces and five-fold CV results for $\alpha = 0.6$ from glmnet().

Table 2.8 Enet estimates, training MSE and test MSE as a function of α for the diabetes data

α	0.0	0.2	0.4	0.6	0.8	1.0
$\lambda_{(opt)}$	5.86	3.35	0.76	0.43	0.8	1.10
AGE	0.29	0	0	0	0	0
SEX	−11.00	−10.35	−11.46	−11.60	−9.18	−10.00
BMI	25.30	25.02	25.65	25.74	25.53	25.70
BP	11.61	11.12	11.67	11.74	10.29	10.71
S1	−5.21	−5.27	−10.66	−11.14	−5.29	−6.37
S2	−4.65	−2.45	0	0	0	0
S3	−10.73	−11.69	−9.36	−9.01	−12.15	−12.34
S4	6.27	2.92	4.86	5.30	0	0
S5	21.91	23.29	25.68	25.88	24.35	25.07
S6	3.6	2.79	2.79	2.78	1.74	2.08
p_o	10	9	8	8	7	7
train MSE	2781.48	2782.29	2770.70	2769.95	2794.52	2784.40
test MSE	3373.13	3363.66	3365.12	3367.00	3364.37	3363.78

obtaining the estimates. We can see that a number of parameters have been set to zero based on the position of λ_{opt}.

The model coefficients for the diabetes data set from various cases are shown in Table 2.8. The `glmnet()` estimates for LS-Enet with $\alpha = 0.6$ was used in Table 2.7 since it has the lowest MSE for the training set. It uses $p_o = 8$ parameters in the model. While there are other models with fewer variables, they had higher MSE values for the training set. The LS-LASSO estimates in Table 2.7 are from $\alpha = 1.0$ while the LS-ridge results are from $\alpha = 0.0$.

2.10 Summary

In this chapter, we have presented properties and practical aspects of penalized rank estimators and its application to machine learning. Several new ideas and concepts have been introduced to enable robust linear regression and model building that deserve further investigation. We have demonstrated the importance of penalty estimators and the inherent value of each type of penalty function in robust data science. We combined rank regression with ridge, LASSO, aLASSO, and aEnet and identified some new capabilities using various combinations. Ridge is a smooth penalty function and the best at shrinkage, whereas aLASSO is the best at variable ordering and hence subset selection. One can view LASSO as being somewhere in between. LASSO typically finds one or more of the least important parameters and may set some of them to zero, but not necessarily in the proper order since it does not have the oracle property. Further investigation of these and other robust methods would provide fruitful directions for future research.

The motivation for the use of rank in machine learning has been well-established for multiple linear regression in the first two chapters through practical applications. It is now time to establish the mathematical foundations for these methods, starting from the ground up. In the upcoming chapters, we will carry out a theoretical treatment of topics in rank theory. Our journey begins with restricted and unrestricted rank estimators for the location and simple linear models. We then move to the analysis of variance, and seemingly unrelated simple linear models, followed by multiple linear models. Next, we cover partially linear models and Liu regression, as well as AR(p) models. High-dimension problems will also be addressed in a subsequent chapter. We conclude the book with practical machine learning topics of rank-based logistic regression and neural networks.

2.11 Problems

2.1. Suppose $p = 2$ for a given centered data set and the least squares estimates are $\hat{\beta}_1^{\text{LS}} = 10.0$ with $\text{Var}(\hat{\beta}_1^{\text{LS}}) = 0.49$, and $\hat{\beta}_2^{\text{LS}} = 5.0$ with $\text{Var}(\hat{\beta}_2^{\text{LS}}) = 0.25$. Assume that $\hat{\sigma}^2 = 10.0$.
   (a) What are the standard errors of each parameter?
   (b) What is the matrix $(\boldsymbol{X}^\top\boldsymbol{X})^{-1}$ assuming the off-diagonals are 0.
   (c) Plot the ridge trace for these two parameters as a function of $\lambda_2 =$ 0 to 10. (Hint: since this is the diagonal case, you can use (2.4.10). Your answer should look similar to Figure 2.1). Comment on the results.
   (d) How would you determine $\hat{\theta}_n^{\text{LS}}$ in general? Given $\hat{\boldsymbol{\beta}}_n^{\text{LS}}$, write the expression using uncentered data, $\boldsymbol{X}_{\text{uc}}$ and $\boldsymbol{y}_{\text{uc}}$. No calculations are needed here.

2.2. Suppose $p = 2$ for a given data set and the R-estimates are $\hat{\beta}_1^{\text{R}} = 10.3$ with $\text{Var}(\hat{\beta}_1^{\text{R}}) = 0.343$, and $\hat{\beta}_2^{\text{R}} = 5.1$ with $\text{Var}(\hat{\beta}_2^{\text{R}}) = 0.175$. Let $\hat{\eta}^2 = 9.0$.
   (a) What are the standard errors of each parameter?
   (b) Assuming the same data set as in Problem 2.1, are there outliers in the data?
   (c) It is interesting to note that the rank intercept is not necessarily 0 even with centered data. Also, we do not estimate the intercept using $D_n(\boldsymbol{\beta})$. Instead, we can use the median to find the intercept. How

would you determine $\hat{\theta}_n^{\text{R}}$ in general, given $\hat{\boldsymbol{\beta}}_n^{\text{R}}$. Write the expression using uncentered data, $\boldsymbol{X}_{\text{uc}}$ and $\boldsymbol{y}_{\text{uc}}$. No calculations are needed here.

2.3. Examine Table 2.3 and Figure 2.3.
   (a) Based on the table and the corresponding figure, which parameters would benefit most from a reduction in their variances (ignoring β_0, of course)?
   (b) Ideally, we would like to shrink only those with large variances. Which parameters would we prefer to keep the same because they already have small variances?
   (c) Next consider the ridge traces of Figure 2.5. What are the first three parameters that are reduced to almost 0 as λ_2 increases in the D-ridge plot? What are the last two parameters to be reduced to almost 0?
   (d) Why are the results from LS-ridge and rank-ridge somewhat different from those of D-ridge? (Hint: consider the effects of using the full $\boldsymbol{C}^{-1}$ matrix compared to a diagonal matrix to obtain the shrinkage estimates.)

2.4. Consider Figure 2.10 and the penalized least squares estimators for $\beta_j^{\text{D-LASSO}}$ and $\beta_j^{\text{D-aLASSO}}$, given below as,

$$\hat{\beta}_j^{\text{D-LASSO}} = \text{sgn}(\hat{\beta}_j^{\text{LS}})(|\hat{\beta}_j^{\text{LS}}| - \lambda_1 \text{ Var}(\hat{\beta}_j^{\text{LS}}))^+$$

and

$$\hat{\beta}_j^{\text{D-aLASSO}} = \text{sgn}(\hat{\beta}_j^{\text{LS}})\left(|\hat{\beta}_j^{\text{LS}}| - \lambda_1 \frac{\text{Var}(\hat{\beta}_j^{\text{LS}})}{|\hat{\beta}_j^{\text{LS}}|}\right)^+,$$

respectively. The only difference is the use of a weighting term in aLASSO. Explain in a qualitative sense why aLASSO possesses the oracle property while LASSO does not, in general, have this feature using two different diamond shapes, one for LASSO and one for aLASSO. Show how the trajectories of the estimates may differ in the two cases. Connect your explanation to the traces shown in Figure 2.12. Why might the two traces differ?

2.5. Ridge and LASSO can be applied in a one-dimensional setting even though they were originally intended for multivariate regression with collinearity. For example, it can be applied to the location problem, with location parameter θ, as will be discussed in detail in the next chapter.
   (a) Devise a simple ridge-type penalty estimator that would shrink the estimate by division using λ_2 as the tuning parameter assuming you are given an unbiased rank estimate, $\hat{\theta}_n^{\text{R}}$. (Hint: see Eq. (2.4.7)).
   (b) Likewise, devise a simple LASSO-type penalty estimator by subtraction using λ_1 in your equation. (Hint: see Eq. (2.7.9))

(c) If $\hat{\theta}_n^{\mathrm{R}} = 2.0$, sketch the traces for ridge and LASSO on the same plot with λ as your x-axis (representing both λ_1 and λ_2, depending on the trace).

(d) How would you define an aLASSO-type estimator in this one-dimensional setting? Plot its trace on this same graph. What can you say about the difference in LASSO and aLASSO in this case?

2.6. Consider the following data set with $p = 2$ to study collinearity:

$\boldsymbol{x}_1$	132.2	501.5	904.0	227.6	66.6	43.4	1253.0
$\boldsymbol{x}_2$	11.7	17.9	21.1	14.7	7.7	8.4	32.8
$\boldsymbol{y}$	404.6	1180.6	1807.5	470.0	151.4	93.8	3293.4

Assume the linear regression model to be

$$\boldsymbol{y} = \beta_0 + \beta_1 \boldsymbol{x}_1 + \beta_2 \boldsymbol{x}_2 + \boldsymbol{\varepsilon}.$$

(a) Plot x_2 vs. x_1. Qualitatively assess whether there is a potential for collinearity, that is, a linear relationship between the two variables. (Hint: if there is a strong correlation between the two variables, there is likely to be collinearity in the data set.)

(b) To quantitatively assess collinearity, we can use the condition number which is the ratio of the largest and smallest eigenvalues of the matrix $\boldsymbol{C} = \boldsymbol{X}^\top \boldsymbol{X}$ using the centered matrix $\boldsymbol{X}$. To find the eigenvalues of a 2×2 matrix, we can use the quadratic equation:

$$\lambda^2 + \mathrm{tr}(\boldsymbol{C})\lambda + \det(\boldsymbol{C}) = 0$$

where λ represents the two resulting eigenvalues, λ_{big} and λ_{small}, $\mathrm{tr}(\boldsymbol{C})$ is the trace of the matrix and $\det(\boldsymbol{C})$ is the determinant of the matrix. Compute the condition number as the ratio $\lambda_{\text{big}}/\lambda_{\text{small}}$. Is this value large (i.e. ill-conditioned) or small (i.e. well-conditioned)?

(c) Find the least squares estimators for β_1 and β_2 using centered data as follows

$$\hat{\boldsymbol{\beta}}_n^{\text{LS}} = (\boldsymbol{X}^\top \boldsymbol{X})^{-1} \boldsymbol{X}^\top \boldsymbol{y}.$$

(d) Find the LS-ridge estimates and plot the ridge trace using the closed-form expression given here ($\boldsymbol{I}_p$ is a 2×2 identity matrix):

$$\hat{\boldsymbol{\beta}}_n^{\text{LS-ridge}} = (\boldsymbol{X}^\top \boldsymbol{X} + \lambda_2 \boldsymbol{I}_p)^{-1} \boldsymbol{X}^\top \boldsymbol{y}.$$

Use $\lambda_2 = 0, 1, 2, 5, 10, 20, 50, 100, 200, 500, 1000$.

2.7. (PROJECT) This problem uses R to obtain results for both least squares and rank estimation using the data set from the previous problem. You will have to run commands in R and write some simple R code to complete this problem[11].

(a) Run the following lines of code in R to obtain LS estimates. Compare your results with the values obtained in the previous problem.

```
x<-cbind(c(11.7,17.9,21.1,14.7,7.7,8.4,32.8),
    c(132.2,501.5,904.0,227.6,66.6,43.4,1253.0))
y<-c(404.6,1180.6,1807.5,470.0,
    151.4,93.8,3293.4)
lmfit <- lm(y ~ x)
summary(lmfit)
```

(b) To obtain R-estimates, run the `rfit()` function. You may have to install the `Rfit` library in order to run the commands below.

```
library("Rfit")
fit <- rfit(y ~ x)
summary(fit)
```

How do your results compare with part (a). That is, are they almost the same or somewhat different?

(c) In this step you will enter R code to obtain the same results as in part (b). For this purpose, write a routine to perform the following operation:

$$\hat{\beta}_n^{\mathrm{R}} = \underset{\beta}{\operatorname{argmin}} \left\{ \sum_{i=1}^{n} (y_i - \boldsymbol{x}_i^\top \boldsymbol{\beta}) \sqrt{12} \left(\frac{R_{n_i}(\boldsymbol{\beta})}{n+1} - \frac{1}{2} \right) \right\}.$$

The `rank()` and `optim()` functions of R will be needed. You may find this starter R code below useful. It requires `x_centered` and `y_centered` to use with this code. The results in `betaR` should match the output of `rfit()`.

```
minimize.DnB <- function(par, X, y) {
    n <- length(y)
    diff <- y-(X %*% par)
    Rni <- rank(diff,ties.method = "random")
    u <- Rni/(n+1.)
    an <- sqrt(12.)*(u-0.5)
    Dnsum <- sum(diff*an)  #Estimator line
    return( Dnsum )
}
a <- t(x)-colMeans(x)
x_centered <- t(a)
```

11 Python and its associated libraries may also be used to solve all programming problems, if desired.

```
y_centered <- y - mean(y)
p = ncol(x_centered)
op.result <- optim(rep(0, p),
        fn = minimize.DnB, method = "BFGS",
        X = x_centered, y = y_centered)
betaR <- op.result$par
betaR
```

(d) In this step you will update your R code to obtain the ridge trace results using a ridge R-estimator. For this purpose, you will need to program the following estimator:

$$\hat{\boldsymbol{\beta}}_n^{\text{R-ridge}} = \underset{\boldsymbol{\beta}}{\text{argmin}} \left\{ \sum_{i=1}^{n} (y_i - \boldsymbol{x}_i^\top \boldsymbol{\beta}) \sqrt{12} \left(\frac{R_{n_i}(\boldsymbol{\beta})}{n+1} - \frac{1}{2} \right) + \lambda \sum_{j=1}^{p} \beta_j^2 \right\}.$$

Choose values in the range of $\lambda = 0.0$ to 1.0 and re-run the code for each value of λ. Plot as a function of $\log(\lambda)$ and compare with the previous problem. Use the existing code but replace the estimator line with

```
lambda <- 0.0
Dnsum <- sum(diff*an) + lambda * (par %*% par)
```

You will need to set lambda each time you run it.

2.8. (PROJECT) This problems concerns the use of R to obtain results for rank estimation with the LASSO penalty.

(a) For the same data set above, write an R program to carry out the following optimization problem:

$$\hat{\boldsymbol{\beta}}_n^{\text{R-LASSO}} = \text{argmin}_{\boldsymbol{\beta}} \left\{ \sum_{i=1}^{n} (y_i - \boldsymbol{x}_i^\top \boldsymbol{\beta}) \sqrt{12} \left(\frac{R_{n_i}(\boldsymbol{\beta})}{n+1} - \frac{1}{2} \right) + \lambda \sum_{j=1}^{p} |\beta_j| \right\}.$$

Use the previous R code but with

```
Dnsum <- sum(diff*an) + lambda * sum(abs(par))
```

as the estimator. Compute the rank estimates for $\lambda = 0$.

(b) Plot the LASSO traces. Use $\lambda = 0$ to 10. Which parameter is more important based on the results?

2.9. (PROJECT) This problem concerns the generation of R-aLASSO traces.

(a) Write a program for rank estimation with an aLASSO penalty as follows.

$$\hat{\boldsymbol{\beta}}_n^{\text{R-aLASSO}} = \text{argmin}_{\boldsymbol{\beta}} \left\{ \sum_{i=1}^{n} (y_i - \boldsymbol{x}_i^\top \boldsymbol{\beta}) \sqrt{12} \left(\frac{R_{n_i}(\boldsymbol{\beta})}{n+1} - \frac{1}{2} \right) + \lambda \sum_{j=1}^{p} \frac{|\beta_j|}{|\beta_j^{\text{R}}|} \right\}.$$

Compute the rank estimates for $\lambda = 0$ and use those values as weights. Use

```
Dnsum <- sum(diff*an) + lambda * sum(abs(par/betaR))
```

(b) Plot the aLASSO traces with $\lambda = 0, 400, 800, 1200$ and 1600, but set $\lambda = 0$ and re-run the code to reset the rank estimates each time. Or use `betaRaLASSO <- op.result$par` to store results from each aLASSO run rather than in `betaR`. Which parameter is more important?

2.10. (PROJECT) This problem concerns the use of R-Enet on the data set of Problem 2.6. Write a program to implement the following minimization problem.

$$\hat{\boldsymbol{\beta}}_n^{\text{R-aEnet}} = \underset{\boldsymbol{\beta}}{\text{argmin}} \left\{ \sum_{i=1}^{n} (y_i - \boldsymbol{x}_i^{\top}\boldsymbol{\beta}) \sqrt{12} \left(\frac{R_{n_i}(\boldsymbol{\beta})}{n+1} - \frac{1}{2} \right) + \lambda \left(\frac{1-\alpha}{2} \sum_{j=1}^{p} \beta_j^2 + \alpha \sum_{j=1}^{p} |\beta_j| \right) \right\}.$$

(a) Determine the R-Enet estimates for $\alpha = 0$, $\lambda = 0$.

(b) Set $\alpha = 0, 0.25, 0.5, 0.75, 1.0$ and then sweep λ from 0 to 10 for each α. You should produce five different plots of the Enet traces.

(c) Which plots are similar to R-ridge and which are similar to R-LASSO obtained in previous problems.

3

Location and Simple Linear Models

In previous chapters, we explored new ideas, concepts and practical aspects of rank-based regression with penalty functions towards robust data science. Our next step is to examine the theoretical underpinnings of shrinkage R-estimators, their asymptotic distributional characteristics, including quadratic risk and hypothesis testing. We will introduce the formalisms, terminology, equations and notation to be used in the rest of this book. Our goal is to provide a strong foundational background for further research in this field as it pertains to rank-based methods.

Readers will likely have different backgrounds in statistics when approaching this chapter. As such, we have included some basic introductory material for those who are unfamiliar with this field, but this chapter is primarily intended for those who are already well-versed in statistics. The chapter begins with the location parameter and simple linear models and then describes R-estimators and R-test for the parameters of the models under consideration. In order to build up the needed knowledge for later chapters, we describe a variety of penalized R-estimators of the location model and of the slope for the simple linear model, and then compare the different R-estimators using an L_2-risk criterion.

3.1 Introduction

A location model is used to estimate the center of a sample of n values, y_1, ..., y_n, for a random variable Y. This could be the sample mean or median or any other similar type of estimate depending on the underlying distribution. We will describe how to analyze this simple case to begin our theoretical study of rank regression because it is much easier to understand the theory in this

Rank-Based Methods for Shrinkage and Selection: With Application to Machine Learning.
First Edition. A. K. Md. Ehsanes Saleh, Mohammad Arashi, Resve A. Saleh, and
Mina Norouzirad.

setting as opposed to the multivariate and high-dimensional settings of later chapters.

Consider the location model, for which we seek to estimate θ, given by

$$y_i = \theta + \varepsilon_i, \quad i = 1, \dots, n$$

where $\varepsilon_1, \dots, \varepsilon_n$ are i.i.d. random variables with a continuous and symmetric c.d.f., $F(\cdot)$, and absolutely continuous p.d.f., $f(\cdot)$, having finite Fisher information

$$I(f) = \int_{-\infty}^{+\infty} \left\{ -\frac{f'(x)}{f(x)} \right\}^2 f(x)\mathrm{d}x < \infty.$$

We will be developing several consistent rank estimators for θ and the required functions for hypothesis testing. We say that an arbitrary estimator, $\hat{\theta}_n$, is asymptotically normal if, as $n \to \infty$, we have convergence in distribution of the form

$$\sqrt{n}(\hat{\theta}_n - \theta) \xrightarrow{\mathcal{D}} \mathcal{N}(0, \eta^2),$$

where η^2 is the variance of the rank estimator, $\hat{\theta}_n$. That is, the R-estimator generally satisfies the law of large numbers and the properties of the central limit theorem.

When comparing different estimators, we consider the asymptotic distributional risk of each estimator under the L_2 loss (also called the quadratic loss and mean squared error), denoted in the rest of this book as the ADL_2-risk, as follows

$$\text{ADL}_2\text{-risk}(\hat{\theta}_n) = \mathbb{E}\left[n\left(\hat{\theta}_n - \theta\right)^2\right]. \tag{3.1.1}$$

We can show that the above expands out to be

$$\text{ADL}_2\text{-risk}(\hat{\theta}_n) = \mathbb{E}[n(\hat{\theta}_n - \mathbb{E}[\hat{\theta}_n])^2] + n(\mathbb{E}[\hat{\theta}_n] - \theta)^2.$$

The asymptotic distributional risk is comprised of two terms:
(1) The asymptotic distributional variance (ADV), defined as

$$\text{ADV}(\hat{\theta}_n) = \mathbb{E}[n(\hat{\theta}_n - \mathbb{E}[\hat{\theta}_n])^2]. \tag{3.1.2}$$

(2) The square of the asymptotic distributional bias (ADB) which is defined as

$$\text{ADB}(\hat{\theta}_n) = \sqrt{n}(\mathbb{E}[\hat{\theta}_n] - \theta). \tag{3.1.3}$$

Hence, Eq. (3.1.1) can be written as

$$\text{ADL}_2\text{-risk}(\hat{\theta}_n) = \text{ADV}(\hat{\theta}_n) + (\text{ADB}(\hat{\theta}_n))^2. \tag{3.1.4}$$

An estimator performs better than another if it has a lower ADL$_2$-risk. We will also be comparing the ADL$_2$-risk efficiency (ADRE) of the various estimators. The relative efficiency of an estimator of interest, $\hat{\theta}_n$, compared to a reference estimator, $\hat{\theta}_n^*$, is defined as

$$\text{ADRE}(\hat{\theta}_n : \hat{\theta}_n^*) = \frac{\text{ADL}_2\text{-risk}(\hat{\theta}_n^*)}{\text{ADL}_2\text{-risk}(\hat{\theta}_n)}. \tag{3.1.5}$$

For R-estimation, we will need the services of a scoring function. We define the score function in the following manner:

$$\psi(u, f) = -\frac{f'(F^{-1}(u))}{f(F^{-1}(u))}, \quad 0 < u < 1,$$

and

$$\psi^+(u, f) = -\frac{f'(F^{-1}(\frac{1+u}{2}))}{f(F^{-1}(\frac{1+u}{2}))}.$$

For instance, consider the logistic distribution,

$$F(x) = \frac{1}{1 + \exp(-x)},$$

which satisfies the required continuity and symmetry conditions. Then

$$f(x) = \frac{\mathrm{d}F(x)}{\mathrm{d}x} = \frac{\exp(-x)}{(1 + \exp(-x))^2}.$$

It follows that

$$f'(x) = \frac{\mathrm{d}f(x)}{\mathrm{d}x} = -\frac{\exp(-x)\,(1 - \exp(-x))}{(1 + \exp(-x))^3},$$

and also

$$F^{-1}(u) = -\ln\left(\frac{1}{u} - 1\right),$$

such that

$$F^{-1}\left(\frac{1+u}{2}\right) = -\ln\left(\frac{2}{1+u} - 1\right) = -\ln\left(\frac{1-u}{1+u}\right).$$

Therefore, the score function is

$$\psi(u, f) = -\frac{f'(F^{-1}(u))}{f(F^{-1}(u))} = -\frac{-u(1-u)(2u-1)}{u(1-u)} = 2u - 1$$

and the corresponding score for signed-rank statistic is

$$\psi^+(u,f) = -\frac{f'(F^{-1}(\frac{1+u}{2}))}{f(F^{-1}(\frac{1+u}{2}))} = -\frac{-\frac{(1-u)(1+u)u}{4}}{\frac{(1-u)(1+u)}{4}} = u.$$

3.2 Location Estimators and Testing

3.2.1 Unrestricted R-estimator of θ

Let us first consider the unrestricted rank estimate of the location parameter, θ. The ranks of $|y_i-\theta|$ among $|y_1-\theta|, \dots, |y_n-\theta|$ are given by $R^+_{n_i}(\theta) = R^+_{n_i}(|y_i-\theta|)$. Then $\boldsymbol{R}^+(\theta) = (R^+_{n_1}(\theta), \dots, R^+_{n_n}(\theta))^\top$ is independent of the ordered $|y_1-\theta|, \dots, |y_n-\theta|$. Let

$$\phi^+(u) = \phi\left(\frac{u+1}{2}\right), \quad \phi(u) + \phi(1-u) = 0 \quad \forall u \in (0,1),$$

where $\phi(u)$ is skew-symmetric about $\frac{1}{2}$ and then let

$$A^2_\phi = \int_0^1 \phi^2(u)du. \tag{3.2.1}$$

Now, we define the scores

$$a^+_n(i) = \mathbb{E}[\phi^+(U_{(i)})] \qquad 1 \le i \le n,$$

where

$$\mathbb{E}[U_{(i)}] = \frac{i}{n+1}$$

such that $0 < U_{(1)} \le \dots \le U_{(n)} < 1$ are the ordered statistics of a random sample of size n from $U(0,1)$. There is a related definition for the score function given by

$$a^+_n(R^+_{n_i}(\theta)) = \phi^+\left(\frac{R^+_{n_i}(\theta)}{n+1}\right)$$

as described in Chapter 1, where $\phi^+(u)$, $0 < u < 1$, is a non-decreasing square integrable function.

We require a formal procedure for estimation and hypothesis testing of the location parameter. For estimation, we minimize, with respect to θ, the rank dispersion function, $D_n(\theta)$, as defined by Jaeckel (1972), and given by

$$D_n(\theta) = \sum_{i=1}^{n} |y_i - \theta| a_n^+(R_{n_i}^+(\theta)).$$

It may be noted that $D_n(\theta)$ is a convex, continuous, and piece-wise linear function of θ. Hence, we define the unrestricted R-estimator (URE), denoted by $\hat{\theta}_n^{\mathrm{UR}}$, as

$$\hat{\theta}_n^{\mathrm{UR}} = \underset{\theta \in \mathbb{R}}{\operatorname{argmin}}\{D_n(\theta)\}.$$

Clearly,

$$\frac{\partial D_n(\theta)}{\partial \theta} = -\sum_{i=1}^{n} \operatorname{sgn}(y_i - \theta) a_n^+(R_{n_i}^+(\theta))$$

$$= -\sqrt{n} T_n(\theta),$$

where

$$T_n(\theta) = \frac{1}{\sqrt{n}} \sum_{i=1}^{n} \operatorname{sgn}(y_i - \theta) a_n^+(R_{n_i}^+(\theta)). \tag{3.2.2}$$

Hence, the minimum of $D_n(\theta)$ occurs when $T_n(\theta) = 0$ which we can now use to define the unrestricted R-estimation of θ as

$$\hat{\theta}_n^{\mathrm{UR}} = \frac{1}{2}(\hat{\theta}_n^{(\mathrm{U})} + \hat{\theta}_n^{(\mathrm{L})}),$$

where $\hat{\theta}_n^{(\mathrm{U})} = \sup\{\theta : T_n(\theta) > 0\}$ and $\hat{\theta}_n^{(\mathrm{L})} = \inf\{\theta : T_n(\theta) < 0\}$.

In fact, we can obtain an explicit expression for $\hat{\theta}_n^{\mathrm{UR}}$ by solving the equation

$$\hat{\theta}_n^{\mathrm{UR}} = \underset{\theta \in \mathbb{R}}{\operatorname{argmin}} \left\{ \sum_{i=1}^{n} |y_i - \theta| R_{n_i}^+(\theta) \right\},$$

based on Wilcoxon signed-rank statistics to obtain

$$\hat{\theta}_n^{\mathrm{UR}} = \underset{1 \leq i \leq j \leq n}{\operatorname{median}} \left\{ \frac{y_i + y_j}{2} \right\}.$$

For the R-test of $\mathcal{H}_o : \theta = 0$ vs. $\mathcal{H}_A : \theta \neq 0$, the test-statistic is given by

$$T_n(0) = \frac{1}{\sqrt{n}} \sum_{i=1}^{n} \operatorname{sgn}(y_i) a_n^+(R_{n_i}^+(0)).$$

It may be shown easily that as $n \to \infty$,

$$T_n(0) \xrightarrow{\mathcal{D}} \mathcal{N}(0, A_\phi^2), \tag{3.2.3}$$

where A_ϕ^2 is given by (3.2.1).

Theorem 3.1 The asymptotic distributional bias, variance and L_2-risk of the unrestricted R-estimator, $\hat{\theta}_n^{UR}$, are respectively given by

$$\text{ADB}(\hat{\theta}_n^{UR}) = 0,$$

$$\text{ADV}(\hat{\theta}_n^{UR}) = \eta^2,$$

$$\text{ADL}_2\text{-risk}(\hat{\theta}_n^{UR}) = \eta^2.$$

Proof: To find the asymptotic distribution of $\hat{\theta}_n^{UR}$, we invoke Jurečková (1971) linearity results as $n \to \infty$, to get

$$\left| T_n\left(n^{-\frac{1}{2}} w\right) - T_n(0) + \gamma(\phi, f) w \right| \xrightarrow{P} 0, \quad w = \sqrt{n}(\hat{\theta}_n^{UR} - \theta),$$

where

$$\gamma(\phi, f) = \int_0^1 \phi(u)\psi(u, f)du.$$

Hence, we obtain the quadratic approximation as

$$\left| D_n\left(n^{-\frac{1}{2}} w\right) - D_n(0) + wT_n(0) + \frac{1}{2}\gamma(\phi, f)w^2 \right| \xrightarrow{P} 0 \quad \text{as} \quad n \to \infty.$$

The convergence is uniform in w: $0 < w^2 < c^2$. Thus, as $n \to \infty$,

$$\left| w - \underset{w}{\text{argmin}} \left\{ \frac{1}{2}\gamma(\phi, f)w^2 + wT_n(0) \right\} \right| \xrightarrow{P} 0$$

or, equivalently,

$$|w - \gamma^{-1}(\phi, f)T_n(0)| \xrightarrow{P} 0.$$

Hence, based on Eq. (3.2.3), we can show that

$$\sqrt{n}(\hat{\theta}_n^{UR} - \theta) \xrightarrow{\mathcal{D}} \mathcal{N}(0, \eta^2), \quad \eta^2 = \frac{A_\phi^2}{\gamma^2(\phi, f)} \quad \text{as} \quad n \to \infty.$$

Now, using Eqs. (3.1.2)–(3.1.4), the proof can be easily completed. □

3.2.2 Restricted R-estimator of θ

If θ is suspected to be 0, the restricted R-estimator (RRE) of θ, denoted by $\hat{\theta}_n^{\mathrm{RR}}$, is clearly given by

$$\hat{\theta}_n^{\mathrm{RR}} = 0.$$

To overcome the difficulty of identical asymptotic distributions of different large-sample estimators under fixed alternatives, $\mathcal{H}_A : \theta \neq 0$, we consider the local alternatives

$$K_{(n)} : \theta_{(n)} = \frac{1}{\sqrt{n}}\delta \tag{3.2.4}$$

where δ is a fixed number.

Theorem 3.2 Under the local alternatives $K_{(n)}$, the asymptotic distributional bias, variance and L_2-risk of the restricted R-estimator, $\hat{\theta}_n^{\mathrm{RR}}$, are respectively given by

$$\begin{aligned}
\mathrm{ADB}(\hat{\theta}_n^{\mathrm{RR}}) &= -\delta, \\
\mathrm{ADV}(\hat{\theta}_n^{\mathrm{RR}}) &= 0, \\
\mathrm{ADL_2\text{-}risk}(\hat{\theta}_n^{\mathrm{RR}}) &= \eta^2\Delta^2, \quad \text{where } \Delta^2 = \frac{\delta^2}{\eta^2}.
\end{aligned}$$

Proof: Under the local alternative hypothesis, it is known that $\theta = \delta/\sqrt{n}$; so, based on Eq. (3.1.3), it is easy to derive that

$$\mathrm{ADB}(\hat{\theta}_n^{\mathrm{RR}}) = \sqrt{n}\left(\mathbb{E}[\hat{\theta}_n^{\mathrm{RR}}] - \frac{1}{\sqrt{n}}\delta\right) = \sqrt{n}\left(0 - \frac{1}{\sqrt{n}}\delta\right) = -\delta$$

and

$$\mathrm{ADL_2\text{-}risk}(\hat{\theta}_n^{\mathrm{RR}}) = n\mathbb{E}\left[\left(0 - \frac{1}{\sqrt{n}}\delta\right)^2\right] = n\left(0 - \frac{1}{\sqrt{n}}\delta\right)^2 = \delta^2.$$

If Δ^2 is defined as δ^2/η^2, then $\mathrm{ADL_2\text{-}risk}(\hat{\theta}_n^{\mathrm{RR}}) = \eta^2\Delta^2$. □

Clearly, Theorem 3.2 results in the ADRE of $\hat{\theta}_n^{\mathrm{RR}}$ given by

$$\mathrm{ADRE}(\hat{\theta}_n^{\mathrm{RR}} : \hat{\theta}_n^{\mathrm{UR}}) = \frac{1}{\Delta^2}.$$

3.3 Shrinkage R-estimators of Location

Penalty estimators can be formulated in a number of different ways. The approach taken here is to start with the unrestricted estimate, $\hat{\theta}_n^{\mathrm{UR}}$, and then explicitly reduce its magnitude in some manner. This leads to a family of such shrinkage estimators that can be compared and contrasted using ADL$_2$-risk. Ideally, we would like to find shrinkage estimators that are uniformly better than the unrestricted estimate but, in practice, there will be intervals over which they are uniformly better, and each one will differ in terms of that interval. The next section provides a conceptual view of the various shrinkage estimators to be covered in this chapter. Then, the ridge-type and LASSO-type R-estimators will be derived. The last section provides a generalized framework for the analysis of shrinkage estimators.

3.3.1 Overview of Shrinkage R-estimators of θ

In the sections to follow, a number of specific shrinkage R-estimators are examined. The nature of the analysis will be detailed, with theorems and proofs that are logically sequenced. As above, the purpose is to compare these estimators in terms of ADRE. The material is rather comprehensive and deliberate due to the fact that subsequent chapters rely on knowledge and understanding of the derivations and proofs.

As such, it is worthwhile, before embarking on the more rigorous process, to understand these estimators at a conceptual level and compare them qualitatively. Once the characteristics of the estimators are clear, the rest of the contents of the chapter will be relatively easy to follow. Further, in many of the proofs, the χ^2-distribution p.d.f and c.d.f will be used extensively to derive the ADL$_2$-risk. The reader should utilize any appropriate resources to supply the necessary background as it will not be reviewed here.

Assume for the moment that we have obtained the unrestricted R-estimate, $\hat{\theta}_n^{\mathrm{UR}}$, and we desire to shrink it in some manner. The efficiency of a shrinkage estimator, $\hat{\theta}_n^{\mathrm{Shrinkage}}$, relative to $\hat{\theta}_n^{\mathrm{UR}}$, is given by ADRE($\hat{\theta}_n^{\mathrm{Shrinkage}}$:$\hat{\theta}_n^{\mathrm{UR}}$). Each shrinkage estimator in Figure 3.1 will be compared in this way. In particular, we will study following R-estimators: hard threshold (HTR), Saleh (SR), positive-rule Saleh (SR+), LASSO (LASSOR), ridge (ridgeR) and the naive elastic net (nEnetR).

First we define a tuning parameter used by all shrinkage estimators, $\lambda \geq 0$. As we have seen in previous chapters, this value can be adjusted to achieve the optimum risk value. The characteristic traces of the estimators as a function of λ are provided in Figure 3.1. The tuning parameter, λ, is swept from 0 to 20.

Next, we establish a threshold quantity using a test statistic, $\mathcal{L}_n$, and use it to define the shrinkage R-estimators. It has the following definition:

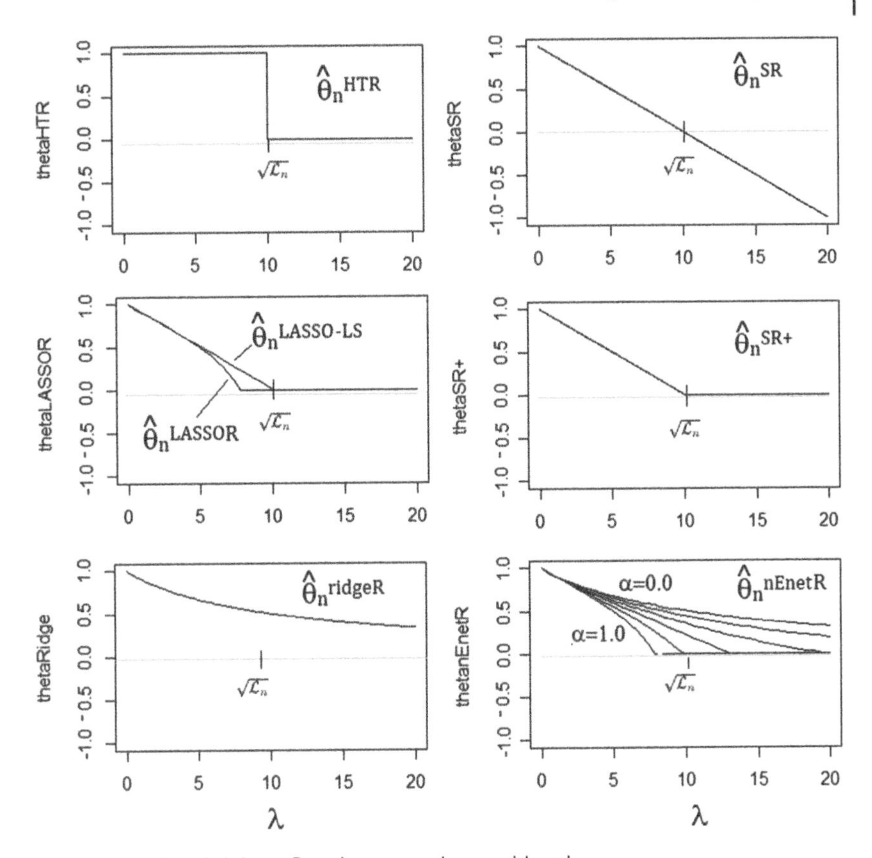

Figure 3.1 Key shrinkage R-estimators to be considered.

$$\mathcal{L}_n = \left(\frac{\sqrt{n}\hat{\theta}_n^{\text{UR}}}{\hat{\eta}}\right)^2.$$

In the figure, we use the values $\hat{\theta}_n^{\text{UR}} = 1.0$, $\hat{\eta} = 0.8$, $n = 64$ and $\mathcal{L}_n = 100$.

We can now begin our cursory look at the shrinkage R-estimators starting with the hard threshold R-estimator, as follows

$$\hat{\theta}_n^{\text{HTR}}(\lambda) = \hat{\theta}_n^{\text{UR}} I(\mathcal{L}_n > \lambda^2).$$

This estimator is also called the preliminary test subset selector R estimator (PTSSRE). Note that the hard-threshold estimator chooses $\hat{\theta}_n^{\text{UR}}$ or 0 as the estimate based on the decision of the indicator function $I(\cdot)$. As we can see in the top-left plot of Figure 3.1, when λ reaches the threshold of $\sqrt{\mathcal{L}_n}$, the estimate shrinks abruptly from $\hat{\theta}_n^{\text{UR}}$ to 0. This is where the name "hard threshold" comes from.

Next, the Saleh-type R-estimator is a continuous version of the hard threshold estimator which removes the step function in favor of a linear function, as follows

$$\hat{\theta}_n^{\mathrm{SR}}(\lambda) = \hat{\theta}_n^{\mathrm{UR}}(1 - \frac{\lambda}{\sqrt{\mathcal{L}_n}}).$$

We see in the top-right plot of Figure 3.1 a line that passes through $\lambda = \sqrt{\mathcal{L}_n}$ and then continues linearly below 0. While this alleviates the discontinuity of the hard threshold estimator, it changes sign and this may not be desirable for a variety of reasons. The positive-rule Saleh-type R-estimator truncates the shrinkage at exactly $\lambda = \sqrt{\mathcal{L}_n}$ such that it can only have positive or negative values, depending on the original sign of $\hat{\theta}_n^{\mathrm{UR}}$, but it cannot switch signs. This can be formulated using the $(\cdot)^+$ notation as follows

$$\hat{\theta}_n^{\mathrm{SR+}}(\lambda) = \hat{\theta}_n^{\mathrm{UR}}(1 - \frac{\lambda}{\sqrt{\mathcal{L}_n}})^+.$$

The estimator trace is shown in the middle-right plot of Figure 3.1. It can be viewed as a 'soft threshold' estimator. It is reminiscent of the LASSO trace seen in earlier chapters so it is worthwhile looking at LASSO in more detail here.

We begin with an L_2 loss and a LASSO penalty for the location problem. The objective function is given by

$$J(\theta) = \frac{1}{2n}||\boldsymbol{y} - \theta 1_n||^2 + \lambda|\theta|. \tag{3.3.1}$$

As described in Chapter 2, we can take the derivative and set it to 0 to obtain the estimator. This results in

$$\hat{\theta}_n^{\mathrm{LASSO\text{-}LS}} = \mathrm{sgn}(\hat{\theta}_n^{\mathrm{LS}})(|\hat{\theta}_n^{\mathrm{LS}}| - \lambda)^+ \tag{3.3.2}$$

where $\hat{\theta}_n^{\mathrm{LS}} = \bar{\boldsymbol{y}}$ is the least-squares (LS) estimate and $\hat{\theta}_n^{\mathrm{LASSO\text{-}LS}}$ is the penalized estimate. Next, let us redefine the objective function using σ, which is assumed to be a known variance quantity, along with a factor of $\sqrt{n}$, as follows

$$J(\theta) = ||\boldsymbol{y} - \theta 1_n||^2 + 2\lambda\sqrt{n}\sigma|\theta|. \tag{3.3.3}$$

The only difference is that we have included additional terms in the penalty function and removed them from the loss function. This type of modification does not introduce any fundamental difference in the resulting estimator except to scale the value of λ. Further, an estimate of σ can easily be obtained from the standard error, $\hat{\sigma} = \sqrt{\mathrm{Var}(\hat{\theta}_n^{\mathrm{LS}})}$. We can use the same steps as above to obtain the estimator for this formulation (see Saleh et al. (2019) for the derivation):

$$\hat{\theta}_n^{\text{LassoLS}} = \text{sgn}(\hat{\theta}_n^{\text{LS}})(|\hat{\theta}_n^{\text{LS}}| - \lambda\frac{\sigma}{\sqrt{n}})^+. \tag{3.3.4}$$

The reason we choose this formulation is the appearance of the $\sigma/\sqrt{n}$ term in the estimator which we will use in our theoretical analysis to draw a connection to the positive-rule Saleh-type R-estimator. In particular, if we define a test statistic

$$\mathcal{L}_n^{\text{LS}} = \frac{n(\hat{\theta}_n^{\text{LS}})^2}{\hat{\sigma}^2},$$

then

$$\sqrt{\mathcal{L}_n^{\text{LS}}} = \frac{\sqrt{n}|\hat{\theta}_n^{\text{LS}}|}{\hat{\sigma}}.$$

Now we can use this term to substitute into the LASSO expression to obtain:

$$\hat{\theta}_n^{\text{LassoLS}} = \text{sgn}(\hat{\theta}_n^{\text{LS}})(|\hat{\theta}_n^{\text{LS}}| - \lambda\frac{|\hat{\theta}_n^{\text{LS}}|}{\sqrt{\mathcal{L}_n^{\text{LS}}}})^+ = \hat{\theta}_n^{\text{LS}}(1 - \lambda\frac{1}{\sqrt{\mathcal{L}_n^{\text{LS}}}})^+. \tag{3.3.5}$$

Clearly, the positive-rule Saleh-type R-estimator and the Lasso-LS estimator are exactly the same for the location problem, as shown in the middle-left plot in Figure 3.1.

The next step is to examine the LASSO R-estimator. The objective function is

$$\mathcal{J}(\theta) = D_n(\theta) + \lambda\sqrt{n}\eta|\theta|. \tag{3.3.6}$$

Although a closed-form solution is not readily available, it is possible to plot the trace for this estimator using

$$\hat{\theta}_n^{\text{LassoR}} = \underset{\theta\in\mathbb{R}}{\text{argmin}}\left\{D_n(\theta) + \lambda\sqrt{n}\,\eta|\theta|\right\}.$$

This is also shown in the middle-left plot. We can observe that the trace is not linear. However, most of the trace has a linear characteristic and, for this reason, the positive-rule Saleh-type R-estimator is a reasonable approximation to the true LASSO R-estimator trace. We propose the use of positive-rule Saleh-type R-estimator using the unrestricted R-estimate to provide a LASSO-type estimator as follows:

$$\hat{\theta}_n^{\text{LassoR}} \approx \text{sgn}(\hat{\theta}_n^{\text{UR}})(|\hat{\theta}_n^{\text{UR}}| - \lambda\frac{\eta}{\sqrt{n}})^+ = \hat{\theta}_n^{\text{SR+}}. \tag{3.3.7}$$

This can also be rewritten in the form we will typically use, as follows

$$\hat{\theta}_n^{\text{SR+}} = \text{sgn}(\hat{\theta}_n^{\text{UR}})(|\hat{\theta}_n^{\text{UR}}| - \lambda\frac{|\hat{\theta}_n^{\text{UR}}|}{\sqrt{\mathcal{L}_n}})^+ = \hat{\theta}_n^{\text{UR}}(1 - \lambda\frac{1}{\sqrt{\mathcal{L}_n}})^+ \tag{3.3.8}$$

By using this positive-rule Saleh-type R-estimator, a LASSO-type of penalty is applied to the unrestricted R-estimates in a manner similar to how LASSO operates on LS estimates.

The ridge R-estimator of the location parameter can be obtained using

$$\hat{\theta}_n^{\text{ridgeR}} = \underset{\theta \in \mathbb{R}}{\text{argmin}}\{D_n(\theta) + \frac{\lambda}{2}\theta^2\} \tag{3.3.9}$$

or in terms of the unrestricted estimate, we define the ridge-type R-estimator as

$$\hat{\theta}_n^{\text{ridgeR}}(\lambda) = \frac{1}{1+\lambda}\hat{\theta}_n^{\text{UR}}.$$

Its trace is illustrated in the bottom-left plot of Figure 3.1. You can observe the decay characteristic that was detailed in Chapter 2.

Finally, the nEnet R-estimator is as follows:

$$\hat{\theta}_n^{\text{nEnetR}}(\lambda, \alpha) = \underset{\theta \in \mathbb{R}}{\text{argmin}}\{D_n(\theta) + \alpha\lambda|\theta| + \frac{\lambda(1-\alpha)}{2}\theta^2\}. \tag{3.3.10}$$

In this case, we have two tuning parameters[1], λ and α. In the bottom right figure, we show six cases corresponding to $\alpha = 0, 0.2, 0.4, 0.6, 0.8, 1.0$. The case for $\alpha = 1.0$ is the same as the LASSO R-estimator, and $\alpha = 0.0$ is the same as the ridge R-estimator. Other values of α produce curves that lie between the two, as one would expect.

The corresponding nEnet-type R-estimator is given by

$$\hat{\theta}_n^{\text{nEnetR}}(\lambda, \alpha) = \frac{\hat{\theta}_n^{\text{UR}}}{1+\lambda(1-\alpha)}\left(1 - \frac{\lambda\alpha}{\sqrt{\mathcal{L}_n}}\right)I(\mathcal{L}_n > \lambda^2\alpha^2). \tag{3.3.11}$$

While not shown in the figure, it would have similar characteristics as the bottom-right plot.

This completes the first pass through all the estimators. In the sections to follow, we will carry out detailed analysis of each estimator. For this purpose, Figure 3.1 can be very useful as a quick reference guide to give context and meaning for the analysis in each section.

1 The α value used in Enet is a tuning parameter, and unrelated to the α-level of confidence intervals or critical values of a distribution.

3.3.2 Derivation of the Ridge-type R-estimator

The estimator for a rank loss function with a ridge penalty can be written as:

$$\hat{\theta}_n^{\text{ridgeR}} = \underset{\omega}{\text{argmin}}\{\mathcal{J}(\omega)\} \tag{3.3.12}$$

where

$$\mathcal{J}(\omega) = \sum_{i=1}^{n} |y_i - n^{-\frac{1}{2}}\omega| a_n^+(R_{n_i}^+(\omega)) + \frac{\lambda}{2\eta}\omega^2. \tag{3.3.13}$$

The ridge-type estimator can be derived starting with

$$\begin{aligned}\frac{\partial \mathcal{J}(\omega)}{\partial \omega} &= -\frac{1}{\sqrt{n}}\sum_{i=1}^{n} \text{sgn}(y_i - n^{-\frac{1}{2}}\omega) a_n^+(R_{n_i}^+(\omega)) + \frac{\lambda}{\eta}\omega \\ &= -T_n(n^{-\frac{1}{2}}\omega) + \frac{\lambda\omega}{\eta}.\end{aligned} \tag{3.3.14}$$

Now

$$|T_n(n^{-\frac{1}{2}}\omega) - T_n(0) + \gamma(\phi, f)\omega| \xrightarrow{\text{P}} 0. \tag{3.3.15}$$

Therefore,

$$\omega = \sqrt{n}(\hat{\theta}_n^{\text{ridgeR}} - \theta) = \gamma^{-1}(\phi, f)T_n(0), \quad |\omega| < c. \tag{3.3.16}$$

Hence, under $\theta = 0$, we have that $\sqrt{n}\hat{\theta}_n^{\text{UR}} = \gamma^{-1}(\phi, f)T_n(0)$, implying that

$$T_n(0) - \left[\gamma(\phi, f) - \lambda\gamma(\phi, f)/A_\phi\right]\omega = 0, \quad A_\phi^2 = \int_0^1 [\phi^+(u)]^2 du. \tag{3.3.17}$$

It follows that

$$\begin{aligned}\sqrt{n}\hat{\theta}_n^{\text{ridgeR}} &= \gamma^{-1}(\phi, f)(1 + \lambda/A_\phi)^{-1}T_n(0) \\ &= (1 + \lambda/A_\phi)^{-1}\sqrt{n}\hat{\theta}_n^{\text{UR}}.\end{aligned} \tag{3.3.18}$$

And finally, we conclude that

$$\hat{\theta}_n^{\text{ridgeR}} = \frac{1}{1 + \lambda/A_\phi}\hat{\theta}_n^{\text{UR}}. \tag{3.3.19}$$

This is the form of the ridge-type estimator acting on $\hat{\theta}_n^{\text{UR}}$.

3.3.3 Derivation of the LASSO-type R-estimator

The estimator for a rank loss function with a LASSO penalty function is given by:

$$\hat{\theta}_n^{\text{LassoR}} = \underset{\omega}{\text{argmin}}\{\mathcal{J}(\omega)\} \tag{3.3.20}$$

where

$$\mathcal{J}(\omega) = \sum_{i=1}^{n} |y_i - n^{-\frac{1}{2}}\omega| a_n^+(R_{n_i}^+(\omega)) + \lambda\eta|\omega|. \tag{3.3.21}$$

The LASSO-type estimator can be derived starting with

$$\begin{aligned} \frac{\partial \mathcal{J}(\omega)}{\partial \omega} &= -\frac{1}{\sqrt{n}} \sum_{i=1}^{n} \text{sgn}(y_i - n^{-\frac{1}{2}}\omega) a_n^+(R_{n_i}^+(\omega)) + \lambda\eta\,\text{sgn}(\omega) \qquad (3.3.22) \\ &= -T_n(n^{-\frac{1}{2}}\omega) + \lambda\eta\,\text{sgn}(\omega). \end{aligned}$$

Now

$$|T_n(n^{-\frac{1}{2}}\omega) - T_n(0) + \gamma(\phi, f)\omega| \overset{\text{P}}{\to} 0. \tag{3.3.23}$$

Therefore,

$$\omega = \sqrt{n}(\hat{\theta}_n^{\text{LassoR}} - \theta) = \gamma^{-1}(\phi, f)T_n(0) - \lambda\eta\,\text{sgn}(\omega), \;\; |\omega| < c. \tag{3.3.24}$$

Hence, under $\theta = 0$, we have that

$$\sqrt{n}\hat{\theta}_n^{\text{LassoR}} = \sqrt{n}\hat{\theta}^{\text{UR}} - \lambda\eta\text{sgn}(\hat{\theta}_n^{\text{LassoR}}) \tag{3.3.25}$$

and rearranging, we obtain

$$\hat{\theta}_n^{\text{LassoR}} = \hat{\theta}^{\text{UR}} - \lambda\frac{\eta}{\sqrt{n}}\text{sgn}(\hat{\theta}_n^{\text{LassoR}}). \tag{3.3.26}$$

And finally, we conclude that

$$\hat{\theta}_n^{\text{LassoR}} = \text{sgn}(\hat{\theta}_n^{\text{UR}})(|\hat{\theta}_n^{\text{UR}}| - \lambda\frac{\eta}{\sqrt{n}})^+ = \hat{\theta}_n^{\text{SR+}}. \tag{3.3.27}$$

This LASSO-type R-estimator is the same as the positive-rule Saleh-type R-estimator.

3.3.4 General Shrinkage R-estimators of θ

We define the shrinkage R-estimator in its most general form as

$$\hat{\theta}_n^{\text{ShrinkageR}}(c) = c\,\hat{\theta}_n^{\text{UR}}, \quad 0 < c < 1,$$

where c may be constant or random.

Theorem 3.3 Under the local alternative hypothesis, $K_{(n)}$, the ADV, ADB and ADL$_2$-risk of $\hat{\theta}_n^{\text{ShrinkageR}}(c)$ are given by

$$\begin{aligned}
\text{ADB}(\hat{\theta}_n^{\text{ShrinkageR}}(c)) &= -\eta(1-c)\Delta, \\
\text{ADV}(\hat{\theta}_n^{\text{ShrinkageR}}(c)) &= \eta^2 c^2, \\
\text{ADL}_2\text{-risk}(\hat{\theta}_n^{\text{ShrinkageR}}(c)) &= \eta^2\left[c^2 + (1-c)^2\Delta^2\right].
\end{aligned}$$

Proof: To calculate the bias of the shrinkage estimator, it is sufficient to use Eq. (3.1.3) as follows:

$$\text{ADB}(\hat{\theta}_n^{\text{ShrinkageR}}(c)) = \sqrt{n}\left(\mathbb{E}[\hat{\theta}_n^{\text{ShrinkageR}}(c)] - \frac{1}{\sqrt{n}}\delta\right) = \sqrt{n}\left(c\hat{\theta}_n^{\text{UR}} - \frac{1}{\sqrt{n}}\delta\right) = (c-1)\delta.$$

This can be rewritten w.r.t. $\Delta^2 = \delta^2/\eta^2$ (hence, $\Delta = \delta/\eta$) as

$$\text{ADB}(\hat{\theta}_n^{\text{ShrinkageR}}(c)) = -(1-c)(\eta\Delta) = -\eta(1-c)\Delta.$$

Using Eq. (3.1.2), under the local alternative hypothesis, it is easily seen that

$$\begin{aligned}
\text{ADV}(\hat{\theta}_n^{\text{ShrinkageR}}(c)) &= \mathbb{E}\left[n\left(\hat{\theta}_n^{\text{ShrinkageR}}(c) - \mathbb{E}[\hat{\theta}_n^{\text{ShrinkageR}}(c)]\right)^2\right] \\
&= \mathbb{E}\left[n\left(c\hat{\theta}_n^{\text{UR}} - \mathbb{E}[c\hat{\theta}_n^{\text{UR}}]\right)^2\right] \\
&= \mathbb{E}\left[nc^2\left(\hat{\theta}_n^{\text{UR}} - \frac{1}{\sqrt{n}}\delta\right)^2\right] \\
&= nc^2\left(\mathbb{E}\left[\hat{\theta}_n^{\text{UR}}\right]^2 - 2\frac{1}{\sqrt{n}}\delta\mathbb{E}\left[\hat{\theta}_n^{\text{UR}}\right] + \frac{1}{n}\delta^2\right).
\end{aligned} \tag{3.3.28}$$

From Eq. (3.2.3), it follows that Eq. (3.3.28) is equivalent to

$$nc^2\left(\left(\frac{1}{n}\eta^2 + \frac{1}{n}\delta^2\right) - 2\frac{1}{\sqrt{n}}\delta\frac{1}{\sqrt{n}}\delta + \frac{1}{n}\delta^2\right) = c^2\eta^2.$$

Now, it is possible to use Eq. (3.1.4) to prove that ADL$_2$-risk$(\hat{\theta}_n^{\text{ShrinkageR}}(c))$ is $\eta^2\left[c^2 + (1-c)^2\Delta^2\right]$. □

Corollary 3.1 The optimum value of c of the shrinkage R-estimator is given by

$$c_{\text{opt}} = \frac{\Delta^2}{1+\Delta^2}.$$

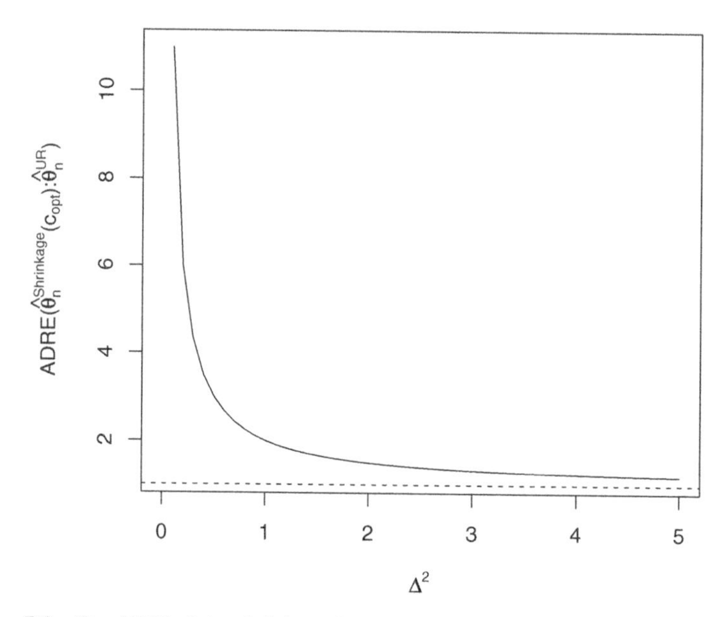

Figure 3.2 The ADRE of the shrinkage R-estimator using the optimal c and URE.

Proof: It is sufficient to find the root of the following:

$$\frac{\partial \text{ADL}_2\text{-risk}(\hat{\theta}_n^{\text{ShrinkageR}}(c))}{\partial c} = \eta^2 \left[2c - 2(1-c)\Delta^2\right] = 0$$

which leads to $c = \Delta^2/(1+\Delta^2)$ as the optimum value. □

Thus,

$$\text{ADL}_2\text{-risk}(\hat{\theta}_n^{\text{ShrinkageR}}(c_{\text{opt}})) = \eta^2 \frac{\Delta^2}{1+\Delta^2},$$

and from Eq. (3.1.5), it can be concluded that

$$\text{ADRE}(\hat{\theta}_n^{\text{ShrinkageR}}(c_{\text{opt}}) : \hat{\theta}_n^{\text{UR}}) = 1 + \frac{1}{\Delta^2}.$$

Further,

$$\text{ADL}_2\text{-risk}(\hat{\theta}_n^{\text{UR}}) - \text{ADL}_2\text{-risk}(\hat{\theta}_n^{\text{ShrinkageR}}(c_{\text{opt}})) = \frac{\eta^2}{1+\Delta^2} \quad \forall \Delta^2.$$

Hence, we conclude that $\hat{\theta}_n^{\text{ShrinkageR}}(c_{\text{opt}})$ is uniformly better than $\hat{\theta}_n^{\text{UR}}$. Figure 3.2 shows the ADRE function for different Δ^2 values that confirm this result.

3.4 Ridge-type R-estimator of θ

We examined general shrinkage estimators that used a factor, c, in the previous section. As a special case of a shrinkage R-estimator, let $c = 1/(1+\lambda), \lambda > 0$ and use this to define a ridge-type R-estimator of the location parameter as follows

$$\hat{\theta}_n^{\text{ridgeR}}(\lambda) = \frac{1}{1+\lambda}\hat{\theta}_n^{\text{UR}}.$$

Theorem 3.4 Under the local alternative hypothesis, $K_{(n)}$, the ADB, ADV, and ADL$_2$-risk of the ridge-type R-estimator of θ, i.e. $\hat{\theta}_n^{\text{ridgeR}}(\lambda)$, are

$$\text{ADB}\left(\hat{\theta}_n^{\text{ridgeR}}(\lambda)\right) = -\eta\frac{\lambda}{1+\lambda}\Delta,$$

$$\text{ADV}\left(\hat{\theta}_n^{\text{ridgeR}}(\lambda)\right) = \eta^2\frac{1}{(1+\lambda)^2},$$

$$\text{ADL}_2\text{-risk}\left(\hat{\theta}_n^{\text{ridgeR}}(\lambda)\right) = \eta^2\frac{1}{(1+\lambda)^2}\left(1+\lambda^2\Delta^2\right).$$

Proof: The result can easily be found by considering $c = 1/(1+\lambda)$ in Theorem 3.3. □

Solving

$$\begin{aligned}\frac{\text{dADL}_2\text{-risk}\left(\hat{\theta}_n^{\text{ridgeR}}(\lambda)\right)}{\text{d}\lambda} &= \eta^2\frac{2\lambda\Delta^2(1+\lambda)-2(1+\lambda^2\Delta^2)}{(1+\lambda)^3}\\ &= \eta^2\frac{2\lambda\Delta^2-2}{(1+\lambda)^3} = 0 \end{aligned} \quad (3.4.1)$$

results in $\lambda_{\text{opt}} = \Delta^{-2}$. Then, for λ_{opt}, the ADL$_2$-risk is given by

$$\text{ADL}_2\text{-risk}\left(\hat{\theta}_n^{\text{ridgeR}}(\lambda_{\text{opt}})\right) = \eta^2\frac{\Delta^2}{1+\Delta^2}.$$

The ADL$_2$-risk-ratio of $\hat{\theta}_n^{\text{UR}}$ and $\hat{\theta}_n^{\text{ridgeR}}(\Delta^{-2})$ is given by $\Delta^2/(1+\Delta^2)$ for $\Delta^2 \in \mathbb{R}^+$. It follows that $\hat{\theta}_n^{\text{ridgeR}}(\lambda_{\text{opt}})$ is uniformly better than $\hat{\theta}_n^{\text{UR}}$ from

$$\text{ADRE}(\hat{\theta}_n^{\text{ridgeR}}(\lambda_{\text{opt}}) : \hat{\theta}_n^{\text{UR}}) = 1 + \frac{1}{\Delta^2}. \quad (3.4.2)$$

Table 3.1 shows the ADRE values for different Δ^2 values. It confirms the better performance of ridge vs. the unrestricted R-estimation.

Table 3.1 The ADRE values of ridge for different values of Δ^2

Δ^2	0	0.01	0.05	0.1	0.5	1	2	5	10	20	50
ADRE	∞	101	21	11	3	2	1.5	1.2	1.1	1.05	1.02

3.5 Preliminary Test R-estimator of θ

In the statistical literature, the preliminary test estimation of parameters was introduced by Bancroft (1944, 1964, 1965) to estimate the parameters of a model when it is suspected that some "uncertain prior information" (UPI) on the parameter(s) of interest are available. The method involves a statistical test of UPI based on the appropriate statistics and a decision on whether the model parameters should be accepted.

Consider the UPI, as denoted here by the null hypothesis, $\mathcal{H}_o : \theta = 0$. If we suspect θ to be 0, then the restricted R-estimator ($\hat{\theta}_n^{\rm RR} = 0$) should be used. In practice, there may be some degree of uncertainty about this prior information, $\theta = 0$. This doubt about the UPI can be removed by using "Fisher's recipe" of testing the hypothesis $\mathcal{H}_o : \theta = 0$ against the alternative $\mathcal{H}_A : \theta \neq 0$, or the local alternative $K_{(n)} : \theta = n^{-1/2}\delta$ with a fixed δ. As a result of this test, we choose $\hat{\theta}_n^{\rm UR}$ or $\hat{\theta}_n^{\rm RR}$ based on the rejection or acceptance of $\mathcal{H}_o$. Accordingly, following Bancroft (1944), we introduce the preliminary test R-estimator (PTRE) denoted by $\hat{\theta}_n^{\rm PTR}(\alpha)$ as

$$\hat{\theta}_n^{\rm PTR}(\alpha) = \hat{\theta}_n^{\rm UR} I\left(\mathcal{L}_n > \chi_1^2(\alpha)\right) \tag{3.5.1}$$

where $\mathcal{L}_n$ is the test statistic for testing $\mathcal{H}_o : \theta = 0$ and defined as

$$\mathcal{L}_n = \frac{n(\hat{\theta}_n^{\rm UR})^2}{\hat{\eta}^2}. \tag{3.5.2}$$

Under the null hypothesis, $\mathcal{L}_n$ has the chi-square distribution with one degree of freedom (d.f.), and $\chi_1^2(\alpha)$ is the α-level upper critical value of a central χ^2 distribution with 1 d.f. If $\alpha = 1, \hat{\theta}_n^{\rm UR}$ is always chosen; if $\alpha = 0, \hat{\theta}_n^{\rm RR}$ is chosen. Since $0 < \alpha < 1, \hat{\theta}_n^{\rm PTR}$ in repeated samples will result in a combination of $\hat{\theta}_n^{\rm UR}$ and $\hat{\theta}_n^{\rm RR}$. Note that the PTRE procedure leads to the choice of only two values, namely, either $\hat{\theta}_n^{\rm UR}$ or $\hat{\theta}_n^{\rm RR}$. Also, the PTRE depends on the level of significance α.

Remark 3.1 For the test statistic, we need $\hat{\eta}$. A consistent estimator is the median absolute deviation (MAD) as discussed in Chapter 1. More generally, we showed earlier that

$$\eta^2 = \frac{A_\phi^2}{\gamma^2(\phi, f)}, \quad \gamma(\phi, f) = \int_{-\infty}^{\infty} f(x)\mathrm{d}\phi(F(x)).$$

To estimate $\gamma(\phi, f)$, we use

$$\hat{\gamma}(\phi, f) = \int_{-\infty}^{\infty} f_n(x)\mathrm{d}\phi(F_n(x)) \tag{3.5.3}$$

where $f_n(x)$ and $F_n(x)$ are the estimates of $f(x)$ and $F(x)$, respectively. Then,

$$\hat{\eta}^2 = \frac{A_n^2}{\hat{\gamma}^2(\phi, f)}, \quad A_n^2 = n^{-1} \sum_{i=1}^{n} a_n^+(i)^2. \tag{3.5.4}$$

Note also that

$$\frac{\sqrt{n}\hat{\theta}_n^{\text{UR}}}{\hat{\eta}} \sim \mathcal{N}(0, 1), \tag{3.5.5}$$

and from Eq. (3.2.3), we know that

$$\frac{T_n(0)}{A_n} \sim \mathcal{N}(0, 1). \tag{3.5.6}$$

Then,

$$\mathcal{L}_n = \frac{T_n^2(0)}{A_n^2} = \left(\frac{\sqrt{n}\,\hat{\theta}_n^{\text{UR}}}{\hat{\eta}}\right)^2 + o_p(1). \tag{3.5.7}$$

This result is used in the theorems to follow.

Theorem 3.5 Under the local alternative hypothesis, the ADB and ADL$_2$-risk of the preliminary test R-estimator of θ are as follows:

$$\text{ADB}(\hat{\theta}_n^{\text{PTR}}(\alpha)) = -\eta\Delta\mathcal{H}_3(\chi_1^2(\alpha), \Delta^2),$$
$$\text{ADL}_2\text{-risk}(\hat{\theta}_n^{\text{PTR}}(\alpha)) = \eta^2\rho_{\text{PT}}(\alpha, \Delta^2),$$

where

$$\rho_{\text{PT}}(\alpha, \Delta^2) = 1 - \mathcal{H}_3(\chi_1^2(\alpha), \Delta^2) + \Delta^2\left(2\mathcal{H}_3(\chi_1^2(\alpha), \Delta^2) - \mathcal{H}_5(\chi_1^2(\alpha), \Delta^2)\right).$$

Here, $\mathcal{H}_\nu(\cdot, \Delta^2)$ is the c.d.f. of a χ^2-distribution with ν d.f. and non-centrality parameter Δ^2.

Proof: If we combine the definition of PTRE with the shrinkage R-estimator, it is easily found that the choice of $c = I(\mathcal{L}_n > \chi_1^2(\alpha))$ confirms that the PTRE is a member of shrinkage R family. One may think that the ADB and ADL$_2$-risk of this estimator can easily be found by substituting this choice of c in Theorem 3.3, but the problem here is that c is not fixed. It is a random variable. So, we have to follow Eqs. (3.1.3) and (3.1.4). The PTRE in Eq. (3.5.1) can be rewritten as

$$\hat{\theta}_n^{\text{PTR}}(\alpha) = \hat{\theta}_n^{\text{UR}} - \hat{\theta}_n^{\text{UR}} I(\mathcal{L}_n \leq \chi_1^2(\alpha)). \tag{3.5.8}$$

Based on Eq. (3.1.3), the ADB of PTRE is

$$\begin{aligned}
\text{ADB}(\hat{\theta}_n^{\text{PTR}}(\alpha)) &= \sqrt{n}\left(\mathbb{E}[\hat{\theta}_n^{\text{PTR}}(\alpha)] - \theta\right) \\
&= \text{ADB}(\hat{\theta}_n^{\text{UR}}) - \mathbb{E}\left[\sqrt{n}\hat{\theta}_n^{\text{UR}} I(\mathcal{L}_n \leq \chi_1^2(\alpha))\right].
\end{aligned}$$

If we consider $Z_n = \sqrt{n}\hat{\theta}_n^{\text{UR}}/\eta$, Eq. (3.2.3) results in $Z_n \sim \mathcal{N}(\Delta, 1)$ under $K_{(n)}$. Knowing that $\text{ADB}(\hat{\theta}_n^{\text{UR}}) = 0$ and using Theorem 4 of Chapter 2 of Saleh (2006) leads to

$$\mathbb{E}\left[\sqrt{n}\hat{\theta}_n^{\text{UR}} I(\mathcal{L}_n \leq \chi_1^2(\alpha))\right] = \eta\mathbb{E}\left[ZI(Z^2 \leq \chi_1^2(\alpha))\right] = \eta\Delta\mathcal{H}_3(\chi_1^2(\alpha), \Delta^2). \tag{3.5.9}$$

This completes the proof of $\text{ADB}(\hat{\theta}_n^{\text{PTR}}(\alpha))$.

To prove ADL$_2$-risk of $\hat{\theta}_n^{\text{PTR}}(\alpha)$, we proceed as follows,

$$\begin{aligned}
\text{ADL}_2\text{-risk}(\hat{\theta}_n^{\text{PTR}}(\alpha)) &= \mathbb{E}\left[n(\hat{\theta}_n^{\text{PTR}}(\alpha) - \theta)^2\right] \\
&= \mathbb{E}\left[n\left(\hat{\theta}_n^{\text{UR}} - \theta\right)^2\right] - 2\mathbb{E}\left[n\left(\hat{\theta}_n^{\text{UR}} - \theta\right)\left(\hat{\theta}_n^{\text{UR}} I(\mathcal{L}_n \leq \chi_1^2(\alpha))\right)\right] \\
&\quad + \mathbb{E}\left[n\left(\hat{\theta}_n^{\text{UR}}\right)^2 I(\mathcal{L}_n \leq \chi_1^2(\alpha))\right] \\
&= \text{ADL}_2\text{-risk}(\hat{\theta}_n^{\text{UR}}) - \mathbb{E}\left[n\left(\hat{\theta}_n^{\text{UR}}\right)^2 I(\mathcal{L}_n \leq \chi_1^2(\alpha))\right] \\
&\quad + 2\sqrt{n}\theta\mathbb{E}\left[\sqrt{n}\hat{\theta}_n^{\text{UR}} I(\mathcal{L}_n \leq \chi_1^2(\alpha))\right].
\end{aligned} \tag{3.5.10}$$

Again, considering $Z = \sqrt{n}\hat{\theta}_n^{\text{UR}}/\eta$, under $K_{(n)}$ and Eq. (3.2.3), it is possible to conclude that $Z \sim \mathcal{N}(\Delta, 1)$. Theorem 5 of Chapter 2 of Saleh (2006) leads to

$$\begin{aligned}
\mathbb{E}\left[n\left(\hat{\theta}_n^{\text{UR}}\right)^2 I(\mathcal{L}_n \leq \chi_1^2(\alpha))\right] &= \eta^2\mathbb{E}\left[Z^2 I(Z^2 \leq \chi_1^2(\alpha))\right] \\
&= \eta^2\left(\mathcal{H}_3(\chi_1^2(\alpha), \Delta^2)\right. \\
&\quad \left. + \Delta^2\mathcal{H}_5(\chi_1^2(\alpha), \Delta^2)\right).
\end{aligned} \tag{3.5.11}$$

Based on Eq. (3.5.9),

$$2\sqrt{n}\theta\mathbb{E}\left[\sqrt{n}\hat{\theta}_n^{\text{UR}} I(\mathcal{L}_n \leq \chi_1^2(\alpha))\right] = 2\eta^2\Delta\left(\Delta\mathcal{H}_3(\chi_1^2(\alpha), \Delta^2)\right). \tag{3.5.12}$$

By substituting Eqs. (3.5.11) and (3.5.12) in Eq. (3.5.10), we have

$$\text{ADL}_2\text{-risk}(\hat{\theta}_n^{\text{PTR}}(\alpha)) = \eta^2 - \eta^2\left(\mathcal{H}_3(\chi_1^2(\alpha), \Delta^2) + \Delta^2\mathcal{H}_5(\chi_1^2(\alpha), \Delta^2)\right)$$

$$+2\eta^2\Delta^2\mathcal{H}_3(\chi_1^2(\alpha),\Delta^2)$$

$$= \eta^2\Big(1 - \mathcal{H}_3(\chi_1^2(\alpha),\Delta^2) + \Delta^2\left(2\mathcal{H}_3(\chi_1^2(\alpha),\Delta^2) - \mathcal{H}_5(\chi_1^2(\alpha),\Delta^2)\right)\Big).$$

The first term is due to Theorem 3.1, and the proof is complete. □

Thus,

$$\text{ADRE}(\hat{\theta}_n^{\text{PTR}}(\alpha) : \hat{\theta}_n^{\text{UR}}) = \rho_{\text{PT}}^{-1}(\alpha,\Delta^2). \tag{3.5.13}$$

3.5.1 Optimum Level of Significance of PTRE

It is possible to rewrite Eq. (3.5.13) as

$$\text{ADRE}(\hat{\theta}_n^{\text{PTR}}(\alpha) : \hat{\theta}_n^{\text{UR}}) = \left[1 + g(\alpha,\Delta^2)\right]^{-1}, \tag{3.5.14}$$

where

$$g(\alpha,\Delta^2) = -\mathcal{H}_3(\chi_1^2(\alpha),\Delta^2) + \Delta^2\left(2\mathcal{H}_3(\chi_1^2(\alpha),\Delta^2) - \mathcal{H}_5(\chi_1^2(\alpha),\Delta^2)\right).$$

The graph of $\text{ADRE}(\hat{\theta}_n^{\text{PTR}}(\alpha) : \hat{\theta}_n^{\text{UR}})$, as a function of Δ^2 for a fixed α, is decreasing until it crosses the 1-line to reach a minimum at some $\Delta^2 = \Delta_0^2(\alpha)$; then it increases toward the 1-line as $\Delta^2 \to \infty$. The maximum value of $\text{ADRE}(\hat{\theta}_n^{\text{PTR}}(\alpha) : \hat{\theta}_n^{\text{UR}})$ occurs at $\Delta^2 = 0$ with the value

$$\text{ADRE}(\hat{\theta}_n^{\text{PTR}}(\alpha) : \hat{\theta}_n^{\text{UR}}) = \left\{1 - \mathcal{H}_3(\chi_1^2(\alpha),0)\right\}^{-1} \geq 1,$$

for all $\alpha \in A$, the set of possible values of α.

The value of $\text{ADRE}(\hat{\theta}_n^{\text{PTR}}(\alpha) : \hat{\theta}_n^{\text{UR}})$ decreases as α increases, i.e, if $\alpha = 0$ and Δ^2 varies, the graphs of $\text{ADRE}(\hat{\theta}_n^{\text{PTR}}(0) : \hat{\theta}_n^{\text{UR}})$ and $\text{ADRE}(\hat{\theta}_n^{\text{PTR}}(1) : \hat{\theta}_n^{\text{UR}})$ intersect at $\Delta^2 = 1$. In general, $\text{ADRE}(\hat{\theta}_n^{\text{PTR}}(\alpha_1) : \hat{\theta}_n^{\text{UR}})$ and $\text{ADRE}(\hat{\theta}_n^{\text{PTR}}(\alpha_2) : \hat{\theta}_n^{\text{UR}})$ intersect within the interval $0 \leq \Delta^2 \leq 1$; the value of Δ^2 at the intersection grows as α increases. Thus, for two different α-values, $\text{ADRE}(\hat{\theta}_n^{\text{PTR}}(\alpha_1) : \hat{\theta}_n^{\text{UR}})$ and $\text{ADRE}(\hat{\theta}_n^{\text{PTR}}(\alpha_2) : \hat{\theta}_n^{\text{UR}})$ will always intersect below the 1-line.

In order to obtain a PTR estimator with a minimum guaranteed efficiency, E_0, we adopt the following procedure: if $0 \leq \Delta^2 \leq 1$, we always choose $\hat{\theta}_n^{\text{UR}}$, because $\text{ADRE}(\hat{\theta}_n^{\text{PTR}}(\alpha) : \hat{\theta}_n^{\text{UR}}) \geq 1$ in this interval. However, since Δ^2 is unknown generally, there is no way to choose an estimate that will be uniformly the best. For this reason, we select an estimator with minimum guaranteed efficiency, and look for a suitable α from the set, $A = \{\alpha |\, \text{ADRE}(\hat{\theta}_n^{\text{PTR}}(\alpha) : \hat{\theta}_n^{\text{UR}}) \geq E_0\}$. The estimator as chosen maximizes $\text{ADRE}(\hat{\theta}_n^{\text{PTR}}(\alpha) : \hat{\theta}_n^{\text{UR}})$ over all $\alpha \in A$, and

Table 3.2 Maximum and minimum guaranteed ADRE of the preliminary test R-estimator for different values of α.

α	0.05	0.10	0.15	0.20	0.25
$E_{\max}$	3.5829	2.2764	1.7936	1.5389	1.3820
$E_{\min}$	0.4058	0.5122	0.5950	0.6646	0.7247
Δ^2	4.6820	3.9692	3.5890	3.3409	3.1632

Δ^2. Thus, we solve the following equation for the optimum α^*:

$$\min_{\Delta^2} \text{ADRE}(\hat{\theta}_n^{\text{PTR}}(\alpha) : \hat{\theta}_n^{\text{UR}}) = E(\alpha, \Delta_0^2(\alpha)) = E_0. \tag{3.5.15}$$

The solution α^* obtained this way results in the PTRE with minimum guaranteed efficiency E_0, which may increase toward $\text{ADRE}(\hat{\theta}_n^{\text{PTR}}(\alpha^*) : \hat{\theta}_n^{\text{UR}})$ for $\Delta^2 = 0$ given by (3.5.15). See Table 3.2 for the optimum values.

The above mentioned approach has been used often in the literature. We refer to Saleh (2006), Safariyan et al. (2019), Karbalaee et al. (2018, 2019) to mention a few.

3.6 Saleh-type R-estimators

In this section, we sequence through a series of R-estimators that eventually lead to the positive-rule Saleh-type R-estimator which is a LASSO-type of estimator. This is used in conjunction with the ridge-type R-estimator to produce a naive elastic net (or nEnet-type) R-estimator. The theorems will be presented without proof to improve readability but they are extensions of the proofs given in earlier sections.

3.6.1 Hard-Threshold R-estimator of θ

Inspired by the PTR estimator, we can effect a simplification by introducing a tuning parameter, λ^2, to play the role of $\chi_1^2(\alpha)$. Then we adjust λ until we achieve the minimum ADL_2-risk. For this purpose, we now define hard threshold R-estimator of θ when one suspects θ may be 0, as

$$\hat{\theta}_n^{\text{HTR}}(\lambda) = \hat{\theta}_n^{\text{UR}} I(\mathcal{L}_n > \lambda^2) \tag{3.6.1}$$

where $I(\cdot)$ is an indicator function taking on two possible values, 0 or 1, depending on the threshold value, $\mathcal{L}_n$, as defined earlier in Eq. (3.5.7). This HTR estimator should be considered as identical in form to PTRE.

Using $Z_n = \sqrt{n}\hat{\theta}_n^{\text{UR}}/\eta$, the definition of the hard threshold R-estimator of θ can be rewritten as

$$\sqrt{n}\hat{\theta}_n^{\text{HTR}}(\lambda) = \eta Z_n I(|Z_n| > \lambda).$$

Theorem 3.6 Under the local alternative hypothesis, $K_{(n)}$, the ADB, ADV, and ADL$_2$-risk of the hard threshold R-estimator of θ, i.e. $\hat{\theta}_n^{\text{HTR}}(\lambda)$, are

$$\begin{aligned}
\text{ADB}(\hat{\theta}_n^{\text{HTR}}(\lambda)) &= -\eta\Delta\mathcal{H}_3(\lambda^2, \Delta^2),\\
\text{ADV}(\hat{\theta}_n^{\text{HTR}}(\lambda)) &= \eta^2\big[1 - \mathcal{H}_3(\lambda^2, \Delta^2) - \Delta^2\mathcal{H}_3^2(\lambda^2, \Delta^2)\\
&\quad + \Delta^2\left(2\mathcal{H}_3(\lambda^2, \Delta^2) - \mathcal{H}_5(\lambda^2, \Delta^2)\right)\big],\\
\text{ADL}_2\text{-risk}(\hat{\theta}_n^{\text{HTR}}(\lambda)) &= \eta^2\rho_{\text{PT}}(\lambda, \Delta^2),
\end{aligned}$$

respectively, where

$$\rho_{\text{PT}}(\lambda, \Delta^2) = 1 - \mathcal{H}_3(\lambda^2, \Delta^2) + \Delta^2\left(2\mathcal{H}_3(\lambda^2, \Delta^2) - \mathcal{H}_5(\lambda^2, \Delta^2)\right)$$

and $\mathcal{H}_\nu(\lambda^2, \Delta^2)$ is the c.d.f. of a χ^2-distribution with ν d.f.

If we replace $\chi_1^2(\alpha)$ with λ^2 in the preliminary test R-estimator, we will obtain the hard threshold R-estimator. Since the $\hat{\theta}_n^{\text{HTR}}(\lambda)$ is effectively the same as $\hat{\theta}_n^{\text{PTR}}(\alpha)$, we skip the proof.

One may use $\lambda^* = \sqrt{2\ln(2)}$ which is the minimax value of λ to compute the ADRE of $\hat{\theta}_n^{\text{HTR}}(\lambda^*)$ w.r.t. $\hat{\theta}_n^{\text{UR}}$, i.e.

$$\text{ADRE}\left(\hat{\theta}_n^{\text{HTR}}(\lambda^*) : \hat{\theta}_n^{\text{UR}}\right) = \rho_{\text{PT}}^{-1}(\lambda^*, \Delta).$$

See Saleh et al. (2019) for details of the choice λ^*. Figure 3.3 shows the ADRE function of the HT (or PT) R-estimator for different Δ^2 values. From this figure, it is obvious that the PTRE has a better performance than the unrestricted R-estimation in the case of $\Delta^2 \leq 0.6$ and vice versa when $\Delta^2 > 0.6$. Also, as Δ^2 goes to ∞, the estimator behaves like the unrestricted case.

3.6.2 Saleh-type R-estimator of θ

Again, following the spirit of Sections 3.3 and 3.5, we define a continuous version of PTRE, namely the Saleh-type R-estimator (SRE), denoted by $\hat{\theta}_n^{\text{SR}}(\lambda)$, and given by

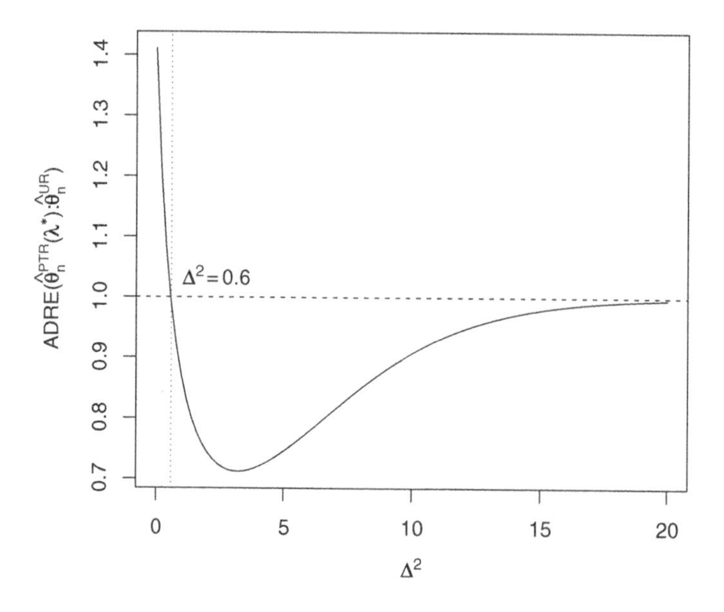

Figure 3.3 The ADRE of the preliminary test (or hard threshold) R-estimator for different Δ^2 based on $\lambda^* = \sqrt{2\ln(2)}$.

$$\hat{\theta}_n^{SR}(\lambda) = \hat{\theta}_n^{UR}(1 - \frac{\lambda}{\sqrt{\mathcal{L}_n}}) = \hat{\theta}_n^{UR} - \lambda\eta\frac{\hat{\theta}_n^{UR}}{|\sqrt{n}\hat{\theta}_n^{UR}|}. \tag{3.6.2}$$

This can be rewritten as

$$\begin{aligned}\sqrt{n}\hat{\theta}_n^{SR}(\lambda) &= \sqrt{n}\hat{\theta}_n^{UR} - \lambda\eta\frac{\sqrt{n}\hat{\theta}_n^{UR}}{|\sqrt{n}\hat{\theta}_n^{UR}|} \\ &= \eta\left(1 - \frac{\lambda}{|Z_n|}\right)Z_n = \eta(Z_n - \lambda\text{sgn}(Z_n))\end{aligned} \tag{3.6.3}$$

with $Z_n = \sqrt{n}\hat{\theta}_n^{UR}/\eta \sim \mathcal{N}(\Delta, 1)$.

Theorem 3.7 Under the local alternative, $K_{(n)}$, the ADB and ADL$_2$-risk of the Saleh-type R-estimator of θ are

$$\text{ADB}(\hat{\theta}_n^{SR}(\lambda)) = -\eta\lambda\,[2\Phi(\Delta) - 1]\,,$$

and

$$\text{ADL}_2\text{-risk}(\hat{\theta}_n^{SR}(\lambda)) = \eta^2\left[1 + \lambda^2 - 2\lambda\sqrt{\frac{2}{\pi}}\exp\left\{-\frac{\Delta^2}{2}\right\}\right].$$

Table 3.3 The ADRE values of the Saleh-type R-estimator for $\lambda^*_{\max} = \sqrt{\frac{2}{\pi}}$ and different Δ^2.

Δ^2	0	0.05	0.1	0.5	1	2 ln(2)	2	5	10
ADRE	2.75	2.53	2.35	1.55	1.16	1.00	0.86	0.65	0.61

Corollary 3.2 The optimum value of λ is

$$\lambda^* = \sqrt{\frac{2}{\pi}} \exp\left\{-\frac{\Delta^2}{2}\right\},$$

with a maximum $\lambda^*_{\max} = \sqrt{\frac{2}{\pi}}$ at $\Delta^2 = 0$.

Thus, the optimum value is

$$\text{ADL}_2\text{-risk}(\hat{\theta}_n^{\text{SR}}(\lambda^*_{\max})) = \eta^2 \left[1 - \frac{2}{\pi}\left(2\exp\left\{-\frac{\Delta^2}{2}\right\} - 1\right)\right].$$

The ADRE of $\hat{\theta}_n^{\text{SR}}(\lambda^*_{\max})$ is then given by

$$\text{ADRE}(\hat{\theta}_n^{\text{SR}}(\lambda^*_{\max}) : \hat{\theta}_n^{\text{UR}}) = \left[1 - \frac{2}{\pi}\left(2\exp\left\{-\frac{\Delta^2}{2}\right\} - 1\right)\right]^{-1}. \quad (3.6.4)$$

Table 3.3 shows the ADRE values of Eq. (3.6.4) for some selected values of Δ^2.

It is easy to see from Table 3.3 that $\hat{\theta}_n^{\text{SR}}(\lambda^*_{\max})$ performs better than $\hat{\theta}_n^{\text{UR}}$ whenever $0 < \Delta^2 \le 2\ln(2)$ and $\hat{\theta}_n^{\text{UR}}$ performs better than $\hat{\theta}_n^{\text{SR}}(\lambda^*_{\max})$ for $\Delta^2 > 2\ln(2)$.

3.6.3 Positive-rule Saleh-type (LASSO-type) R-estimator of θ

From Section 3.6.2, consider the Saleh-type R-estimator of θ which was defined as

$$\hat{\theta}_n^{\text{SR}}(\lambda) = \hat{\theta}_n^{\text{UR}}(1 - \frac{\lambda}{\sqrt{\mathcal{L}_n}}).$$

Notice that as $|\hat{\theta}_n^{\text{UR}}| \to \infty$, $\hat{\theta}_n^{\text{SR}}(\lambda)$ retains its sign. However, if $|\hat{\theta}_n^{\text{UR}}| \to 0$, then the sign of $\hat{\theta}_n^{\text{SR}}(\lambda)$ changes. To correct this anomaly, we define the positive-rule Saleh-type R-estimator as

$$\hat{\theta}_n^{\text{SR+}}(\lambda) = \hat{\theta}_n^{\text{UR}}(1 - \frac{\lambda}{\sqrt{\mathcal{L}_n}})^+ = \left(\hat{\theta}_n^{\text{UR}} - \frac{\lambda\eta}{\sqrt{n}}\text{sgn}(\hat{\theta}_n^{\text{UR}}(\lambda))\right) I(\mathcal{L}_n > \lambda^2). \quad (3.6.5)$$

A LASSO-type R-estimator, based on the derivation given earlier is:

$$\hat{\theta}_n^{\text{LassoR}}(\lambda) = \text{sgn}(\hat{\theta}_n^{\text{UR}})\left(|\hat{\theta}_n^{\text{UR}}| - \lambda\frac{\eta}{\sqrt{n}}\right)^+.$$

Since $\mathcal{L}_n = n(\hat{\theta}_n^{\text{UR}})^2/\eta^2$, we find that it can be rewritten as

$$\hat{\theta}_n^{\text{LassoR}}(\lambda) = \text{sgn}(\hat{\theta}_n^{\text{UR}})\left(|\hat{\theta}_n^{\text{UR}}| - \lambda\frac{|\hat{\theta}_n^{\text{UR}}|}{\sqrt{\mathcal{L}_n}}\right)^+.$$

This is the same as (3.6.5), and was shown to be a linearized version of true LASSO R-estimator. Hence, this is referred to as a LASSO-type R-estimator of θ. Since it is different from the actual LASSO R-estimator, we refer to it as a positive-rule Saleh-type R-estimator. Considering $Z_n = \sqrt{n}\hat{\theta}_n^{\text{UR}}/\eta$, the positive-rule Saleh-type R-estimator of θ can be written as

$$\begin{aligned}\sqrt{n}\,\hat{\theta}_n^{\text{SR+}}(\lambda) &= \eta Z_n\left(1 - \frac{\lambda}{|Z_n|}\right)^+ = \eta Z_n\left(1 - \frac{\lambda}{|Z_n|}\right)I(|Z_n| > \lambda)\\ &= \eta\,\text{sgn}(Z_n)(|Z_n| - \lambda)^+ = \eta(Z_n - \lambda\text{sgn}(Z_n))I(|Z_n| > \lambda). \quad (3.6.6)\end{aligned}$$

Theorem 3.8 Under the local alternative hypothesis, $K_{(n)}$, the ADB and ADL$_2$-risk of $\hat{\theta}_n^{\text{SR+}}(\lambda)$ are given by

$$\text{ADB}(\hat{\theta}_n^{\text{SR+}}(\lambda)) = -\eta\left\{\Delta\mathcal{H}_3(\lambda^2, \Delta^2) - \lambda(\Phi(\lambda - \Delta) - \Phi(\lambda + \Delta))\right\}, \quad (3.6.7)$$

and

$$\text{ADL}_2\text{-risk}(\hat{\theta}_n^{\text{SR+}}(\lambda)) = \eta^2\rho_{\text{ST}}(\lambda, \Delta), \quad (3.6.8)$$

where

$$\begin{aligned}\rho_{\text{ST}}(\lambda, \Delta) &= 1 - \mathcal{H}_3(\lambda^2, \Delta^2) + \Delta^2(2\mathcal{H}_3(\lambda^2, \Delta^2) - \mathcal{H}_5(\lambda^2, \Delta^2))\\ &\quad + \lambda^2(1 - \mathcal{H}_1(\lambda^2, \Delta^2)) - 2\lambda(\phi(\lambda - \Delta) + \phi(\lambda + \Delta))\\ &= \rho_{\text{PT}}(\lambda, \Delta) + \lambda^2(1 - \mathcal{H}_1(\lambda^2, \Delta^2))\\ &\quad - 2\lambda(\phi(\lambda - \Delta) + \phi(\lambda + \Delta)). \quad (3.6.9)\end{aligned}$$

Hence,

$$\text{ADRE}(\hat{\theta}_n^{\text{SR+}}(\lambda) : \hat{\theta}_n^{\text{UR}}) = \rho_{\text{ST}}^{-1}(\lambda, \Delta). \quad (3.6.10)$$

Table 3.4 shows the ADRE values of Eq. (3.6.10) for a range of Δ^2.

Further, it may be seen that ADL$_2$-risk$(\hat{\theta}_n^{\text{SR+}}(\lambda^*_{\max})) \leq$ ADL$_2$-risk$(\hat{\theta}_n^{\text{SR}}(\lambda^*_{\max}))$ uniformly in $\Delta^2 \in \mathbb{R}^+$. See Tables 3.3 and 3.4.

Table 3.4 The ADRE values of the positive-rule Saleh-type R-estimator for $\lambda^*_{\max} = \sqrt{\frac{2}{\pi}}$ and different Δ^2.

Δ^2	0	0.05	0.1	0.5	1	2 ln(2)	2	5	10
ADRE	4.30	3.83	3.46	2.03	1.42	1.19	0.98	0.68	0.62

Comparing Tables 3.3 and 3.4 confirms that the positive-rule Saleh-type R-estimator has less risk than the Saleh-type R-estimator. So, this estimator is an improvement on the Saleh-type R-estimator.

3.6.4 Elastic Net-type R-estimator of θ

If we combine the ridge and LASSO R-estimators, we obtain the elastic net-type R-estimator (EnetRE) of θ. This estimator is defined by

$$\hat{\theta}_n^{\text{EnetR}}(\lambda,\alpha) = \underset{\theta \in \mathbb{R}}{\text{argmin}} \left\{ \sum_{i=1}^{n} |y_i - \theta| a_n^+(R_{n_i}^+(\theta)) + \lambda \left[\eta\alpha|\theta| + \frac{1-\alpha}{2}\theta^2 \right] \right\}.$$

Correspondingly, the naive Enet-type R-estimator (nEnetRE) is given by

$$\hat{\theta}_n^{\text{nEnetR}}(\lambda,\alpha) = \frac{\hat{\theta}_n^{\text{UR}}}{1+\lambda(1-\alpha)} \left(1 - \frac{\lambda\alpha}{\sqrt{\mathcal{L}_n}} \right) I(\mathcal{L}_n > \lambda^2\alpha^2) \tag{3.6.11}$$

or equivalently

$$\sqrt{n}\hat{\theta}_n^{\text{nEnetR}}(\lambda,\alpha) = \eta \frac{\text{sgn}(Z_n)}{1+\lambda(1-\alpha)} (|Z_n| - \lambda\alpha)^+, \quad Z_n = \frac{\sqrt{n}\hat{\theta}_n^{\text{UR}}}{\eta},$$

where $0 < \alpha < 1$, and $\lambda > 0$. If $\alpha = 0$, then the nEnetRE becomes the ridge-type R-estimator, $(\eta Z_n)/(1+\lambda)$. On the other hand, taking $\alpha = 1$, it reduces to the positive-rule Saleh-type (i.e. LASSO-type) R-estimator, $\eta\text{sgn}(Z_n)(|Z_n| - \lambda)^+$.

Theorem 3.9 Under the local alternative hypothesis, $K_{(n)}$, the ADB and ADL$_2$-risk of the naive elastic net-type R-estimator of θ are

$$\begin{aligned} \text{ADB}(\hat{\theta}_n^{\text{nEnetR}}(\lambda,\alpha)) &= -\eta \frac{1}{1+\lambda(1-\alpha)} \Big\{ \Delta\mathcal{H}_3(\lambda^2\alpha^2, \Delta^2) \\ &\quad -\lambda\alpha \left(\Phi(\lambda\alpha - \Delta) - \Phi(\lambda\alpha + \Delta)\right) - \lambda(1-\alpha)\Delta \Big\} \end{aligned}$$

and

$$\text{ADL}_2\text{-risk}(\hat{\theta}_n^{\text{nEnetR}}(\lambda,\alpha)) = \eta^2 \frac{1}{[1+\lambda(1-\alpha)]^2} \rho_{\text{EN}}(\lambda,\alpha,\Delta)$$

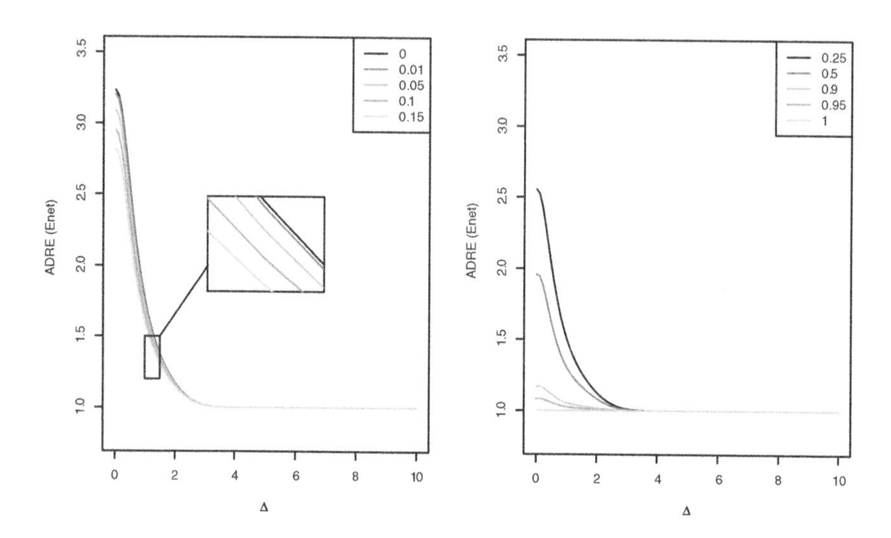

Figure 3.4 The ADRE of nEnet-type R-estimators.

where

$$\begin{aligned}\rho_{\rm EN}(\lambda,\alpha,\Delta) &= \Big(1-\mathcal{H}_3(\lambda^2\alpha^2,\Delta^2)+\Delta^2\big(2\mathcal{H}_3(\lambda^2\alpha^2,\Delta^2)-\mathcal{H}_5(\lambda^2\alpha^2,\Delta^2)\big)\\
&\quad+\lambda^2\alpha^2\big(1-\mathcal{H}_1(\lambda^2\alpha^2,\Delta^2)\big)-2\lambda\alpha\big(\,[\phi(\lambda\alpha+\Delta)+\phi(\lambda\alpha-\Delta)]\,\big)\Big)\\
&\quad+2\lambda(1-\alpha)\Delta\big(\Delta\mathcal{H}_3(\lambda^2\alpha^2,\Delta^2)-\lambda\alpha\,(\Phi(\lambda\alpha-\Delta)-\Phi(\lambda\alpha+\Delta))\,\big)\\
&\quad+\lambda^2(1-\alpha)^2\Delta^2\\
&= \rho_{\rm ST}(\lambda\alpha,\Delta)+\lambda^2(1-\alpha)^2\Delta^2+2\lambda(1-\alpha)\Delta\Big\{\Delta\mathcal{H}_3(\lambda^2\alpha^2;\Delta^2)\\
&\quad-\lambda\alpha\,(\Phi(\lambda\alpha-\Delta)-\Phi(\lambda\alpha+\Delta))\Big\}.\end{aligned}$$

Hence,

$$\text{ADRE}(\hat{\theta}_n^{\rm nEnetR}(\lambda,\alpha):\hat{\theta}_n^{\rm UR}) \;=\; [1+\lambda(1-\alpha)]^2\,(\rho_{\rm EN}(\lambda,\alpha,\Delta))^{-1}.$$

Figure 3.4 shows that the ADRE of nEnet-type R-estimators for different α values.

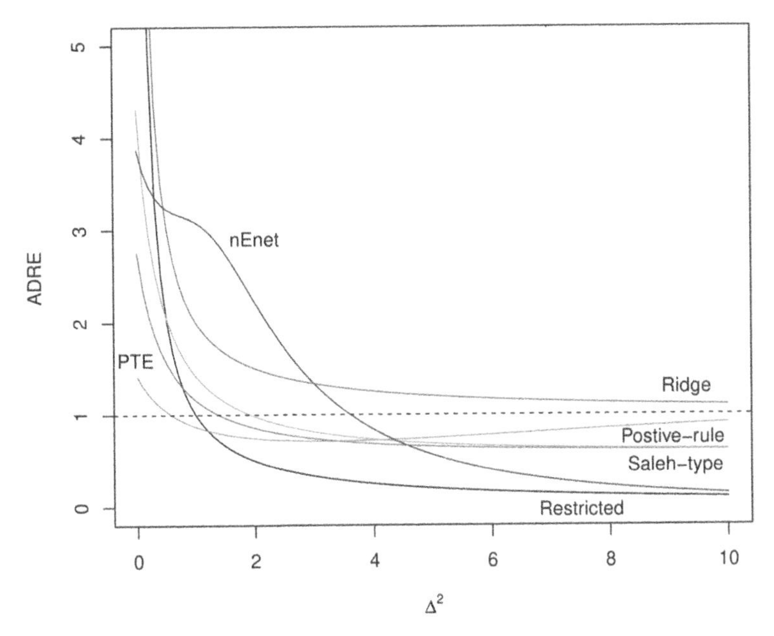

Figure 3.5 Figure of the ADRE of all R-estimators for different Δ^2.

3.7 Comparative Study of the R-estimators of Location

This section provides a summary of the results obtained using all the R-estimators of the location model. Figure 3.5 and Table 3.5 provide the corresponding graphical and tabular data, respectively. Note that for PTRE, we use $\lambda = \lambda^* = \sqrt{2\ln(2)}$; for the Saleh-type and its positive-rule Saleh-type, we set $\lambda = \lambda_{\text{opt}} = \sqrt{\frac{2}{\pi}}$; and, finally, for the nEnet R-estimator, we use $\lambda = \sqrt{2\ln(2)}$ and $\alpha = 0.5$.

(a) RRE vs. URE. We first consider the comparison of unrestricted and restricted R-estimators of θ from Section 3.2.2. The ADL_2-risk difference of the two R-estimators is given by $\eta^2(1 - \Delta^2)$. Thus, the restricted R-estimator, $\hat{\theta}_n^{\text{RR}}$, outperforms the unrestricted R-estimator, $\hat{\theta}_n^{\text{UR}}$, when $0 \leq \Delta^2 \leq 1$ and $\hat{\theta}_n^{\text{UR}}$ outperforms $\hat{\theta}_n^{\text{RR}}$ whenever $\Delta^2 > 1$, as seen in the 1st column of the table derived from

$$\text{ADRE}(\hat{\theta}_n^{\text{RR}} : \hat{\theta}_n^{\text{UR}}) = \Delta^{-2}.$$

(b) General shrinkage R-estimator vs. URE. A comparative study of the general shrinkage R-estimator of the location parameter, $\hat{\theta}^{\text{ShrinkageR}}(c_{\text{opt}})$, and URE,

Table 3.5 The ADRE of all R-estimators for different Δ^2

Δ^2	RR	ridgeR	PTR/HTR	SR	SR+	nEnetR
0	∞	∞	1.411	2.752	4.303	3.869
0.01	100.000	101.000	1.399	2.705	4.199	3.842
0.02	50.000	51.000	1.387	2.659	4.101	3.816
0.03	33.333	34.333	1.375	2.615	4.007	3.791
0.04	25.000	26.000	1.364	2.573	3.918	3.766
0.05	20.000	21.000	1.353	2.533	3.833	3.743
0.1	10.000	11.000	1.301	2.350	3.463	3.636
0.2	5.000	6.000	1.213	2.064	2.916	3.471
0.3	3.333	4.333	1.141	1.849	2.532	3.355
0.4	2.500	3.500	1.082	1.683	2.248	3.277
0.5	2.000	3.000	1.031	1.550	2.029	3.224
1	1.000	2.000	0.869	1.157	1.417	3.084
$2\ln(2)$	0.721	1.721	0.801	1.000	1.187	2.854
2	0.500	1.500	0.743	0.856	0.980	2.224
3	0.333	1.333	0.713	0.739	0.813	1.332
4	0.250	1.250	0.720	0.683	0.730	0.839
5	0.200	1.200	0.746	0.653	0.685	0.561
10	0.100	1.100	0.908	0.614	0.619	0.141
20	0.050	1.050	0.996	0.611	0.611	0.033
30	0.033	1.033	1.000	0.611	0.611	0.015
40	0.025	1.025	1.000	0.611	0.611	0.008
50	0.020	1.020	1.000	0.611	0.611	0.005

$\hat{\theta}_n^{\rm UR}$, produces an ADL$_2$-risk difference given by

$$\eta^2\left[1-\frac{\Delta^2}{1+\Delta^2}\right]=\frac{\eta^2}{1+\Delta^2}.$$

Hence,

$$\text{ADRE}(\hat{\theta}^{\text{SrinkageR}}(c_{\text{opt}}):\hat{\theta}_n^{\rm UR})=1+\frac{1}{\Delta^2}.$$

The general shrinkage R-estimator of θ outperforms $\hat{\theta}_n^{\rm UR}$ uniformly.

(c) Ridge-type R-estimator vs. URE. Here, the ADL$_2$-risk difference of $\hat{\theta}_n^{\rm ridgeR}(\lambda_{\rm opt})$ and $\hat{\theta}_n^{\rm UR}$ is again $\frac{\eta^2}{1+\Delta^2}$. Therefore,

$$\text{ADRE}(\hat{\theta}^{\text{ridgeR}}(\lambda_{\text{opt}}) : \hat{\theta}_n^{\text{UR}}) = 1 + \frac{1}{\Delta^2}.$$

Hence, ridge-type R-estimator outperforms URE uniformly, as seen in the second column of the table.

(d) HTRE (or PTRE) vs. URE. In this case, the ADL_2-risk difference of $\hat{\theta}_n^{\text{HTR}}(\lambda)$ (i.e. $\hat{\theta}_n^{\text{PTR}}(\lambda)$) and $\hat{\theta}_n^{\text{UR}}$ is given by

$$\mathcal{H}_3(\lambda^2, \Delta^2) - \Delta^2 \left(2\mathcal{H}_3(\lambda^2, \Delta^2) - \mathcal{H}_5(\lambda^2, \Delta^2)\right).$$

The difference is non-negative whenever

$$0 \le \Delta^2 \le \frac{\mathcal{H}_3(\lambda^2, \Delta^2)}{2\mathcal{H}_3(\lambda^2, \Delta^2) - \mathcal{H}_5(\lambda^2, \Delta^2)}.$$

Therefore, $\hat{\theta}_n^{\text{HTR}}(\lambda)$ outperforms $\hat{\theta}_n^{\text{UR}}$; otherwise, $\hat{\theta}_n^{\text{UR}}$ outperforms $\hat{\theta}_n^{\text{HTR}}(\lambda)$ for Δ^2 in the complimentary interval, as in the third column of the table. Finally,

$$\text{ADRE}(\hat{\theta}_n^{\text{HTR}}(\lambda) : \hat{\theta}_n^{\text{UR}}) = \left[1 - \mathcal{H}_3(\lambda^2, \Delta^2) + \Delta^2 \left(2\mathcal{H}_3(\lambda^2, \Delta^2) - \mathcal{H}_5(\lambda^2, \Delta^2)\right)\right]^{-1}.$$

(e) Saleh-type R-estimator vs. URE. The ADL_2-risk difference is

$$\frac{2}{\pi}\left[2\exp\left\{-\frac{\Delta^2}{2}\right\} - 1\right].$$

Then for $0 \le \Delta^2 \le 2\ln 2$, the Saleh-type R-estimator is better than URE; otherwise, $\hat{\theta}^{\text{UR}}(\lambda)$ is better. This is due to the fact that

$$\text{ADRE}(\hat{\theta}^{\text{SR}}(\lambda) : \hat{\theta}_n^{\text{UR}}) = \left[\frac{\pi + 2}{\pi} - \frac{4}{\pi}\exp\left\{-\frac{\Delta^2}{2}\right\}\right]^{-1}.$$

(f) Positive-rule Saleh-type vs. Saleh-type R-estimators. The ADL_2-risk of $\hat{\theta}_n^{\text{SR+}}$ is given by

$$\begin{aligned}\text{ADL}_2\text{-risk}(\hat{\theta}_n^{\text{SR+}}) \quad = \quad & \eta^2\Big[(1 - \mathcal{H}_3(\lambda^2, \Delta^2)) + \Delta^2 \left(2\mathcal{H}_3(\lambda^2, \Delta^2) - \mathcal{H}_5(\lambda^2, \Delta^2)\right) \\ & + \lambda^2 \left(1 - \mathcal{H}_3(\lambda^2, \Delta^2)\right) - 2\lambda \left(\phi(\lambda + \Delta) + \phi(\lambda - \Delta)\right)\Big].\end{aligned}$$

If $\Delta^2 = 0$, then ADL_2-risk of $\hat{\theta}_n^{\text{SR+}}(\lambda)$ is

$$\eta^2\left[(1 - \mathcal{H}_3(\lambda^2, 0)) + \lambda^2\left(1 - \mathcal{H}_1(\lambda^2, 0)\right) - 4\lambda\phi(\lambda)\right].$$

On the other hand,

$$\text{ADL}_2\text{-risk}(\hat{\theta}_n^{\text{SR+}}(\lambda)) \le \eta^2\left[(1 + \lambda)^2\right].$$

We note that ADL$_2$-risk of $\hat{\theta}_n^{\text{SR}+}(\lambda)$ is better than the Saleh-type R-estimator, $\hat{\theta}_n^{\text{SR}}(\lambda)$, uniformly in $\Delta^2 \in \mathbb{R}^+$, as seen in the fourth and fifth columns of the table. It maybe noted that $\hat{\theta}_n^{\text{SR}+}(\lambda)$ is also better than $\hat{\theta}_n^{\text{UR}}$ for $0 < \Delta^2 < 2\ln(2)$; otherwise, $\hat{\theta}_n^{\text{UR}}$ is better than $\hat{\theta}_n^{\text{SR}+}(\lambda)$. See column 5 of the table.

(g) nEnet-type vs. postive-rule Saleh-type R-estimators. Consider ADL$_2$-risk difference of $\hat{\theta}_n^{\text{nEnetR}}(\lambda,\alpha)$ and $\hat{\theta}_n^{\text{SR}+}(\lambda)$ given by

$$\begin{aligned}
&\eta^2\Bigg(\rho_{\text{ST}}(\lambda\alpha,\Delta) - \frac{1}{[1+\lambda(1-\alpha)]^2}\Big[\rho_{\text{ST}}(\lambda\alpha,\Delta) + \lambda^2(1-\alpha)^2\Delta^2 \\
&\quad +2\lambda(1-\alpha)\Delta\left\{\Delta\mathcal{H}_3(\lambda\alpha,\Delta^2) - \lambda\alpha\left(\Phi(\lambda\alpha-\Delta) - \Phi(\lambda\alpha+\Delta)\right)\right\}\Big]\Bigg) \\
&= \eta^2\rho_{\text{ST}}(\lambda\alpha,\Delta)\left[1 - \frac{1}{[1+\lambda(1-\alpha)]^2}\right] - \frac{1}{[1+\lambda(1-\alpha)]^2}\Big[\lambda^2(1-\alpha)^2\Delta^2 \\
&\quad +2\lambda(1-\alpha)\Delta\left\{\Delta\mathcal{H}_3(\lambda\alpha,\Delta) - \lambda\alpha\left(\Phi(\lambda\alpha-\Delta) - \Phi(\lambda\alpha+\Delta)\right)\right\}\Big].
\end{aligned}$$

This expression may be "positive", "negative" or zero. Thus, the positive-rule Saleh nor nEnetRE dominates the other uniformly in Δ^2 and, likewise, neither ridge nor nEnetRE dominates the other. This is shown in columns 2, 5 and 6. These results are expected since nEnetRE is a combination of the two R-estimators.

3.8 Simple Linear Model

In this section, we consider the simple linear model,

$$y_i = \theta + \beta x_i + \varepsilon_i,\ i = 1,\dots,n \tag{3.8.1}$$

where $\varepsilon_1, \dots, \varepsilon_n$ are i.i.d. random variables with symmetric c.d.f. $F(\cdot)$ and a p.d.f. $f(\cdot)$ having a finite Fisher information, $I(f) < \infty$. Let $Q_n = \sum_{i=1}^{n}(x_i - \bar{x}_n)^2$. We may rewrite (3.8.1) as

$$y_i = \theta^* + \beta^* Q_n^{-\frac{1}{2}}(x_i - \bar{x}_n) + \varepsilon_i,\ i = 1,\dots,n \tag{3.8.2}$$

where $\theta^* = \theta + \beta\bar{x}_n$ and $\beta^* = Q_n^{\frac{1}{2}}\beta$.

Clearly,

$$Q_n^{-\frac{1}{2}}\sum_{i=1}^{n}(x_i - \bar{x}_n) = 0 \quad \text{and} \quad Q_n^{-1}\sum_{i=1}^{n}(x_i - \bar{x}_n)^2 = 1.$$

Let $c_i = Q_n^{-\frac{1}{2}}(x_i - \bar{x}_n)$, $i = 1,\dots,n$. Also, suppose $R_{n_i}^+(\theta^*,\beta^*)$ is the rank of $|y_i - \theta^* - \beta^* c_i|$ among $|y_1 - \theta^* - \beta^* c_1|, \dots, |y_n - \theta^* - \beta^* c_n|$. Then, we may write the rank dispersion function as

$$D_n(\theta,\beta) = \sum_{i=1}^{n} |y_i - \theta - \beta c_i| a_n^+(R_{n_i}^+(\theta,\beta))$$

which is convex, continuous and piece-wise linear in (θ, β). Thus, we define the R-estimator of $(\theta^*, \beta^*)^\top$ as

$$(\hat{\theta}_n^{*\text{UR}}, \hat{\beta}_n^{*\text{UR}})^\top = \underset{(\theta,\beta)^\top \in \mathbb{R}^2}{\text{argmin}} \{D_n(\theta,\beta)\}.$$

It is easy to verify that at the point of continuity of $D_n(\theta, \beta)$, we have

$$\frac{\partial D_n(\theta,\beta)}{\partial \theta} = -\sum_{i=1}^{n} \text{sgn}(y_i - \theta - \beta c_i) a_n^+(R_{n_i}^+(\theta,\beta)) = -n^{-\frac{1}{2}} T_n(\theta,\beta)$$

and

$$\frac{\partial D_n(\theta,\beta)}{\partial \beta} = -\sum_{i=1}^{n} c_i \text{sgn}(y_i - \theta - \beta c_i) a_n^+(R_{n_i}^+(\theta,\beta)) = -n^{-\frac{1}{2}} L_n(\theta,\beta).$$

This result provides the motivation and justification for incorporation of weighted signed-rank statistics for simultaneous R-estimation of (θ^*, β^*) or (θ, β) based on (3.8.1) or (3.8.2), respectively. Further, using the asymptotic linearity results, we obtain that for any (fixed) $(\theta, \beta)^\top \in \mathbb{R}^2$, as $n \to \infty$,

$$|T_n(n^{-\frac{1}{2}}\omega_1, \omega_2) - T_n(0,0) + \omega_1 \gamma(\phi, f)| \xrightarrow{P} 0, \tag{3.8.3}$$

and

$$|L_n(\omega_1, n^{-\frac{1}{2}}\omega_2) - L_n(0,0) + \omega_2 \gamma(\phi, f)| \xrightarrow{P} 0. \tag{3.8.4}$$

Hence,

$$\begin{aligned} D_n(n^{-\frac{1}{2}}\omega_1, n^{-\frac{1}{2}}\omega_2) - D_n(0,0) + \omega_1 T_n(0,0) \\ + \omega_2 L_n(0,0) + \frac{\gamma(\phi,f)}{2}(\omega_1^2 + \omega_2^2) \xrightarrow{P} 0. \end{aligned} \tag{3.8.5}$$

The above mentioned convexity of $D_n(\theta, \beta)$, and point-wise stochastic convergence in (3.8.5), is uniform in $(\theta, \beta), \theta^2 + \beta^2 < k^2$ with same uniform convergence as for (3.8.3) and (3.8.4). Following the proof of Theorem 5.5.2 of Jurečková and Sen (1996), we conclude as $n \to \infty$,

$$\left\| \begin{pmatrix} \sqrt{n}(\hat{\theta}_n^{*\text{UR}} - \theta^*) \\ (\hat{\beta}_n^{*\text{UR}} - \beta^*) \end{pmatrix} - \underset{(\theta,\beta) \in \mathbb{R}^2}{\text{argmin}} \left\{ \frac{\gamma(\phi,f)}{2}(\omega_1^2 + \omega_2^2) - \omega_1 T_n(0,0) - \omega_2 L_n(0,0) \right\} \right\| \xrightarrow{P} 0.$$

Therefore,

$$\begin{pmatrix} \sqrt{n}(\hat{\theta}_n^{*\mathrm{UR}} - \theta^*) \\ (\hat{\beta}_n^{*\mathrm{UR}} - \beta^*) \end{pmatrix} - \gamma^{-1}(\phi, f)\begin{pmatrix} T_n(0,0) \\ L_n(0,0) \end{pmatrix} \xrightarrow{P} 0. \tag{3.8.6}$$

Because of the transformation scale-equivariance properties of the R-estimator, we have from (3.8.6),

$$\begin{pmatrix} \sqrt{n}(\hat{\theta}_n^{\mathrm{UR}} - \theta) \\ Q_n^{\frac{1}{2}}(\hat{\beta}_n^{\mathrm{UR}} - \beta) \end{pmatrix} - \gamma^{-1}(\phi, f)\begin{pmatrix} 1 & -\sqrt{n}\bar{x}_n/Q_n^{\frac{1}{2}} \\ 0 & 1 \end{pmatrix}\begin{pmatrix} T_n(0,0) \\ L_n(0,0) \end{pmatrix} \xrightarrow{P} 0$$

where $(\hat{\theta}_n^{\mathrm{UR}}, \hat{\beta}_n^{\mathrm{UR}})^\top$ refers to the R-estimator of $(\theta, \beta)^\top$ using $D_n(\theta, \beta)$. Hence,

$$\begin{pmatrix} \sqrt{n}(\hat{\theta}_n^{\mathrm{UR}} - \theta) \\ \sqrt{n}(\hat{\beta}_n^{\mathrm{UR}} - \beta) \end{pmatrix} \xrightarrow{\mathcal{D}} \mathcal{N}_2\left(\begin{pmatrix} 0 \\ 0 \end{pmatrix}, \eta^2\begin{pmatrix} 1 + \frac{\bar{x}^2}{Q} & -\frac{\bar{x}}{Q} \\ -\frac{\bar{x}}{Q} & \frac{1}{Q} \end{pmatrix}\right) \quad \text{as} \quad n \to \infty,$$

where $Q = \lim_{n\to\infty} n^{-1}Q_n$.

3.8.1 Restricted R-estimator of Slope

Clearly, if β is suspected to be 0, the restricted R-estimator (RRE) of β, denoted by $\hat{\beta}_n^{\mathrm{RR}}$, is

$$\hat{\beta}_n^{\mathrm{RR}} = 0.$$

To overcome the difficulty of identical asymptotic distributions of different large-sample estimators under fixed alternatives, $\mathcal{H}_A : \beta \neq 0$, we consider the local alternatives

$$K_{(n)} : \beta_{(n)} = \frac{1}{\sqrt{n}}\delta, \tag{3.8.7}$$

where δ is a fixed number.

Theorem 3.10 Under the local alternatives, $K_{(n)}$, the asymptotic distributional bias, and ADL$_2$-risk of the restricted R-estimator, $\hat{\beta}_n^{\mathrm{RR}}$, are respectively given by

$$\begin{aligned} \mathrm{ADB}(\hat{\beta}_n^{\mathrm{RR}}) &= -\delta, \\ \mathrm{ADL}_2\text{-risk}(\hat{\beta}_n^{\mathrm{RR}}) &= \eta^2\Delta^2, \quad \Delta^2 = \frac{Q\delta^2}{\eta^2}. \end{aligned}$$

In what follows, we define the shrinkage and penalty estimators for the slope parameter in the simple linear model (3.8.1).

3.8.2 Shrinkage R-estimator of Slope

Consider the shrinkage R-estimator of the slope given by

$$\hat{\beta}_n^{\text{ShrinkageR}}(c) = c\hat{\beta}_n^{\text{UR}}, \quad 0 \le c \le 1. \tag{3.8.8}$$

Theorem 3.11 Under the local alternative hypothesis, $K_{(n)} : \beta = n^{-\frac{1}{2}}\delta$, the ADB and ADL_2-risk of $\hat{\theta}_n^{\text{ShrinkageR}}(c)$ are given by

$$\begin{aligned}
\text{ADB}(\hat{\theta}_n^{\text{ShrinkageR}}(c)) &= -(1-c)\delta = -\frac{\eta}{\sqrt{Q}}(1-c)\Delta, \\
\text{ADL}_2\text{-risk}(\hat{\theta}_n^{\text{ShrinkageR}}(c)) &= \frac{\eta^2}{Q}[c^2 + (1-c)^2\Delta^2], \quad \Delta^2 = \frac{Q\delta^2}{\eta^2}.
\end{aligned} \tag{3.8.9}$$

Corollary 3.3 The optimum value of c of the shrinkage R-estimator of β is given by

$$c_{\text{opt}} = \frac{\Delta^2}{1+\Delta^2}.$$

Proof: Minimizing (3.8.9) w.r.t. c, one finds

$$\frac{\partial \text{ADL}_2\text{-risk}(\hat{\beta}_n^{\text{ShrinkageR}}(c))}{\partial c} = \frac{\eta^2}{Q}\left[2c - 2(1-c)\Delta^2\right] = 0$$

which leads to $\Delta^2/(1+\Delta^2)$ as the optimum value. □

Using this optimum value in (3.8.9) yields

$$\text{ADL}_2\text{-risk}(\hat{\beta}_n^{\text{ShrinkageR}}(c_{\text{opt}})) = \eta^2\frac{\Delta^2}{1+\Delta^2}.$$

Hence,

$$\text{ADRE}(\hat{\beta}_n^{\text{ShrinkageR}}(c_{\text{opt}}) : \hat{\beta}_n^{\text{UR}}) = 1 + \frac{1}{\Delta^2}.$$

3.8.3 Ridge-type R-estimation of Slope

Now, consider (3.8.8) with $c = 1/(1+\lambda)$, $\lambda > 0$. Then, the ridge-type R-estimator of β is given by

$$\hat{\beta}_n^{\text{ridgeR}}(\lambda) = \frac{1}{1+\lambda}\hat{\beta}_n^{\text{UR}}.$$

Theorem 3.12 Under the local alternative hypothesis, $K_{(n)}$, the ADB and ADL_2-risk of the ridge-type R-estimator of β, i.e. $\hat{\beta}_n^{\text{ridgeR}}(\lambda)$, are

$$\text{ADB}\left(\hat{\beta}_n^{\text{ridgeR}}(\lambda)\right) = -\frac{\lambda\delta}{1+\lambda} = -\frac{\eta}{\sqrt{Q}}\frac{\lambda}{1+\lambda}\Delta$$

$$\text{ADL}_2\text{-risk}\left(\hat{\beta}_n^{\text{ridgeR}}(\lambda)\right) = \frac{\eta^2}{\sqrt{Q}}\frac{1+\lambda^2\Delta^2}{(1+\lambda)^2}.$$

The optimum value of λ is $\lambda_{\text{opt}} = \Delta^{-2}$. Hence,

$$\text{ADL}_2\text{-risk}(\hat{\beta}_n^{\text{ridgeR}}(\Delta^{-2})) = \frac{\eta^2}{Q}\frac{\Delta^2}{1+\Delta^2}.$$

Therefore,

$$\text{ADRE}(\hat{\beta}_n^{\text{ridgeR}}(\Delta^{-2}) : \hat{\beta}_n^{\text{UR}}) = 1 + \frac{1}{\Delta^2}.$$

3.8.4 Hard-Threshold R-estimator of Slope

Using Section 3.6.1, we define the hard threshold R-estimator (HTRE) of the slope as follows,

$$\hat{\beta}_n^{\text{HTR}}(\lambda) = \hat{\beta}_n^{\text{UR}} I(|\hat{\beta}_n^{\text{UR}}| > \frac{\lambda}{\sqrt{n}}\frac{\eta}{\sqrt{Q_n}}). \tag{3.8.10}$$

Thus,

$$\begin{aligned}\sqrt{n}\hat{\beta}_n^{\text{HTR}}(\lambda) &= \sqrt{n}\hat{\beta}_n^{\text{UR}} I(|\sqrt{n}\hat{\beta}_n^{\text{UR}}| > \frac{\lambda\eta}{\sqrt{Q_n}}) \\ &= \frac{\eta}{\sqrt{Q_n}} Z_n I(|Z_n| > \lambda).\end{aligned}$$

Theorem 3.13 Under the local alternative hypothesis, $K_{(n)}$, the ADB and ADL_2-risk of the hard threshold R-estimator of β are

$$\begin{aligned}\text{ADB}(\hat{\beta}_n^{\text{HTR}}(\lambda)) &= -\frac{\eta}{\sqrt{Q}}\Delta\mathcal{H}_3(\lambda^2, \Delta^2) \\ \text{ADL}_2\text{-risk}(\hat{\beta}_n^{\text{HTR}}(\lambda)) &= \frac{\eta^2}{Q}\rho_{\text{PT}}(\lambda, \Delta^2),\end{aligned}$$

respectively, where

$$\rho_{\text{PT}}(\lambda, \Delta^2) = 1 - \mathcal{H}_3(\lambda^2, \Delta^2) + \Delta^2\left(2\mathcal{H}_3(\lambda^2, \Delta^2) - \mathcal{H}_5(\lambda^2, \Delta^2)\right)$$

and $\mathcal{H}_\nu(\lambda^2, \Delta^2)$ is the c.d.f. of a χ^2-distribution with ν d.f.

The minimax value is $\lambda^* = \sqrt{2\ln(2)}$. So for the HTRE of β in case of λ^*, the ADRE is calculated as

$$\text{ADRE}(\hat{\beta}_n^{\text{HTR}}(\lambda^*) : \hat{\beta}_n^{\text{UR}}) = \left(\rho_{\text{PT}}(\lambda^*, \Delta^2)\right)^{-1}.$$

3.8.5 Saleh-type R-estimator of Slope

Following Saleh (2006, chapter 3), we define the Saleh-type R-estimator as

$$\hat{\beta}_n^{\text{SR}}(\lambda) = \hat{\beta}_n^{\text{UR}} - \frac{\lambda}{\sqrt{n}} \frac{\eta}{\sqrt{Q_n}} \text{sgn}(\hat{\beta}_n^{\text{UR}}). \tag{3.8.11}$$

Thus,

$$\sqrt{n}\hat{\beta}_n^{\text{SR}}(\lambda) = \sqrt{n}\hat{\beta}_n^{\text{UR}} - \frac{\lambda\eta}{\sqrt{Q_n}} \frac{\sqrt{n}\hat{\beta}_n^{\text{UR}}}{|\sqrt{n}\hat{\beta}_n^{\text{UR}}|} = \frac{\eta}{\sqrt{Q_n}} \left[Z_n - \lambda \text{sgn}(Z_n)\right],$$

since $Z_n \sim \mathcal{N}(\Delta, 1)$ under the local alternatives $K_{(n)} : \beta = n^{-\frac{1}{2}}\delta$, with $\Delta = Q\delta^2/\eta^2$.

Theorem 3.14 Under the local alternative, $K_{(n)}$, the ADB and ADL$_2$-risk of Saleh-type R-estimator of β are

$$\text{ADB}(\hat{\beta}_n^{\text{SR}}(\lambda)) = -\frac{\eta\lambda}{\sqrt{Q}} \left[2\Phi(\Delta) - 1\right]$$

and

$$\text{ADL}_2\text{-risk}(\hat{\beta}_n^{\text{SR}}(\lambda)) = \frac{\eta^2}{Q} \left[1 + \lambda^2 - 2\lambda\sqrt{\frac{2}{\pi}} \exp\left\{-\frac{\Delta^2}{2}\right\}\right].$$

Using $\lambda_{\text{opt}} = \max\limits_{\Delta^2 \in \mathbb{R}^+} \sqrt{\frac{2}{\pi}} \exp\left\{-\frac{\Delta^2}{2}\right\} = \sqrt{\frac{2}{\pi}}$, the ADL$_2$-risk of the Saleh-type R-estimator is defined as

$$\text{ADL}_2\text{-risk}(\hat{\beta}_n^{\text{SR}}(\lambda_{\text{opt}})) = \frac{\eta^2}{Q} \left[1 - \frac{2}{\pi}\left(2\exp\left\{-\frac{\Delta^2}{2}\right\} - 1\right)\right].$$

Thus, the ADRE is

$$\text{ADRE}(\hat{\beta}_n^{\text{SR}}(\lambda_{\text{opt}}) : \hat{\beta}_n^{\text{UR}}) = \left[1 - \frac{2}{\pi}\left(2\exp\left\{-\frac{\Delta^2}{2}\right\} - 1\right)\right]^{-1}.$$

3.8.6 Positive-rule Saleh-type (LASSO-type) R-estimator of Slope

The positive-rule Saleh-type R-estimator of β is defined as

$$\hat{\beta}_n^{\mathrm{SR+}}(\lambda) = \sqrt{n}\mathrm{sgn}(\hat{\beta}_n^{\mathrm{UR}})\left(|\hat{\beta}_n^{\mathrm{UR}}| - \frac{\lambda}{\sqrt{n}}\frac{\eta}{\sqrt{Q_n}}\right)^{+}.$$

It can be rewritten as

$$\sqrt{n}\hat{\beta}_n^{\mathrm{SR+}}(\lambda) = \frac{\eta}{\sqrt{Q_n}}\mathrm{sgn}(Z_n)(|Z_n| - \lambda)^{+}, \quad Z_n = \eta^{-1}Q_n^{\frac{1}{2}}\sqrt{n}\hat{\beta}_n^{\mathrm{UR}}.$$

Theorem 3.15 Under the local alternative hypothesis, $K_{(n)}$, the ADB and ADL$_2$-risk of $\hat{\beta}_n^{\mathrm{SR+}}(\lambda)$ are given by

$$\mathrm{ADB}(\hat{\beta}_n^{\mathrm{SR+}}(\lambda)) = -\frac{\eta}{\sqrt{Q}}\left\{\Delta\mathcal{H}_3(\lambda^2, \Delta^2) - \lambda(\Phi(\lambda - \Delta) - \Phi(\lambda + \Delta))\right\},$$

and

$$\mathrm{ADL_2\text{-}risk}(\hat{\beta}_n^{\mathrm{SR+}}(\lambda)) = \frac{\eta^2}{Q}\rho_{\mathrm{ST}}(\lambda, \Delta),$$

where

$$\begin{aligned}\rho_{\mathrm{ST}}(\lambda, \Delta) &= 1 - \mathcal{H}_3(\lambda^2, \Delta^2) + \Delta^2(2\mathcal{H}_3(\lambda^2, \Delta^2) - \mathcal{H}_5(\lambda^2, \Delta^2)) \\ &\quad + \lambda^2(1 - \mathcal{H}_1(\lambda^2, \Delta^2)) - 2\lambda(\phi(\lambda - \Delta) + \phi(\lambda + \Delta)).\end{aligned}$$

Finally,

$$\mathrm{ADRE}(\hat{\beta}_n^{\mathrm{SR+}}(\lambda) : \hat{\beta}_n^{\mathrm{UR}}) = \rho_{\mathrm{ST}}^{-1}(\lambda, \Delta).$$

3.8.7 The Adaptive LASSO (aLASSO-type) R-estimator

The aLASSO-type R-estimator is defined as

$$\hat{\beta}_n^{\mathrm{aLassoR}}(\lambda) = \hat{\beta}_n^{\mathrm{UR}}\left(1 - \frac{\lambda_n}{\sqrt{n}}\frac{\hat{\omega}}{|\hat{\beta}_n^{\mathrm{UR}}|}\right)I\left(|\sqrt{n}\hat{\beta}_n^{\mathrm{UR}}| > \lambda_n\hat{\omega}\right), \quad \hat{\omega} = |\hat{\beta}_n^{\mathrm{UR}}|^{-\gamma},$$

where γ determines the weighting factor and $\hat{\beta}_n^{\mathrm{UR}}$ is the unrestricted R-estimator of β based on

$$\hat{\beta}_n^{\mathrm{UR}} = \underset{\beta\in\mathbb{R}}{\mathrm{argmin}}\left\{\sum_{j=1}^{n} |y_j - \beta(x_j - \bar{x}_n)|a_n^{+}(R_{n_j}(\beta))\right\}.$$

Further, the true adaptive LASSO of β is obtained as

$$\hat{\beta}_n^{\text{aLassoR}} = \underset{\beta \in \mathbb{R}}{\text{argmin}} \left\{ \sum_{j=1}^{n} |y_j - \beta(x_j - \bar{x}_n)| a_n^+(R_{n_j}(\beta)) + \lambda_n \hat{w} |\beta| \right\}.$$

If $\hat{\beta}_n^{\text{aLassoR}} \neq 0$, it belongs to the event A_n^* and if $\hat{\beta}_n^{\text{aLassoR}} = 0$, it belongs to $A_n^{*^c}$. The aLASSO R-estimator is consistent if $P(A_n^* = A) = 1$ where A is the event if $\beta \neq 0$.

3.8.8 nEnet-type R-estimator of Slope

Following Zou and Hastie (2005), we may write the expression for naive Enet-type R-estimator of β as

$$\hat{\beta}_n^{\text{nEnetR}}(\lambda, \alpha) = \frac{1}{1 + \lambda(1 - \alpha)} \text{sgn}(\hat{\beta}_n^{\text{UR}}) \left(|\hat{\beta}_n^{\text{UR}}| - \frac{\lambda\alpha}{\sqrt{n}} \frac{\eta}{\sqrt{Q_n}} \right)^+$$

which it can be rewritten as

$$\hat{\beta}_n^{\text{nEnetR}}(\lambda, \alpha) = \frac{\eta}{\sqrt{Q}} \frac{\text{sgn}(Z_n)}{[1 + \lambda(1 - \alpha)]} (|Z_n| - \lambda\alpha)^+,$$

where $Z_n \sim \mathcal{N}(\Delta, 1)$.

Theorem 3.16 Under the local alternative hypothesis, $K_{(n)}$, the ADB and ADL_2-risk of the naive elastic net-type R-estimator of β are

$$\begin{aligned} \text{ADB}(\hat{\beta}_n^{\text{nEnetR}}(\lambda, \alpha)) &= -\frac{\eta}{\sqrt{Q}} \frac{1}{1 + \lambda(1 - \alpha)} \Big\{ \Delta \mathcal{H}_3(\lambda^2\alpha^2, \Delta^2) \\ &\quad - \lambda\alpha \left(\Phi(\lambda\alpha - \Delta) - \Phi(\lambda\alpha + \Delta) \right) + \lambda(1 - \alpha)\Delta \Big\} \\ \text{ADL}_2\text{-risk}(\hat{\theta}_n^{\text{nEnetR}}(\lambda, \alpha)) &= \frac{\eta^2}{Q} \frac{1}{[1 + \lambda(1 - \alpha)]^2} \rho_{\text{EN}}(\lambda, \alpha, \Delta) \end{aligned}$$

where

$$\begin{aligned} \rho_{\text{EN}}(\lambda, \alpha, \Delta) &= \Big(1 - \mathcal{H}_3(\lambda^2\alpha^2, \Delta^2) + \Delta^2 \big(2\mathcal{H}_3(\lambda^2\alpha^2, \Delta^2) - \mathcal{H}_5(\lambda^2\alpha^2, \Delta^2) \big) \\ &\quad + \lambda^2\alpha^2 \big(1 - \mathcal{H}_1(\lambda^2\alpha^2, \Delta^2) \big) - 2\lambda\alpha \big([\phi(\lambda\alpha + \Delta) + \phi(\lambda\alpha - \Delta)] \big) \Big) \\ &\quad + 2\lambda(1 - \alpha)\Delta \big(\Delta\mathcal{H}_3(\lambda^2\alpha^2, \Delta^2) - \lambda\alpha \left(\Phi(\lambda\alpha - \Delta) - \Phi(\lambda\alpha + \Delta) \right) \big) \\ &\quad + \lambda^2(1 - \alpha)^2\Delta^2. \end{aligned}$$

Therefore,

$$\text{ADRE}(\hat{\beta}_n^{\text{nEnetR}}(\lambda,\alpha) : \hat{\beta}_n^{\text{UR}}) = [1+\lambda(1-\alpha)]^2 \left(\rho_{\text{EN}}(\lambda,\alpha,\Delta)\right)^{-1}. \qquad (3.8.12)$$

Table 3.5 contains the tabular values for $\alpha = 0.5$ and $\lambda = \sqrt{2\ln 2}$.

3.8.9 Comparative Study of R-estimators of Slope

(a) Unrestricted vs. restricted R-estimators of slope. We consider the ADL$_2$-risk-difference of $\hat{\beta}_n^{\text{UR}}$ and $\hat{\beta}_n^{\text{RR}} = 0$ given by $\eta^2\left[\frac{1}{Q} - \Delta^2\right]$. This difference is non-negative whenever $0 \le \Delta^2 \le \frac{1}{Q}$ and $\Delta^2 \in \left[\frac{1}{Q}, \infty\right)$. Hence, the RRE outperforms URE when $0 \le \Delta^2 \le \frac{1}{Q}$, otherwise $\hat{\beta}_n^{\text{UR}}$ outperforms $\hat{\beta}_n^{\text{RR}}$.

(b) Shrinkage R-estimator of slope vs. URE. The ADL$_2$-risk difference for the slope R-estimator is $\eta^2/(1+\Delta^2)$ and the ADL$_2$-risk difference for the intercept parameter is

$$\eta^2\left[\left(1+\frac{\bar{x}_n^2}{Q}\right) - \left(1+\frac{\bar{x}_n^2}{Q}\frac{\Delta^2}{1+\Delta^2}\right)\right] = \eta^2\frac{\bar{x}_n^2}{Q}\frac{1}{1+\Delta^2}.$$

Hence, in both cases the shrinkage R-estimators outperform the URE uniformly. The same thing holds for the ridge-type R-estimators.

(c) HTRE vs. URE of slope. The ADL$_2$-risk differences in this case are

$$\text{ADL}_2\text{-risk}(\hat{\beta}_n^{\text{UR}}) - \text{ADL}_2\text{-risk}(\hat{\beta}_n^{\text{HTR}}(\lambda)) = \eta^2[\mathcal{H}_3(\lambda^2,\Delta^2) \qquad (3.8.13)$$

$$-\Delta^2\{2\mathcal{H}_3(\lambda^2,\Delta^2) - \mathcal{H}_5(\lambda^2,\Delta^2)\}].$$

Hence, $\hat{\beta}_n^{\text{HTR}}(\lambda)$ outperforms $\hat{\beta}_n^{\text{UR}}$ whenever

$$0 \le \Delta^2 \le \frac{\mathcal{H}_3(\lambda^2,\Delta^2)}{2\mathcal{H}_3(\lambda^2,\Delta^2) - \mathcal{H}_5(\lambda^2,\Delta^2)}.$$

Hence, neither $\hat{\beta}_n^{\text{UR}}$ nor $\hat{\beta}_n^{\text{HTR}}(\lambda)$ outperforms the other uniformly.

(d) Saleh-type R-estimator vs. URE of slope. In this case, ADL$_2$-risk difference is given by

$$\eta^2\frac{\bar{x}_n^2}{Q}\frac{2}{\pi}\left(2\exp\left\{-\frac{\Delta^2}{2} - 1\right\}\right).$$

Hence, the Saleh-type R-estimator outperforms the URE whenever $0 \le \Delta^2 \le 2\ln 2$; otherwise, URE outperforms Saleh-type R-estimator.

(e) Positive-rule Saleh-type R-estimator vs. URE. Similarly, in this case, the ADL$_2$-risk difference shows that $\hat{\beta}_n^{\text{SR+}}(\lambda)$ dominates $\hat{\beta}_n^{\text{UR}}$ over $0 \le \Delta^2 \le 2\ln(2)$.

However, the positive-rule Saleh-type R-estimator dominates over the Saleh-type R-estimator uniformly.

(f) The naive elastic-net vs. positive-rule Saleh-type R-estimators. Now, consider ADL$_2$-risk difference of $\hat{\beta}_n^{\text{nEnetR}}(\lambda, \alpha)$ and $\hat{\beta}_n^{\text{SR+}}(\lambda)$ given by

$$\eta^2 \frac{\bar{x}_n^2}{Q} \Bigg(\rho_{\text{ST}}(\lambda\alpha, \Delta) - \frac{1}{[1+\lambda(1-\alpha)]^2} \Big[\rho_{\text{ST}}(\lambda\alpha, \Delta) + \lambda^2(1-\alpha)^2\Delta^2$$
$$+ 2\lambda(1-\alpha)\Delta \left\{ \Delta\mathcal{H}_3(\lambda\alpha, \Delta^2) - \lambda\alpha\left(\Phi(\lambda\alpha - \Delta) - \Phi(\lambda\alpha + \Delta)\right)\right\} \Big] \Bigg)$$
$$= \eta^2 \frac{\bar{x}_n^2}{Q} \rho_{\text{ST}}(\lambda\alpha, \Delta) \left[1 - \frac{1}{[1+\lambda(1-\alpha)]^2} \right] - \frac{1}{[1+\lambda(1-\alpha)]^2} \left[\lambda^2(1-\alpha)^2\Delta^2 \right.$$
$$\left. + 2\lambda(1-\alpha)\Delta \left\{ \Delta\mathcal{H}_3(\lambda\alpha, \Delta) - \lambda\alpha\left(\Phi(\lambda\alpha - \Delta) - \Phi(\lambda\alpha + \Delta)\right)\right\} \right].$$

This quantity can be either positive or negative, or perhaps zero if there is perfect cancellation of all the terms. Hence, neither the positive-rule Saleh-type nor nEnet dominates the other uniformly.

3.9 Summary

The goal of this chapter was to use the location and simple linear models to lay the foundation of the theoretical analysis to follow in upcoming chapters. A number of rank penalty estimators were introduced for these models as they are easiest to understand in a one-dimensional setting. In particular, the ridge-type, positive-rule Saleh-type (LASSO-type) and naive elastic net-type estimators were introduced. In contrast with standard penalty estimators that minimize an objective function, these estimators act to directly shrink the value of the unrestricted estimate, $\hat{\theta}_n^{\text{UR}}$, in a manner that is based on the key shrinkage property of each of the original estimators. A comparison of the ADL$_2$-risk risk for both the location and slope parameters were carried out for the different estimators. From the results in Table 3.5, we found that

(i) $\hat{\theta}_n^{\text{ridgeR}}$ is superior among all the R-estimators in the sense that it is the only one that is uniformly better than $\hat{\theta}_n^{\text{UR}}$ in terms of Δ^2.
(ii) ADL$_2$-risk($\hat{\theta}_n^{\text{SR+}}$ or $\hat{\beta}_n^{\text{SR+}}$) $\leq$ ADL$_2$-risk($\hat{\theta}_n^{\text{SR}}$ or $\hat{\beta}_n^{\text{SR}}$) $\leq$ ADL$_2$-risk($\hat{\theta}_n^{\text{UR}}$ or $\hat{\beta}_n^{\text{UR}}$) uniformly in $\Delta^2 \in R^+$.
(iii) Neither ($\hat{\theta}_n^{\text{HTR}}$ or $\hat{\beta}_n^{\text{HTR}}$) nor ($\hat{\theta}_n^{\text{SR+}}$ or $\hat{\beta}_n^{\text{SR+}}$) dominate the other uniformly.
(iv) Neither $\hat{\theta}_n^{\text{nEnetR}}$ nor $\hat{\theta}_n^{\text{SR+}}$ nor $\hat{\theta}_n^{\text{ridgeR}}$ dominate each other uniformly.

The groundwork is now in place to study these estimators in a multivariate setting and to explore other advanced topics in the next seven chapters intended mainly for statisticians.

3.10 Problems

3.1. The Wilcoxon score functions are given by

$$\phi(u) = \sqrt{12}(u - 1/2) \quad \text{and} \quad \phi^+(u) = \sqrt{3}u.$$

(a) Show that $\phi(u) = \sqrt{12}(u-1/2)$ is skew symmetric around $\frac{1}{2}$. That is, $\phi(u) + \phi(1-u) = 0 \quad \forall u \in (0,1)$.
(b) Scores are standardized such that $\int_0^1 \phi(u)du = 0$ and $\int_0^1 \phi^2(u)du = 1$. Show that $\phi(u)$ satisfies these requirements. Then use the fact that $\phi^+(u) = \phi((u+1)/2)$ to obtain the score function, $\phi^+(u)$.

3.2. Show under Jurečková linearity results, as $n \to \infty$, as $n \to \infty$,

$$\left| T_n(n^{-\frac{1}{2}}\omega) - T_n(0) + \gamma(\phi, f)\omega \right| \xrightarrow{P} 0,$$

where $T_n(\cdot)$ is defined by (3.2.2).

3.3. Derive ADV $\left(\hat{\beta}_n^{\text{ridgeR}}(\lambda)\right)$in Theorem 3.12. (Hint: use the results of ADB and ADL$_2$-risk, if necessary).

3.4. In Section 3.8, the simple linear model given by

$$y_i = \theta + \beta x_i + \varepsilon_i, \; i = 1, \dots, n \tag{3.10.1}$$

was considered. After defining $Q_n = \sum_{i=1}^{n}(x_i - \bar{x}_n)^2$, it was rewritten as

$$y_i = \theta^* + \beta^* Q_n^{-\frac{1}{2}}(x_i - \bar{x}_n) + \varepsilon_i, \; i = 1, \dots, n$$

where $\theta^* = \theta + \beta\bar{x}_n$ and $\beta^* = Q_n^{\frac{1}{2}}\beta$. Given the unrestricted intercept estimate is $\hat{\theta}_n^{\text{UR}}$ with variance $\eta^2(1 + \bar{x}_n/Q_n)$, we can improve the estimator using, for example, ridge as follows:

$$\hat{\theta}_n^{*\text{ridgeR}}(\lambda) = \hat{\theta}_n^{*\text{UR}} - \bar{x}_n \frac{1}{1+\lambda}\hat{\beta}_n^{*\text{UR}}.$$

In a similar manner, define the following improved estimators:
(i) Preliminary test (or hard threshold), $\hat{\theta}_n^{*\text{PTR}}$, when θ may be 0.
(ii) Saleh-type R-estimator $\hat{\theta}_n^{*\text{SR}}$.
(iii) Positive-rule Saleh-type R-estimator $\hat{\theta}_n^{*\text{SR+}}$.

3.5. The ADL$_2$-risk for the intercept parameter using ridge as described in Problem 3.4 is given by:

$$\text{ADB}(\hat{\theta}_n^{*\text{ridgeR}}(\lambda)) \quad = \quad \eta \frac{\lambda}{1+\lambda}\frac{\bar{x}_n}{\sqrt{Q}}\Delta$$

$$\text{ADL}_2\text{-risk}(\hat{\theta}_n^{*\text{ridgeR}}(\lambda)) = \eta^2\left(1 + \frac{\bar{x}_n^2}{Q}\frac{1}{(1+\lambda)^2}(1+\lambda^2\Delta^2)\right).$$

Derive the ADL_2-risk for the intercept parameter $\hat{\theta}_n^{*\text{PTR}}$.

3.6. Derive the ADL_2-risk for the intercept parameter $\hat{\theta}_n^{*\text{SR}}$.

3.7. Derive the ADL_2-risk for the intercept parameter $\hat{\theta}_n^{*\text{SR+}}$.

3.8. Consider the results shown in Table 3.5 and Figure 3.5. Noting the similarity of the results of the location and intercept parameters, compare the different estimators in Problems 3.4, 3.5, 3.6 and 3.7. Which is the best and which is the worst of the three?

3.9. (PROJECT) The dispersion function for the location parameter, θ, was given as

$$D_n(\theta) = \sum_{i=1}^{n} |y_i - \theta| a_n^+(R_{n_i}^+(\theta)).$$

The estimator is given by

$$\hat{\theta}_n^{\text{UR}} = \underset{\theta \in \mathbb{R}}{\text{argmin}}\{D_n(\theta)\}.$$

(a) Use the following R code to generate a data set with $n = 64$:

```
set.seed(123)
#Pick 64 random points
n <- 64
y <- rnorm(n)+0.97194 #adjust to make median = 1
```

(b) Use the following unrestricted rank estimator to find the estimate. It uses the dispersion function for the location parameter. What is $\hat{\theta}_n^{\text{UR}}$ based on the results of the code?

```
minimize.Dn <- function(theta, y) {
  n <- length(y)
  abs_diff <- abs(y-theta)
  rank_abs_diff <- rank(abs_diff,
                      ties.method = "first")
  an_plus <- sqrt(3.)*(rank_abs_diff)/(n+1.)
  Dnsum <- sum(abs_diff*an_plus)/n
  return(Dnsum)
}
op.result <- optim(2.0,
                   fn = minimize.Dn,
                   method = "BFGS",
                   y = y)
```

```
thetaUR <- op.result$par
thetaUR
```

3.10. (PROJECT) The function $T_n(\theta)$ was described to be used for a gradient test for $\mathcal{H}_0 : \theta = 0$ vs. $\mathcal{H}_A : \theta \neq 0$. The test statistic is $T_n(0)/A_n$. Compute this statistic for the data set of the previous problem and perform the hypothesis test for an alpha level of 0.05. This R code should be useful to compute the test statistic:

```
Tn <- function(theta, y ) {
   n <- length(y)
   diff <- y-theta
   abs_diff <- abs(diff)
   sgn_diff <- sign(diff)
   rank_abs_diff <- rank(abs_diff,
           ties.method = "first")
   an_plus <- sqrt(3.)*(rank_abs_diff)/(n+1.)
   Tnsum <- sum(sgn_diff*an_plus)/sqrt(n)
   return(Tnsum)
}
# Obtain Tn(0)
Tn0 <- Tn(0.0,y = y)
Tn0
```

Modify the above R code to obtain $A_n^2 = n^{-1} \sum_{i=1}^{n} [a_n^+(i)]^2$. Compute the test statistic and decide whether or not to reject the null hypothesis.

3.11. (PROJECT) In this problem, the goal is to implement ridge, LASSO and Enet R-estimators for the location parameter and plot traces vs. λ using the same data set given in Problem 3.9(a).

```
set.seed(123)
#Pick 64 random points
n <- 64
y <- rnorm(n)+0.97194 #adjust to make median = 1
```

(a) Use the following R code to implement the ridge penalty function and plot the ridge trace as a function of λ.

```
minimize.penaltyR <- function(theta, y, lambda ) {
  n <- length(y)
  abs_diff <- abs(y-theta)
  rank_abs_diff <- rank(abs_diff,
                     ties.method = "first")
  an_plus <- sqrt(3.)*(rank_abs_diff)/(n+1.)
  Dnsum <- sum(abs_diff*an_plus)
  eta <- 0.8
```

```
    #ridge penalty
    penalty <- sqrt(n)*eta*lambda* theta*theta/2
    return(Dnsum + penalty)
  }
  p <- 1
  lambda_grid = seq(0,20, length = 50)
  penalty_mat_opt = matrix(0,nr = p,
                           nc = length(lambda_grid))
  for(i in 1:length(lambda_grid)){
    op.result <- optim(0.0,
                       fn = minimize.penaltyR,
                       method = "BFGS",
                       y = y,
                       lambda = lambda_grid[i])
  penalty_mat_opt[,i] <- op.result$par
  }
  plot(lambda_grid,penalty_mat_opt[1,],
       type = "l", col = 1, ylim = c(-1,1),
       xlab = "lambda", ylab = "thetapenaltyR")
```

(b) Replace the penalty function to implement LASSO and re-run the R code, as follows:

```
#ridge penalty
#penalty <- sqrt(n)*eta*lambda*  theta*theta/2
#LASSO penalty
penalty <- lambda * abs(theta)*eta*sqrt(n)
```

(c) Modify the penalty function to implement elastic net and re-run the R code, as follows:

```
#ridge penalty
#penalty <- sqrt(n)*eta*lambda* theta*theta/2
#LASSO penalty
#penalty <- lambda * abs(theta)*eta*sqrt(n)
#Enet penalty
alpha = 1.0 #adjust alpha from 0 to 1 manually
penalty <- lambda *alpha*abs(theta)*eta*sqrt(n)+
eta*sqrt(n)*lambda *((1-alpha)* theta*theta/2)
```

3.12. (PROJECT) Implement the Saleh-type estimators using the same data set used earlier.

```
set.seed(123)
#Pick 64 random points
n <- 64
y <- rnorm(n)+0.97194 #adjust to make median = 1
```

(a) Begin by running the following code for the hard threshold estimator.

```
pts <- 100
eta <- 0.8
n = 64
thetaUR <- 1.0
Ln <- n*thetaUR^2/eta^2
thetaShrink <- rep(0,pts)
lambda <- rep(0,pts)
#HTR
for (lambdax in 0:pts) {
  lambda2 <- lambdax/5
  thetaShrink[lambdax+1] <- thetaUR*
    ( Ln > lambda2*lambda2)
  lambda[lambdax+1] <- lambda2
}
plot(lambda,thetaShrink, ylab="thetaHTR",
     xlab="lambda",ylim=c(-1,1),
     col="black",type="l")
points(c(-0.1,20.1),c(-0.0,-0.0),
       col="gray",type="l")
```

(b) Next, run the R code that generates a trace for the Saleh R-estimator.

```
#SR
for (lambdax in 0:pts) {
  lambda2 <- lambdax/5
  thetaShrink[lambdax+1] <- thetaUR*
       (1- lambda2/sqrt(Ln) )
  lambda[lambdax+1] <- lambda2
}
plot(lambda,thetaShrink, ylab="thetaSR",
    xlab="lambda",ylim=c(-1,1),
    col="black",type="l")
points(c(-0.1,20.1),c(-0.0,-0.0),
    col="gray",type="l")
```

(c) Finally, run the R code that generates a trace for the positive-rule Saleh-type R-estimator.

```
# SR+
for (lambdax in 0:pts) {
  lambda2 <- lambdax/5
  thetaShrink[lambdax+1] <- thetaUR*
       (1- lambda2/sqrt(Ln) )*
       ( Ln > lambda2*lambda2)
  lambda[lambdax+1] <- lambda2
```

```
}
plot(lambda,thetaShrink, ylab="thetaSR+",
    xlab="lambda",ylim=c(-1,1),
    col="black",type="l")
points(c(-0.1,20.1),c(-0.05,-0.05),
    col="gray",type="l")
```

(d) Comment on the different traces obtained in Problems 3.9 and 3.10 and compare them with the results of Figure 3.1.

4

Analysis of Variance (ANOVA)

4.1 Introduction

An important model belonging to the class of general linear hypothesis is the analysis of variance (ANOVA) model. In this chapter, we consider the ANOVA model in its simplest form, namely, we collect p samples of differing sample sizes $n_1, \ldots, n_p$ from p dimensions,

$$F(y_{ij} - \theta_i), \quad j = 1, \ldots, n_i, \quad i = 1, \ldots, p$$

where $\theta = (\theta_1, \ldots, \theta_p)^\top$ is the unknown vector of p treatment effects. In the traditional methodology, one tests the hypothesis of equality of p treatments, i.e.

$$\theta_1 = \ldots = \theta_p = \theta_0 \quad \text{(unknown)}.$$

In our case, we attempt to discover a subset of treatments which contains significant treatments relative to the complimentary subset.

In this chapter, we consider (i) the unrestricted R-estimator, and (ii) the restricted R-estimator, along with some penalty estimators, namely, (iii) the subset selection rule, (iv) the LASSO-type, (v) the aLASSO-type, (vi) the ridge-type, and (vii) the elastic net-type estimators together with the traditional shrinkage estimators such as (viii) preliminary test, and (ix) the Stein–Saleh estimators in a nonparametric setup to study their asymptotic distributional L_2-risk properties.

4.2 Model, Estimation and Tests

Consider the ANOVA model

$$Y = B\theta + \varepsilon, \tag{4.2.1}$$

Rank-Based Methods for Shrinkage and Selection: With Application to Machine Learning.
First Edition. A. K. Md. Ehsanes Saleh, Mohammad Arashi, Resve A. Saleh, and Mina Norouzirad.

where

$$\begin{aligned}
\boldsymbol{Y} &= (y_{11}, \dots, y_{1n_1}, y_{21}, \dots, y_{2n_2}, \dots, y_{p1}, \dots, y_{pn_p})^\top, \\
\boldsymbol{\varepsilon} &= (\varepsilon_{11}, \dots, \varepsilon_{1n_1}, \varepsilon_{21}, \dots, \varepsilon_{2n_2}, \dots, \varepsilon_{p1}, \dots, \varepsilon_{pn_p})^\top, \\
\boldsymbol{B} &= \text{block-diagonal vectors}(\mathbf{1}_{n_1}, \dots, \mathbf{1}_{n_p}), \mathbf{1}_{n_i} = (1, \dots, 1)\text{- an } n_i \text{ tuple of 1s}, \\
n &= n_1 + \dots + n_p \\
\mathcal{K}_n &= \text{Diag}(\frac{n_1}{n}, \dots, \frac{n_p}{n}).
\end{aligned}$$

We assume that

$$F(\varepsilon) = \prod_{i=1}^{p} \prod_{j=1}^{n_i} F_0(y_{ij} - \theta_i), \quad \boldsymbol{\theta} = (\theta_1, \dots, \theta_p)^\top,$$

where
(i) F_0 belongs to the class of absolutely continuous, symmetric c.d.f. with absolutely continuous p.d.f., f_0, having finite Fisher information, $I(f_0)$.

(ii) $I(f_0) = \int_{-\infty}^{\infty} \left\{ -\frac{f_0'(x)}{f_0(x)} \right\}^2 f_0(x)\mathrm{d}x < \infty.$

(iii) $\lim_{n\to\infty} \frac{n_i}{n} = \kappa_i \ (0 < \kappa_i < 1), \ \mathcal{K} = \text{Diag}(\kappa_1, \dots, \kappa_p) = \lim_{n\to\infty} \mathcal{K}_n.$

(iv) Score functions $a_n(i)$ are generated by a technique $\phi(u)$, $0 < u < 1$ that is non-constant, non-decreasing and squared integrable, where

$$\phi^+(u) = \phi\left(\frac{1+u}{2}\right), \quad \phi(u) + \phi(1-u) = 0.$$

Then, $a_{n_i}^+(i) = \mathbb{E}\left[\phi^+(U_{n_i i})\right]$ or $\phi\left(\frac{i}{n_i+1}\right)$, $1 \le i \le n_i$ and $0 < U_{n_i 1} \le \dots \le U_{n_i n_i} < 1$ are the order statistics of a sample of size n_i from $U(0,1)$.

4.3 Overview of Multiple Location Models

The setup in this chapter is equivalent to solving a number of related location problems. The vector of location estimates can be used for hypothesis testing, as in ANOVA, or for shrinkage estimation, as will be described in this chapter. The results herein are an extension of the material covered in Chapter 3.

4.3.1 Example: Corn Fertilizers

To give the upcoming material some context and to develop intuition, consider the following hypothetical application of ANOVA. A company seeks to compare a few of its corn crop fertilizers currently in production. If some fertilizers do not improve the yield, then they can be removed from production. The target yield is approximately 1500 bushels per 10 acres of farmland. The treatments involve six different fertilizers, each requiring 10 acres per test, with any number of tests per fertilizer. The fertilizers are applied early in the year with all other factors such as irrigation, sunlight, soil type, etc., being equal. Corn production results at the end of the growing season are entered in Table 4.1. They are all relative to the 1500 bushels/10 acre target. Those below the target will have a negative number; those above, a positive number. At the end of the columns are listed the median and mean for each treatment, as well as the median absolute deviation, $\text{mad}(\boldsymbol{y}_j), j = 1, \ldots, 6$. The MAD value in each column is a robust measure of variability and increases in value from fertilizer 1 to fertilizer 6.

4.3.2 One-way ANOVA

We first want to test the null hypothesis that there are no differences between the fertilizers. Typically, a one-way ANOVA is performed using a least-squares methodology to compute the F-statistic and a corresponding p-value using the sum of squares of the means within columns and between columns. This type of analysis is available in R using the `aov()` function. When executed on the data table, it produces an F-statistic of 1.538 based on a least-squares approach and a p-value of 0.194. These two values inform us that we cannot reject the null hypothesis at the $\alpha = 0.05$ level of significance and therefore the means are not statistically different from one another. As a result, we do not need to carry out a post hoc comparison of the different fertilizer means. The implication is that we should perhaps keep all the fertilizers, or randomly select some for removal.

However, the method described above may not take into account the effect of outliers in the data set. Consider the boxplots of the fertilizers on the left of Figure 4.1 which shows the median as a bold line, the inter-quartile range (first to third quartile) as a box and the data range as whiskers. We can observe that fertilizers 1 and 3 have noticeably higher medians than the others. Furthermore, a Q–Q plot on the right indicates that the randomized errors have a skewed distribution. Therefore, the results from the standard ANOVA may not be reliable in this case.

A robust ANOVA is appropriate in this case and is available in the `Rfit` package using a function called `oneway.rfit()` (Kloke and McKean, 2012). We will

Table 4.1 Table of (hypothetical) corn crop yield from six different fertilizers. Units: bushels per 10 acre allotments relative to a 1500 bushels per 10 acres target production level. Production of exactly 1500 bushels/10 acres would produce an entry of 0 in the table.

Field	Fertilizer 1	Fertilizer 2	Fertilizer 3	Fertilizer 4	Fertilizer 5	Fertilizer 6
1	138.8	43.2	79.7	−54.0	−102.1	−379.5
2	145.4	53.1	135.9	−4.5	−36.0	−206.1
3	181.2	106.8	440.0	263.8	24.1	−168.1
4	151.4	62.1	187.0	40.6	35.9	−69.0
5	152.6	63.9	197.0	49.4	353.0	21.2
6	184.3	111.5	466.6	287.3	102.2	38.8
7	–	73.8	253.3	99.1	321.7	138.8
8	–	22.0	−40.1	−159.8	−243.0	467.6
9	–	39.4	–	−73.0	−127.4	514.5
10	–	46.6	–	−36.8	−79.1	–
11	–	96.7	–	213.6	254.8	–
12	–	70.8	–	–	82.0	–
13	–	72.0	–	–	–	–
14	–	63.3	–	–	–	–
n_j	$n_1 = 6$	$n_2 = 14$	$n_3 = 8$	$n_3 = 11$	$n_9 = 12$	$n_6 = 9$
Median	152.0	63.6	192.0	40.6	30.0	21.2
Mean	158.9	66.1	214.9	56.9	48.8	39.7
MAD(y_j)	14.7	20.4	128.7	140.3	178.8	280.6

not cover this topic in this book as it is detailed in Hettmansperger and McKean (2011), but we will show the results of this feature of `Rfit` and its importance in robust data science. In Hettmansperger and McKean (2011), they develop a robust F-statistic using the dispersion function, which is itself a measure of variability.

When the `oneway.rfit()` function is executed on the ANOVA data table, it produces a robust F-statistic of 3.731 and a p-value of 0.0057. The null hypothesis is rejected implying that there are differences in the fertilizers at any reasonable level of significance. In that case, we should carry out a post hoc pairwise comparison which is automatically performed in `Rfit`. The results are shown in Table 4.2. If we examine the p-values in the table with an $\alpha = 0.1$ level, we see that fertilizers 1 and 3 form one group while fertilizers 2, 4, 5 and 6 form another group. Therefore, it is important to consider the use of robust ANOVA

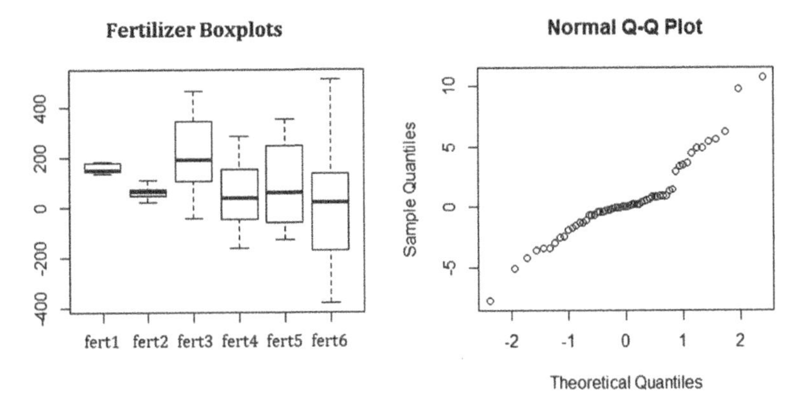

Figure 4.1 Boxplot and Q–Q plot using ANOVA table data.

Table 4.2 Table of p-values from pairwise comparisons of fertilizers.

	Fertilizer 1	Fertilizer 2	Fertilizer 3	Fertilizer 4	Fertilizer 5	Fertilizer 6
Fertilizer 1	–	–	–	–	–	–
Fertilizer 2	0.08766	–	–	–	–	–
Fertilizer 3	0.52868	0.00954	–	–	–	–
Fertilizer 4	0.03351	0.52388	0.00290	–	–	–
Fertilizer 5	0.02101	0.39134	0.00146	0.84601	–	–
Fertilizer 6	0.01071	0.20787	0.00075	0.52707	0.64440	–

methods since the least-squares methodology may produce misleading results in some cases with outliers.

4.3.3 Effect of Variance on Shrinkage Estimators

We now use the data from the ANOVA table to examine the effect of variance on the shrinkage characteristics of rank penalty estimators as compared to LS penalty estimators. The analysis will be used to construct suitable penalized rank estimators for the multiple location model. For this purpose, the columns in the ANOVA table can be treated as a related set of location problems and solved using linear regression. To begin our study, we examine the ridge penalty function. For this multiple location model, the LS-ridge estimator is given by

$$\hat{\theta}_n^{\text{LS-ridge}}(\lambda) = \underset{\theta_i \in \mathbb{R}}{\text{argmin}} \left\{ \sum_{i=1}^{p} \sum_{j=1}^{n_i} (y_{ij} - \theta_i)^2 + \frac{\lambda}{2} \sum_{i=1}^{p} \theta_i^2 \right\}.$$

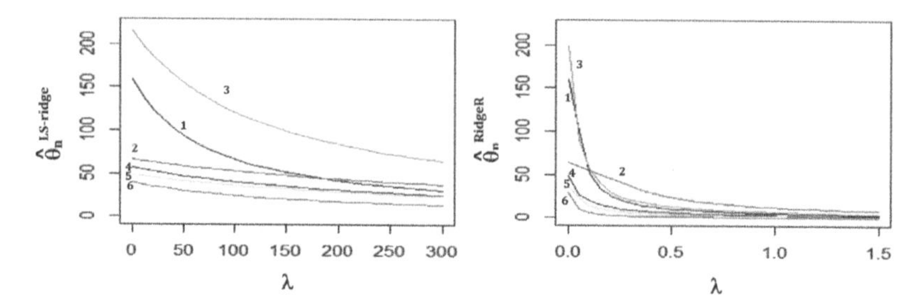

Figure 4.2 LS-ridge and ridge R traces for fertilizer problem from ANOVA table data.

Note that each column (treatments 1–6) of the table is a separate location problem as described in Chapter 3 but here we are combining all columns into one linear regression for each different value of λ.

The ridge R-estimator is given by

$$\hat{\theta}_n^{\text{ridgeR}}(\lambda) = \underset{\theta_i \in \mathbb{R}}{\operatorname{argmin}} \left\{ \sum_{i=1}^{p} \sum_{j=1}^{n_i} |y_{ij} - \theta_i| a_n^+(R_{n_j}(\theta_i)) + \frac{\lambda}{2} \sum_{i=1}^{p} \theta_i^2 \right\}.$$

If we examine the resulting ridge traces in Figure 4.2, we note that they both exhibit the decay characteristics of a typical ridge shrinkage estimator. However, the curves for the LS estimation and R-estimation are quite different in their shrinkage rates. We observe a sharper drop to 0 in the case of rank as compared to LS.

Next, we examine the LASSO shrinkage characteristics. We saw in previous chapters that the LASSO is very effective when addressing sparsity in the estimates, establishing subgroups of parameters and identifying parameters that should be set to zero. The least-squares loss with a LASSO penalty is given by

$$\hat{\theta}_n^{\text{LS-LASSO}}(\lambda) = \underset{\theta_i \in \mathbb{R}}{\operatorname{argmin}} \left\{ \sum_{i=1}^{p} \sum_{j=1}^{n_i} (y_{ij} - \theta_i)^2 + \lambda \sum_{i=1}^{p} |\theta_i| \right\}.$$

The LASSO R-estimator is given by

$$\hat{\theta}_n^{\text{LASSOR}}(\lambda) = \underset{\theta_i \in \mathbb{R}}{\operatorname{argmin}} \left\{ \sum_{i=1}^{p} \sum_{j=1}^{n_i} |y_{ij} - \theta_i| a_n^+(R_{n_j}(\theta_i)) + \lambda \sum_{i=1}^{p} |\theta_i| \right\}.$$

The LASSO traces are shown in Figure 4.3 as a function of λ. This time, the shrinkage characteristics are dramatically different. In the least-squares case, the shrinkage is linear. In the rank case, some parameters appear to shrink linearly (although not strictly linear) while others remain constant and then

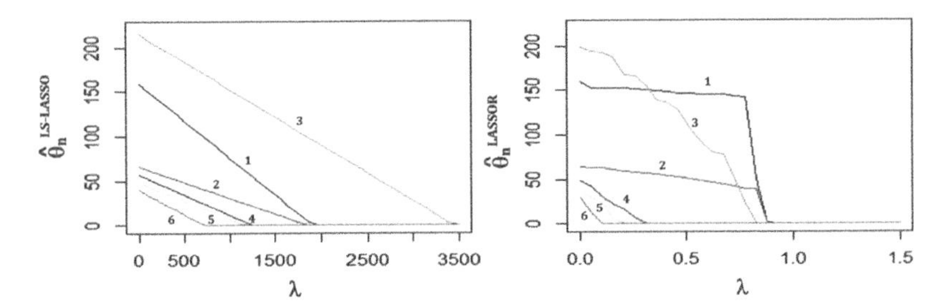

Figure 4.3 LS-LASSO and LASSOR traces for the fertilizer problem from the ANOVA table data.

sharply drop to 0. This reminds us of the hard threshold estimator of Chapter 3. The traces in the rank case are dependent on the variance whereas, for LS, it is independent of the variance. This will be made clear shortly by standardizing the comparison.

Before doing that, we return to the ANOVA problem described earlier. The LASSOR traces of Figure 4.3 can be used to address the issue of which fertilizers to remove. We note that as λ is increased, the estimates shrink in value. Fertilizers 4, 5 and 6 are set to zero when $\lambda \geq 0.3$. Hence, we may choose to remove any or all of these three fertilizers from production, while keeping all of fertilizers 1, 2 and 3 since they have yields well above the 1500 bushels/10 acres target.

Returning to the issue at hand, it is somewhat difficult to notice the effect of the variance on the traces in the figures discussed thus far. This is because the unrestricted estimate is different for each location problem. If we now utilize a data set with essentially equal unrestricted estimates (albeit not exactly the same), but different variances (taken from the MAD($\boldsymbol{y}_j$) row of Table 4.1), we can determine if the variance has a direct influence on the traces. Figure 4.4 shows the resulting traces. We note that for the least-squares case, variance does not have a significant effect on the traces in both ridge and LASSO for the location problem. In fact, the ridge traces are all defined by one equation as follows:

$$\hat{\theta}_n^{\text{LS-ridge}} = \frac{\hat{\theta}_n^{\text{LS}}}{1 + \lambda/2n} \tag{4.3.1}$$

while the LASSO traces are all given by:

$$\hat{\theta}_n^{\text{LS-LASSO}} = \text{sgn}(\hat{\theta}_n^{\text{LS}})(|\hat{\theta}_n^{\text{LS}}| - \lambda/2n)^+. \tag{4.3.2}$$

The labels 1 and 6 are provided on the traces on the outer edges of the plot. The remaining traces are sandwiched in between in sequential order. The slight

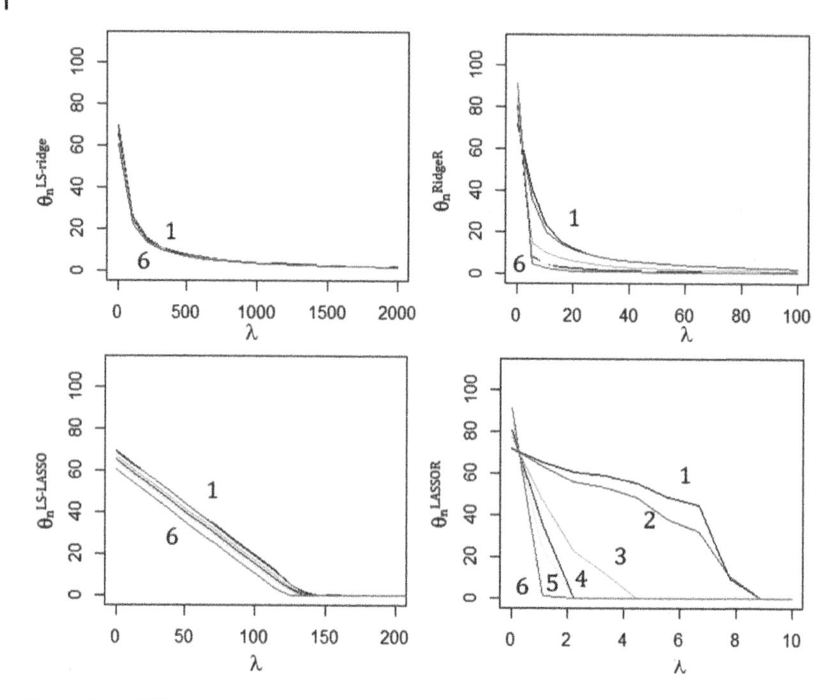

Figure 4.4 Effect of variance on shrinkage using ridge and LASSO traces.

differences in the least-squares estimates with $\lambda = 0$ is responsible for the discrepancy between the curves.

Variance is not important when using the quadratic loss function. However, it clearly plays a significant role in the rank case. We see that higher variances cause rapid shrinkage whereas a lower variance causes much less shrinkage, especially in the case of LASSO where each curve is visibly further apart. A similar characteristic was described in Chapter 2. Therefore, the trace for a given location parameter clearly depends on its variance (or MAD) in the rank case. To summarize, the shrinkage characteristics in the least-squares location problem is not greatly affected by variance whereas in the rank case there is a direct relationship between the variance and the shrinkage characteristics. This intuition is important for developing R-estimators for the multiple location model.

4.3.4 Shrinkage Estimators for Multiple Location

In order to construct and evaluate different shrinkage R-estimators for the multiple location model, we can re-examine the traces of ridge and LASSO

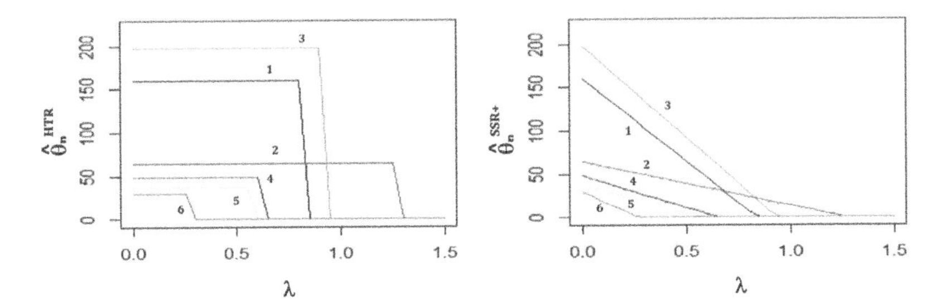

Figure 4.5 Hard threshold and positive-rule Stein–Saleh traces for ANOVA table data.

given above. There are two considerations to keep in mind. First, there may be sparsity in the parameters in which case we want to be able to partition the parameters into two groups. Second, there are parameters with high variance and others with low variance so we may also want to split them into two groups. Combining these two objectives, we define $\theta = (\theta_{1n}^{\top}, \theta_{2n}^{\top})^{\top}$, where θ_{1n} contains non-zero values with low variance and θ_{2n} contains the parameters that we suspect might be 0 or have high variance.

Consider the LASSO R trace on the right in Figure 4.3. Two of the traces with low variance look like the hard threshold estimator from Chapter 3 while the remaining four resemble the positive-rule Saleh R-estimator. To specify these two estimators, we need to define a test statistic, $\mathcal{L}_{n_j}, j = 1, \dots, p$, which will be elaborated later in the chapter. For now, consider it a threshold quantity. Then, the hard threshold R-estimator is given by

$$\hat{\theta}_n^{\mathrm{HTR}}(\lambda) = \hat{\theta}_{n_j}^{\mathrm{UR}} I(\mathcal{L}_{n_j} > \lambda^2), \ j = 1, \dots, p.$$

The positive-rule Stein–Saleh R-estimator[1] is given by

$$\hat{\theta}_n^{\mathrm{SSR+}}(\lambda) = \hat{\theta}_n^{\mathrm{UR}}(1 - \lambda/\sqrt{\mathcal{L}_n}) I(\sqrt{\mathcal{L}_n} > \lambda).$$

In Figure 4.5, we show traces for these two R-estimators for the multiple location model. Note the resemblance between two of the LASSOR traces (1 and 2) and the corresponding traces of hard threshold estimator, and four of the LASSOR traces (3, 4, 5 and 6) and the corresponding traces of the positive-rule Stein–Saleh estimator. The LASSOR traces of Figure 4.3 indicate that the unrestricted value should be used for some parameters while a shrinkage penalty should be applied to others. Since our objective in the remainder of this chapter is to compare different estimators using ADL$_2$-risk, when we suspect sparsity,

1 When the multiple location model is considered, we use the term positive-rule Stein–Saleh R-estimator.

we can define the following R-estimator:

$$\hat{\theta}_n^{\text{SSR+}}(\lambda) = \left(\hat{\theta}_{1n}^{\text{UR}^\top}, \hat{\theta}_{2n}^{\text{UR}^\top}(1 - \lambda\mathcal{L}_n^{-1})I(\mathcal{L}_n > \lambda)\right)$$

where $\hat{\theta}_n^{\text{SSR+}}$ is the positive-rule Stein–Saleh R-estimator. Here, $\hat{\theta}_{1n}^{\text{UR}}$ are the unrestricted estimates that will not experience shrinkage and $\hat{\theta}_{2n}^{\text{UR}}$ are the parameters that will experience shrinkage. There is an assumption that there is low variance for group 1 and therefore their values do not shrink rapidly. However, for group 2 with possible sparsity, the positive-rule Stein–Saleh R-estimator is used, which applies shrinkage at a linear rate.

If we do not suspect sparsity in the parameter set, then no groups are defined and we can apply a full shrinkage model using, for example, ridge-type regression as follows

$$\hat{\theta}_n^{\text{ridgeR}}(\lambda) = \frac{1}{1+\lambda}\hat{\theta}_n^{\text{UR}}.$$

In the remainder of this chapter, we define an evaluate a number of penalty R-estimators for the multiple location problem. We assume that partitioning is based on some suitable criteria such as variance or the marginal theory due to Saleh et al. (2019) when we suspect sparsity. In the sections to follow, we will define the unrestricted and restricted estimators, and then use ADL$_2$-risk to compare the different rank-based penalty estimators.

4.4 Unrestricted R-estimator

Let $R_{n_ij}^+(\theta_j)$ be the rank of $|y_{ij} - \theta_j|$ among $|y_{i1} - \theta_1|, \ldots, |y_{in_p} - \theta_p|$. For the unrestricted R-estimator (URE) of θ_j, we minimize the Jaeckel rank dispersion function w.r.t. θ_j, that is to say, minimize

$$D_{n_j}(\theta_j) = \sum_{i=1}^{n_j} |y_{ij} - \theta_j| a_{n_j}^+(R_{n_i}^+(\theta_j)), \quad j = 1, \ldots, p, \tag{4.4.1}$$

w.r.t. θ_j. This can be accomplished by solving the equation

$$\frac{\partial D_{n_j}(\theta_j)}{\partial \theta_j} = -n_j^{\frac{1}{2}} T_{n_j}(\theta_j) = -\sum_{j=1}^{n_i} \text{sgn}(y_{ij} - \theta_j) a_{n_j}^+(R_{n_i}^+(\theta_j)) = 0, \tag{4.4.2}$$

and then the URE of θ_j is given by

$$\hat{\theta}_{n_j}^{\text{UR}} = \frac{1}{2}\left[\sup\{\theta_j : T_{n_j}(\theta_j) > 0\} + \inf\left\{\theta_j : T_{n_j}(\theta_j) < 0\right\}\right].$$

Thus, the URE of $\boldsymbol{\theta}$ is given by

$$\hat{\boldsymbol{\theta}}_n^{\text{UR}} = (\hat{\theta}_{n_1}^{\text{UR}}, \dots, \hat{\theta}_{n_p}^{\text{UR}})^\top.$$

Using Eqs. (4.4.1) and (4.4.2), let

$$\boldsymbol{D}_n(\boldsymbol{\theta}) = (D_{n_1}(\theta_1), \dots, D_{n_p}(\theta_p))^\top$$

and

$$\boldsymbol{T}_n(\boldsymbol{\theta}) = (T_{n_1}(\theta_1), \dots, T_{n_p}(\theta_p))^\top.$$

Then, using the Jurečková linearity result and the asymptotic uniform quadratic result due to Jaeckel (1972), we have as $n \to \infty$,

$$\sup_{\|\boldsymbol{w}\|<n} n^{-\frac{1}{2}} \left\| \boldsymbol{T}_n(n^{-\frac{1}{2}}\boldsymbol{w}) - \boldsymbol{T}_n(0) + \gamma(\phi, \psi)\mathcal{K}\boldsymbol{w} \right\| \xrightarrow{P} 0,$$

$$\sup_{\|\boldsymbol{w}\|<n} n^{-\frac{1}{2}} \left\| \boldsymbol{D}_n(n^{-\frac{1}{2}}\boldsymbol{w}) - \boldsymbol{D}_n(0) + \boldsymbol{w}^\top \boldsymbol{T}_n(0) + \frac{1}{2}\gamma(\phi, \psi)\boldsymbol{w}^\top \mathcal{K}\boldsymbol{w} \right\| \xrightarrow{P} 0$$

where $\|\boldsymbol{a}\| = \sum_{j=1}^{p} |a_j|$ and $\mathcal{K} = \text{Diag}(\kappa_1, \dots, \kappa_p)$.

Thus, we conclude that, as $n \to \infty$, for

$$W_n(\omega) = \omega^\top \boldsymbol{T}_n(\boldsymbol{0}) + \frac{1}{2}\gamma(\phi, \psi)\omega^\top \mathcal{K}\omega,$$

we have

$$\left\| \sqrt{n}(\hat{\boldsymbol{\theta}}_n^{\text{UR}} - \boldsymbol{\theta}) - \underset{\boldsymbol{w}\in\mathbb{R}^p}{\text{argmin}}\{W_n(\boldsymbol{w})\} \right\| \xrightarrow{P} 0.$$

Hence, as $n \to \infty$,

$$\boldsymbol{T}_n(\boldsymbol{0}) \xrightarrow{\mathcal{D}} \mathcal{N}_p(\boldsymbol{0}, A_\phi^2 \mathcal{K}^{-1}),$$

and

$$|\sqrt{n}(\hat{\boldsymbol{\theta}}_n^{\text{UR}} - \boldsymbol{\theta}) - \gamma^{-1}(\phi, \psi)\mathcal{K}\boldsymbol{T}_n(\boldsymbol{0})| \xrightarrow{P} \boldsymbol{0},$$

which leads to

$$\sqrt{n}(\hat{\boldsymbol{\theta}}_n^{\text{UR}} - \boldsymbol{\theta}) \xrightarrow{\mathcal{D}} \mathcal{N}_p(\boldsymbol{0}, \eta^2 \mathcal{K}), \quad \eta^2 = \frac{A_\phi^2}{\gamma^2(\phi, \psi)}, \tag{4.4.3}$$

with

$$\gamma(\phi, \psi) = \int_0^1 \phi(u)\psi(u)\text{d}u, \; \psi(u) = -\frac{f_0'(F_0^{-1}(u))}{f_0(F_0^{-1}(u))}. \tag{4.4.4}$$

We consider the expressions for the asymptotic distributional bias (ADB), L_2-risk (ADL_2-risk) and the weighted ADL_2-risk of various R-estimators. The ADB and ADL_2-risk expressions are defined by

$$\mathrm{ADB}(\hat{\theta}_n^{*}) = \left(\lim_{n\to\infty} \mathbb{E}\left[\sqrt{n}(\hat{\theta}_{1n}^{*} - \theta_1)^{\top}\right], \lim_{n\to\infty} \mathbb{E}\left[\sqrt{n}(\hat{\theta}_{2n}^{*} - \theta_2)^{\top}\right] \right)^{\top}, \tag{4.4.5}$$

and

$$\begin{aligned} \mathrm{ADL_2\text{-}risk}(\hat{\theta}_n^{*} : \boldsymbol{W}_1, \boldsymbol{W}_2) &= \lim_{n\to\infty} \mathbb{E}\left[n(\hat{\theta}_{1n}^{*} - \theta)^{\top}\boldsymbol{W}_1(\hat{\theta}_{1n}^{*} - \theta_1)\right] \\ &\quad + \lim_{n\to\infty} \mathbb{E}\left[n(\hat{\theta}_{n2}^{*} - \theta)^{\top}\boldsymbol{W}_2(\hat{\theta}_{2n}^{*} - \theta_2)\right] \\ &= \mathrm{tr}(\boldsymbol{W}_1\boldsymbol{M}_1) + \mathrm{tr}(\boldsymbol{W}_2\boldsymbol{M}_2) \end{aligned} \tag{4.4.6}$$

where $\boldsymbol{W}_1$ and $\boldsymbol{W}_2$ are the weight matrices of p_1 and p_2 dimensions, respectively.

If $\boldsymbol{W}_1 = \boldsymbol{I}_{p_1}$ and $\boldsymbol{W}_2 = \boldsymbol{I}_{p_2}$, then we obtain the trace of the MSE matrices of $\boldsymbol{M}_1$ and $\boldsymbol{M}_2$ as $\mathrm{tr}(\boldsymbol{M}_1) = \sum_{i=1}^{p_1} \kappa_i$ and $\mathrm{tr}(\boldsymbol{M}_2) = \sum_{i=p_1+1}^{p} \kappa_i$, respectively, where

$$\boldsymbol{M}_i = \lim_{n\to\infty} n\mathbb{E}\left[(\hat{\theta}_{in}^{*} - \theta_i)(\hat{\theta}_{in}^{*} - \theta_i)^{\top}\right], \quad \forall i = 1, 2.$$

Thus,

$$\begin{aligned} \mathrm{ADL_2\text{-}risk}(\hat{\theta}_n^{*} : \boldsymbol{W}_1, \boldsymbol{W}_2) &= \mathrm{ADL_2\text{-}risk}(\hat{\theta}_{1n}^{*} : \boldsymbol{W}_{p_1}) \\ &\quad + \mathrm{ADL_2\text{-}risk}(\hat{\theta}_{2n}^{*} : \boldsymbol{W}_{p_2}). \end{aligned}$$

Briefly, the unrestricted R-estimator of θ may be written as $(\hat{\theta}_{1n}^{\mathrm{UR}^{\top}}, \hat{\theta}_{2n}^{\mathrm{UR}^{\top}})^{\top}$ with the marginal asymptotic distribution of $\hat{\theta}_{in}^{\mathrm{UR}}$ given by $\mathcal{N}_{p_i}(\theta_i, \eta^2\mathcal{K}_i)$, where $\mathcal{K} = \mathrm{blockdiagonal}(\mathcal{K}_1, \mathcal{K}_2)$.

Theorem 4.1 Under the local alternatives, $K_{(n)}$, the asymptotic distributional bias, and ADL_2-risk of the unrestricted R-estimator, $\hat{\theta}_n^{\mathrm{UR}}$, are respectively given by

$$\mathrm{ADB}(\hat{\theta}_n^{\mathrm{UR}}) = (\mathbf{0}_{p_1}^{\top}, \mathbf{0}_{p_2}^{\top})^{\top},$$

$$\mathrm{ADL_2\text{-}risk}(\hat{\theta}_n^{\mathrm{UR}}) = \eta^2 \left(\sum_{i=1}^{p_1} \kappa_i + \sum_{i=p_1+1}^{p} \kappa_i \right).$$

4.5 Test of Significance

From Eq. (4.4.2), the statistic

$$T_{n_j}(0) = \frac{1}{\sqrt{n_j}} \sum_{i=1}^{n_j} \operatorname{sgn}(y_{ij}) a_{n_i}^{+}(R_{n_i}^{+}(0))$$

may be used to test the null hypothesis $\mathcal{H}_o : \theta_j = 0$ vs. $\mathcal{H}_A : \theta_j \neq 0$. Note that $\mathbb{E}[T_{n_j}(0)] = 0$ and

$$\mathrm{Var}(T_{n_j}(0)) = \frac{1}{n_j} \sum_{i=1}^{n_j} [a_{n_j}^{+}(i)]^2 = A_{n_j}^2.$$

It may be shown that, as $n \to \infty$,

$$T_{n_j}(0) \xrightarrow{\mathcal{D}} \mathcal{N}(\mathbf{0}, A_{\phi}^2 \kappa_j^{-1}) \tag{4.5.1}$$

and we can use $\sqrt{\kappa_j} T_{n_j}(0)/A_{n_j}$ as the test statistic for $\mathcal{H}_o : \theta_j = 0$ vs. $\mathcal{H}_A : \theta_j \neq 0$. Using Hajek and Sidák (1967), as $n \to \infty$ and $\theta_j = 0$, we have that

$$\left| \sqrt{n} \hat{\theta}_{n_j}^{\mathrm{UR}} - \gamma^{-1}(\phi, \psi) \kappa_j T_{n_j}(0) \right| \xrightarrow{P} 0. \tag{4.5.2}$$

Then, after some rearrangement and substitutions, we obtain the statistic

$$\mathcal{L}_{n_j} = \frac{\kappa_j T_{n_j}^2(0)}{A_{n_j}^2}, \quad j = 1, \dots, p. \tag{4.5.3}$$

Alternatively, we may define the statistic as

$$\mathcal{L}_{n_j} = \frac{n \left(\hat{\theta}_n^{\mathrm{UR}} \right)^2}{\hat{\eta}^2 \kappa_j}, \quad j = 1, \dots, p \tag{4.5.4}$$

where $\hat{\eta} = \mathrm{MAD}(\boldsymbol{y} - \boldsymbol{B} \hat{\theta}_n^{\mathrm{UR}})$ for the ANOVA model of Eq. (4.2.1).

For the whole vector, we have

$$\left| \sqrt{n} \hat{\theta}_n^{\mathrm{UR}} - \gamma^{-1}(\phi, \psi) \mathcal{K} \boldsymbol{T}_n(\mathbf{0}) \right| \xrightarrow{P} 0. \tag{4.5.5}$$

Next, we consider the test of any sub-hypothesis

$$\mathcal{H}_o : \theta_2 = (\theta_{p_1+1}, \dots, \theta_p)^{\top} = \mathbf{0} \quad \text{vs.} \quad \mathcal{H}_A : \theta_2 \neq \mathbf{0}.$$

In this case, one may use the statistic, $\mathcal{L}_{n(2)}$, given by

$$\mathcal{L}_{n(2)} = \boldsymbol{A}_{n(2)}^{-2} \boldsymbol{T}_{n(2)}^{\top}(\mathbf{0}) \mathcal{K}_2 \boldsymbol{T}_{n(2)}(\mathbf{0}),$$

where

$$T_n(\mathbf{0}) = (T_{n(1)}^\top(\mathbf{0}), T_{n(2)}^\top(\mathbf{0}))^\top,$$
$$A_n = \mathrm{Diag}(A_{n(1)}, A_{n(2)}),$$
$$\mathcal{K} = \mathrm{Diag}(\mathcal{K}_1, \mathcal{K}_2),$$

and $T_{n(2)}(0)$ is p_2-vector. We also use the definition

$$A_{n(2)}^2 = \sum_{j=p_1+1}^{p} (n_j - 1)A_{n_j}^2. \tag{4.5.6}$$

Under the local alternatives,

$$K_{(n)} : \theta_{n_j} = n^{-\frac{1}{2}}\delta_j, \quad \text{for} \quad j = 1, \dots, p.$$

From chapter 5, section 5.2 of Hajek and Sidák (1967), we have

$$\forall j = 1, \dots, p, \qquad Z_j = \frac{\sqrt{n}\hat{\theta}_{n_j}^{\mathrm{UR}}}{\eta\sqrt{\kappa_j}} \xrightarrow{\mathcal{D}} \mathcal{N}(\Delta_j, 1), \qquad \frac{\delta_j}{\eta\sqrt{\kappa_j}} = \Delta_j.$$

Under the assumed regularity conditions, as $n \to \infty$, $\mathcal{L}_{n(2)}$ is asymptotically distributed as a non-central χ^2-distribution with p_2 degrees of freedom (d.f.) and non-centrality parameter

$$\Delta^2 = \frac{1}{\eta^2}\delta_2^\top \mathcal{K}_2^{-1}\delta_2, \quad \delta_2 = (\delta_{p_1+1}, \dots, \delta_p)^\top$$

using the local alternatives $\theta_{2(n)} = n^{-\frac{1}{2}}\delta_2$.

4.6 Restricted R-estimator

The restricted R-estimator (RRE) of $(\theta_1^\top, \mathbf{0}^\top)^\top$ is $(\hat{\theta}_{1n}^{\mathrm{UR}^\top}, \mathbf{0}^\top)^\top$.

Theorem 4.2 Under the local alternatives, $K_{(n)}$, the asymptotic distributional bias, and ADL$_2$-risk of the restricted R-estimator, $\hat{\theta}_n^{\mathrm{RR}}$, are respectively given by

$$\mathrm{ADB}(\hat{\theta}_n^{\mathrm{RR}}) = (\mathbf{0}_{p_1}^\top, -\delta_2^\top)^\top \text{ and}$$

$$\mathrm{ADL}_2\text{-risk}(\hat{\theta}_n^{\mathrm{RR}}) = \eta^2 \sum_{i=1}^{p_1} \kappa_i + \delta_2^\top \delta_2.$$

The lower bound of ADL$_2$-risk is given by

$$\mathrm{ADL}_2^*\text{-risk}(\hat{\theta}_n^{\mathrm{RR}}) = \eta^2 \left\{\mathrm{tr}(\mathcal{K}_1) + \Delta^2 \mathrm{Ch}_{\min}(\mathcal{K}_2)\right\}, \quad \Delta^2 = \frac{1}{\eta^2}\delta_2^\top \mathcal{K}_2^{-1}\delta_2.$$

Hence,

$$\text{ADRE}(\hat{\theta}_n^{\text{RR}} : \hat{\theta}_n^{\text{UR}}) = \left(1 + \frac{\text{tr}(\mathcal{K}_2)}{\text{tr}(\mathcal{K}_1)}\right)\left(1 + \frac{\Delta^2 \text{Ch}_{\min}(\mathcal{K}_2)}{\text{tr}(\mathcal{K}_1)}\right)^{-1}.$$

4.7 Shrinkage Estimators

When the problem is sparse, that is, we suspect that some estimates may be 0, we can partition the problem into two subsets, $i = 1, 2$, and retain subset 1 while applying shrinkage estimators of various forms to the potentially null subset 2.

4.7.1 Preliminary Test R-estimator

We partition the parameters as $\theta = (\theta_1^\top, \theta_2^\top)^\top$, where θ_i is a p_i-vector then $p_1 + p_2 = p$. Now, we may define the preliminary test R-estimator of θ as follows

$$\hat{\theta}_n^{\text{PTR}}(\lambda) = \left(\hat{\theta}_{1n}^{\text{UR}^\top}, \hat{\theta}_{2n}^{\text{UR}^\top} I(\mathcal{L}_n > \lambda^2)\right)^\top$$

where $I(A)$ is the indicator function of the set A and $\mathcal{L}_n$ is given by

$$\mathcal{L}_n = \boldsymbol{A}_{n(2)}^{-2} \boldsymbol{T}_{n(2)}(\boldsymbol{0}) \mathcal{K}_2^{-1} \boldsymbol{T}_{n(2)}(\boldsymbol{0})$$

and $\chi_1^2(\lambda^2)$ is the λ-level upper critical value of a central χ^2 distribution with p_2 d.f.

Theorem 4.3 Under the local alternatives, $K_{(n)}$, the asymptotic distributional bias, and ADL$_2$-risk of the preliminary test R-estimator, $\hat{\theta}_n^{\text{PTR}}(\lambda)$, are respectively given by

$$\text{ADB}(\hat{\theta}_n^{\text{PTR}}(\lambda)) = (\boldsymbol{0}_{p_1} - \delta_2^\top \mathcal{H}_{p_2+2}(\chi_{p_2}^2(\lambda^2), \Delta^2))^\top$$

and

$$\begin{aligned}\text{ADL}_2\text{-risk}(\hat{\theta}_n^{\text{PTR}}(\lambda)) \quad &= \quad \eta^2(\text{tr}(\mathcal{K}_1) + \text{tr}(\mathcal{K}_2)) - \eta^2 \, \text{tr}(\mathcal{K}_2)\mathcal{H}_{p_2+2}(\chi_{p_2}^2(\lambda), \Delta^2) \\ &\quad + \theta_2^\top \theta_2 \{2\mathcal{H}_{p_2+2}(\chi_{p_2}^2(\lambda^2), \Delta^2) - \mathcal{H}_{p_2+4}(\chi_{p_2}^2(\lambda^2), \Delta^2)\}.\end{aligned}$$

The lower bound of the ADL$_2$-risk is given by

$$\begin{aligned}\text{ADL}_2^*\text{-risk}(\hat{\theta}_n^{\text{PTR}}(\lambda)) \quad &= \quad \eta^2 \Big[\text{tr}(\mathcal{K}_1) + \text{tr}(\mathcal{K}_2)(1 - \mathcal{H}_{p_2+2}(\chi_{p_2}^2(\lambda^2), \Delta^2)) \\ &\quad + \Delta^2 \text{Ch}_{\min}(\mathcal{K}_2)\{2\mathcal{H}_{p_2+2}(\chi_{p_2}^2(\lambda^2), \Delta^2) - \mathcal{H}_{p_2+4}(\chi_{p_2}^2(\lambda^2), \Delta^2)\} \Big].\end{aligned}$$

Hence,

$$\mathrm{ADRE}(\hat{\theta}_n^{\mathrm{PTR}}(\lambda) : \hat{\theta}_n^{\mathrm{UR}}) = \left(1 + \frac{\mathrm{tr}(\mathcal{K}_2)}{\mathrm{tr}(\mathcal{K}_1)}\right)(A(\lambda, \Delta))^{-1} \tag{4.7.1}$$

where

$$A(\lambda, \Delta) = 1 + \frac{\mathrm{tr}(\mathcal{K}_2)}{\mathrm{tr}(\mathcal{K}_1)}\{1 - \mathcal{H}_{p_2+2}(\chi^2_{p_2}(\lambda^2), \Delta^2)\}$$

$$+\frac{\Delta^2\mathrm{Ch}_{\min}(\mathcal{K}_2)}{\mathrm{tr}(\mathcal{K}_1)}\left(2\mathcal{H}_{p_2+2}(\chi^2_{p_2}(\lambda^2), \Delta^2) - \mathcal{H}_{p_2+4}(\chi^2_{p_2}(\lambda^2), \Delta^2)\right). \tag{4.7.2}$$

4.7.2 The Stein–Saleh-type R-estimator

Note that $\hat{\theta}_n^{\mathrm{PTR}}(\lambda)$ is a discrete function and some desirable properties of the estimator may lost in the process. Hence, we consider the continuous version of $\hat{\theta}_n^{\mathrm{PTR}}(\lambda)$ to be called the Stein–Saleh-type R-estimator and defined by

$$\hat{\theta}_n^{\mathrm{SSR}}(\lambda) = \left(\hat{\theta}_{1n}^{\mathrm{UR}^\top}, \hat{\theta}_{2n}^{\mathrm{UR}^\top}(1 - \lambda\mathcal{L}_n^{-1})\right)^\top \quad \text{with} \quad p_2 - 2 \geq 3. \tag{4.7.3}$$

Theorem 4.4 Under the local alternatives, $K_{(n)}$, the asymptotic distributional bias, and ADL$_2$-risk of the Stein–Saleh-type R-estimator, $\hat{\theta}_n^{\mathrm{SSR}}(\lambda)$, are respectively given by

$$\mathrm{ADB}(\hat{\theta}_n^{\mathrm{SSR}}(\lambda)) = \left(\mathbf{0}_{p_1}^\top, -\boldsymbol{\delta}_2^\top(p_2 - 2)\mathbb{E}\left[\chi^{-2}_{p_2+2}(\Delta^2)\right]\right)^\top$$

and

$$\begin{aligned}\mathrm{ADL}_2\text{-risk}(\hat{\theta}_n^{\mathrm{SSR}}(\lambda)) &= \eta^2\Big\{(\mathrm{tr}(\mathcal{K}_1) + \mathrm{tr}(\mathcal{K}_2)) - \lambda\,\mathrm{tr}(\mathcal{K}_2)\Big(\mathbb{E}\left[\chi^{-2}_{p_2+2}(\Delta^2)\right] \\ &\quad +\Delta^2\mathbb{E}\left[\chi^{-4}_{p_2+2}(\Delta^2)\right]\Big) + \lambda(\lambda + 4)\boldsymbol{\delta}_2^\top\boldsymbol{\delta}_2\mathbb{E}\left[\chi^{-4}_{p_2+4}(\Delta^2)\right]\Big\}.\end{aligned}$$

The lower bound of ADL$_2$-risk is given by

$$\begin{aligned}\mathrm{ADL}_2^*\text{-risk}(\hat{\theta}_n^{\mathrm{SSR}}(\lambda)) &= \eta^2\Big[\mathrm{tr}(\mathcal{K}_1) + \mathrm{tr}(\mathcal{K}_2) - (p_2 - 2)\,\mathrm{tr}(\mathcal{K}_2) \\ \times\Big(\mathbb{E}\left[\chi^{-2}_{p_2+2}(\Delta^2)\right] + \Delta^2\mathbb{E}\left[\chi^{-4}_{p_2+2}(\Delta^2)\right]\Big) &+ (p_2^2 - 4)\Delta^2\mathrm{Ch}_{\min}(\mathcal{K}_2)\mathbb{E}[\chi^{-4}_{p_2+4}(\Delta^2)]\Big].\end{aligned}$$

Hence,

$$\text{ADRE}(\hat{\theta}_n^{\text{SSR}}(\lambda) : \hat{\theta}_n^{\text{UR}}) = \left(1 + \frac{\text{tr}(\mathcal{K}_2)}{\text{tr}(\mathcal{K}_1)}\right)(B(\Delta))^{-1}$$

where

$$\begin{aligned} B(\Delta) &= 1 + \frac{\text{tr}(\mathcal{K}_2)}{\text{tr}(\mathcal{K}_1)}\left\{1 - (p_2 - 2)\left(\mathbb{E}\left[\chi_{p_2+2}^{-2}(\Delta^2)\right] + \Delta^2\mathbb{E}\left[\chi_{p_2+2}^{-4}(\Delta^2)\right]\right)\right\} \\ &\quad + (p_2^2 - 4)\frac{\Delta^2 \text{Ch}_{\min}(\mathcal{K}_2)}{\text{tr}(\mathcal{K}_1)}\mathbb{E}\left[\chi_{p_2+4}^{-4}(\Delta^2)\right] \end{aligned}$$

and

$$\mathbb{E}\left[\chi_{p+2}^{-2}(\Delta^2)\right] = \exp\left\{-\frac{\Delta^2}{2}\right\}\sum_{r\geq 0}\frac{1}{r!}\left(\frac{\Delta^2}{2}\right)^r \frac{1}{p+2r} = \mathbb{E}_R\left[\frac{1}{p+2R}\right].$$

In the above, $\mathbb{E}_R$ stands for the expectation with respect to a Poisson variable R with mean $\frac{\Delta^2}{2}$. We may obtain the following useful identities:

$$\begin{aligned} \mathbb{E}\left[\chi_{p+2}^{-4}(\Delta^2)\right] &= \exp\left\{-\frac{\Delta^2}{2}\right\}\sum_{r\geq 0}\frac{1}{r!}\left(\frac{\Delta^2}{2}\right)^r \frac{1}{(p+2r)(p+2r-2)} \\ &= \mathbb{E}_R\left[\frac{1}{(p+2R)(p+2R-2)}\right], \\ \Delta^2\mathbb{E}\left[\chi_{p+2}^{-2}(\Delta^2)\right] &= 1 - (p-2)\mathbb{E}\left[\chi_p^{-2}(\Delta^2)\right], \\ \Delta^2\mathbb{E}\left[\chi_{p+4}^{-4}(\Delta^2)\right] &= \mathbb{E}\left[\chi_{p+2}^{-2}(\Delta^2)\right] - (p-2)\mathbb{E}\left[\chi_{p+2}^{-4}(\Delta^2)\right], \\ 2\mathbb{E}\left[\chi_p^{-2}(\Delta^2)\right] &= \mathbb{E}\left[\chi_p^{-2}(\Delta^2)\right] - \mathbb{E}\left[\chi_{p+2}^{-2}(\Delta^2)\right]. \end{aligned}$$

We note that the above Stein–Saleh R-estimator depends on $p_2 - 2 \geq 3$. Saleh (2006) considers the case of $p_2 - 2 < 3$ in Chapter 2.

4.7.3 The Positive-rule Stein–Saleh-type R-estimator

The estimator in the previous section may change signs in some cases. As a remedy, we consider the positive-rule Stein–Saleh-type R-estimator which is denoted by $\hat{\theta}_n^{\text{SSR+}}(\lambda)$ and defined by

$$\hat{\theta}_n^{\text{SSR+}}(\lambda) = \left(\hat{\theta}_{1n}^{\text{UR}^\top}, \hat{\theta}_{2n}^{\text{UR}^\top}(1 - \lambda\mathcal{L}_n^{-1})I(\mathcal{L}_n > \lambda)\right) \quad \text{with} \quad p_2 - 2 \geq 3.$$

Theorem 4.5 Under the local alternatives, $K_{(n)}$, the asymptotic distributional bias, and ADL$_2$-risk of the positive-rule Stein–Saleh-type R-estimator, $\hat{\theta}_n^{\text{SSR+}}(\lambda)$, are respectively given by

$$\text{ADB}(\hat{\theta}_n^{\text{SSR+}}(\lambda)) = \begin{pmatrix} \mathbf{0}_{p_1} \\ -\boldsymbol{\delta}_2^\top \mathbb{E}\left[(1-(p_2-2)\chi_{p_2+2}^{-2}(\Delta^2))I(\chi_{p_2+2}^2(\Delta^2) < p_2-2)\right] \end{pmatrix}$$

and

$$\begin{aligned}\text{ADL}_2\text{-risk}(\hat{\theta}_n^{\text{SSR+}}(\lambda)) &= \text{ADL}_2\text{-risk}(\hat{\theta}_n^{\text{SSR}}(\lambda)) \\ &-\eta^2\,\text{tr}(\mathcal{K}_2)\mathbb{E}\left[\left(1-(p_2-2)\chi_{p_2+2}^{-2}(\Delta^2)\right)^2 I(\chi_{p_2+2}^2(\Delta^2) < p_2-2)\right] \\ &+\boldsymbol{\delta}_2^\top\boldsymbol{\delta}_2\Big\{2\mathbb{E}\left[(1-(p_2-2)\chi_{p_2+2}^{-2}(\Delta^2))I(\chi_{p_2+2}^2(\Delta^2) < p_2-2)\right] \\ &-\mathbb{E}\left[\left(1-(p_2-2)\chi_{p_2+4}^{-2}(\Delta^2)\right)^2 I(\chi_{p_2+4}^2(\Delta^2) < p_2-2)\right]\Big\}.\end{aligned}$$

The lower bound of ADL$_2$-risk is given by

$$\begin{aligned}\text{ADL}_2^*\text{-risk}(\hat{\theta}_n^{\text{SSR+}}(\lambda)) &= \text{ADL}_2\text{-risk}(\hat{\theta}_n^{\text{SSR}}(\lambda)) \\ &-\frac{\eta^2}{p_2-2}\text{tr}(\mathcal{K}_2^{-1})\mathbb{E}\left[\left(1-(p_2-2)\chi_{p_2+2}^{-2}(\Delta^2)\right)^2 I(\chi_{p_2+2}^2(\Delta^2) < p_2-2)\right] \\ &+\Delta^2\text{Ch}_{\min}(\mathcal{K}_2)\Big\{2\mathbb{E}\left[\left(1-(p_2-2)\chi_{p_2+2}^{-2}(\Delta^2)\right)I(\chi_{p_2+2}^2(\Delta^2) < p_2-2)\right] \\ &-\mathbb{E}\left[\left(1-(p_2-2)\chi_{p_2+4}^{-2}(\Delta^2)\right)^2 I(\chi_{p_2+4}^2(\Delta^2) < p_2-2)\right]\Big\}.\end{aligned}$$

Hence,

$$\text{ADRE}(\hat{\theta}_n^{\text{SSR+}}(\lambda) : \hat{\theta}_n^{\text{UR}}) = \left(1+\frac{\text{tr}(\mathcal{K}_2)}{\text{tr}(\mathcal{K}_1)}\right)(C(\Delta))^{-1}$$

where

$$\begin{aligned}C(\Delta) &= \frac{\text{ADL}_2\text{-risk}(\hat{\theta}_n^{\text{SSR}})}{\eta^2\,\text{tr}(\mathcal{K}_1)} \\ &-\frac{\text{tr}(\mathcal{K}_2)}{\text{tr}(\mathcal{K}_1)}\mathbb{E}\left[\left(1-(p_2-2)\chi_{p_2+2}^{-2}(\Delta^2)\right)^2 I(\chi_{p_2+2}^2(\Delta^2) < (p_2-2))\right] \\ &+\frac{\Delta^2\text{Ch}_{\min}(\mathcal{K}_2)}{\text{tr}(\mathcal{K}_1)}\Big\{2\mathbb{E}\left[\left(1-(p_2-2)\chi_{p_2+2}^{-2}(\Delta^2)\right)I(\chi_{p_2+2}^2(\Delta^2) < (p_2-2))\right]\end{aligned}$$

$$-\mathbb{E}\left[\left(1-(p_2-2)\chi^{-2}_{p_2+4}(\Delta^2)\right)^2 I(\chi^2_{p_2+4}(\Delta^2)<(p_2-2))\right]\Big\},$$

$$\mathbb{E}\left[\chi^{-2}_{p+2}(\Delta^2)I(\chi^2_{p+2}(\Delta^2)<c)\right] = \mathbb{E}_R\left[\frac{\mathcal{H}_{p+2R}(c;0)}{(p+2R)}\right],$$

and

$$\mathbb{E}\left[\chi^{-4}_{p+2}(\Delta^2)I(\chi^2_{p+2}(\Delta^2)<c)\right] = \mathbb{E}_R\left[\frac{\mathcal{H}_{p-2+2R}(c;0)}{(p+2R)(p-2+2R)}\right].$$

In the above, $\mathbb{E}_R$ stands for the expectation with respect to a Poisson variable R with mean $\frac{\Delta^2}{2}$.

4.7.4 The Ridge-type R-estimator

As we described earlier in Section 4.3, LASSO facilitated the partitioning of the R-estimator of θ into two sub-vectors $(\theta_1^\top, \theta_2^\top)^\top$ as follows

$$\hat{\theta}_{1n}^{\text{UR}} = (\hat{\theta}_{n_1}^{\text{UR}}, \dots, \hat{\theta}_{n_{p_1}}^{\text{UR}})^\top \quad \text{and} \quad \hat{\theta}_{2n} = (0, \dots, 0)^\top$$

where $\hat{\theta}_{1n}^{\text{UR}}$ contains p_1 non-zero treatment effects and $\hat{\theta}_{2n}$ contains p_2 'zero' treatment effects so that $p_1 + p_2 = p$. Hence, in general, we may suspect θ_2 to be $\mathbf{0}$. In this case, one may use the R-estimator

$$\hat{\theta}_n^{\text{ridgeR}}(\lambda) = \underset{\theta_2\in\mathbb{R}^2}{\text{argmin}}\left\{\sum_{i=1}^{p}\sum_{j=1}^{n_i}|y_{ij}-\theta_i|a_{n_i}^+(R_{n_i}^+(\theta_i)) + \frac{\lambda}{2}\theta_2^\top\theta_2\right\}.$$

Thus, we define

$$\hat{\theta}_{n_j}^{\text{ridgeR}}(\lambda) = \hat{\theta}_{n_j}^{\text{UR}} \qquad j = 1, \dots, p_1, \text{ and}$$

$$\hat{\theta}_{n_j}^{\text{ridgeR}}(\lambda) = \frac{\hat{\theta}_{n_j}^{\text{UR}}}{1+\lambda} \qquad j = p_1+1, \dots, p, \tag{4.7.4}$$

to obtain the ridge-type R-estimator

$$\hat{\theta}_n^{\text{ridgeR}}(\lambda) = \begin{pmatrix} \hat{\theta}_{n_1}^{\text{UR}} \\ \vdots \\ \hat{\theta}_{n_{p_1}}^{\text{UR}} \\ \cdots\cdots\cdots \\ \frac{1}{1+\lambda}\hat{\theta}_{n_{p_1+1}}^{\text{UR}} \\ \vdots \\ \frac{1}{1+\lambda}\hat{\theta}_{n_p}^{\text{UR}} \end{pmatrix} = \left(\hat{\theta}_{1n}^{\text{UR}^\top}, \frac{1}{1+\lambda}\hat{\theta}_{2n}^{\text{UR}^\top}\right)^\top. \tag{4.7.5}$$

Theorem 4.6 Under the local alternatives, $K_{(n)}$, the asymptotic distributional bias, and L_2-risk of the ridge-type R-estimator, $\hat{\theta}_n^{\text{ridgeR}}(\lambda)$, are respectively given by

$$\text{ADB}(\hat{\theta}_n^{\text{ridgeR}}(\lambda)) = \left(\mathbf{0}^\top, -\frac{\lambda}{1+\lambda}\delta_2^\top\right)^\top$$

and

$$\text{ADL}_2\text{-risk}(\hat{\theta}_n^{\text{ridgeR}}(\lambda)) = \eta^2\left[\text{tr}(\mathcal{K}_1) + \frac{\text{tr}(\mathcal{K}_2) + \lambda^2\delta_2^\top\delta_2}{(1+\lambda)^2}\right].$$

The lower-bound of ADL_2-risk is

$$\text{ADL}_2^*\text{-risk}(\hat{\theta}_n^{\text{ridgeR}}(\lambda)) = \eta^2\left[\text{tr}(\mathcal{K}_1) + \frac{\text{tr}(\mathcal{K}_2) + \lambda^2\Delta^2\text{Ch}_{\min}(\mathcal{K}_2)}{(1+\lambda)^2}\right].$$

We may note that the optimum value of λ for $\text{ADL}_2^*\text{-risk}(\hat{\theta}_n^{\text{ridgeR}}(\lambda))$ equals

$$\lambda_{\text{opt}} = \text{tr}(\mathcal{K}_2)\Delta^{-2}(\text{Ch}_{\min}(\mathcal{K}_2))^{-1}.$$

Thus,

$$\text{ADL}_2^*\text{-risk}(\hat{\theta}_n^{\text{ridgeR}}(\lambda_{\text{opt}})) = \eta^2\left[\text{tr}(\mathcal{K}_1) + \frac{\text{tr}(\mathcal{K}_2)\Delta^2\text{Ch}_{\min}(\mathcal{K}_2)}{\text{tr}(\mathcal{K}_2) + \Delta^2\text{Ch}_{\min}(\mathcal{K}_2)}\right],$$

and

$$\begin{aligned}\text{ADRE}(\hat{\theta}_n^{\text{ridgeR}}(\lambda) : \hat{\theta}_n^{\text{UR}}) &= \left(1 + \frac{\text{tr}(\mathcal{K}_2)}{\text{tr}(\mathcal{K}_1)}\right) \\ &\times \left(1 + \frac{\text{tr}(\mathcal{K}_2)}{\text{tr}(\mathcal{K}_1)}\frac{\Delta^2\text{Ch}_{\min}(\mathcal{K}_2)}{[\text{tr}(\mathcal{K}_2) + \Delta^2\text{Ch}_{\min}(\mathcal{K}_2)]}\right)^{-1}\end{aligned} \quad (4.7.6)$$

which is a decreasing function of Δ^2. At $\Delta^2 = 0$, the value is $\left(1 + \frac{\text{tr}(\mathcal{K}_2)}{\text{tr}(\mathcal{K}_1)}\right)$ and for $\Delta^2 \to \infty$, it is 1. Clearly,

$$\begin{aligned}0 &\le \left(1 + \frac{\text{tr}(\mathcal{K}_2)}{\text{tr}(\mathcal{K}_1)}\right)\left(1 + \frac{(\text{tr}(\mathcal{K}_2))\Delta^2\text{Ch}_{\min}(\mathcal{K}_2)}{(\text{tr}(\mathcal{K}_1))(\text{tr}(\mathcal{K}_2) + \Delta^2\text{Ch}_{\min}(\mathcal{K}_2))}\right)^{-1} \\ &< 1 + \frac{\text{tr}(\mathcal{K}_2)}{\text{tr}(\mathcal{K}_1)} \qquad \forall \Delta^2 \in \mathbb{R}^+.\end{aligned}$$

4.8 Subset Selection Penalty R-estimators

In the prior section, we described methods to handle sparsity in the model. In this section, we discuss several penalty R-estimators that seek to find the parameters that lead to sparsity in the model, such as

(i) Preliminary test R-estimator (PTRE) generally designated as a hard threshold estimators (HTE) due to Donoho and Johnstone (1994).
(ii) LASSO proposed by Tibshirani (1996) designated as a soft threshold estimator (STE) due to Donoho and Johnstone (1994). Saleh et al. (2019) identified LASSO as equivalent to the positive-rule Saleh selector (PRSS) for the location model, as noted in Chapter 3.
(iii) The adaptive LASSO-type estimator.
(iv) The elastic-net-type estimator.

As mentioned before, we use the "-type" designation since we are applying the shrinkage factor directly to the unrestricted estimate to obtain a closed-form estimator rather than minimizing a penalized loss function.

4.8.1 Preliminary Test Subset Selector R-estimator

The preliminary test subset selector R estimator (PTSSRE) is defined by

$$\hat{\theta}_n^{\text{PTSSR}}(\lambda) = \left(\hat{\theta}_{n_j}^{\text{UR}} I(\mathcal{L}_{n_j} > \lambda^2) | j = 1, \dots, p\right)^\top \tag{4.8.1}$$

where

$$\mathcal{L}_{n_j} = \frac{n\left(\hat{\theta}_{n_j}^{\text{UR}}\right)^2}{\hat{\eta}^2 \kappa_j}, \quad j = 1, \dots, p$$

is the test statistic.

Each component is a PTSSRE of θ_j, $j = 1, \dots, p$.

Theorem 4.7 Under the local alternatives, $K_{(n)}$, the asymptotic distributional bias, and ADL$_2$-risk of the preliminary test subset selector R-estimator, $\hat{\theta}^{\text{PTSSR}}$, are respectively given by

$$\text{ADB}(\hat{\theta}_n^{\text{PTSSR}}(\lambda)) = -\eta \left(\delta_j \mathcal{H}_3(\lambda^2, \Delta_j) | j = 1, \dots, p\right)^\top,$$

and

$$\text{ADL}_2\text{-risk}(\hat{\theta}_n^{\text{PTSSR}}(\lambda)) = \eta^2\left\{\sum_{j=1}^{p}\kappa_j(1-\mathcal{H}_3(\lambda^2,\Delta_j^2)) + \sum_{j=1}^{p}\Delta_j^2\left(2\mathcal{H}_3(\lambda^2,\Delta_j^2)-\mathcal{H}_5(\lambda^2,\Delta_j^2)\right)\right\}.$$

Proof: We begin by showing that

$$\sqrt{n}\hat{\theta}_n^{\text{PTSSR}}(\lambda) = \left(\sqrt{n}\hat{\theta}_{n_j}^{\text{UR}} I(|\sqrt{n}\hat{\theta}_{n_j}^{\text{UR}}| > \lambda\kappa_j^{\frac{1}{2}}\eta) | j=1,\dots,p\right)^{\top}$$
$$= \left(\eta\kappa_j^{\frac{1}{2}} Z_{n_j} I(|Z_{n_j}| > \lambda) | j=1,\dots,p\right)^{\top},$$

where

$$Z_{n_j} = \frac{\sqrt{n}\hat{\theta}_{n_j}^{\text{UR}}}{\eta\sqrt{\kappa_j}}, \quad j=1,\dots,p.$$

Clearly, $Z_{n_j} \sim \mathcal{N}(\Delta_j, 1)$, $j = 1,\dots,p$, as $n \to \infty$, and $\Delta_j = \delta_j/(\eta\sqrt{\kappa_j})$. Since the proof of this theorem is similar to Theorem 3.5, we leave the details to the reader. □

4.8.2 Saleh-type R-estimator

We noted that the hard threshold R-estimator is discrete in nature and may be extremely variable and unstable due to the fact that small changes in the data can result in a very different model and may reduce the prediction accuracy. The Saleh R-estimator is a continuous version of the hard threshold estimator and is defined as

$$\hat{\theta}_n^{\text{SR}}(\lambda) = \left(\hat{\theta}_{n_j}^{\text{UR}}\left(1-\frac{\lambda}{\mathcal{L}_{n_j}}\right)\middle| j=1,\dots,p\right)^{\top}, \quad \mathcal{L}_{n_j} = |Z_{n_j}|.$$

Theorem 4.8 Under the local alternatives, $K_{(n)}$, the asymptotic distributional bias, and ADL_2-risk of the Saleh R-estimator, $\hat{\theta}_n^{\text{SR}}$, are respectively given by

$$\text{ADB}(\hat{\theta}_n^{\text{SR}}(\lambda)) = \left(-\lambda\eta\sqrt{\kappa_j}(2\Phi(\Delta_j)-1) | j=1,\dots,p\right)^{\top},$$

and

$$\text{ADL}_2\text{-risk}(\hat{\theta}_n^{\text{SR}}(\lambda)) = \eta^2 \sum_{j=1}^{p} \left(1 + \lambda^2 - 2\lambda\sqrt{\frac{2}{\pi}} \exp\left\{ -\frac{\Delta_j^2}{2} \right\} \right).$$

Proof: We first show that

$$\sqrt{n}\hat{\theta}_n^{\text{SR}}(\lambda) = \left(\sqrt{n}\hat{\theta}_{n_j}^{\text{UR}} - \lambda\eta\sqrt{\kappa_j} \frac{\sqrt{n}\hat{\theta}_{n_j}^{\text{UR}}}{|\sqrt{n}\hat{\theta}_{n_j}^{\text{UR}}|} \middle| j = 1, \dots, p \right)^{\top}$$

$$= \left(\eta\sqrt{\kappa_j} \left(1 - \frac{\lambda}{|Z_{n_j}|} \right) Z_{n_j} \middle| j = 1, \dots, p \right)^{\top}$$

where $Z_{n_j} = \sqrt{n}\hat{\theta}_{n_j}^{\text{UR}}/(\eta\sqrt{\kappa_j})$, $j = 1, \dots, p$. Clearly, $Z_{n_j} \sim \mathcal{N}(\Delta_j, 1)$, as $n \to \infty$, and $\Delta_j = \delta_j/(\eta\sqrt{\kappa_j})$. Since the proof of this theorem is similar to Theorem 3.7, we leave the remainder to the reader. □

Hence, for $\lambda_{\max}^* = \sqrt{\frac{2}{\pi}}$, we obtain

$$\text{ADL}_2\text{-risk}(\hat{\theta}_n^{\text{SR}}(\lambda^*)) = \eta^2 \sum_{j=1}^{n} \left[1 - \frac{2}{\pi} \left(2\exp\left\{ -\frac{\Delta_j}{2} \right\} - 1 \right) \right].$$

Finally,

$$\text{ADRE}(\hat{\theta}_n^{\text{SR}}(\lambda^*) : \hat{\theta}_n^{\text{UR}}) = \left(\sum_{j=1}^{n} \left[1 - \frac{2}{\pi} \left(2\exp\left\{ -\frac{\Delta_j}{2} \right\} - 1 \right) \right] \right)^{-1}.$$

4.8.3 Positive-rule Saleh Subset Selector (PRSS)

The Saleh R-estimator described in the previous section changes sign and this may not be desirable. However, the LASSO R-estimator, which is given by

$$\hat{\theta}_n^{\text{LassoR}}(\lambda) = \underset{\theta_i \in \mathbb{R}}{\text{argmin}} \left\{ \sum_{i=1}^{p} \sum_{j=1}^{n_i} |y_{ij} - \theta_i| a_n^+(R_{n_j}(\theta_i)) + \lambda \sum_{i=1}^{p} |\theta_i| \right\}$$

does not have this issue. Motivated by LASSO and soft threshold estimator (Donoho and Johnstone, 1994), we consider the positive-rule of Saleh subset selector (PRSS) to remedy this problem. It can be defined as

$$\hat{\theta}_n^{\text{SR+}}(\lambda) = \left(\text{sgn}(\hat{\theta}_{n_j}^{\text{UR}}) \left(|\hat{\theta}_{n_j}^{\text{UR}}| - \lambda \right) I(\mathcal{L}_{n_j} > \lambda^2) \middle| j = 1, 2, \dots, p \right)^{\top}$$

where

$$\mathcal{L}_{n_j} = \frac{(\sqrt{n}\hat{\theta}_j^{\text{UR}})^2}{\hat{\eta}^2\kappa_j}. \tag{4.8.2}$$

Theorem 4.9 Under the local alternatives, $K_{(n)}$, the asymptotic distributional bias, and ADL$_2$-risk of the positive-rule Saleh R-estimator, $\hat{\theta}_n^{\text{SR+}}$, are respectively given by

$$\begin{aligned}\text{ADB}(\hat{\theta}_n^{\text{SR+}}(\lambda)) &= -\eta\left(\Delta_j\mathcal{H}_3(\lambda^2,\Delta^2) - \lambda\left(\Phi(\lambda-\Delta_j)\right.\right.\\ &\qquad \left.\left.-\Phi(\lambda+\Delta_j)\right)\middle| j=1,\dots,p\right)^{\top},\end{aligned}$$

and

$$\begin{aligned}\text{ADL}_2\text{-risk}(\hat{\theta}_n^{\text{SR+}}(\lambda)) =&\eta^2\Bigg(\sum_{j=1}^{p}\Big\{\kappa_j\left(1-\mathcal{H}_3(\lambda^2,\Delta_j^2)\right)\\ &+\Delta_j^2\left(2\mathcal{H}_3(\lambda^2,\Delta_j^2)-\mathcal{H}_5(\lambda^2,\Delta_j^2)\right)\\ &+\lambda^2(1-\mathcal{H}_1(\lambda^2,\Delta_j^2))-2\lambda\left(\phi(\lambda-\Delta_j)+\phi(\lambda+\Delta_j)\right)\Big\}\Bigg).\end{aligned}$$

To obtain the LASSO R-estimator, one minimizes the rank dispersion function (Jaeckel, 1972) with an L$_1$ penalty as follows,

$$\hat{\theta}_{n_j}^{\text{LassoR}}(\lambda) = \left(\underset{\theta}{\text{argmin}}\left\{\sum_{j=1}^{n_i}|y_{i_j}-\theta|a_n^+(R_{n_j}(\theta_i))+\lambda\kappa_j^{\frac{1}{2}}\eta|\theta_i|\right\}\middle| j=1,\dots,p\right)^{\top}.$$

Now, consider a family of diagonal linear projections,

$$T_{\text{DP}}(\theta,\boldsymbol{\lambda}) = \left(\lambda_1\hat{\theta}_{n_1}^{\text{UR}},\dots,\lambda_p\hat{\theta}_{n_p}^{\text{UR}}\right)^{\top}.$$

Such estimators *keep* or *kill* each coordinate. Suppose we have available an oracle which would supply for us the optimal coefficients, $(\lambda_j|\, j = 1,\dots,p)$, for use in the diagonally linear projection scheme. These ideal coefficients are $I(|\delta_j| > \eta\sqrt{\kappa_j})$, $j = 1,\dots,p$. These yield the ideal asymptotic distributional L$_2$-risk as

$$\text{ADL}_2^*\text{-risk}(T_{\text{DP}}(\theta,\boldsymbol{\lambda})) = \sum_{j=1}^{p}\min(\delta_j^2,\eta^2\kappa_j).$$

Now suppose p_1 of them satisfy $\{\delta_j^2 < \eta^2|\, j = 1,\dots,p\}$ and p_2 of them satisfy $\{\delta_j^2 > \eta^2|\, j = 1,\dots,p\}$. Then we obtain ADL$_2^*$-risk$(T_{\text{DP}}(\theta,\boldsymbol{\lambda}))$ as

$$\text{ADL}_2^*\text{-risk}(\hat{\theta}_n^{\text{LassoR}}(\lambda)) = \eta^2(\text{tr}(\mathcal{K}_1)+\Delta^2\text{Ch}_{\min}(\mathcal{K}_2)),$$

where

$$\Delta^2 = \eta^{-2}\delta_2^{\top}\mathcal{K}_2^{-1}\delta_2$$

and $\mathcal{K} = \text{Diag}(\mathcal{K}_1, \mathcal{K}_2)$ and $\text{Ch}_{\min}(\mathcal{K}_2^{-1})$ is the minimum eigenvalue of $\mathcal{K}_2^{-1}$. Here, we assume that p_1 parameters exceed the noise level, $\eta\sqrt{\kappa_j}$, and p_2 of them do not. Hence, the lower bound of $\text{ADL}_2^*\text{-risk}(\hat{\theta}_n^{\text{LassoR}}(\lambda))$ is

$$\text{ADL}_2^*\text{-risk}(\hat{\theta}_n^{\text{LassoR}}(\lambda)) = \eta^2(\text{tr}(\mathcal{K}_1) + \Delta_L^2),$$

where $\Delta_L^2 = \Delta^2\text{Ch}_{\min}(\mathcal{K}_2)$. Thus,

$$\text{ADRE}(\hat{\theta}_n^{\text{LassoR}}(\lambda) : \hat{\theta}_n^{\text{UR}}) = \left(1 + \frac{\text{tr}(\mathcal{K}_2)}{\text{tr}(\mathcal{K}_1)}\right)\left(1 + \frac{\Delta^2\text{Ch}_{\min}(\mathcal{K}_2)}{\text{tr}(\mathcal{K}_1)}\right)^{-1}$$

using the lower bound.

4.8.4 The Adaptive LASSO (aLASSO)

Adaptive LASSO is a variation of LASSO whereby

$$\hat{\theta}_n^{\text{aLassoR}}(\lambda_n) = \underset{\theta_i \in \mathbb{R}}{\text{argmin}}\left\{\sum_{i=1}^{p}\sum_{j=1}^{n_i}|y_{ij} - \theta_i|a_n^+(R_{n_j}(\theta_i)) + \lambda_n\sum_{i=1}^{p}\hat{\omega}_i|\theta_i|\right\}$$

and $\hat{\omega}_i$, $(i = 1, \dots, p)$ are the weight functions, $\hat{\omega}_i = |\hat{\theta}_i^{\text{UR}}|^{-\gamma}$, $\gamma > 0$, $i = 1, \dots, p$. Like the positive-rule Saleh R-estimator, we can define an aLASSO-type R-estimator as

$$\hat{\theta}_n^{\text{aLassoR}}(\lambda_n) = \left(\left(\text{sgn}(\hat{\theta}_{n_j}^{\text{UR}})\left(|\hat{\theta}_{n_j}^{\text{UR}}| - \lambda_n\hat{\omega}_j\right)I(\mathcal{L}_{n_j} > \lambda_n^2\hat{\omega}_j^2)\right)\Big| j = 1, \dots, p\right)^{\top}. \tag{4.8.3}$$

We can rewrite this estimator based on $Z_{n_i} = \sqrt{n}\hat{\theta}_{n_i}^{\text{UR}}/\eta\sqrt{\kappa_i}$ as

$$\begin{aligned}\sqrt{n}\hat{\theta}_{n_i}^{\text{aLassoR}}(\lambda_n) &= \text{sgn}(\sqrt{n}\hat{\theta}_{n_i}^{\text{UR}})\left(|\sqrt{n}\hat{\theta}_{n_i}^{\text{UR}}| - \lambda_n\hat{\omega}_i\sqrt{\kappa_i}\eta\right)^+ \\ &= \eta\sqrt{\kappa_i}\text{sgn}(Z_{n_i})\left(|Z_{n_i}| - \lambda_n\hat{\omega}_i\right)^+.\end{aligned}$$

Let $\mathcal{A}_n^* = \{j : \hat{\theta}_{jn}^{\text{aLassoR}} \neq 0\}$ and $\mathcal{A} = \{j : \theta_j^* \neq 0\}$ and the size of $\mathcal{A}$ be p_o $(< p)$.

Theorem 4.10 (Oracle property) Let $\lambda_n/\sqrt{n} \to 0$ and $\lambda_n n^{\frac{\gamma-1}{2}} \to \infty$. Then, adaptive LASSO must satisfy the following:

(a) Asymptotic normality: $\sqrt{n}(\hat{\theta}_{\mathcal{A}}^{*(n)} - \hat{\theta}_{\mathcal{A}}^*) \xrightarrow{\mathcal{D}} \mathcal{N}_{p_o}(\mathbf{0}, \eta^2\mathcal{K}_1)$.

(b) If $\sqrt{n}\lambda_n \to \lambda_0 \in (0, \infty)$, then $\lim_{n\to\infty} P(\hat{\theta}^*_{2n} = \mathbf{0}) = 1$, and

$$\sqrt{n}\gamma(\phi, f)\mathcal{K}_{p_0}\left\{(\hat{\theta}^*_{1n} - \theta^*_1) + n^{-\frac{1}{2}}\gamma^{-1}(\phi, f)\mathcal{K}_{p_0}^{-1}\lambda_0\mathbf{b}\right\} \to \mathcal{N}_{p_0}(\mathbf{0}, A^2_\phi\mathcal{K}_1) \tag{4.8.4}$$

where $\mathcal{K}_{p_0} = \text{Diag}(\kappa_1, \dots, \kappa_{p_0})$ and $\boldsymbol{b} = \left(\frac{\text{sgn}(\theta^*_j)}{|\theta^*_j|}\middle| j = 1, \dots, p_0\right)$.

Proof: To prove the asymptotic normality, let $\theta_i = \theta^*_i + \frac{u_i}{\sqrt{n}}$ and

$$\psi_n(u) = \sum_{i=1}^{p}\sum_{j=1}^{n_i}\left|y_{ij} - (\theta^*_i + \frac{u_i}{\sqrt{n}})\right| a^+_n(R_{n_j}(\theta^*_i)) + \frac{\lambda_n}{\sqrt{n}}\sum_{i=1}^{p}\hat{\omega}_i\sqrt{n}\left(|\theta^*_i + \frac{u_j}{\sqrt{n}}|\right)$$

and

$$\psi_n(0) = \sum_{i=1}^{p}\sum_{j=1}^{n_i}|y_{ij} - \theta^*_i|\, a^+_n(R_{n_j}(\theta^*_i)) + \frac{\lambda_n}{\sqrt{n}}\sum_{i=1}^{p}\hat{\omega}_i\sqrt{n}|\theta^*_i|.$$

Let $\hat{\boldsymbol{u}}_n = \text{argmin}\{\psi_n(\boldsymbol{u})\}$; then $\hat{\boldsymbol{u}}_n = \sqrt{n}\left(\hat{\theta}^*_n - \theta^*\right)$. Further, note that $\Psi_n(\boldsymbol{u}) - \Psi_n(\mathbf{0}) = V^{(n)}_0(\boldsymbol{u})$ where

$$V^{(n)}_0(\boldsymbol{u}) = \boldsymbol{u}^\top_{\mathcal{A}}\mathcal{K}^{-1}_{p_0}\boldsymbol{u}_{\mathcal{A}} - 2\boldsymbol{u}^\top_{\mathcal{A}}\boldsymbol{W} + \frac{\lambda_n}{\sqrt{n}}\sum_{j=1}^{p}\hat{\omega}_j\sqrt{n}\left(|\theta^*_j + \frac{u_j}{\sqrt{n}}| - |\theta^*_j|\right).$$

We know that $\mathcal{K}_n \to \mathcal{K}_{p_0}$ and $\boldsymbol{W} \xrightarrow{\mathcal{D}} \mathcal{N}_p(\mathbf{0}, \eta^2\mathcal{K}^{-1}_{p_0})$. Consider the limiting behavior of the third term if $\theta^*_j \neq 0$. Then, we have that

$$\hat{\omega}_j \xrightarrow{P} |\theta^*_j|^{-\gamma},$$

and

$$\sqrt{n}\left(\left|\theta^*_j + \frac{u_j}{\sqrt{n}}\right| - \left|\theta^*_j\right|\right) \xrightarrow{P} |u_j|\text{sgn}(\theta^*_j).$$

By Slutsky's theorem, we can show that

$$\frac{\lambda_n}{\sqrt{n}}\hat{\omega}_j\sqrt{n}\left(\left|\theta^*_j + \frac{u_j}{\sqrt{n}}\right| - \left|\theta^*_j\right|\right) \xrightarrow{P} 0. \tag{4.8.5}$$

If $\theta^*_j = 0$, then $\sqrt{n}\left(\left|\theta^*_j + \frac{u_j}{\sqrt{n}}\right| - \left|\theta^*_j\right|\right) = |u_j|$ and

$$\frac{\lambda_n}{\sqrt{n}}\hat{\omega}_j = \frac{\lambda_n}{\sqrt{n}}n^{\frac{\gamma}{2}}\left(\left|\sqrt{n}\hat{\theta}^{\text{UR}}_j\right|\right)^{-\gamma}$$

where $\sqrt{n}\hat{\theta}_j^{\text{UR}} = O_p(1)$. Then by using Slutsky's theorem again, we obtain that $V_{\mathcal{A}}^{(n)}(\boldsymbol{u}) \xrightarrow{\mathcal{D}} V_{\mathcal{A}}(\boldsymbol{u})$ for every $\boldsymbol{u}$, where

$$V_{\mathcal{A}}(\boldsymbol{u}) = \begin{cases} \boldsymbol{u}_{\mathcal{A}}^{\top}\mathcal{K}_{p_0}^{-1}\boldsymbol{u}_{\mathcal{A}} - 2\boldsymbol{u}_{\mathcal{A}}^{\top}\boldsymbol{W}_{\mathcal{A}} & u_j = 0\ \forall j \notin \mathcal{A} \\ \infty & \text{otherwise.} \end{cases} \tag{4.8.6}$$

Note that $V_{\mathcal{A}}(\boldsymbol{u})$ is convex, and with a unique minimum of $V_{\mathcal{A}}(\boldsymbol{u})$ is $(\mathcal{K}_1^{-1}\boldsymbol{W}_{\mathcal{A}}, \boldsymbol{0})^{\top}$. Following the epi-convergence[2] result by Geyer (1994) and Knight and Fu (2000), we have

$$u_{\mathcal{A}} \xrightarrow{\mathcal{D}} \mathcal{K}_1^{-1}\boldsymbol{W}_{\mathcal{A}} \quad \text{and} \quad u_{\mathcal{A}^c} \xrightarrow{\mathcal{D}} \boldsymbol{0}. \tag{4.8.7}$$

Finally, $\boldsymbol{W}_{\mathcal{A}} \sim \mathcal{N}_{p_0}(\boldsymbol{0}, \eta^2\mathcal{K}_1)$, and the proof of asymptotic normality follows.

Before we discuss the "oracle" property of aLASSO, we partition true location vector $\theta = (\theta_1, \dots, \theta_p)^{\top}$ into two subset $(\theta_{p_0}^{\top}, \theta_{p-p_0}^{\top})^{\top}$ where $\theta_j \neq 0$ for $j \leq p_0$ and $\theta_{p-p_0} = \boldsymbol{0}$.

Let $\hat{\theta}_n^{\text{UR}} = (\hat{\theta}_{p_0}^{\text{UR}^{\top}}, \hat{\theta}_{p-p_0}^{\text{UR}^{\top}})^{\top}$ is the partition of R-estimators of θ. Basic assumption of the analysis is based on Jurečková (1969, 1971).

Accordingly, the p-dimensional rank-statistic is given by

$$\nabla D(\theta) = (-1)\boldsymbol{T}_n(\theta), \qquad \boldsymbol{T}_n(\theta) = \left(T_{n_1}(\theta_1), \dots, T_{n_p}(\theta_p)\right) \tag{4.8.8}$$

where $T_{n_i}(\theta_i) = n^{-\frac{1}{2}} \sum_{j=1}^{n_i} \text{sgn}(y_{ij} - \theta_i)a_{n_i}^{+}(R_{n_i}(\theta_i))$.

First, we consider the usual asymptotic linearity for rank-statistics, i.e. for any $M > 0$,

$$\sup_{\left\|\theta^* - \theta^*\right\| < Mn^{-\frac{1}{2}}} \left\| \boldsymbol{T}_n(\theta^* + \frac{\boldsymbol{u}}{\sqrt{n}}) - \boldsymbol{T}_n(\theta^*) + \gamma(\phi, f)\mathcal{K}^{-1}\boldsymbol{u} \right\| \xrightarrow{P} 0 \tag{4.8.9}$$

where $\mathcal{K} = \text{Diag}(\kappa_1, \dots, \kappa_p)$. Let $\boldsymbol{u} = \sqrt{n}(\hat{\theta}_n^* - \theta^*)$.

Secondly, $D_n(\theta)$ has quadratic approximation in any root-n neighborhood of θ_0,

$$\sup_{\left\|\theta^* - \theta^*\right\| < Mn^{-\frac{1}{2}}} \left| D_n(\theta^* + \frac{\boldsymbol{u}}{\sqrt{n}}) - D_n(\theta^*) + \boldsymbol{T}_n(\theta^*)^{\top}\boldsymbol{u} - \frac{1}{2}\boldsymbol{u}^{\top}\mathcal{K}\boldsymbol{u} \right| \xrightarrow{P} 0. \tag{4.8.10}$$

2 Epi-convergence is a type of convergence for real-valued functions that is important in approximating minimization problems.

Equivalently, we may write

$$\sup_{\left\|\hat{\theta}^{*}-\theta^{*}\right\|<Mn^{-\frac{1}{2}}} \left\| \boldsymbol{T}_n(\theta^{*} + \frac{\boldsymbol{u}}{\sqrt{n}}) - \boldsymbol{T}_n(\theta^{*}) + \gamma(\phi, f)\mathcal{K}\boldsymbol{u} \right\| \xrightarrow{P} 0 \tag{4.8.11}$$

and

$$\sup_{\left\|\hat{\theta}^{*}-\theta^{*}\right\|<Mn^{-\frac{1}{2}}} n^{-\frac{1}{2}} \left| D_n(\theta^{*} + \frac{\boldsymbol{u}}{\sqrt{n}}) - D_n(\theta^{*}) - \boldsymbol{u}^{\top}\boldsymbol{T}_n(\boldsymbol{0}) + \frac{1}{2}\gamma(\phi, f)\boldsymbol{u}^{\top}\mathcal{K}\boldsymbol{u} \right| \xrightarrow{P} 0 \tag{4.8.12}$$

as $n \to \infty$,

$$\mathcal{W}_n(\boldsymbol{u}) = \boldsymbol{u}^{\top}\boldsymbol{T}_n(\boldsymbol{0}) + \frac{1}{2}\gamma(\phi, f)\boldsymbol{u}^{\top}\mathcal{K}_{p_0}\boldsymbol{u}$$

$$n^{-\frac{1}{2}}\boldsymbol{T}_n(\boldsymbol{0}) \xrightarrow{\mathcal{D}} \mathcal{N}_p(\boldsymbol{0}, A_{\phi}^2, \mathcal{K}^{-1}).$$

Hence, we have

$$\left| \sqrt{n}(\hat{\theta}_n^{\mathrm{UR}} - \theta) - \operatorname{argmin}\{\mathcal{W}(\boldsymbol{u})\} \right| \xrightarrow{P} 0 \tag{4.8.13}$$

which leads to

$$\sqrt{n}\left(\hat{\theta}_n^{\mathrm{UR}} - \theta\right) \xrightarrow{\mathcal{D}} \mathcal{N}_p(\boldsymbol{0}, \eta^2\mathcal{K}). \tag{4.8.14}$$

Also,

$$n^{\frac{1}{2}}\mathcal{K}_{p_0}\gamma(\phi, f)\left\{(\hat{\theta}_{1n}^{*} - \theta_1) + n^{-\frac{1}{2}}\gamma^{-1}(\phi, f)\mathcal{K}_{p_0}^{-1}\lambda_0\boldsymbol{b}\right\} = n^{-\frac{1}{2}}\boldsymbol{T}_{n(1)}(\theta) + o_p(1). \tag{4.8.15}$$

Given that

$$n^{-\frac{1}{2}}\boldsymbol{T}_{n_1}(\theta) \xrightarrow{\mathcal{D}} \mathcal{N}_{p_0}(\boldsymbol{0}, A_{\phi}^2\mathcal{K}_{p_0}^{-1}), \tag{4.8.16}$$

we can infer that

$$\sqrt{n}(\hat{\theta}_{1n}^{*} - \theta_1) \xrightarrow{\mathcal{D}} \mathcal{N}_{p_0}(\boldsymbol{0}, A_{\phi}^2\mathcal{K}_{p_0}), \tag{4.8.17}$$

where $\mathcal{K}_{p_0} = \mathrm{Diag}(\kappa_1, \dots, \kappa_{p_0})$. □

4.8.5 Elastic-net-type R-estimator

Consider the R-estimation of $\boldsymbol{\theta}$ by using a combination of the L_1 and L_2 penalty functions. That is, if ridge and LASSO are used together, it is possible to retain the benefits of both shrinkage and subset selection in one penalized R-estimator. We require the introduction of two tuning parameters, λ and α, to formulate the estimator as follows:

$$\hat{\boldsymbol{\theta}}_n^{\text{EnetR}}(\lambda,\alpha) = \underset{\boldsymbol{\theta}\in\mathbb{R}^p}{\text{argmin}}\left\{\sum_{i=1}^{p}\sum_{j=1}^{n_i}|y_{ij}-\theta_i|a_{n_i}^{+}(R_{n_i}(\theta_i)) + \lambda\sum_{i=1}^{p}\{\eta\sqrt{\kappa_i}(\alpha|\theta_i| + \frac{1}{2}(1-\alpha)\theta_i^2)\}\right\}. \tag{4.8.18}$$

By the same token, the combination of ridge-type and LASSO-type R-estimators can be used to formulate a naive Enet-type (nEnet) R-estimator as follows

$$\hat{\boldsymbol{\theta}}_n^{\text{nEnetR}}(\lambda,\alpha) = \begin{pmatrix} \hat{\theta}_{n_1}^{\text{UR}}\left(1-\frac{\eta\lambda\alpha\sqrt{\kappa_1}}{\sqrt{\mathcal{L}_{n_1}}}\right)\frac{I(\mathcal{L}_{n_1}>\lambda^2\alpha^2)}{1+\lambda(1-\alpha)\kappa_1} \\ \vdots \\ \hat{\theta}_{n_p}^{\text{UR}}\left(1-\frac{\eta\lambda\alpha\sqrt{\kappa_p}}{\sqrt{\mathcal{L}_{n_p}}}\right)\frac{I(\mathcal{L}_{n_p}>\lambda^2\alpha^2)}{1+\lambda(1-\alpha)\kappa_p} \end{pmatrix}.$$

Theorem 4.11 Under the local alternatives, $K_{(n)}$, the asymptotic distributional bias, and L_2-risk of the nEnet R-estimator, $\hat{\boldsymbol{\theta}}_n^{\text{nEnetR}}(\lambda)$, are respectively given by

$$\text{ADB}(\hat{\boldsymbol{\theta}}_n^{\text{nEnetR}}(\lambda,\alpha)) = \left(\frac{-\eta\sqrt{\kappa_j}}{[1+\lambda(1-\alpha)\kappa_j]}\Big[\Delta_j\mathcal{H}_3(\lambda^2\alpha^2,\Delta_j^2) - \lambda\alpha\left(\Phi(\lambda\alpha-\Delta_j)-\Phi(\lambda\alpha+\Delta_j)\right) - \lambda(1-\alpha)\Delta_j\Big]\Big| j=1,2,\ldots,p\right)^{\top}$$

and

$$\begin{aligned}\text{ADL}_2\text{-risk}(\hat{\boldsymbol{\theta}}_n^{\text{nEnetR}}(\lambda,\alpha)) \;=\; &\eta^2\Big[\sum_{j=1}^{p}\frac{1}{[1+\lambda(1-\alpha)\kappa_j]^2}\Big\{\kappa_j\left(1-\mathcal{H}_3(\lambda^2\alpha^2,\Delta_j^2)\right)\\ &+\Delta_j^2\Big(2\mathcal{H}_3(\lambda^2\alpha^2,\Delta_j^2)-\mathcal{H}_5(\lambda^2\alpha^2,\Delta^2)\Big)+\lambda^2\left(1-\mathcal{H}_1(\lambda^2\alpha^2,\Delta_j^2)\right)\\ &-2\lambda\Big(\phi(\lambda\alpha-\Delta_j)+\phi(\lambda\alpha+\Delta_j)\Big)+\lambda^2(1-\alpha)^2\Delta_j^2\\ &-2\lambda(1-\alpha)\Delta_j\Big(\Delta_j\mathcal{H}_3(\lambda^2\alpha^2,\Delta_j^2)-\lambda\alpha\Big(\Phi(\lambda\alpha-\Delta_j)-\Phi(\lambda\alpha+\Delta_j)\Big)\Big)\Big\}\Big].\end{aligned}$$

If $\alpha = 1$,

$$\text{ADL}_2\text{-risk}(\hat{\theta}_n^{\text{nEnetR}}(\lambda, \alpha = 1)) = \eta^2\Bigg[\sum_{j=1}^{p}\kappa_j\left(1 - \mathcal{H}_3(\lambda^2, \Delta_j^2)\right)$$

$$+ \Delta_j^2\Big(2\mathcal{H}_3(\lambda^2, \Delta_j^2) - \mathcal{H}_5(\lambda^2, \Delta^2)\Big) + \lambda^2\left(1 - \mathcal{H}_1(\lambda^2, \Delta_j^2)\right)$$

$$- 2\lambda\Big(\phi(\lambda - \Delta_j) + \phi(\lambda + \Delta_j)\Big)\Bigg].$$

If $\alpha = 0$,

$$\text{ADL}_2\text{-risk}(\hat{\theta}_n^{\text{nEnetR}}(\lambda, \alpha = 0)) = \eta^2\sum_{j=1}^{p}\frac{1}{\left[1 + \lambda\kappa_j\right]^2}$$

$$\times\left\{\kappa_j + \lambda^2 - 4\lambda\phi(\Delta_j) + \lambda^2\Delta_j^2\right\}.$$

Now, using the family of linear projections, we obtain the lower bound of the $\text{ADL}_2\text{-risk}(\hat{\theta}_n^{\text{nEnetR}}(\lambda, \alpha))$ as

$$\frac{\eta^2}{[1 + \lambda(1-\alpha)]^2}\Bigg[\text{tr}(\mathcal{K}_1) + \Delta^2\text{Ch}_{\min}(\mathcal{K}_2)$$

$$- 2\lambda(1-\alpha)\Delta^2\text{Ch}_{\min}(\mathcal{K}_2)\sum_{j=1}^{p}\mathcal{H}_3(\lambda^2\alpha^2, \Delta_j^2)$$

$$+ 2\lambda(1-\alpha)\sum_{j=1}^{p}\left(\Phi(\lambda\alpha - \Delta_j) - \Phi(\lambda\alpha + \Delta_j)\right)^2\Bigg]$$

where $\mathcal{K} = \text{Diag}(\mathcal{K}_1, \mathcal{K}_2)$ and $\Delta^2 = (\delta_2^\top\mathcal{K}_2^{-1}\delta)/\eta^2$. Hence,

$$\text{ADRE}(\hat{\theta}_n^{\text{nEnetR}}(\kappa, \alpha); \hat{\theta}_n^{\text{UR}}) = \begin{cases} (1+\lambda)^2\Bigg[\text{tr}(\mathcal{K}_1) + \Delta^2\text{Ch}_{\min}(\mathcal{K}_2) \\ \qquad + 2\lambda\sum_{j=1}^{p}(2\Phi(\Delta_j) - 1)^2\Bigg]^{-1} & \alpha = 0 \\ \left[\text{tr}(\mathcal{K}_1) + \Delta^2\text{Ch}_{\min}(\mathcal{K}_2)\right]^{-1} & \alpha = 1. \end{cases}$$

4.9 Comparison of the R-estimators

In this section, we compare all the proposed R-estimators discussed with respect to the unrestricted R-estimator of $\theta = (\theta_1, \dots, \theta_p)^\top$.

4.9.1 Comparison of URE and RRE

From Sections 4.4 and 4.6, we have

$$\begin{aligned}\text{ADL}_2\text{-risk}(\hat{\theta}_n^{\text{RR}}) - \text{ADL}_2\text{-risk}(\hat{\theta}_n^{\text{UR}}) &= \delta_2^\top\delta_2 - \eta^2\,\text{tr}(\mathcal{K}_2) \\ &\geq \eta^2\left[\Delta^2\text{Ch}_{\min}(\mathcal{K}_2) - \text{tr}(\mathcal{K}_2)\right] \quad (4.9.1)\end{aligned}$$

Based on Eq. (4.9.1), we find that $\hat{\theta}_n^{\text{UR}}$ outperforms $\hat{\theta}_n^{\text{RR}}$ whenever

$$\Delta^2 \geq \frac{\text{tr}(\mathcal{K}_2)}{\text{Ch}_{\min}(\mathcal{K}_2)}$$

and $\hat{\theta}_n^{\text{RR}}$ outperforms $\hat{\theta}_n^{\text{UR}}$ whenever

$$0 \leq \Delta^2 \leq \frac{\text{tr}(\mathcal{K}_2)}{\text{Ch}_{\min}(\mathcal{K}_2)}.$$

4.9.2 Comparison of URE and Stein–Saleh-type R-estimators

From Section 4.7.2, we find that ADL_2-risk difference of $\hat{\theta}_n^{\text{UR}}$ and $\hat{\theta}_n^{\text{SSR}}$ is given by

$$\begin{aligned}\eta^2\Big\{&(p_2-2)\,\text{tr}(\mathcal{K}_2)\left(\mathbb{E}\left[\chi^{-2}_{p_2+2}(\Delta^2)\right] + \Delta^2\mathbb{E}\left[\chi^{-4}_{p_2+2}(\Delta^2)\right]\right) \\ &-(p_2-2)(p_2+2)\Delta^2\text{Ch}_{\min}(\mathcal{K}_2)\mathbb{E}\left[\chi^{-4}_{p_2+4}(\Delta^2)\right]\Big\} \geq 0\end{aligned}$$

for all $\Delta^2 \in \mathbb{R}^2$. Thus, $\hat{\theta}_n^{\text{SSR}}$ outperforms $\hat{\theta}_n^{\text{UR}}$ uniformly. Furthermore,

$$\text{ADL}_2\text{-risk}(\hat{\theta}_n^{\text{SSR+}}) \leq \text{ADL}_2\text{-risk}(\hat{\theta}_n^{\text{SSR}}) \leq \text{ADL}_2\text{-risk}(\hat{\theta}_n^{\text{UR}}).$$

4.9.3 Comparison of URE and Ridge-type R-estimators

We consider ADL_2-risk difference expressions of $\hat{\theta}_n^{\text{UR}}$ and $\hat{\theta}_n^{\text{ridgeR}}(\lambda_{\text{opt}})$, with

$$\lambda_{\text{opt}} = \text{tr}(\mathcal{K}_2)\Delta^{-2}\left(\text{Ch}_{\min}(\mathcal{K}_2)\right)^{-1},$$

given by

$$\begin{aligned}&\eta^2\left[(\text{tr}(\mathcal{K}_1) + \text{tr}(\mathcal{K}_2)) - \text{tr}(\mathcal{K}_1) - \frac{\text{tr}(\mathcal{K}_2)\Delta^2\text{Ch}_{\min}(\mathcal{K}_2)}{\text{tr}(\mathcal{K}_2) + \Delta^2\text{Ch}_{\min}(\mathcal{K}_2)}\right] \\ &= \eta^2\,\text{tr}(\mathcal{K}_2)\left[1 - \frac{\Delta^2\text{Ch}_{\min}(\mathcal{K}_2)}{\text{tr}(\mathcal{K}_2) + \Delta^2\text{Ch}_{\min}(\mathcal{K}_2)}\right]\end{aligned}$$

$$= \eta^2 \frac{(\text{tr}(\mathcal{K}_2))}{\text{tr}(\mathcal{K}_2) + \Delta^2 \text{Ch}_{\max}(\mathcal{K}_2)} > 0, \qquad \forall \Delta^2 \in \mathbb{R}^+.$$

Hence, $\hat{\theta}_n^{\text{ridgeR}}(\lambda_{\text{opt}})$ uniformly outperforms $\hat{\theta}_n^{\text{UR}}$.

4.9.4 Comparison of URE and PTSSRE

From Sections 4.4 and 4.8.1, we have the ADL$_2$-risk difference of $\hat{\theta}_n^{\text{UR}}$ and $\hat{\theta}_n^{\text{PTSSR}}(\lambda)$ as

$$\begin{aligned}
&\eta^2 \Big\{ \text{tr}(\mathcal{K}_1) + \text{tr}(\mathcal{K}_2) - \text{tr}(\mathcal{K}_1) - \text{tr}(\mathcal{K}_2) \\
&\times (1 - \mathcal{H}_{p_2+2}(\lambda^2, \Delta^2)) - \eta^{-2} \delta_2^\top \delta_2 \left(2\mathcal{H}_{p_2+2}(\lambda^2, \Delta^2) - \mathcal{H}_{p_2+4}(\lambda^2, \Delta^2) \right) \Big\} \\
&= \eta^2 \left\{ \text{tr}(\mathcal{K}_2) \mathcal{H}_{p_2+2}(\lambda^2, \Delta^2) - \eta^{-2} \delta_2^\top \delta \left(2\mathcal{H}_{p_2+2}(\lambda^2, \Delta^2) - \mathcal{H}_{p_2+4}(\lambda^2, \Delta^2) \right) \right\} \\
&\geq \eta^2 \Big\{ \text{tr}(\mathcal{K}_2) \mathcal{H}_{p_2+2}(\lambda^2, \Delta^2) \\
&\qquad - \Delta^2 \text{Ch}_{\min}(\mathcal{K}_2) \left(2\mathcal{H}_{p_2+2}(\lambda^2, \Delta^2) - \mathcal{H}_{p_2+4}(\lambda^2, \Delta^2) \right) \Big\}.
\end{aligned}$$

Clearly, $\hat{\theta}_n^{\text{PTSSR}}(\lambda)$ outperforms $\hat{\theta}_n^{\text{UR}}$ whenever

$$0 \leq \Delta^2 \leq \frac{\text{tr}(\mathcal{K}_2) \mathcal{H}_{p_2+2}(\lambda^2, \Delta^2)}{\text{Ch}_{\min}(\mathcal{K}_2) \left[2\mathcal{H}_{p_2+2}(\lambda^2, \Delta^2) - \mathcal{H}_{p_2+4}(\lambda^2, \Delta^2) \right]} \tag{4.9.2}$$

and $\hat{\theta}_n^{\text{UR}}$ outperforms $\hat{\theta}_n^{\text{PTSSR}}(\lambda)$ outside Eq. (4.9.2).

We know that the lower bound of ADL$_2$-risk($\hat{\theta}_n^{\text{LassoR}}(\lambda)$) is $\eta^2(\text{tr}(\mathcal{K}_1) + \Delta_L^2)$, $\Delta_L^2 = \Delta^2 \text{Ch}_{\min}(\mathcal{K}_2)$. Hence, we compare all R-estimators with LASSO relative to this lower bound.

4.9.5 Comparison of LASSO-type and Ridge-type R-estimators

From Section 4.8.3, we know that

$$\text{ADL}_2^*\text{-risk}(\hat{\theta}_n^{\text{LassoR}}(\lambda)) = \eta^2(\text{tr}(\mathcal{K}_1) + \Delta_L^2),$$

where $\Delta_L^2 = \Delta^2 \text{Ch}_{\min}(\mathcal{K}_2)$. Thus, the lower bound of ADL$_2$-risk difference is given by

$$\eta^2 \left[\text{tr}(\mathcal{K}_1) + \Delta^2 \text{Ch}_{\min}(\mathcal{K}_2) - \left(\text{tr}(\mathcal{K}_1) + \frac{(\text{tr}(\mathcal{K}_2))\Delta^2 \text{Ch}_{\min}(\mathcal{K}_2)}{\text{tr}(\mathcal{K}_2) + \Delta^2 \text{Ch}_{\min}(\mathcal{K}_2)} \right) \right]$$

$$= \eta^2 \frac{\Delta^2 \text{Ch}_{\min}(\mathcal{K}_2)}{\text{tr}(\mathcal{K}_2) + \Delta^2 \text{Ch}_{\min}(\mathcal{K}_2)} \geq 0, \quad \forall \Delta^2 \in \mathbb{R}^+. \tag{4.9.3}$$

Therefore, the ridge-type R-estimator uniformly outperforms the LASSO-type R-estimator when using the optimal ridge parameter, λ_{opt}.

4.9.6 Comparison of URE, RRE and LASSO

To compare URE and LASSO R-estimator, the ADL$_2$-risk difference is given by

$$\eta^2 \left[(\text{tr}(\mathcal{K}_1) + \text{tr}(\mathcal{K}_2) - (\text{tr}(\mathcal{K}_1 + \Delta_L^2) \right] = \eta^2 (\text{tr}(\mathcal{K}_2) - \Delta_L^2) \tag{4.9.4}$$

where $\Delta_L^2 = \Delta^2 \text{Ch}_{\min}(\mathcal{K}_2)$. Hence, if $\Delta_L^2 \in (0, \text{tr}(\mathcal{K}_2))$, LASSO outperforms URE, while URE outperforms LASSO for $\Delta_L^2 \in (\text{tr}(\mathcal{K}_2), \infty)$. Neither LASSO R-estimator nor URE outperforms the other uniformly.

As far as the LASSO and RRE are concerned, they perform equally well since the ADL$_2$-risk of the RRE is the lower bound of the ADL$_2$-risk of LASSO R-estimator.

4.9.7 Comparison of LASSO with PTRE

The ADL$_2$-risk-risk difference in this case is given by

$$\text{ADL}_2\text{-risk}(\hat{\theta}_n^{\text{PTR}}(\lambda)) - \text{ADL}_2\text{-risk}(\hat{\theta}_n^{\text{LassoR}}(\lambda))$$

$$= \eta^2 \left[\text{tr}(\mathcal{K}_2)(1 - \mathcal{H}_{p_2+2}(\lambda^2, \Delta^2)) - \Delta_L^2 \left(2\mathcal{H}_{p_2+2}(\lambda^2, \Delta^2) \right.\right.$$

$$\left.\left. - \mathcal{H}_{p_2+4}(\alpha, \Delta^2) \right) \right] - \Delta^2 \text{Ch}_{\min}(\mathcal{K}_2)$$

$$\geq \eta^2 \, \text{tr}(\mathcal{K}_2)(1 - \mathcal{H}_{p_2+2}(\lambda^2, 0)) \geq 0 \quad \text{if} \quad \Delta^2 = 0.$$

However, if $\Delta^2 \neq 0$, LASSO R-estimator outperforms PTRE for

$$0 \leq \Delta^2 \leq \frac{(\text{tr}(\mathcal{K}_2))(1 - \mathcal{H}_{p_2+2}(\alpha, \Delta^2))}{\text{Ch}_{\min}(\mathcal{K}_2^{-1}) \left[2\mathcal{H}_{p_2+2}(\lambda^2, \Delta^2) - \mathcal{H}_{p_2+4}(\lambda^2, \Delta^2) + 1 \right]}. \tag{4.9.5}$$

Otherwise, PTSSRE outperforms LASSO R-estimator and neither LASSO R-estimator nor PTSSRE outperforms the other uniformly.

4.9.8 Comparison of LASSO with SSRE

Consider the ADL$_2$-risk difference of the Stein–Saleh and LASSO-type R-estimators given by

$$\begin{aligned}\text{ADL}_2\text{-risk}(\hat{\theta}_n^{\text{SSR}}(\lambda)) - \text{ADL}_2\text{-risk}(\hat{\theta}_n^{\text{LassoR}}(\lambda)) &= \eta^2\,[\text{tr}(\mathcal{K}_2) \\ &-(p_2-2)\,\text{tr}(\mathcal{K}_2)\left(\mathbb{E}\left[\chi^{-2}_{p_2+2}(\Delta^2)\right] - \Delta^2\mathbb{E}\left[\chi^{-4}_{p_2+2}(\Delta^2)\right]\right) \\ &+(p_2^2-4)\Delta_L^2\mathbb{E}\left[\chi^{-4}_{p_2+4}(\Delta^2)\right] - \Delta_L^2\Big].\end{aligned}$$

Thus, the LASSO R-estimator outperforms the Stein–Saleh R-estimator whenever

$$0 < \Delta^2 \leq \frac{(\text{tr}(\mathcal{K}_2))\left\{1-(p_2-2)\left(\mathbb{E}\left[\chi^{-2}_{p_2+2}(\Delta^2)\right] + \Delta^2\mathbb{E}\left[\chi^{-4}_{p_2+2}(\Delta^2)\right]\right)\right\}}{\text{Ch}_{\min}(\mathcal{K}_2)\left(1-(p_2^2-4)\mathbb{E}\left[\chi^{-4}_{p_2+4}(\Delta^2)\right]\right)}. \tag{4.9.6}$$

Otherwise, the Stein–Saleh R-estimator outperforms LASSO. Neither LASSO R-estimator nor the Stein–Saleh R-estimator outperforms the other uniformly.

4.9.9 Comparison of LASSO with PRSSRE

To compare LASSO R-estimator vs. positive-rule Stein–Saleh-type R-estimator, consider the lower bound of the ADRE difference

$$\begin{aligned}&\eta^2\Big[\text{tr}(\mathcal{K}_1) + \Delta^2\text{Ch}_{\min}(\mathcal{K}_2) - \Big\{\frac{1}{\eta^2}\text{ADL}_2\text{-risk}(\hat{\theta}_n^{\text{SSR}}) \\ &- \text{tr}(\mathcal{K}_2)\mathbb{E}\left[\left(1-(p_2-2)\chi^{-2}_{p_2+1}(\Delta^2)\right)^2 I(\chi^2_{p_2+2}(\Delta^2) < p_2-2)\right] \\ &+ \Delta^2\text{Ch}_{\min}(\mathcal{K}_2)\Big\{2\mathbb{E}\left[\left(1-(p_2-2)\chi^{-2}_{p_2+2}\right) I(\chi^2_{p_2+2}(\Delta^2) \leq p_2-2)\right] \\ &- \mathbb{E}\left[\left(1-(p_2-2)\chi^{-2}_{p_2+4}(\Delta^2)\right)^2 I(\chi^2_{p_2+4}(\Delta^2) \leq p_2-2)\right]\Big\}.\end{aligned}$$

Since ADL$_2$-risk($\hat{\theta}_n^{\text{SSR+}}$) $\leq$ ADL$_2$-risk($\hat{\theta}_n^{\text{SSR}}$) $\leq$ ADL$_2$-risk($\hat{\theta}_n^{\text{UR}}$), the first term is a non-negative decreasing function of Δ^2 towards 0. As far as the second term is concerned, it is also a decreasing function of Δ^2. For Δ^2 in the range defined by Eq. (4.9.6), the term is non-positive and it is non-negative, otherwise. Thus, LASSO outperforms PRSSRE for Δ^2 in the interval of Eq. (4.9.6) and PRSSRE outperforms LASSO otherwise, and neither LASSO nor PRSSRE uniformly outperforms the other.

4.9.10 Comparison of nEnetRE with URE

In this case, we consider the ADL$_2$-risk difference of the URE, RRE and nEnetRE given by

$$\eta^2\left[(\mathrm{tr}(\mathcal{K}_1)+\mathrm{tr}(\mathcal{K}_2))-\frac{1}{\eta^2}\mathrm{ADL}_2\text{-risk}(\hat{\theta}_n^{\mathrm{nEnetR}}(\lambda,\alpha))\right]\geq 0.$$

Thus, the nEnetRE outperforms URE, uniformly.

4.9.11 Comparison of nEnetRE with RRE

In this case, the ADL$_2$-risk difference is given by

$$\eta^2\left[(\mathrm{tr}(\mathcal{K}_1)+\Delta^2\mathrm{Ch}_{\min}(\mathcal{K}_2))-\frac{1}{\eta^2}\mathrm{ADL}_2\text{-risk}(\hat{\theta}_n^{\mathrm{nEnetR}}(\lambda,\alpha))\right]\geq 0.$$

The nEnetRE uniformly outperforms RRE.

4.9.12 Comparison of nEnetRE with HTRE

In this case, the ADL$_2$-risk difference is given by

$$\begin{aligned}
&\eta^2\Big[(\mathrm{tr}(\mathcal{K}_1)+\mathrm{tr}(\mathcal{K}_2))-\mathrm{tr}(\mathcal{K}_2)\mathcal{H}_{p_2+2}(\lambda^2,\Delta^2)\\
&+\Delta_L^2\left(2\mathcal{H}_{p_2+2}(\lambda^2,\Delta^2)-\mathcal{H}_{p_2+4}(\lambda^2,\Delta^2)\right)-\Big\{\mathrm{tr}(\boldsymbol{I}_{p_1}+\lambda(1-\alpha)\mathcal{K}_1)^{-2}\mathcal{K}_1\\
&+\lambda^2(1-\alpha)^2\boldsymbol{\delta}_1^\top\left[\mathcal{K}_1+\lambda(1-\alpha)\boldsymbol{I}_{p_1}\right]^{-2}\boldsymbol{\delta}_1+\Delta^2\mathrm{Ch}_{\min}(\mathcal{K}_2)\Big\}\Big]\\
=\;&\eta^2\Big[\mathrm{tr}(\mathcal{K}_2)\left(1-\mathcal{H}_{p_2+2}(\lambda^2,\Delta^2)\right)\\
&-\Delta_L^2\left(1-2\mathcal{H}_{p_2+2}(\lambda^2,\Delta^2)+\mathcal{H}_{p_2+4}(\lambda^2,\Delta^2)\right)\Big]\\
&+\eta^2\Big[\mathrm{tr}(\mathcal{K}_1)-\mathrm{tr}(\boldsymbol{I}_{p_1}+\lambda(1-\alpha)\mathcal{K}_1)^{-2}\mathcal{K}_1\\
&+\lambda^2(1-\alpha)^2\boldsymbol{\delta}_1^\top\left[\mathcal{K}_1+\lambda(1-\alpha)\boldsymbol{I}_{p_1}\right]^{-2}\boldsymbol{\delta}_1\Big].
\end{aligned}$$

Then, the HTRE outperforms nEnetRE, whenever

$$0\leq\Delta^2\leq\frac{\mathrm{tr}(\mathcal{K}_2)}{\mathrm{Ch}_{\min}(\mathcal{K}_2)}\frac{f(\lambda,\Delta^2)}{\left[1-2\mathcal{H}_{p_2+2}(\lambda^2,\Delta^2)+\mathcal{H}_{p_2+4}(\lambda^2,\Delta^2)\right]}\tag{4.9.7}$$

where

$$f(\lambda,\Delta^2)=(1-\mathcal{H}_{p_2+2}(\lambda^2,\Delta^2))+\mathrm{tr}(\mathcal{K}_1)-\mathrm{tr}(\boldsymbol{I}_{p_1}+\lambda(1-\alpha)\mathcal{K}_1)^{-2}\mathcal{K}_1$$

$$+\lambda^2(1-\alpha)^2\boldsymbol{\delta}_1^\top\left[\mathcal{K}_1+\lambda(1-\alpha)\boldsymbol{I}_{p_1}\right]^{-2}\boldsymbol{\delta}_1. \tag{4.9.8}$$

Otherwise, the nEnetRE is superior to HTRE. Thus, neither HTRE nor nEnetRE outperforms the other uniformly.

4.9.13 Comparison of nEnetRE with SSRE

The ADL_2-risk difference of the Stein–Saleh-type R-estimator and nEnetRE is given by

$$\begin{aligned}
&\eta^2\Big[\text{tr}(\mathcal{K}_2)-(p_2-2)\,\text{tr}(\mathcal{K}_2)\Big(\mathbb{E}\left[\chi^{-2}_{p_2+2}(\Delta^2)\right]\\
&+\Delta^2\mathbb{E}\left[\chi^{-4}_{p_2+2}(\Delta^2)\right]\Big)+(p_2^2-4)\Delta_L^2\mathbb{E}\left[\chi^{-4}_{p_2+4}(\Delta^2)\right]\Big]\\
&-\eta^2\Big[\text{tr}(\mathcal{K}_1)-\Big\{\text{tr}(\boldsymbol{I}_{p_1}+\lambda(1-\alpha)\mathcal{K}_1)^{-2}\mathcal{K}_1\\
&+\lambda^2(1-\alpha)^2\boldsymbol{\delta}_1^\top\left[\mathcal{K}_1^{-1}+\lambda(1-\alpha)\boldsymbol{I}_{p_1}\right]^{-2}\boldsymbol{\delta}_1+\Delta^2\text{Ch}_{\min}(\mathcal{K}_2)\Big\}\Big]\geq 0.
\end{aligned}$$

That means

$$\begin{aligned}
\text{ADL}_2\text{-risk}(\hat{\theta}_{in}^{\text{SSR}})+\eta^2\Delta_L^2 &\geq \eta^2\Big[\text{tr}(\mathcal{K}_1)-\text{tr}(\boldsymbol{I}_{p_1}+\lambda(1-\alpha)\mathcal{K}_1)^{-2}\mathcal{K}_1\\
&+\lambda^2(1-\alpha)^2\boldsymbol{\delta}_1^\top\left[\mathcal{K}_1^{-1}+\lambda(1-\alpha)\boldsymbol{I}_{p_1}\right]^{-2}\boldsymbol{\delta}_1\Big]\geq 0.
\end{aligned}$$

Hence, the nEnetRE performs better than SSRE uniformly.

4.9.14 Comparison of Ridge-type vs. nEnetRE

In this case, the ADL_2-risk difference is given by

$$\begin{aligned}
&\eta^2\Big[\Big\{\text{tr}(\mathcal{K}_1)+\frac{(\text{tr}(\mathcal{K}_2))\Delta_L^2}{\text{tr}(\mathcal{K}_2)+\Delta_L^2}\Big\}-\Big\{\text{tr}(\boldsymbol{I}_{p_1}+\lambda(1-\alpha)\mathcal{K}_1)^{-2}\mathcal{K}_1\\
&+\lambda^2(1-\alpha)^2\boldsymbol{\delta}_1^\top\left[\mathcal{K}_1^{-1}+\lambda(1-\alpha)\boldsymbol{I}_{p_1}\right]^{-2}\boldsymbol{\delta}_1+\Delta_L^2\Big\}\Big]\\
&=\eta^2\Big[\text{tr}(\mathcal{K}_1)-\Big(\text{tr}\left((\boldsymbol{I}_{p_1}+\lambda(1-\alpha)\mathcal{K}_1)^{-2}\mathcal{K}_1\right)\\
&+\lambda^2(1-\alpha)^2\boldsymbol{\delta}_1^\top[\mathcal{K}_1^{-1}+\lambda(1-\alpha)]^{-2}\boldsymbol{\delta}_1\Big)\\
&+\Delta_L^2\left(1+\frac{\text{tr}(\mathcal{K}_1)}{\text{tr}(\mathcal{K}_2)+\Delta_L^2}\right)\Big]\geq 0.
\end{aligned} \tag{4.9.9}$$

Hence, the ridge-type R-estimator outperforms nEnetRE uniformly.

4.10 Summary

This chapter opened with the ANOVA model and described issues of robustness and partitioning. In particular, rank and LASSO can be effectively used to partition the treatments into subgroups, if the null hypothesis is rejected using a robust ANOVA. The variance of the treatments are instrumental in separating the different parameters into subgroups. If sparsity is suspected, then the parameters can be divided into two subgroups: "zero" and "non-zero". To identify the sparse parameters, one of the subset selection penalties can be applied. After partitioning, the estimation can be carried out using one of several shrinkage estimators.

A comparison of the different shrinkage estimators using ADL_2-risk produced the following results:

(i) Ridge regression as a shrinkage estimator dominates all other estimators uniformly.
(ii) The multi-dimensional preliminary test estimator and the Stein–Saleh estimator have similar performance characteristics to the one-dimensional PTE and Stein–Saleh estimators. That is, $\text{ADL}_2\text{-risk}(\hat{\theta}_n^{\text{SSR+}}) \leq \text{ADL}_2\text{-risk}(\hat{\theta}_n^{\text{SSR}}) \leq \text{ADL}_2\text{-risk}(\hat{\theta}_n^{\text{UR}})$ uniformly.
(iii) None of the other estimators dominate over each other uniformly as seen in the previous chapter.

These results are similar in nature and consistent with the results presented in Chapter 3 where the single location problem was studied. Chapter 3 also addressed the simple linear model in detail. The next step is to solve a number of simple linear problems simultaneously and this is the subject of Chapter 5.

4.11 Problems

4.1. Derive $\text{ADV}(\hat{\theta}_n^{\text{UR}})$ to complete Theorem 4.1. (Hint: see Eq. (4.4.3)).
4.2. Show the steps needed to derive $\mathcal{L}_{n_j} = \kappa_j T_{n_j}^2(0)/A_{n_j}^2$.
4.3. Qualitatively compare and contrast the preliminary test, Stein–Saleh, positive-rule Stein–Saleh and ridge R-estimators. Quantitatively compare them by plotting graphs of the ADL_2-risk as a function of Δ^2. Compare your results with those given in Section 4.9.
4.4. Tabulate the ADRE values in Eq. (4.7.6) against Δ^2 based on $\lambda^2 = 2\ln(\text{tr}(\mathcal{K}_2))$.

4.5. (PROJECT) In this problem, the LASSO R-estimator will be used to partition the ANOVA table in Section 4.3.

(a) Generate the data shown in Table 4.1 using the following R code.

```
set.seed(123)
n <- 6
fert1 <- rnorm(n)*20+150
set.seed(123)
n <- 14
fert2 <- rnorm(n)*30+60
set.seed(123)
n <- 8
fert3 <- rnorm(n)*170+175
set.seed(123)
n <- 11
fert4 <- rnorm(n)*150+30
set.seed(123)
n <- 12
fert5 <- rnorm(n)*200+10
set.seed(123)
n <- 9
fert6 <- rnorm(n)*300+0
```

(b) Enter the R code for the functions needed to compute the estimates as a function of λ.

```
minimize.DnLassoV <- function(theta,lambda) {
  sum1 <- minimize.DnLasso(theta[1],fert1,lambda)
  sum2 <- minimize.DnLasso(theta[2],fert2,lambda)
  sum3 <- minimize.DnLasso(theta[3],fert3,lambda)
  sum4 <- minimize.DnLasso(theta[4],fert4,lambda)
  sum5 <- minimize.DnLasso(theta[5],fert5,lambda)
  sum6 <- minimize.DnLasso(theta[6],fert6,lambda)
  Dnsum <- sum1+sum2+sum3+sum4+sum5+sum6
  return(Dnsum)
}
minimize.DnLasso <- function(theta, y,lambda) {
  n <- length(y)
  abs_diff <- abs(y-theta)
  rank_abs_diff <- rank(abs_diff,
                        ties.method = "first")
  an_plus <- sqrt(3.)*((rank_abs_diff)/(n+1.))
  Dnsum <- sum(abs_diff*an_plus)/n
  penalty <- lambda * abs(theta)
```

```
    return(Dnsum+penalty)
  }
```

(c) Run the optimizer and generate the plots.

```
p <- 6
lambda_grid = seq(0,1.5, length = 30)
penalty_mat_opt = matrix(0,nr = p,
    nc = length(lambda_grid))
for(i in 1:length(lambda_grid)){
  op.result <- optim(par= c(1,2,3,4,5,6),
                     fn = minimize.DnLassoV,
                     method = "BFGS",
                     lambda = lambda_grid[i])
  penalty_mat_opt[,i] <- op.result$par
}
plot(lambda_grid,penalty_mat_opt[1,], type = "l",
    col = 1,  ylim = c(0,220), xlab = "lambda",
    ylab = "thetaLASSOR")
points(lambda_grid,penalty_mat_opt[2,], type = "l",
    col = 2,  xlab = "lambda", ylab = "thetaLASSOR")
points(lambda_grid,penalty_mat_opt[3,], type = "l",
    col = 3,  xlab = "lambda", ylab = "thetaLASSOR")
points(lambda_grid,penalty_mat_opt[4,], type = "l",
    col = 4,  xlab = "lambda", ylab = "thetaLASSOR")
points(lambda_grid,penalty_mat_opt[5,], type = "l",
    col = 5,  xlab = "lambda", ylab = "thetaLASSOR")
points(lambda_grid,penalty_mat_opt[6,], type = "l",
    col = 6,  xlab = "lambda", ylab = "thetaLASSOR")
```

(d) Identify the "non-zero" and "zero" subgroups.

4.6. (PROJECT) Assuming that the previous problem has been completed, it is rather straightforward to run ANOVA in R, which is the purpose of this problem. Run the following R code to obtain the comparison of standard ANOVA using **aov()** and the robust ANOVA using **oneway.rfit()**.

```
yield <- c(fert1,fert2,fert3,fert4,fert5,fert6)
fertilizer <- c(rep("fert1",6),rep("fert2",14),
                rep("fert3",8),rep("fert4",11),
                rep("fert5",12),rep("fert6",9))
anova.mod <- aov(yield ~ fertilizer)
```

```
summary(anova.mod)
boxplot(yield ~ fertilizer)
library(Rfit)
anovafit<- oneway.rfit(yield,fertilizer)
qqnorm(rstudent(anovafit$fit))
oneway.rfit(yield,fertilizer)
```

Comment on the results based on an examination of the boxplot , Q-Q plot and the p-values of the two approaches.

4.7. (PROJECT) In this problem, we will compare the LASSO trace of fertilizer 3 of the ANOVA table using $\mathcal{L}_{n_3}$ and the positive-rule Stein–Saleh estimator. The code below assumes that the code in Problem 4.6 has been executed.

(a) Compute the unrestricted estimates using the following R code.

```
op.result <- optim(par= c(1,2,3,4,5,6),
                   fn = minimize.DnLassoV,
                   method = "BFGS",
                   lambda = 0.0)
thetaUR <- op.result$par
```

(b) Use the following codes to obtain $T_n(\theta)$ and A_n^2.

```
Tn <- function(theta, y ) {
  n <- length(y)
  diff <- y-theta
  abs_diff <- abs(diff)
  sgn_diff <- sign(diff)
  rank_abs_diff <- rank(abs_diff,
           ties.method = "first")
  an_plus <- sqrt(3.)*(rank_abs_diff)/(n+1.)
  Tnsum <- sum(sgn_diff*an_plus)/sqrt(n)
  return(Tnsum)
}

An2 <- function(theta, y ) {
  n <- length(y)
  diff <- y-theta
  abs_diff <- abs(diff)
  sgn_diff <- sign(diff)
  rank_abs_diff <- rank(abs_diff,
       ties.method = "first")
  an_plus <- sqrt(3.)*(rank_abs_diff)/(n+1.)
  An2sum <- sum((an_plus)^2)/(n)
```

```
    return(An2sum)
  }
```

(c) Compute $T_n(0)$, A_n^2 and $\mathcal{L}_n$ and plot the SSR+ trace overlayed on the LASSO trace.

```
n <- length(fert1)+length(fert2)+length(fert3)+
  length(fert4)+length(fert5)+length(fert6)
#Try fert3
nj <- length(fert3)
theta1UR <- thetaUR[3]
kj <- nj/n
Tn0 <- Tn(0.0,y = fert3)
An <- sqrt(An2(0,fert3))
Ln <- sqrt(kj)*Tn0/An
# SR+
pts <- 30
thetaShrink <- rep(0,pts)
lambda1 <- rep(0,pts)
for (lambdax in 0:pts) {
  lambda2 <- lambdax/20
  thetaShrink[lambdax+1] <- theta1UR*
    (1- lambda2/sqrt(Ln) )*( 1> lambda2/sqrt(Ln))
  lambda1[lambdax+1] <- lambda2
}
plot(lambda1,thetaShrink, ylim=c(0,200),ylab="thetaSR+",
    xlab="lambda", col="black",type="l")
#From previous problem
points(lambda_grid,penalty_mat_opt[3,], type = "l",
    col = 3,  xlab = "lambda", ylab = "thetaLASSOR")
```

(d) Plot the remaining traces in the same manner and compare the results.

4.8. The ridge-type R-estimator when there is sparsity is given by

$$\hat{\theta}_n^{\text{ridgeR}}(\lambda) = \begin{pmatrix} \hat{\theta}_{n_1}^{\text{UR}} \\ \hat{\theta}_{n_2}^{\text{UR}} \\ \hat{\theta}_{n_3}^{\text{UR}} \\ \cdots\cdots\cdots\cdots \\ \frac{1}{1+\lambda}\hat{\theta}_{n_4}^{\text{UR}} \\ \frac{1}{1+\lambda}\hat{\theta}_{n_5}^{\text{UR}} \\ \frac{1}{1+\lambda}\hat{\theta}_{n_6}^{\text{UR}} \end{pmatrix} = \left(\hat{\theta}_{1n}^{\text{UR}^\top}, \frac{1}{1+\lambda}\hat{\theta}_{2n}^{\text{UR}^\top}\right)^\top.$$

Sketch the trace of each parameter as a function of λ (assuming the setup of ANOVA Table 4.1) where the first three parameters are "non-zero" and the last three are "zero".

4.9. In Section 4.7.4, we defined the simplest ridge regression R-estimator of θ. Now, we define it in its full generality as

$$\hat{\theta}_n^{\text{ridgeR}}(\lambda) = (\hat{\theta}_{1n}^{\top}, \hat{\theta}_{2n}^{\top}(I_{p_2} + \lambda\mathcal{K}_2)^{-1})^{\top}$$

Verify the following results for the general ridge R-estimator:

$$\text{ADB}(\hat{\theta}_n^{\text{ridgeR}}(\lambda)) = (\mathbf{0}^{\top}, -\lambda\theta_2^{\top}(I_{p_2} + \lambda\mathcal{K}_2)^{-1})^{\top}.$$

$$\text{ADL}_2\text{-risk}(\hat{\theta}_n^{\text{ridgeR}}(\lambda)) = \eta^2\Big[\text{tr}(\mathcal{K}_1) + \text{tr}\{(\mathcal{K}_2 + \lambda I_{p_2})^{-2}\mathcal{K}_2\}$$

$$+\lambda^2\theta_2^{\top}(\mathcal{K}_2 + \lambda I_{p_2})^{-2}\theta_2\Big].$$

5

Seemingly Unrelated Simple Linear Models

5.1 Introduction

In this chapter, we consider a class of models used in econometrics, namely, the seemingly unrelated simple linear models. These models belong to the class of linear hypothesis, useful in the analysis of bio-assay data, shelf-life determination of pharmaceutical products, profitability analysis of factory products in terms of costs and outputs, among other applications. In the previous chapter, we performed regression on multiple location problems simultaneously. Here, the same approach is applied to simple linear models that are seemingly unrelated to each other but, rather than solve them separately, they will be solved together as one simultaneous regression problem. At first, this may appear to be pointless since they could, in principle, be solved separately. However, there may in fact be correlations between the error terms that could be used to produce better estimates afterwards. In other words, after simultaneous regression, we could carry out a second step using the linear regression results and use the residuals to make corrections to the estimates when we suspect that there are indeed correlations.

More concretely, consider the number of fruits or vegetables sold as a function of price. For example, assume that we keep track of prices for Hawaiian bananas, Mexican avocados, Washington apples and Valencia oranges sold in different parts of California. Then our model for each would be as follows:

$$\text{bananas: } \boldsymbol{y}_1 = \mathbf{1}_1\theta_1 + \beta_1\boldsymbol{x}_1 + \boldsymbol{\varepsilon}_1$$

$$\text{avocados: } \boldsymbol{y}_2 = \mathbf{1}_2\theta_2 + \beta_2\boldsymbol{x}_2 + \boldsymbol{\varepsilon}_2$$

$$\text{apples: } \quad \boldsymbol{y}_3 = \mathbf{1}_3\theta_3 + \beta_3\boldsymbol{x}_3 + \boldsymbol{\varepsilon}_3$$

Rank-Based Methods for Shrinkage and Selection: With Application to Machine Learning. First Edition. A. K. Md. Ehsanes Saleh, Mohammad Arashi, Resve A. Saleh, and Mina Norouzirad.

$$\text{oranges:} \quad \boldsymbol{y}_4 = \mathbf{1}_4\theta_4 + \beta_4\boldsymbol{x}_4 + \boldsymbol{\varepsilon}_4$$

where $\boldsymbol{x}_j$ is the price of each item, $\boldsymbol{y}_j$ is the number items sold in data set j, with $j = 1, \ldots, 4$, the term $\mathbf{1}_j$ is a suitable vector of 1s, and $(\theta_j, \beta_j)^\top$ are the parameters of interest.

Since these are all different fruits or vegetables with different prices and different amounts sold, it would seem that they should be solved as four separate linear regressions because there does not appear to be any correlations between them. However, there are many factors that determine the relationship between the different prices: income of buyers, type of store, placement within the store, region of the state, demographics, season, location grown, cost of transport, price of other items, etc. Many of these factors cannot be easily incorporated into the model and, furthermore, simple models are generally preferred over complex ones. Some of these factors may be minor but still they contribute to the overall relationships of price and demand in each case.

While there may not be obvious correlations between the variables, the hidden correlations between them can be found in the error terms and utilized to improve the estimates. In the seemingly unrelated regression methodology, we first solve for the ordinary least-squares (OLS) estimates and then use the residual errors to construct a covariance matrix that is used to produce better estimates with smaller variances.

The first step is to construct one large regression model, given in matrix form as:

$$\begin{pmatrix} \boldsymbol{y}_{n_1} \\ \boldsymbol{y}_{n_2} \\ \boldsymbol{y}_{n_3} \\ \boldsymbol{y}_{n_4} \end{pmatrix} = \begin{pmatrix} \mathbf{1}_{n_1} & 0 & 0 & 0 \\ 0 & \mathbf{1}_{n_2} & 0 & 0 \\ 0 & 0 & \mathbf{1}_{n_3} & 0 \\ 0 & 0 & 0 & \mathbf{1}_{n_4} \end{pmatrix} \begin{pmatrix} \theta_1 \\ \theta_2 \\ \theta_3 \\ \theta_4 \end{pmatrix} + \begin{pmatrix} \boldsymbol{x}_{n_1} & 0 & 0 & 0 \\ 0 & \boldsymbol{x}_{n_2} & 0 & 0 \\ 0 & 0 & \boldsymbol{x}_{n_3} & 0 \\ 0 & 0 & 0 & \boldsymbol{x}_{n_4} \end{pmatrix} \begin{pmatrix} \beta_1 \\ \beta_2 \\ \beta_3 \\ \beta_4 \end{pmatrix} + \begin{pmatrix} \boldsymbol{\varepsilon}_{n_1} \\ \boldsymbol{\varepsilon}_{n_2} \\ \boldsymbol{\varepsilon}_{n_3} \\ \boldsymbol{\varepsilon}_{n_4} \end{pmatrix} \tag{5.1.1}$$

where $\boldsymbol{y}_{n_j}$, $\boldsymbol{x}_{n_j}$ and $\boldsymbol{\varepsilon}_{n_j}$ are all $n_j \times 1$ vectors, $\mathbf{1}_{n_j}$ are vectors of 1s of length n_j, and n_j is the number of observations in data set j. Then the OLS estimation of the whole system produces the uncorrelated estimates for the parameters. Next, the residuals are computed as $\boldsymbol{\varepsilon}_{n_j} = \boldsymbol{y}_{n_j} - (\hat{\theta}_j\mathbf{1}_{n_j} + \hat{\beta}_j\boldsymbol{x}_{n_j})$. Finally, a generalized least-squares method is used to obtain the improved estimates as described in Greene (2012). Our objective in this chapter is to replace the first step with a rank-based method and study the ADL$_2$-risk of various penalty estimators in this context.

5.1.1 Problem Formulation

We consider p simple linear models of the form

$$\boldsymbol{y}_{n_j} = \theta_j \mathbf{1}_{n_j} + \beta_j \boldsymbol{x}_{n_j} + \boldsymbol{\varepsilon}_{n_j}, \ j = 1, \dots, p, \tag{5.1.2}$$

where $\boldsymbol{y}_{n_j} = (y_{j1}, \dots, y_{jn_j})^\top$,

$\mathbf{1}_{n_j} = (1, \dots, 1)^\top$ - an n_j-tuple of 1s,

$\boldsymbol{x}_{n_j} = (x_{j1}, \dots, x_{jn_j})^\top$, and

$\boldsymbol{\varepsilon}_{n_j} = (\varepsilon_{j1}, \dots, \varepsilon_{jn_j})^\top$ with a c.d.f. defined by

$$F(\boldsymbol{\varepsilon}_j) = \prod_{i=1}^{n_j} F_0(y_{ji} - \theta_j - \beta_j x_{ji}) \tag{5.1.3}$$

and

$$F(\boldsymbol{\varepsilon}_1, \dots, \boldsymbol{\varepsilon}_p) = \prod_{j=1}^{p} \prod_{i=1}^{n_j} F_0(y_{ji} - \theta_j - \beta_j x_{ji}). \tag{5.1.4}$$

As defined above, the errors ε_{ji} are mutually i.i.d. random variables, $F_0(\cdot)$ belonging to the class of absolutely, continuous c.d.f. with absolutely continuous p.d.f., $f_0(\cdot)$ having finite Fisher information along with the stated conditions below:

(i) $\int_{-\infty}^{\infty} \left\{ \frac{f_0'(x)}{f_0(x)} \right\}^2 f_0(x) dx < \infty$.

(ii) $\lim_{n_j \to \infty} n_j^{-1} \sum_{i=1}^{n_j} x_{ji} = \bar{x}_j \ (|\bar{x}_j| < \infty)$, and $\lim_{n_j \to \infty} n_j^{-1} \sum_{i=1}^{n_j} (x_{ji} - \bar{x}_j)^2 = \lambda_j Q_j$.

(iii) $\lim_{n \to \infty} \frac{n_j}{n} = \kappa_j$, where $\mathcal{K}_n = \mathrm{Diag}(\frac{n_1}{n}, \dots, \frac{n_p}{n})$ and $\mathcal{K} = \mathrm{Diag}(\kappa_1, \dots, \kappa_p)$, then $\lim_{n \to \infty} \mathcal{K}_n = \mathcal{K}$, where $n = n_1 + \dots + n_p$.

(iv) Score functions of the form $a_n^+(i)$ are generated by a function $\phi(u)$, $u \in (0, 1)$, which is non-decreasing, skew symmetric, i.e. $\phi(u) + \phi(1 - u) = 0$, for all $u \in (0, 1)$ and is square integrable.

Let $\phi^+(u) = \phi(\frac{1+u}{2})$, $u \in (0, 1)$, and

$$a_{n_j}^+(k) = \mathbb{E}\left[\phi^+(U_{n_j k})\right] \quad \text{or} \quad U_{n_j} = \frac{k}{n_j + 1}, \ k = 1, \dots, n_j, \tag{5.1.5}$$

where $0 \leq U_{n_j 1} \leq \dots \leq U_{n_j n_j} \leq 1$ are order statistics of sample size n_j from $U(0, 1)$.

5.2 Signed and Signed Rank Estimators of Parameters

We consider sign rank statistics for the parameters (θ_j, β_j), $j = 1, \dots, p$. For this, we adopt a re-parameterization:

$$(\theta_j, \beta_j) \to (\theta_j^*, \beta_j^*),$$

where $\theta_j^* = \theta_j + \beta_j \bar{x}_j$ and $\beta_j^* = Q_{n_j}^{-\frac{1}{2}} \beta_j$, $Q_{n_j} = \sum^{n_j}_{i=1} (x_{ij} - \bar{x}_j)^2$. Thus, we have

$$\theta_j + \beta_j x_{n_{ji}} = \theta_j^* + \beta_j^* c_{n_{ji}} \quad \text{for} \quad i = 1, \dots, n_j, \tag{5.2.1}$$

where

$$c_{n_{ji}} = Q_{n_j}^{-\frac{1}{2}} (x_{n_{ji}} - \bar{x}_j),$$

with

$$\sum_{i=1}^{n_j} c_{n_{ji}} = 0, \quad \text{and} \quad \sum_{i=1}^{n_j} c_{n_{ji}}^2 = 1.$$

Also, we use $R_{n_{ji}}^+(\theta_j^*, \beta_j^*)$ to denote the rank of $|y_{n_{ji}} - \theta_j^* - \beta_j^* c_{n_{ji}}|$ among

$$|y_{n_{ji}} - \theta_j^* - \beta_j^* c_{n_{ji}}|, \quad 1 \le i \le n_j$$

for $j = 1, \dots, p$. For $(\theta_j^*, \beta_j^*) \in \mathbb{R}^2$, let

$$D_{n_j}(\theta_j^*, \beta_j^*) = \sum_{i=1}^{n_j} |y_{n_j i} - \theta_j^* - \beta_j^* c_{n_j i}| a_{n_j}^+(R_{n_j i}^+(\theta_j^*, \beta_j^*)). \tag{5.2.2}$$

It is known that $D_{n_j}(\theta_j^*, \beta_j^*)$ is a convex, continuous and piece-wise linear function of (θ_j^*, β_j^*) and we define the R-estimate of $(\hat{\theta}_j^{*\mathrm{UR}}, \hat{\beta}_j^{*\mathrm{UR}})$ as

$$(\hat{\theta}_j^{*\mathrm{UR}}, \hat{\beta}_j^{*\mathrm{UR}}) = \underset{(\theta_j^*, \beta_j^*) \in \mathbb{R}^2}{\operatorname{argmin}} \left\{ D_{n_j}(\theta_j^*, \beta_j^*) \right\}. \tag{5.2.3}$$

Clearly, we have at the continuity points of $D_{n_j}(\theta_j^*, \beta_j^*)$,

$$\begin{aligned} \frac{\partial}{\partial \theta_j^*} D_{n_j}(\theta_j^*, \beta_j^*) &= -\sum_{i=1}^{n_j} \operatorname{sgn}(y_{n_{ji}} - \theta_j^* - \beta_j^* c_{n_{ji}}) a_{n_j}^+(R_{n_{ji}}(\theta_j^*, \beta_j^*)) \\ &= -\sqrt{n_j}\ T_{n_j}(\theta_j^*, \beta_j^*) \end{aligned} \tag{5.2.4}$$

and

$$\frac{\partial}{\partial \beta_j^*} D_{n_j}(\theta_j^*, \beta_j^*) = -\sum_{i=1}^{n_j} c_{n_{ji}} \operatorname{sgn}(y_{n_j i} - \theta_j^* - \beta_j^* c_{n_{ji}}) a_{n_j}^+(R_{n_{ji}}(\theta_j^*, \beta_j^*))$$

$$= \quad -L_{n_j}(\theta_j^*, \beta_j^*). \tag{5.2.5}$$

We use the following quantities for our expressions of ADB and ADL$_2$-risk of the R-estimators,

$$A_\phi^2 = \int_0^1 \phi^2(u)\mathrm{d}u, \qquad A_\psi^2 = \int_0^1 \psi^2(u)\mathrm{d}u,$$

$$\psi(u) = -\frac{f_0'(F_0^{-1}(u))}{f_0(F^{-1}(u))},$$

and

$$\gamma(\phi,\psi) = \int_0^1 \psi(u)\phi(u)\mathrm{d}u.$$

Also, let

$$A_{n_j} = \frac{1}{n_j - 1}\sum_{i=1}^{n_j}(a_{n_j}^+(i) - \bar{a}_j^+)^2,$$

$$\bar{a}_j^+ = \frac{1}{n_j}\sum_{i=1}^{n_j} a_j^+(i),$$

$$A_n^2 = \sum_{j=1}^{p}(n_j - 1)A_{n_j}^2.$$

Note that for a given x_j,

$$L_{n_j}(\theta_j^*, \beta_j^*) \searrow \beta_j^* \quad \text{for} \quad \beta_j^* \in \mathbb{R},$$

and

$$T_{n_j}(\theta_j^*, \beta_j^*) \searrow \theta_j^* \quad \text{for a fixed} \quad \beta_j^*.$$

Under the model of Eq. (5.1.2) with $\theta_j^* = \beta_j^* = 0$, both $T_{n_j}(0,0)$ and $L_{n_j}(0,0)$ have symmetric distributions about $(0,0)$. Then, as $n \to \infty$, we have

$$\begin{pmatrix} T_{n_j}(0,0) \\ \\ L_{n_j}(0,0) \end{pmatrix} \xrightarrow{\mathcal{D}} \mathcal{N}_2\left(\begin{pmatrix} 0 \\ \\ 0 \end{pmatrix}, A_\phi^2 \kappa_j^{-1}\begin{pmatrix} 1 & 0 \\ \\ 0 & 1 \end{pmatrix}\right). \tag{5.2.6}$$

Now, from Section 3.2.1, the R-estimation of $(\theta_j^*, \beta_j^*)^\top$ is given by

$$\hat{\beta}_j^{*\mathrm{UR}} = \frac{1}{2}(\hat{\beta}_j^{*(\mathrm{U})} + \hat{\beta}_j^{*(\mathrm{L})}) \tag{5.2.7}$$

where

$$\hat{\beta}_j^{*(U)} = \sup\left\{\beta_j^* : L_{n_j}(0, \beta_j^*) > 0\right\}$$
$$\hat{\beta}_j^{*(L)} = \inf\left\{\beta_j^* : L_{n_j}(0, \beta_j^*) < 0\right\}$$

and

$$\hat{\theta}_j^{*UR} = \frac{1}{2}(\hat{\theta}_j^{*(U)} + \hat{\theta}_j^{*(L)}) \tag{5.2.8}$$

where

$$\hat{\theta}_j^{*(U)} = \sup\left\{\theta_j^* : T_{n_j}(\theta_j^*, \hat{\beta}_j^{*UR}) > 0\right\},$$
$$\hat{\theta}_j^{*(L)} = \inf\left\{\theta_j^* : T_{n_j}(\theta_j^*, \hat{\beta}_j^{*UR}) < 0\right\}.$$

Using asymptotic linearity results of Jurečková (1971), we may obtain for a fixed $(\hat{\theta}_j^{*UR}, \hat{\beta}_j^{*UR})^\top \in \mathbb{R}^2$, as $n \to \infty$ (Hajek et al., 1999, p. 325),

$$\left|T_{n_j}(n^{-\frac{1}{2}}\hat{\theta}_j^{*UR}, \hat{\beta}_j^{*UR}) - T_{n_j}(0,0) + \gamma(\phi, f)\hat{\theta}_j^{*UR}\right| \xrightarrow{P} 0, \tag{5.2.9}$$

and

$$\left|L_{n_j}(\theta_j^{*UR}, n^{-\frac{1}{2}}\beta_j^{*UR}) - L_{n_j}(0,0) + \gamma(\phi, f)\hat{\beta}_j^{*UR}\right| \xrightarrow{P} 0. \tag{5.2.10}$$

Hence, using the asymptotic quadratic approximation result by Jaeckel (1972), we have as $n \to \infty$,

$$D_{n_j}(n^{-\frac{1}{2}}\hat{\theta}_j^{*UR}, n^{-\frac{1}{2}}\hat{\beta}_j^{*UR}) - D_{n_j}(0,0) + \hat{\theta}_j^{*UR}T_{n_j}(0,0)$$
$$+ \hat{\beta}_j^{*UR}L_{n_j}(0,0) + \frac{1}{2}\gamma(\phi,\psi)\left[\hat{\theta}_j^{*UR^2} + \hat{\beta}_j^{*UR^2}\right] \xrightarrow{P} 0.$$

As a result, we have as $n \to \infty$,

$$\begin{pmatrix} \sqrt{n}(\hat{\theta}_j^{*UR} - \theta_j) \\ \sqrt{n}(\hat{\beta}_j^{*UR} - \beta_j) \end{pmatrix} - \gamma^{-1}(\phi,\psi)\begin{pmatrix} T_{n_j}(0,0) \\ L_{n_j}^*(0,0) \end{pmatrix} \xrightarrow{P} 0. \tag{5.2.11}$$

Further, using translation and scale equivalence properties of $(\theta, \beta)^\top$, we may write the estimation (θ_j, β_j) as

$$\hat{\theta}_j^{UR} = \hat{\theta}_j^{*UR} - \hat{\beta}_j^{*UR}\bar{x}_j \text{ and } \hat{\beta}_j^{UR} = Q_{n_j}^{-\frac{1}{2}}\hat{\beta}_j^{*UR}, \tag{5.2.12}$$

where $Q_{n_j} = \sum_{i=1}^{n_j}(x_{ji} - \bar{x}_j)^2$, which leads to (as $n \to \infty$),

$$\begin{pmatrix} \sqrt{n}(\hat{\theta}_j^{\text{UR}} - \theta_j) \\ \sqrt{n}(\hat{\beta}_j^{\text{UR}} - \beta_j) \end{pmatrix} - \gamma^{-1}(\phi, \psi) \begin{pmatrix} 1 & -\frac{\sqrt{n_j}\bar{x}_j}{Q_{n_j}} \\ 0 & 1 \end{pmatrix} \begin{pmatrix} T_{n_j}(0,0) \\ L_{n_j}(0,0) \end{pmatrix} \xrightarrow{P} 0. \tag{5.2.13}$$

Then, we have following Theorem 5.1.

Theorem 5.1 Under the class of local alternatives

$$K_{(n)} : \boldsymbol{\beta}_{(n)} = n^{-\frac{1}{2}}\boldsymbol{\delta} = n^{-\frac{1}{2}}(\boldsymbol{\delta}_1^{\top}, \boldsymbol{\delta}_2^{\top})^{\top}$$

and assumed regularity conditions on the model of Eq. (5.1.2) as $n \to \infty$, we have for $\hat{\boldsymbol{\theta}}^{\text{UR}} = (\hat{\theta}_1^{\text{UR}}, \dots, \hat{\theta}_p^{\text{UR}})^{\top}$ and $\hat{\boldsymbol{\beta}}^{\text{UR}} = (\hat{\beta}_1^{\text{UR}}, \dots, \hat{\beta}_p^{\text{UR}})^{\top}$ the following results:

$$\begin{pmatrix} \sqrt{n}(\hat{\boldsymbol{\theta}}^{\text{UR}} - \boldsymbol{\theta}) \\ \sqrt{n}(\hat{\boldsymbol{\beta}}^{\text{UR}} - \boldsymbol{\beta}) \end{pmatrix} \xrightarrow{\mathcal{D}} \mathcal{N}_{2p}\left(\begin{pmatrix} \mathbf{0} \\ \mathbf{0} \end{pmatrix}, \eta^2 \begin{pmatrix} \mathcal{K}^{-1} + \boldsymbol{T}_0\boldsymbol{D}_{22}\boldsymbol{T}_0 & -\boldsymbol{T}_0\boldsymbol{D}_{22} \\ -\boldsymbol{T}_0\boldsymbol{D}_{22} & \boldsymbol{D}_{22} \end{pmatrix}\right)$$

where

$$\boldsymbol{D}_{22}^{-1} = \text{Diag}(\kappa_j Q_j \,|\, j = 1, \dots, p), \tag{5.2.14}$$

$$\boldsymbol{D}_{11} = \mathcal{K}^{-1} + \boldsymbol{T}_0\boldsymbol{D}_{22}\boldsymbol{T}_0$$

and

$$\boldsymbol{T}_0 = \text{Diag}(\bar{x}_1, \dots, \bar{x}_p).$$

For the test of a sub-hypothesis on $\boldsymbol{\beta}$, we may consider the hypothesis $\boldsymbol{\beta}_2 = \mathbf{0}$, where $\boldsymbol{\beta} = (\boldsymbol{\beta}_1^{\top}, \boldsymbol{\beta}_2^{\top})^{\top}$. We may use the statistics, $\mathcal{L}_n$ defined by

$$\mathcal{L}_n = A_{n(2)}^{-2} L_{n(2)}^{\top}(\mathbf{0})\boldsymbol{D}_{22}^{[2]} L_{n(2)}(\mathbf{0}), \tag{5.2.15}$$

where $L_{n(2)}(\mathbf{0}) = L_{n(2)}(\hat{\boldsymbol{\beta}}_{1n_1}^{\text{UR}}, \mathbf{0})$ is the p_2-vector test statistics,

$$\boldsymbol{D}_{22} = \text{Diag}(\boldsymbol{D}_{22}^{[1]}, \boldsymbol{D}_{22}^{[2]}),$$

$A_n^2 = (A_{n(1)}^2, A_{n(2)}^2)$ with p_1 and p_2 dimensions, respectively. Thus, under $\mathcal{H}_0 : \boldsymbol{\beta}_2 = \mathbf{0}$, $\mathcal{L}_n$ follows $\chi_{p_2}^2$ as $n \to \infty$ and under the local alternatives, $K_{(n)} : \boldsymbol{\beta}_{2(n)} = n^{-\frac{1}{2}}\boldsymbol{\delta}_2$, $\boldsymbol{\delta}_2 = (\delta_{21}, \dots, \delta_{2p_2})^{\top}$, $\mathcal{L}_n$ follows a non-central χ^2-distribution with p_2

d.f. and non-centrality parameter $\Delta^2 = \boldsymbol{\delta}_2^\top \boldsymbol{D}_{22}^{[2]} \boldsymbol{\delta}_2$. Also, we have

$$\lim_{n\to\infty} P(\mathcal{L}_n \le x) = \mathcal{H}_p(x, \Delta^2).$$

These results allow us to define the preliminary test, Stein–Saleh-type and positive-rule Stein–Saleh-type R-estimators as follows. First apply LASSO to partition $\boldsymbol{\beta}$ into two subsets, namely, $(\boldsymbol{\beta}_1^\top, \boldsymbol{\beta}_2^\top)^\top$ using the unrestricted estimator given by the p-vector, $(\hat{\boldsymbol{\beta}}_{1n_1}^{\mathrm{UR}}, \ldots, \hat{\boldsymbol{\beta}}_{pn_p}^{\mathrm{UR}})^\top$ with independent components using the selector,

$$\begin{aligned}
\hat{\boldsymbol{\beta}}_n^{\mathrm{LassoR}} &= \left(\operatorname{sgn}(\hat{\beta}_{jn}^{\mathrm{UR}}) \left(|\hat{\beta}_{jn}^{\mathrm{UR}}| - \frac{\lambda\eta}{\sqrt{\kappa_j Q_j}} \right)^+ \middle| j = 1, \ldots, p \right) \\
&= \left(\hat{\boldsymbol{\beta}}_{1n}^{\mathrm{UR}^\top}, \hat{\boldsymbol{\beta}}_{2n}^{\mathrm{UR}^\top} \right)^\top \quad \text{using diagonally linear projections}
\end{aligned}$$

where $\hat{\boldsymbol{\beta}}_{2n_2}^{\mathrm{UR}}$ is the sparse set suspected to be $\mathbf{0}$. Then, define

(a) Preliminary test R-estimator

$$\hat{\boldsymbol{\beta}}_n^{\mathrm{PTR}}(\lambda) = \left(\hat{\boldsymbol{\beta}}_{1n}^{\mathrm{UR}^\top}, \hat{\boldsymbol{\beta}}_{2n}^{\mathrm{UR}^\top} I(\mathcal{L}_n \ge \lambda^2) \right)^\top.$$

(b) Stein–Saleh-type R-estimator

$$\hat{\boldsymbol{\beta}}_n^{\mathrm{SSR}} = \left(\hat{\boldsymbol{\beta}}_{1n}^{\mathrm{UR}^\top}, \hat{\boldsymbol{\beta}}_{2n}^{\mathrm{UR}^\top} (1 - c\mathcal{L}_n^{-1}) \right)^\top, \quad c = p_2 - 2.$$

(c) Positive-rule Stein–Saleh-type R-estimator

$$\hat{\boldsymbol{\beta}}_n^{\mathrm{SSR+}} = \left(\hat{\boldsymbol{\beta}}_{1n}^{\mathrm{UR}^\top}, \hat{\boldsymbol{\beta}}_{2n}^{\mathrm{UR}^\top} (1 - c\mathcal{L}_n^{-1}) I(\mathcal{L}_n \ge (p_2 - 2)) \right)^\top, \quad c = p_2 - 2.$$

Similarly, we may define various other R-estimators of slope.

5.2.1 General Shrinkage R-estimator of $\boldsymbol{\beta}$

We consider the general shrinkage R-estimator of $\boldsymbol{\beta}$ as

$$\hat{\boldsymbol{\beta}}_n^{\mathrm{shrinkageR}}(c) = (\hat{\boldsymbol{\beta}}_{1n}^{\mathrm{UR}^\top}, c\hat{\boldsymbol{\beta}}_{2n}^{\mathrm{UR}^\top})^\top. \tag{5.2.16}$$

Theorem 5.2 Under the local alternatives $K_{(n)}$, the asymptotic distributional bias, and ADL$_2$-risk of the shrinkage R-estimator, $\hat{\boldsymbol{\beta}}_n^{\mathrm{shrinkageR}}(c)$, are respectively given by

$$\text{ADB}(\hat{\boldsymbol{\beta}}_n^{\text{shrinkageR}}(c)) = \begin{pmatrix} \mathbf{0}_{p_1} \\ -(1-c)\boldsymbol{\delta}_2 \end{pmatrix}$$

and

$$\begin{aligned} \text{ADL}_2\text{-risk}(\hat{\boldsymbol{\beta}}_n^{\text{shrinkageR}}(c)) &= \text{tr}(\text{Var}(\hat{\boldsymbol{\beta}}_1^{\text{UR}})) + c^2\,\text{tr}(\text{Var}(\hat{\boldsymbol{\beta}}_2^{\text{UR}})) + (1-c)^2\boldsymbol{\delta}_2^\top\boldsymbol{\delta}_2 \\ &= \eta^2\left[\text{tr}(\boldsymbol{D}_{22}^{[1]}) + c^2\,\text{tr}(\boldsymbol{D}_{22}^{[2]}) + (1-c)^2\boldsymbol{\delta}_2^\top\boldsymbol{\delta}_2\right], \end{aligned}$$

where $\boldsymbol{D}_{22} = \text{Diag}(\boldsymbol{D}_{22}^{[1]}, \boldsymbol{D}_{22}^{[2]})$.

The lower bound of ADL_2-risk is

$$\text{ADL}_2^*\text{-risk}(\hat{\boldsymbol{\beta}}_n^{\text{shrinkageR}}(c)) = \eta^2\left[\text{tr}(\boldsymbol{D}_{22}^{[1]}) + c^2\,\text{tr}(\boldsymbol{D}_{22}^{[2]}) + (1-c)^2\Delta^2\text{Ch}_{\min}(\boldsymbol{D}_{22}^{[2]})\right].$$

Minimizing the R.H.S. w.r.t. c, we have

$$c_{\text{opt}} = \frac{\Delta^2\text{Ch}_{\min}(\boldsymbol{D}_{22}^{[2]})}{\text{tr}(\boldsymbol{D}_{22}^{[2]}) + \Delta^2\text{Ch}_{\min}(\boldsymbol{D}_{22}^{[2]})}. \tag{5.2.17}$$

Consequently, the lower bound of the ADL_2-risk of $\hat{\boldsymbol{\beta}}_n^{\text{shrinkageR}}(c_{\text{opt}})$ is given by

$$\text{ADL}_2^*\text{-risk}(\hat{\boldsymbol{\beta}}_n^{\text{shrinkageR}}(c_{\text{opt}})) = \eta^2\left[\text{tr}(\boldsymbol{D}_{22}^{[1]}) + \frac{\text{tr}(\boldsymbol{D}_{22}^{[2]})\Delta^2\text{Ch}_{\min}(\boldsymbol{D}_{22}^{[2]})}{\text{tr}(\boldsymbol{D}_{22}^{[2]}) + \Delta^2\text{Ch}_{\min}(\boldsymbol{D}_{22}^{[2]})}\right]. \tag{5.2.18}$$

The ADRE is then given by

$$\begin{aligned} \text{ADRE}(\hat{\boldsymbol{\beta}}_n^{\text{shrinkageR}}(c_{\text{opt}}) : \hat{\boldsymbol{\beta}}_n^{\text{UR}}) &= \left(1 + \frac{\text{tr}(\boldsymbol{D}_{22}^{[2]})}{\text{tr}(\boldsymbol{D}_{22}^{[1]})}\right) \\ &\quad \times \left[1 + \frac{\text{tr}(\boldsymbol{D}_{22}^{[2]})}{\text{tr}(\boldsymbol{D}_{22}^{[1]})}\left(\frac{\Delta^2\text{Ch}_{\min}(\boldsymbol{D}_{22}^{[2]})}{\text{tr}(\boldsymbol{D}_{22}^{[2]}) + \Delta^2\text{Ch}_{\min}(\boldsymbol{D}_{22}^{[2]})}\right)\right]^{-1} \geq 1. \end{aligned} \tag{5.2.19}$$

5.2.2 Ridge-type R-estimator of $\boldsymbol{\beta}$

Now, let $c = 1/(1+\kappa)$ in Section 5.2.1. Thus, the ridge-type R-estimator of $\boldsymbol{\beta}$ is given by

$$\hat{\boldsymbol{\beta}}_n^{\text{ridgeR}}(\kappa) = \left(\hat{\boldsymbol{\beta}}_{1n}^{\text{UR}}, \frac{1}{1+\kappa}\hat{\boldsymbol{\beta}}_{2n}^{\text{UR}}\right)^\top.$$

Theorem 5.3 Under the local alternatives $K_{(n)}$, the asymptotic distributional bias, and ADL$_2$-risk of the ridge R-estimator, $\hat{\boldsymbol{\beta}}_n^{\text{ridgeR}}(\kappa)$, are respectively given by

$$\text{ADB}(\hat{\boldsymbol{\beta}}_n^{\text{ridgeR}}(\kappa)) = \begin{pmatrix} \mathbf{0}_{p_1} \\ -\frac{\kappa}{1+\kappa}\boldsymbol{\delta}_2 \end{pmatrix}$$

and

$$\text{ADL}_2\text{-risk}(\hat{\boldsymbol{\beta}}_n^{\text{ridgeR}}(\kappa)) = \eta^2 \left[\text{tr}(\boldsymbol{D}_{22}^{[1]}) + \frac{1}{(1+\kappa)^2}\left(\text{tr}(\boldsymbol{D}_{22}^{[2]}) + \kappa^2 \frac{\boldsymbol{\delta}_2^{\top}\boldsymbol{\delta}_2}{\eta^2} \right) \right]$$

where $\boldsymbol{D}_{22} = \text{Diag}(\boldsymbol{D}_{22}^{[1]}, \boldsymbol{D}_{22}^{[2]})$.

The lower bound of ADL$_2$-risk is

$$\text{ADL}_2^*\text{-risk}(\hat{\boldsymbol{\beta}}_n^{\text{ridgeR}}(\kappa)) = \eta^2 \left[\text{tr}(\boldsymbol{D}_{22}^{[1]}) + \frac{1}{(1+\kappa)^2}\left(\text{tr}(\boldsymbol{D}_{22}^{[2]}) + \kappa^2 \Delta^2 \text{Ch}_{\min}(\boldsymbol{D}_{22}^{[2]}) \right) \right]. \tag{5.2.20}$$

The optimum value of κ for the lower bound given in Eq. (5.2.20) for the ADL$_2$-risk$(\hat{\boldsymbol{\beta}}_n^{\text{ridgeR}}(\kappa))$ is

$$\kappa^* = \frac{\text{tr}(\boldsymbol{D}_{22}^{[2]})}{\Delta^2 \text{Ch}_{\min}(\boldsymbol{D}_{22}^{[2]})}.$$

Hence,

$$\text{ADL}_2\text{-risk}(\hat{\boldsymbol{\beta}}_n^{\text{ridgeR}}(\kappa^*)) = \eta^2 \left[\text{tr}(\boldsymbol{D}_{22}^{[1]}) + \frac{\text{tr}(\boldsymbol{D}_{22}^{[2]})\Delta^2 \text{Ch}_{\min}(\boldsymbol{D}_{22}^{[2]})}{\text{tr}(\boldsymbol{D}_{22}^{[2]}) + \Delta^2 \text{Ch}_{\min}(\boldsymbol{D}_{22}^{[2]})} \right].$$

As a result, the lower bound ADL$_2$-risk efficiency of $\hat{\boldsymbol{\beta}}_n^{\text{ridgeR}}(\kappa)$ is given by

$$\text{ADRE}(\hat{\boldsymbol{\beta}}_n^{\text{ridgeR}}(\kappa^*) : \hat{\boldsymbol{\beta}}_n^{\text{UR}}) = \left(1 + \frac{\text{tr}(\boldsymbol{D}_{22}^{[2]})}{\text{tr}(\boldsymbol{D}_{22}^{[1]})} \right) \times \left(1 + \frac{\text{tr}(\boldsymbol{D}_{22}^{[2]})\Delta^2 \text{Ch}_{\min}(\boldsymbol{D}_{22}^{[2]})}{\text{tr}(\boldsymbol{D}_{22}^{[1]})(\text{tr}(\boldsymbol{D}_{22}^{[1]}) + \Delta^2 \text{Ch}_{\min}(\boldsymbol{D}_{22}^{[1]}))} \right)^{-1}$$

which is similar to Eq. (5.2.19).

5.2.3 Preliminary Test R-estimator of β

The PTRE of the slope vector is given by

$$\begin{aligned}\hat{\boldsymbol{\beta}}_n^{\text{PTR}}(\lambda) &= \left(\hat{\boldsymbol{\beta}}_{1n}^{\text{UR}^\top}, \hat{\boldsymbol{\beta}}_{2n}^{\text{UR}^\top} I(\mathcal{L}_n \geq \lambda^2)\right)^\top \\ &= \left(\hat{\boldsymbol{\beta}}_{1n}^{\text{UR}^\top}, \hat{\boldsymbol{\beta}}_{2n}^{\text{UR}^\top} - \hat{\boldsymbol{\beta}}_{2n}^{\text{UR}^\top} I(\mathcal{L}_n \leq \lambda^2)\right)^\top. \end{aligned} \tag{5.2.21}$$

Then, using Theorem 5.1, we may establish the following theorem.

Theorem 5.4 Under the local alternatives, $K_{(n)}$, the asymptotic distributional bias, and ADL$_2$-risk of the preliminary test R-estimator, $\hat{\boldsymbol{\beta}}_n^{\text{PTR}}(\lambda)$, are respectively given by

$$\text{ADB}(\hat{\boldsymbol{\beta}}_n^{\text{PTR}}(\lambda)) = \begin{pmatrix} \mathbf{0}_{p_1} \\ -\boldsymbol{\delta}_2 \mathcal{H}_{p_2+2}(\lambda^2, \Delta^2) \end{pmatrix}$$

and

$$\begin{aligned}\text{ADL}_2\text{-risk}(\hat{\boldsymbol{\beta}}_n^{\text{PTR}}(\lambda)) &= \eta^2 \Big[\text{tr}(\boldsymbol{D}_{22}^{[1]}) + \text{tr}(\boldsymbol{D}_{22}^{[2]})\left(1 - \mathcal{H}_{p_2+2}(\lambda^2, \Delta^2)\right) \\ &\quad + \eta^{-1}\boldsymbol{\delta}_2^\top \boldsymbol{\delta}_2 \left(2\mathcal{H}_{p_2+2}(\lambda^2, \Delta^2) - \mathcal{H}_{p_2+4}(\lambda^2, \Delta^2)\right)\Big]\end{aligned}$$

where $\boldsymbol{D}_{22} = \text{Diag}(\boldsymbol{D}_{22}^{[1]}, \boldsymbol{D}_{22}^{[2]})$.

Hence, the lower bound is given by

$$\begin{aligned}\text{ADL}_2^*\text{-risk}(\hat{\boldsymbol{\beta}}_n^{\text{PTR}}(\lambda)) &= \eta^2 \Big[\text{tr}(\boldsymbol{D}_{22}^{[1]}) + \text{tr}(\boldsymbol{D}_{22}^{[2]})(1 - \mathcal{H}_{p_2+2}(\lambda^2, \Delta^2)) \\ &\quad + \Delta^2 \text{Ch}_{\min}(\boldsymbol{D}_{22}^{[2]}) \left(2\mathcal{H}_{p_2+2}(\lambda^2, \Delta^2) - \mathcal{H}_{p_2+4}(\lambda^2, \Delta^2)\right)\Big].\end{aligned}$$

Therefore, the lower bound ADRE of $\hat{\boldsymbol{\beta}}_n^{\text{PTR}}(\lambda)$ is given by

$$\begin{aligned}\text{ADRE}(\hat{\boldsymbol{\beta}}_n^{\text{PTR}}(\lambda) : \hat{\boldsymbol{\beta}}_n^{\text{UR}}) &= \left(1 + \frac{\text{tr}(\boldsymbol{D}_{22}^{[2]})}{\text{tr}(\boldsymbol{D}_{22}^{[1]})}\right)\Bigg\{1 + \frac{\text{tr}(\boldsymbol{D}_{22}^{[2]})}{\text{tr}(\boldsymbol{D}_{22}^{[1]})} \\ &\quad + \frac{\Delta^2 \text{Ch}_{\min}(\boldsymbol{D}_{22}^{[2]})}{\text{tr}(\boldsymbol{D}_{22}^{[1]})} \left(2\mathcal{H}_{p_2+2}(\lambda^2, \Delta^2) - \mathcal{H}_{p_2+4}(\lambda^2, \Delta^2)\right)\Bigg\}^{-1} \geq 1.\end{aligned}$$

5.3 Stein–Saleh-type R-estimator of β

The Stein–Saleh-type R-estimator of $\boldsymbol{\beta} = (\boldsymbol{\beta}_1^\top, \boldsymbol{\beta}_2^\top)^\top$ is given by

$$\hat{\boldsymbol{\beta}}_n^{\text{SSR}} = \left(\hat{\boldsymbol{\beta}}_{1n}^{\text{UR}^\top}, (1-(p_2-2)\mathcal{L}_n^{-1})\hat{\boldsymbol{\beta}}_{2n}^{\text{UR}^\top}\right)^\top.$$

Theorem 5.5 Under the local alternatives, $K_{(n)}$, the asymptotic distributional bias, and the lower-bound of ADL$_2$-risk of the Stein–Saleh-type R-estimator, $\hat{\boldsymbol{\beta}}_n^{\text{SSR}}$, are respectively given by

$$\text{ADB}(\hat{\boldsymbol{\beta}}_n^{\text{SSR}}) = \begin{pmatrix} \mathbf{0}_{p_1} \\ -\boldsymbol{\delta}_2(p_2-2)\mathbb{E}\left[\chi_{p_2+2}^{-2}(\Delta^2)\right] \end{pmatrix}$$

and

$$\begin{aligned} \text{ADL}_2^*\text{-risk}(\hat{\boldsymbol{\beta}}_n^{\text{SSR}}) &= \eta^2\Big[\operatorname{tr}(\boldsymbol{D}_{22}^{[1]}) + \operatorname{tr}(\boldsymbol{D}_{22}^{[2]}) - (p_2-2)\operatorname{tr}(\boldsymbol{D}_{22}^{[(2)]}) \\ &\quad \Big(2\mathbb{E}\left[\chi_{p_2+2}^{-2}(\Delta^2)\right] - (p_2-2)\mathbb{E}\left[\chi_{p_2+2}^{-4}(\Delta^2)\right] \\ &\quad +(p_2^2-4)\Delta^2\text{Ch}_{\min}(\boldsymbol{D}_{22}^{[2]})\mathbb{E}\left[\chi_{p_2+4}^{-4}(\Delta^2)\right]\Big)\Big], \end{aligned}$$

where $\boldsymbol{D}_{22} = \text{Diag}(\boldsymbol{D}_{22}^{[1]}, \boldsymbol{D}_{22}^{[2]})$.

Hence, the ADRE expression is given by

$$\begin{aligned} \text{ADRE}(\hat{\boldsymbol{\beta}}_n^{\text{SSR}} : \hat{\boldsymbol{\beta}}_n^{\text{UR}}) &= \left(1 + \frac{\operatorname{tr}(\boldsymbol{D}_{22}^{[2]})}{\operatorname{tr}(\boldsymbol{D}_{22}^{[1]})}\right)\Bigg[1 + \frac{\operatorname{tr}(\boldsymbol{D}_{22}^{[2]})}{\operatorname{tr}(\boldsymbol{D}_{22}^{[1]})} \\ &\quad -(p_2-2)\frac{\operatorname{tr}(\boldsymbol{D}_{22}^{[2]})}{\operatorname{tr}(\boldsymbol{D}_{22}^{[1]})}\left(2\mathbb{E}\left[\chi_{p_2+2}^{-2}(\Delta^2)\right] - (p_2-2)\mathbb{E}\left[\chi_{p_2+2}^{-4}(\Delta^2)\right]\right) \\ &\quad +(p_2^2-4)\frac{\Delta^2\text{Ch}_{\min}(\mathcal{D}_{22}^{[2]})}{\operatorname{tr}(\boldsymbol{D}_{22}^{[1]})}\mathbb{E}\left[\chi_{p_2+4}^{-4}(\Delta^2)\right]\Bigg]^{-1} \geq 1. \end{aligned}$$

5.3.1 Positive-rule Stein–Saleh R-estimators of β

The positive-rule-Stein–Saleh R-estimator (PRSSE) is defined by

$$\hat{\boldsymbol{\beta}}_n^{\text{SSR+}} = \left(\hat{\boldsymbol{\beta}}_{1n}^{\text{UR}^\top}, (1-(p_2-2)\mathcal{L}_n^{-1})I(\mathcal{L}_n > (p_2-2))\hat{\boldsymbol{\beta}}_{2n}^{\text{UR}^\top}\right)^\top.$$

Consequently, we have the following theorem.

Theorem 5.6 Under the local alternatives $K_{(n)}$, the asymptotic distributional bias, and the lower-bound of ADL$_2$-risk of the Stein–Saleh R-estimator, $\hat{\boldsymbol{\beta}}_n^{\text{SSR}}$, are respectively given by

$$\text{ADB}(\hat{\boldsymbol{\beta}}_n^{\text{SSR+}}) = \begin{pmatrix} \mathbf{0}_{p_1} \\ \text{ADB}(\hat{\boldsymbol{\beta}}_n^{\text{SSR}}) - \boldsymbol{\delta}_2 \mathbb{E}\left[\left(1-(p_2-2)\chi^{-2}_{p_2+2}(\Delta^2)\right) I(\chi^2_{p_2+2}(\Delta^2) \le (p_2-2))\right] \end{pmatrix}$$

and

$$\begin{aligned}
\text{ADL}_2^*\text{-risk}(\hat{\boldsymbol{\beta}}_n^{\text{SSR+}}) &= \text{ADL}_2^*\text{-risk}(\hat{\boldsymbol{\beta}}_n^{\text{SSR}}) \\
&-\eta^2 \operatorname{tr}(\boldsymbol{D}_{22}^{[2]}) \mathbb{E}\left[\left(1-(p_2-2)\chi^2_{p_2+2}(\Delta^2)\right)^2 I(\chi^2_{p_2+2}(\Delta^2) \le (p_2-2))\right] \\
&+\Delta^2 \text{Ch}_{\min}(\boldsymbol{D}_{22}^{[2]}) \Big\{ 2\mathbb{E}\left[(1-(p_2-2)\chi^{-2}_{p_2+2}(\Delta^2)) I(\chi^2_{p_2+2}(\Delta^2) \le (p_2-2))\right] \\
&-\mathbb{E}\left[\left(1-(p_2-2)\chi^{-2}_{p_2+4}(\Delta^2)\right)^2 I(\chi^2_{p_2+4}(\Delta^2) \le (p_2-2))\right] \Big\}. \qquad (5.3.1)
\end{aligned}$$

Hence,

$$\begin{aligned}
\text{ADRE}(\hat{\boldsymbol{\beta}}_n^{\text{SSR+}} : \hat{\boldsymbol{\beta}}_n^{\text{UR}}) &= \left(1 + \frac{\operatorname{tr}(\boldsymbol{D}_{22}^{[2]})}{\operatorname{tr}(\boldsymbol{D}_{22}^{[1]})}\right) \Bigg[\frac{\text{ADL}_2\text{-risk}(\hat{\boldsymbol{\beta}}_n^{\text{SSR}})}{\operatorname{tr}(\boldsymbol{D}_{22}^{[1]})} \\
&- \frac{\operatorname{tr}(\boldsymbol{D}_{22}^{[2]})}{\operatorname{tr}(\boldsymbol{D}_{22}^{[1]})} \mathbb{E}\left[\left(1-(p_2-2)\chi^2_{p_2+2}(\Delta^2)\right)^2 I(\chi^2_{p_2+2}(\Delta^2) \le (p_2-2))\right] \\
&+ \frac{\Delta^2 \text{Ch}_{\min}(\boldsymbol{D}_{22}^{[2]})}{\operatorname{tr}(\boldsymbol{D}_{22}^{[1]})} \Big\{ 2\mathbb{E}\left[(1-(p_2-2)\chi^{-2}_{p_2+2}(\Delta^2)) I(\chi^2_{p_2+2}(\Delta^2) \le (p_2-2))\right] \\
&- \mathbb{E}\left[\left(1-(p_2-2)\chi^{-2}_{p_2+4}(\Delta^2)\right)^2 I(\chi^2_{p_2+4}(\Delta^2) \le (p_2-2))\right] \Big\} \Bigg]^{-1} \ge 1.
\end{aligned}$$

5.4 Saleh-type R-estimator of β

We may define the Saleh-type R-estimator of the slope vector as

$$\hat{\boldsymbol{\beta}}_n^{\text{SR}}(\lambda) = \left(\hat{\beta}_j^{\text{UR}} - \frac{\lambda \eta}{\sqrt{\kappa_j Q_j}} \operatorname{sgn}(\hat{\beta}_j^{\text{UR}}) \,\Big|\, j = 1, \ldots, p \right)^{\text{T}}$$

$$= \left(\frac{\eta}{\sqrt{\kappa_j Q_j}} (\text{sgn}(Z_j)(|Z_j| - \lambda)) \middle| j = 1, \ldots, p \right)^{\top} \quad (5.4.1)$$

where

$$Z_j = \sqrt{n \kappa_j Q_j} \hat{\beta}_j^{\text{UR}} / \eta.$$

Using Eq. (5.4.1), we have the following theorem.

Theorem 5.7 Under the local alternatives, $K_{(n)}$, the asymptotic distributional bias, and the lower bound of ADL$_2$-risk of the Saleh-type R-estimator, $\hat{\beta}_n^{\text{SR}}(\lambda)$, are respectively given by

$$\text{ADB}(\hat{\boldsymbol{\beta}}_n^{\text{SR}}(\lambda)) = \left(-\frac{\eta}{\sqrt{\kappa_j Q_j}} [2\Phi(\Delta_j) - 1] \middle| j = 1, \ldots, p \right)^{\top} \quad (5.4.2)$$

and

$$\text{ADL}_2\text{-risk}(\hat{\boldsymbol{\beta}}_n^{\text{SR}}(\lambda)) = \eta^2 \sum_{j=1}^{p} \frac{1}{\kappa_j Q_j} \left[1 + \lambda^2 - 2\lambda \sqrt{\frac{2}{n}} \exp\left\{ -\frac{\Delta_j^2}{2} \right\} \right].$$

For $\lambda_{\text{opt}} = \sqrt{\frac{2}{\pi}}$, where $\Delta_j^2 = 0$, we have

$$\text{ADL}_2\text{-risk}(\hat{\boldsymbol{\beta}}_n^{\text{SR}}(\lambda_{\text{opt}})) = \eta^2 \sum_{j=1}^{p} \frac{1}{\kappa_j Q_j} \left[1 - \frac{2}{\pi} \left\{ 2\exp\left\{ -\frac{\Delta_j^2}{2} \right\} - 1 \right\} \right].$$

Hence,

$$\text{ADRE}(\hat{\boldsymbol{\beta}}_n^{\text{SR}}(\lambda_{\text{opt}}) : \hat{\boldsymbol{\beta}}_n^{\text{UR}}) = \frac{\sum_{j=1}^{p} \frac{1}{\kappa_j Q_j}}{\sum_{j=1}^{p} \frac{1}{\kappa_j Q_j} \left[1 - \frac{2}{\pi} \left\{ 2\exp\left\{ -\frac{\Delta_\alpha^2}{2} \right\} - 1 \right\} \right]}.$$

If $\Delta_j = 0, \forall j$, then it is equal to $(\pi - 2)/\pi$ and as $\Delta_\alpha \to \infty, \forall \alpha$, then it equals $(\pi + 2)/\pi$, crossing the surface at $(2\ln 2, \ldots, 2\ln 2)^{\top}$.

5.4.1 LASSO-type R-estimator of the β

We obtain the LASSO-type R-estimator of the slope-vector as

$$\hat{\boldsymbol{\beta}}_n^{\text{LassoR}}(\lambda) = \begin{pmatrix} \hat{\beta}_1^{\text{UR}}\left(1 - \frac{\lambda}{\sqrt{\mathcal{L}_n}}\right)^+ \\ \vdots \\ \hat{\beta}_p^{\text{UR}}\left(1 - \frac{\lambda}{\sqrt{\mathcal{L}_p}}\right)^+ \end{pmatrix}$$

where $\mathcal{L}_j^2 = |\sqrt{\kappa_j}Q_j\hat{\beta}_j^{\text{UR}}/\eta|^2$. It can be rewritten based on Z_j as

$$\sqrt{n}\hat{\boldsymbol{\beta}}_n^{\text{LassoR}}(\lambda) = \left(\frac{\eta}{\sqrt{\kappa_j Q_j}}\text{sgn}(Z_j)\left(|Z_j| - \lambda\right)^+ |j = 1, \dots, p\right)^\top \tag{5.4.3}$$

based on Eq. (5.4.1). Hence, we have the following theorem.

Theorem 5.8 Under the local alternatives, $K_{(n)}$, the asymptotic distributional bias, and the lower bound of ADL$_2$-risk of the LASSO-type R-estimator, $\hat{\boldsymbol{\beta}}_n^{\text{LassoR}}(\lambda)$, are respectively given by

$$\begin{aligned} \text{ADB}(\hat{\boldsymbol{\beta}}_n^{\text{LassoR}}(\lambda)) &= \left(\frac{-\eta}{\sqrt{\kappa_j Q_j}}\left[\Delta_j\mathcal{H}_3(\lambda^2 : \Delta_j^2)\right.\right. \\ &\left.\left. -\lambda\left(\Phi(\lambda - \Delta_j) - \Phi(\lambda + \Delta_j)\right)\right] \middle| j = 1, \dots, p\right)^\top \end{aligned}$$

and

$$\text{ADL}_2\text{-risk}(\hat{\boldsymbol{\beta}}_n^{\text{LassoR}}(\lambda)) = \eta^2 \sum_{j=1}^{p} \frac{1}{\kappa_j Q_j}\rho_{\text{ST}}(\lambda, \Delta_j),$$

where

$$\begin{aligned} \rho_{\text{ST}}(\lambda, \Delta_j) &= 1 - \mathcal{H}_3(\lambda^2, \Delta_j^2) + \Delta_j^2(2\mathcal{H}_3(\lambda^2, \Delta_j^2) - \mathcal{H}_5(\lambda^2, \Delta_j^2)) \\ &\quad + \lambda^2(1 - \mathcal{H}_1(\lambda^2, \Delta_j^2)) - 2\lambda(\phi(\lambda - \Delta_j) + \phi(\lambda + \Delta_j)). \end{aligned}$$

The lower bound of ADL$_2$-risk($\hat{\boldsymbol{\beta}}_n^{\text{LassoR}}(\lambda)$) is given by

$$\text{ADL}_2^*\text{-risk}(\hat{\boldsymbol{\beta}}_n^{\text{LassoR}}(\lambda)) = \eta^2\left(\text{tr}(\boldsymbol{D}_{22}^{[1]}) + \Delta^2\text{Ch}_{\min}(\boldsymbol{D}_{22}^{[2]})\right).$$

Thus, with this lower bound, we have the ADRE expression as

$$\text{ADRE}(\hat{\boldsymbol{\beta}}_n^{\text{LassoR}}(\lambda) : \hat{\boldsymbol{\beta}}_n^{\text{UR}}) = \left(1 + \frac{\text{tr}(\boldsymbol{D}_{22}^{[2]})}{\text{tr}(\boldsymbol{D}_{22}^{[1]})}\right)\left(1 + \frac{\Delta^2 \text{Ch}_{\min}(\boldsymbol{D}_{22}^{[2]})}{\text{tr}(\boldsymbol{D}_{22}^{[1]})}\right)^{-1} \geq 1. \tag{5.4.4}$$

5.5 Elastic-net-type R-estimators

Following Chapter 3, Section 3.8.8, we may write the nEnet R-estimator of $\boldsymbol{\beta}$ as

$$\begin{aligned}
\hat{\boldsymbol{\beta}}_n^{\text{nEnetR}}(\lambda, \alpha) &= \left(\frac{1}{[1 + \lambda(1-\alpha)\kappa_j Q_j]} \text{sgn}(\hat{\boldsymbol{\beta}}_{jn_j}^{\text{UR}}) \left(|\hat{\beta}_{jn_j}^{\text{UR}}| - \frac{\lambda\alpha\eta}{\sqrt{\kappa_j Q_j}} \right)^+ \middle| j = 1, \dots, p \right)^\top \\
&= \left(\frac{\eta}{\sqrt{\kappa_j Q_j}} \frac{\text{sgn}(Z_{jn_j})}{[1 + \lambda(1-\alpha)\kappa_j Q_j]} \left(|Z_{jn_j}| - \lambda\alpha \right)^+ \middle| j = 1, \dots, p \right)^\top
\end{aligned}$$

where $Z_{jn_j} \sim \mathcal{N}(\Delta_j, 1)$.

Then, we have the following theorem.

Theorem 5.9 Under the assumed condition and local alternatives $K_{(n)}$, the asymptotic distributional bias, and ADL$_2$-risk of the nEnet R-estimator, $\hat{\boldsymbol{\beta}}_n^{\text{nEnetR}}(\lambda, \alpha)$, are respectively given by

$$\text{ADB}(\hat{\boldsymbol{\beta}}_n^{\text{nEnetR}}(\lambda, \alpha)) = \left(-\frac{\eta}{\sqrt{\kappa_j Q_j}} A(\lambda, \alpha, \Delta_j) \middle| j = 1, \dots, p \right)^\top \tag{5.5.1}$$

where

$$\begin{aligned}
A(\lambda, \alpha, \Delta_j) &= \frac{1}{[1 + \lambda(1-\alpha)\kappa_j Q_j]} \Big\{ \Delta_j \mathcal{H}_3(\lambda^2\alpha^2, \Delta_j^2) \\
&\quad - \lambda\alpha \left(\Phi(\lambda\alpha - \Delta_j) - \Phi(\lambda\alpha + \Delta_j) \right) - \lambda(1-\alpha) \Big\}
\end{aligned}$$

and

$$\begin{aligned}
\text{ADL}_2\text{-risk}(\hat{\boldsymbol{\beta}}_n^{\text{nEnetR}}(\lambda, \alpha)) &= \eta^2 \Bigg[\sum_{j=1}^{p} \frac{1}{\kappa_j Q_j} \frac{1}{[1 + \lambda(1-\alpha)\kappa_j Q_j]^2} \\
&= \times \Big\{ \left(1 - \mathcal{H}_3(\lambda^2\alpha^2, \Delta_j^2) \right) \\
&\quad + \Delta_j \left(2\mathcal{H}_3(\lambda\alpha, \Delta_j^2) - \mathcal{H}_5(\lambda\alpha, \Delta_j^2) \right)
\end{aligned}$$

$$+\lambda^2\alpha\left(1-\mathcal{H}_1(\lambda^2\alpha^2,\Delta_j^2)\right)+\lambda^2(1-\alpha)^2$$

$$-2\lambda\alpha\left(\phi(\lambda\alpha-\Delta_j)+\phi(\lambda\alpha+\Delta_j)\right)$$

$$+2\lambda(1-\alpha)\Delta_j\Big\{\Delta_j\mathcal{H}_3(\lambda\alpha,\Delta_j^2)$$

$$-\lambda\alpha\left(\Phi(\lambda\alpha-\Delta_j)-\Phi(\lambda\alpha+\Delta_j)\right)\Big\}\Big\}\Big]\geq 1.$$

For $\alpha = 0$,

$$\text{ADL}_2\text{-risk}(\hat{\boldsymbol{\beta}}_n^{\text{nEnetR}}(\lambda,0)) = \eta^2\left[\sum_{j=1}^{p}\frac{1+\lambda^2}{\kappa_j Q_j(1+\lambda\kappa_j Q_j)^2}\right]$$

and for $\alpha = 1$,

$$\begin{aligned}\text{ADL}_2\text{-risk}(\hat{\boldsymbol{\beta}}_n^{\text{nEnetR}}(\lambda,1)) = \eta^2\Big[\sum_{j=1}^{p}\frac{1}{\kappa_j Q_j}\Big\{&\left(1-\mathcal{H}_3(\lambda^2,\Delta_j^2)\right)\\ &+\Delta_j\left(2\mathcal{H}_3(\lambda,\Delta_j^2)-\mathcal{H}_5(\lambda,\Delta_j^2)\right)\\ &+\lambda^2\left(1-\mathcal{H}_1(\lambda^2,\Delta_j^2)\right)\\ &-2\lambda\left(\phi(\lambda-\Delta_j)+\phi(\lambda+\Delta_j)\right)\Big\}\Big].\end{aligned}$$

Alternatively, we can use the lower bound of ADL_2-risk of $\hat{\boldsymbol{\beta}}_n^{\text{nEnet}}(\lambda,\alpha)$ given by

$$\text{ADL}_2^*\text{-risk}(\hat{\boldsymbol{\beta}}_n^{\text{nEnetR}}(\lambda,\alpha)) = \eta^2\left(\text{tr}(\boldsymbol{D}_{22}^{[1]})+\Delta^2\text{Ch}_{\min}(\boldsymbol{D}_{22}^{[2]})\right). \tag{5.5.2}$$

Then,

$$\text{ADRE}(\hat{\boldsymbol{\beta}}_n^{\text{nEnetR}}(\lambda,\alpha):\hat{\boldsymbol{\beta}}_n^{\text{UR}}) = \left(1+\frac{\text{tr}(\boldsymbol{D}_{22}^{[2]})}{\text{tr}(\boldsymbol{D}_{22}^{[1]})}\right)\left(1+\frac{\Delta^2\text{Ch}_{\min}(\boldsymbol{D}_{22}^{[2]})}{\text{tr}(\boldsymbol{D}_{22}^{[1]})}\right)^{-1}.$$

5.6 R-estimator of Intercept When Slope Has Sparse Subset

From Section 5.2, we know that

$$\begin{pmatrix}\sqrt{n}(\hat{\theta}_n^{\text{UR}}-\theta)\\ \sqrt{n}(\hat{\boldsymbol{\beta}}_n^{\text{UR}}-\boldsymbol{\beta})\end{pmatrix}\sim\mathcal{N}_{2p}\left(\begin{pmatrix}\mathbf{0}\\ \mathbf{0}\end{pmatrix},\eta^2\begin{pmatrix}\boldsymbol{D}_{11} & -\boldsymbol{T}_0\boldsymbol{D}_{22}\\ -\boldsymbol{T}_0\boldsymbol{D}_{22} & \boldsymbol{D}_{22}\end{pmatrix}\right) \tag{5.6.1}$$

where the term $\boldsymbol{D}_{11} = \mathcal{K}^{-1} + \boldsymbol{T}_0\boldsymbol{D}_{22}\boldsymbol{T}_0$, $\boldsymbol{T}_0 = \text{Diag}(\bar{x}_1, \dots, \bar{x}_p)$, and the term $\boldsymbol{D}_{22}^{-1} = \text{Diag}(\kappa_1 Q_1, \dots, \kappa_p Q_p)$.

5.6.1 General Shrinkage R-estimator of Intercept

We consider the general shrinkage R-estimator of $\boldsymbol{\theta}$ when the R-estimator of $\boldsymbol{\beta}$ is

$$\hat{\boldsymbol{\beta}}_n^{\text{shrinkageR}}(c) = \left(\hat{\boldsymbol{\beta}}_{1n}^{\text{UR}^\top}, c\hat{\boldsymbol{\beta}}_{2n}^{\text{UR}^\top}\right)^\top$$

given by

$$\hat{\boldsymbol{\theta}}_n^{\text{shrinkageR}}(c) = \left(\hat{\boldsymbol{\theta}}_{1n}^{\text{UR}^\top}, \hat{\boldsymbol{\theta}}_{2n}^{\text{shrinkageR}^\top}(c)\right)^\top, \tag{5.6.2}$$

where

$$\hat{\boldsymbol{\theta}}_{2n}^{\text{shrinkageR}}(c) = \hat{\boldsymbol{\theta}}_{2n}^{\text{UR}} + (1-c)\boldsymbol{T}_2\hat{\boldsymbol{\beta}}_{2n}^{\text{UR}}, \quad 0 < c < 1.$$

Theorem 5.10 Under the local alternatives, $K_{(n)}$, the asymptotic distributional bias, and the lower bound of ADL$_2$-risk of the shrinkage R-estimator, $\hat{\boldsymbol{\theta}}_n^{\text{shrinkageR}}(c)$, are respectively given by

$$\text{ADB}(\hat{\boldsymbol{\theta}}_n^{\text{shrinkageR}}(c)) = \begin{pmatrix} \mathbf{0}_{p_1} \\ (1-c)\boldsymbol{T}_2\boldsymbol{\delta}_2 \end{pmatrix} \tag{5.6.3}$$

where $\boldsymbol{T}_0 = (\boldsymbol{T}_1, \boldsymbol{T}_2)$ and

$$\begin{aligned} \text{ADL}_2\text{-risk}(\hat{\boldsymbol{\theta}}_n^{\text{shrinkageR}}(c)) &= \eta^2 \Big[\text{tr}(\mathcal{K}_1^{-1} + \boldsymbol{T}_1\boldsymbol{D}_{22}^{[1]}\boldsymbol{T}_1) + c^2\,\text{tr}(\boldsymbol{T}_2\boldsymbol{D}_{22}^{[2]}\boldsymbol{T}_2) \\ &\quad + (1-c)^2\boldsymbol{\delta}_2^\top\boldsymbol{T}_2\boldsymbol{\delta}_2 + 2c(1-c)\,\text{tr}(\boldsymbol{T}_2\boldsymbol{D}_{22}^{[2]}\boldsymbol{T}_2)\Big]. \end{aligned}$$

The lower bound of ADL$_2$-risk is given by

$$\begin{aligned} \text{ADL}_2^*\text{-risk}(\hat{\boldsymbol{\theta}}_n^{\text{shrinkageR}}(c)) &= \eta^2 \Big[\text{tr}(\mathcal{K}_1^{-1} + \boldsymbol{T}_1\boldsymbol{D}_{22}^{[1]}\boldsymbol{T}_1) + c^2\,\text{tr}(\boldsymbol{T}_2\boldsymbol{D}_{22}^{[2]}\boldsymbol{T}_2) \\ &\quad + (1-c)^2\Delta^2\text{Ch}_{\min}(\boldsymbol{T}_2^\top\boldsymbol{D}_{22}^{[2]}\boldsymbol{T}_2) \\ &\quad + 2c(1-c)\,\text{tr}(\boldsymbol{T}_2\boldsymbol{D}_{22}^{[2]}\boldsymbol{T}_2)\Big]. \end{aligned}$$

Hence,

$$\text{ADRE}(\hat{\boldsymbol{\theta}}_n^{\text{shrinkageR}}(c) : \hat{\boldsymbol{\theta}}_n^{\text{UR}}) = \left(1 + \frac{\text{tr}(\mathcal{K}_2^{-1} + \boldsymbol{T}_2\boldsymbol{D}_{22}^{[2]}\boldsymbol{T}_2)}{\text{tr}(\mathcal{K}_1^{-1} + \boldsymbol{T}_2\boldsymbol{D}_{22}^{[2]}\boldsymbol{T}_2)}\right)$$

$$\times \left\{ 1 + \frac{\text{tr}(\boldsymbol{T}_2\boldsymbol{D}_{22}^{[2]}\boldsymbol{T}_2)\Delta^2\text{Ch}_{\min}(\boldsymbol{T}_2\boldsymbol{D}_{22}^{[2]}\boldsymbol{T}_2)}{\text{tr}(\mathcal{K}_1^{-1} + \boldsymbol{T}_1\boldsymbol{D}_{22}^{[1]}\boldsymbol{T}_1)\left[\text{tr}(\boldsymbol{T}_2\boldsymbol{D}_{22}^{[2]}\boldsymbol{T}_2) + \Delta^2\text{Ch}_{\min}(\boldsymbol{T}_2\boldsymbol{D}_{22}^{[2]}\boldsymbol{T}_2)\right]} \right\}^{-1} \geq 1.$$

5.6.2 Ridge-type R-estimator of θ

From Section 5.6.1, we use $c = 1/(1+\kappa)$. Thus, the ridge-type R-estimator of θ is given by

$$\hat{\theta}_n^{\text{ridgeR}}(\kappa) = \left(\hat{\theta}_{1n}^{\text{UR}^\top}, \hat{\theta}_{2n}^{\text{ridgeR}^\top}(\kappa) \right)^\top, \tag{5.6.4}$$

where

$$\hat{\theta}_{2n}^{\text{ridgeR}}(\kappa) = \hat{\theta}_{2n}^{\text{UR}} - \frac{\kappa}{1+\kappa}\boldsymbol{T}_2\hat{\boldsymbol{\beta}}_{2n}^{\text{UR}}.$$

In the case of

$$\kappa_{\text{opt}} = \frac{\Delta^2\text{Ch}_{\min}(\boldsymbol{T}_2\boldsymbol{D}_{22}\boldsymbol{T}_2)}{\text{tr}(\boldsymbol{T}_2\boldsymbol{D}_{22}\boldsymbol{T}_2)},$$

all expressions of ADB, ADL$_2$-risk and ADRE are similar to those in Section 5.6.1.

5.6.3 Preliminary Test R-estimators of θ

In this case, we define the PTRE of θ as

$$\hat{\theta}_n^{\text{PTR}}(\lambda) = \begin{pmatrix} \hat{\theta}_{1n}^{\text{UR}} - \boldsymbol{T}_1\hat{\boldsymbol{\beta}}_{1n}^{\text{UR}} \\ \hat{\theta}_{2n}^{\text{UR}} + \boldsymbol{T}_2\hat{\boldsymbol{\beta}}_{2n}^{\text{UR}} I(\mathcal{L}_n \leq \lambda^2) \end{pmatrix}. \tag{5.6.5}$$

Theorem 5.11 Under the local alternatives, $K_{(n)}$, the asymptotic distributional bias, and ADL$_2$-risk of the preliminary test R-estimator, $\hat{\theta}_n^{\text{PTR}}(\lambda)$, are respectively given by

$$\text{ADB}(\hat{\theta}_n^{\text{PTR}}(\lambda)) = \begin{pmatrix} \boldsymbol{0}_{p_1} \\ \boldsymbol{T}_2\boldsymbol{\delta}_2\mathcal{H}_{p_2+2}(\lambda^2, \Delta^2) \end{pmatrix} \tag{5.6.6}$$

and

$$\text{ADL}_2\text{-risk}(\hat{\theta}_n^{\text{PTR}}(\lambda)) = \eta^2\left[\text{tr}(\mathcal{K}_1^{-1} + \boldsymbol{T}_1\boldsymbol{D}_{22}^{[1]}\boldsymbol{T}_1) + \text{tr}(\mathcal{K}_2^{-1} + \boldsymbol{T}_2\boldsymbol{D}_{22}^{[2]}\boldsymbol{T}_2)\right.$$

$$\times\mathcal{H}_{p_2+2}(\lambda^2,\Delta^2)+\boldsymbol{\delta}_2^\top\boldsymbol{T}_2^2\boldsymbol{\delta}_2\left(2\mathcal{H}_{p_2+2}(\lambda^2,\Delta^2)-\mathcal{H}_{p_2+4}(\lambda^2,\Delta^2)\right)\Big].$$

The lower bound of the asymptotic risk of $\hat{\boldsymbol{\theta}}_n^{\mathrm{PTR}}(\lambda)$ is

$$\begin{aligned}\mathrm{ADL}_2^*\text{-risk}(\hat{\boldsymbol{\theta}}_n^{\mathrm{PTR}}(\lambda)) &= \eta^2\Big[\mathrm{tr}(\mathcal{K}_1^{-1}+\boldsymbol{T}_1\boldsymbol{D}_{22}^{[1]}\boldsymbol{T}_1)\\ &\quad+\mathrm{tr}(\mathcal{K}_2^{-1}+\boldsymbol{T}_2\boldsymbol{D}_{22}^{[2]}\boldsymbol{T}_2)\mathcal{H}_{p_2+2}(\lambda^2,\Delta^2)\\ &\quad+\Delta^2\mathrm{Ch}_{\min}(\boldsymbol{T}_2\boldsymbol{D}_{22}^{[2]}\boldsymbol{T}_2)\left(2\mathcal{H}_{p_2+2}(\lambda^2,\Delta^2)-\mathcal{H}_{p_2+4}(\lambda^2,\Delta^2)\right)\Big].\end{aligned}$$

Hence,

$$\begin{aligned}\mathrm{ADRE}(\hat{\boldsymbol{\theta}}_n^{\mathrm{PTR}}(\lambda):\hat{\boldsymbol{\theta}}_n^{\mathrm{UR}}) &= \left(1+\frac{\mathrm{tr}(\mathcal{K}_2^{-1}+\boldsymbol{T}_2\boldsymbol{D}_{22}^{[2]}\boldsymbol{T}_2)}{\mathrm{tr}(\mathcal{K}_1^{-1}+\boldsymbol{T}_1\boldsymbol{D}_{22}^{[1]}\boldsymbol{T}_1)}\right)\\ &\quad\times\left\{1+\frac{\mathrm{tr}(\mathcal{K}_2^{-1}+\boldsymbol{T}_2\boldsymbol{D}_{22}^{[2]}\boldsymbol{T}_2)}{\mathrm{tr}(\mathcal{K}_1^{-1}+\boldsymbol{T}_1\boldsymbol{D}_{22}^{[1]}\boldsymbol{T}_1)}(1-\mathcal{H}_{p_2+2}(\lambda^2,\Delta^2))\right.\\ &\quad\left.-\frac{\Delta^2\mathrm{Ch}_{\min}(\boldsymbol{T}_2\boldsymbol{D}_{22}^{[2]}\boldsymbol{T}_2)}{\mathrm{tr}(\mathcal{K}_1^{-1}+\boldsymbol{T}_1\boldsymbol{D}_{22}^{[1]}\boldsymbol{T}_1)}\left(2\mathcal{H}_{p_2+2}(\lambda^2,\Delta^2)-\mathcal{H}_{p_2+4}(\lambda^2,\Delta^2)\right)\right\}^{-1}\geq 1.\end{aligned}$$

5.7 Stein–Saleh-type R-estimator of θ

In this case, we define the Stein–Saleh-type R-estimator of the intercept vector as

$$\hat{\boldsymbol{\theta}}_n^{\mathrm{SSR}}=\begin{pmatrix}\hat{\boldsymbol{\theta}}_{1n}^{*\mathrm{UR}}-\boldsymbol{T}_1\hat{\boldsymbol{\beta}}_{1n}^{\mathrm{UR}}\\ \hat{\boldsymbol{\theta}}_{2n}^{*\mathrm{UR}}-\boldsymbol{T}_2\hat{\boldsymbol{\beta}}_{2n}^{\mathrm{UR}}(1-(p_2-2)\mathcal{L}_n^{-1})\end{pmatrix}. \tag{5.7.1}$$

Theorem 5.12 Under the local alternatives, $K_{(n)}$, the asymptotic distributional bias, and the lower bound of ADL_2-risk of the Stein–Saleh-type R-estimator, $\hat{\boldsymbol{\theta}}_n^{\mathrm{SSR}}$, are respectively given by

$$\mathrm{ADB}(\hat{\boldsymbol{\theta}}_n^{\mathrm{SSR}})=\begin{pmatrix}\mathbf{0}_{p_1}\\ -(p_2-2)\boldsymbol{T}_2\boldsymbol{\delta}_2\mathbb{E}\left[\chi_{p_2+2}^{-2}(\Delta^2)\right]\end{pmatrix}$$

and

$$\begin{aligned}
\text{ADL}_2^*\text{-risk}(\hat{\theta}_n^{\text{SSR}}) &= \eta^2 \Big[\text{tr}(\mathcal{K}_1^{-1} + \boldsymbol{T}_1 \boldsymbol{D}_{22}^{[1]} \boldsymbol{T}_1) \\
&-(p_2 - 2)\,\text{tr}(\boldsymbol{T}_2 \boldsymbol{D}_{22}^{[(2)]} \boldsymbol{T}_2) \left(2\mathbb{E}\left[\chi_{p_2+2}^{-2}(\Delta^2)\right] - (p_2 - 2)\mathbb{E}\left[\chi_{p_2+2}^{-4}(\Delta^2)\right] \right. \\
&\left. -(p_2^2 - 4)\Delta^2 \text{Ch}_{\min}(\boldsymbol{T}_2 \boldsymbol{D}_{22}^{[2]} \boldsymbol{T}_2) \mathbb{E}\left[\chi_{p_2+4}^{-4}(\Delta^2)\right] \right) \Big].
\end{aligned}$$

Hence, the ADRE expression is given by

$$\begin{aligned}
\text{ADRE}(\hat{\theta}_n^{\text{SSR}} : \hat{\theta}_n^{\text{UR}}) &= \left(1 + \frac{\text{tr}(\mathcal{K}_2^{-1} + \boldsymbol{T}_2 \boldsymbol{D}_{22}^{[2]} \boldsymbol{T}_2)}{\text{tr}(\mathcal{K}_1^{-1} + \boldsymbol{T}_1 \boldsymbol{D}_{22}^{[1]} \boldsymbol{T}_1)} \right) \\
&\times \left[1 + \frac{\text{tr}(\boldsymbol{T}_2 \boldsymbol{D}_{22}^{[2]} \boldsymbol{T}_2)}{\text{tr}(\mathcal{K}_1^{-1} + \boldsymbol{T}_1 \boldsymbol{D}_{22}^{[1]} \boldsymbol{T}_1)} \right. \\
&\times (p_2 - 2) \left(2\mathbb{E}\left[\chi_{p_2+2}^{-2}(\Delta^2)\right] - (p_2 - 2)\mathbb{E}\left[\chi_{p_2+2}^{-4}(\Delta^2)\right] \right) \\
&\left. - \frac{(p_2^2 - 4)\Delta^2 \text{Ch}_{\min}(\boldsymbol{T}_2 \boldsymbol{D}_{22}^{[2]} \boldsymbol{T}_2)}{\text{tr}(\mathcal{K}_1^{-1} + \boldsymbol{T}_1 \boldsymbol{D}_{22}^{[1]} \boldsymbol{T}_1)} \mathbb{E}\left[\chi_{p_2+4}^{-4}(\Delta^2)\right] \right]^{-1} \geq 1. \qquad (5.7.2)
\end{aligned}$$

5.7.1 Positive-rule Stein–Saleh-type R-estimator of θ

We define the PRSSRE as

$$\hat{\theta}_n^{\text{SSR+}} = \begin{pmatrix} \hat{\theta}_{1n}^{*\text{UR}} - \boldsymbol{T}_1 \hat{\boldsymbol{\beta}}_{1n}^{\text{UR}} \\ \hat{\theta}_{2n}^{*\text{UR}} - \boldsymbol{T}_2 \hat{\boldsymbol{\beta}}_{2n}^{\text{UR}} (1 - (p_2 - 2)\mathcal{L}_n^{-1}) I(\mathcal{L}_n > (p_2 - 2)) \end{pmatrix}.$$

Theorem 5.13 Under the local alternatives, $K_{(n)}$, the asymptotic distributional bias, and the lower bound of ADL$_2$-risk of the positive-rule Stein–Saleh-type R-estimator, $\hat{\theta}_n^{\text{SSR+}}$, are respectively given by

$$\text{ADB}(\hat{\theta}_n^{\text{SSR+}}) = \begin{pmatrix} \mathbf{0}_{p_1} \\ \text{ADB}(\hat{\theta}_{2n}^{\text{SSR+}}) \end{pmatrix}$$

where

$$\begin{aligned}\text{ADB}(\hat{\theta}_{2n}^{\text{SSR+}}) &= \boldsymbol{T}_2\boldsymbol{\delta}_2\Big\{(p_2-2)\mathbb{E}\left[\chi_{p_2+2}^{-2}(\Delta^2)\right] \\ &\quad +\mathbb{E}\left[\left(1-(p_2-2)\chi_{p_2+2}^{-2}(\Delta^2)\right)I(\chi_{p_2+2}^{2}(\Delta^2)\leq(p_2-2))\right]\Big\}\end{aligned}$$

and

$$\begin{aligned}\text{ADL}_2^*\text{-risk}(\hat{\theta}_n^{\text{SSR+}}) &= \text{ADL}_2^*\text{-risk}(\hat{\theta}_n^{\text{SSR}}) - \eta^2(p_2-2)\operatorname{tr}(\boldsymbol{T}_2\boldsymbol{D}_{22}^{[2]}\boldsymbol{T}_2)B(\Delta^2) \\ &\quad +\Delta^2\left(C(\Delta^2)-B^*(\Delta^2)\right)\end{aligned}$$

where

$$\begin{aligned}B(\Delta^2) &= \mathbb{E}\left[\left(1-(p_2-2)\chi_{p_2+2}^{2}(\Delta^2)\right)^2 I(\chi_{p_2+2}^{2}(\Delta^2)\leq(p_2-2))\right], \\ C(\Delta^2) &= 2\mathbb{E}\left[(1-(p_2-2)\chi_{p_2+2}^{-2}(\Delta^2))I(\chi_{p_2+2}^{2}(\Delta^2)\leq(p_2-2))\right], \\ B^*(\Delta^2) &= \mathbb{E}\left[\left(1-(p_2-2)\chi_{p_2+4}^{-2}(\Delta^2)\right)^2 I(\chi_{p_2+4}^{2}(\Delta^2)\leq(p_2-2))\right].\end{aligned}$$

Hence, $\text{ADRE}(\hat{\theta}_n^{\text{SSR+}}:\hat{\theta}_n^{\text{UR}})$ is given by

$$\begin{aligned}\text{ADRE}(\hat{\theta}_n^{\text{SSR+}}:\hat{\theta}_n^{\text{UR}}) &= \left(1+\frac{\operatorname{tr}(\mathcal{K}_2^{-1}+\boldsymbol{T}_2\boldsymbol{D}_{22}^{[2]}\boldsymbol{T}_2)}{\operatorname{tr}(\mathcal{K}_1^{-1}+\boldsymbol{T}_1\boldsymbol{D}_{22}^{[1]}\boldsymbol{T}_1)}\right) \\ &\quad\times\left[1+\frac{\operatorname{tr}(\boldsymbol{T}_2\boldsymbol{D}_{22}^{[2]}\boldsymbol{T}_2)}{\operatorname{tr}(\mathcal{K}_1^{-1}+\boldsymbol{T}_1\boldsymbol{D}_{22}^{[1]}\boldsymbol{T}_1)}\right. \\ &\quad\times\Big\{(p_2-2)\left(2\mathbb{E}\left[\chi_{p_2+2}^{-2}(\Delta^2)\right]-(p_2-2)\mathbb{E}\left[\chi_{p_2+2}^{-4}\right]\right) \\ &\quad -(p_2^2-4)\frac{\Delta^2\text{Ch}_{\min}(\boldsymbol{T}_2\boldsymbol{D}_{22}^{[2]}\boldsymbol{T}_2)}{\operatorname{tr}(\mathcal{K}_1^{-1}+\boldsymbol{T}_1\boldsymbol{D}_{22}^{[1]}\boldsymbol{T}_1)}\mathbb{E}\left[\chi_{p_2+4}^{-4}(\Delta^2)\right]\Big\} \\ &\quad -(p_2-2)\frac{\operatorname{tr}(\boldsymbol{T}_2\boldsymbol{D}_{22}^{[2]}\boldsymbol{T}_2)}{\operatorname{tr}(\mathcal{K}_1^{-1}+\boldsymbol{T}_1\boldsymbol{D}_{22}^{[1]}\boldsymbol{T}_1)}B(\Delta^2) \\ &\quad \left.+\frac{\Delta^2\left(C(\Delta^2)-B^*(\Delta^2)\right)}{\operatorname{tr}(\mathcal{K}_1^{-1}+\boldsymbol{T}_1\boldsymbol{D}_{22}^{[1]}\boldsymbol{T}_1)}\right]^{-1}\geq 1.\end{aligned}$$

5.7.2 LASSO-type R-estimator of θ

In this case, we define the LASSO-type R-estimator of the intercept vector as

$$\begin{aligned}\hat{\theta}_n^{\text{LassoR}}(\lambda) &= \hat{\theta}_n^{*\text{UR}} - T_0\hat{\beta}_n^{\text{LassoR}}(\lambda) \\ &= \left(\hat{\theta}_{1n}^{*\text{UR}^\top} - T_1\hat{\beta}_{1n}^{\text{UR}}, \hat{\theta}_{2n}^{*\text{UR}^\top}\right)^\top. \end{aligned} \tag{5.7.3}$$

Hence,

Theorem 5.14 Under the local alternatives, $K_{(n)}$, the asymptotic distributional bias, and ADL$_2$-risk of the LASSO R-estimator, $\hat{\theta}_n^{\text{LassoR}}(\lambda)$, are respectively given by

$$\text{ADB}(\hat{\theta}_n^{\text{LassoR}}(\lambda)) = \begin{pmatrix} \mathbf{0}_{p_1} \\ T_2\delta_2 \end{pmatrix} \tag{5.7.4}$$

and

$$\text{ADL}_2\text{-risk}(\hat{\theta}_n^{\text{LassoR}}(\lambda)) = \eta^2\left[\text{tr}(\mathcal{K}_1^{-1} + T_1 D_{22}^{[1]} T_1) + \frac{1}{\eta^2}\delta_2^\top T_2 \delta_2\right].$$

The lower bound of the ADL$_2$-risk of LASSO R-estimator is

$$\text{ADL}_2^*\text{-risk}(\hat{\beta}_n^{\text{LassoR}}(\kappa)) = \eta^2\left[\text{tr}(\mathcal{K}_1^{-1} + T_1 D_{22}^{[1]} T_1) + \Delta^2 \text{Ch}_{\min}(T_2 D_{22}^{[2]} T_2)\right].$$

Consequently,

$$\begin{aligned}\text{ADRE}(\hat{\theta}_n^{\text{LassoR}}(\kappa) : \hat{\theta}_n^{\text{UR}}) &= \left(1 + \frac{\text{tr}(\mathcal{K}_2^{-1} + T_2 D_{22}^{[2]} T_2)}{\text{tr}(\mathcal{K}_1^{-1} + T_1 D_{22}^{[1]} T_1)}\right) \\ &\quad \times \left(1 + \frac{\Delta^2 \text{Ch}_{\min}(T_2 D_{22}^{[2]} T_2)}{\text{tr}(\mathcal{K}_1^{-1} + T_1 D_{22}^{[2]} T_1)}\right)^{-1} \geq 1. \end{aligned}$$

5.8 Summary

In this chapter, we have discussed selection and estimation of the intercept and slope parameters of several simple linear models solved simultaneously. The properties of their R-estimators are discussed based on the ADL$_2$-risk of the R-estimators with ADRE expressions. It may be observed that Chapter 3 is crucial to the understanding of the results of this chapter.

5.8.1 Problems

5.1. Prove Eq. (5.2.6).
5.2. Establish Eq. (5.2.13).
5.3. Prove Theorem 5.1.
5.4. Verify Theorem 5.2.
5.5. Verify Theorem 5.3.
5.6. Verify Theorem 5.4.
5.7. Verify Theorem 5.5.
5.8. Verify Theorem 5.8.
5.9. The elastic net R-estimator of intercept, $\boldsymbol{\theta} = (\theta_1, \dots, \theta_p)^\top$ has the form

$$\begin{aligned}
\hat{\boldsymbol{\theta}}_n^{\text{nEnetR}}(\lambda, \alpha) &= \hat{\boldsymbol{\theta}}_n^{*\text{UR}} - \boldsymbol{T}_0 \hat{\boldsymbol{\beta}}_n^{\text{nEnetR}}(\lambda, \alpha) \\
&= \left(\hat{\boldsymbol{\theta}}_{1n}^{*\text{UR}} - \boldsymbol{T}_1 \hat{\boldsymbol{\beta}}_{1n}^{\text{UR}} [\boldsymbol{I}_{p_1} + \lambda(1-\alpha) \left(\boldsymbol{D}_{22}^{[1]} \right)^{-1}]^{-1}, \hat{\boldsymbol{\theta}}_{2n}^{*} \right)^\top \\
&= \left(\hat{\boldsymbol{\theta}}_{1n}^{\text{UR}} + \lambda(1-\alpha) \boldsymbol{T}_1 \hat{\boldsymbol{\beta}}_{1n}^{\text{UR}} [\boldsymbol{D}_{22}^{[1]} + \lambda(1-\alpha) \boldsymbol{I}_{p_1}]^{-1}, \hat{\boldsymbol{\theta}}_{2n}^{\text{UR}} + \boldsymbol{T}_2 \hat{\boldsymbol{\beta}}_{2n}^{\text{UR}} \right)^\top .
\end{aligned}$$

(a) Verify that the above three steps are correct.
(b) Derive the $\text{ADB}(\hat{\boldsymbol{\theta}}_n^{\text{nEnetR}}(\lambda, \alpha))$.
(c) Derive the expression for $\text{ADL}_2\text{-risk}(\hat{\boldsymbol{\theta}}_n^{\text{nEnetR}}(\lambda, \alpha))$.

6

Multiple Linear Regression Models

6.1 Introduction

This chapter considers rank-based shrinkage estimation for the non-orthogonal multiple linear model. This subject was studied extensively in Chapter 2 where the practical aspects for machine learning with penalized rank estimators were described. Here, we provide a theoretical treatment and investigate the comparative performance characteristics of the primary penalty estimators, namely, ridge, LASSO, elastic net and aLASSO, together with the restricted, unrestricted, preliminary test and Stein–Saleh-type R-estimators when the dimension of the parameter space is less than the dimension of the sample space. We use marginal distribution theory to evaluate the asymptotic distributional L_2-risks of the R-estimators.

6.2 Multiple Linear Model and R-estimation

Consider the multiple regression model

$$\boldsymbol{y} = \boldsymbol{X}\boldsymbol{\beta} + \boldsymbol{\varepsilon} \tag{6.2.1}$$

where $\boldsymbol{y} = (y_1, \dots, y_n)^\top$ is the response vector, $\boldsymbol{X}$ is an $n \times p$ $(n > p)$ matrix of real numbers, and $\boldsymbol{\beta}$ is a p-vector of regression coefficients. We assume the following conditions due to Jurečková (1971):

(i) $\boldsymbol{\varepsilon} = (\varepsilon_1, \dots, \varepsilon_n)^\top$ are i.i.d. components with c.d.f. F, having absolutely continuous p.d.f. f with finite Fisher information

$$0 < I(f) = \int_{-\infty}^{+\infty} \left(\frac{f'(x)}{f(x)}\right)^2 f(x)\mathrm{d}x < \infty. \tag{6.2.2}$$

Rank-Based Methods for Shrinkage and Selection: With Application to Machine Learning. First Edition. A. K. Md. Ehsanes Saleh, Mohammad Arashi, Resve A. Saleh, and Mina Norouzirad.

(ii) We consider the score generating function $\phi : (0,1) \to \mathbb{R}$ which is assumed to be non-constant, non-decreasing, and square integrable on $(0,1)$ such that

$$A_\phi^2 = \int_0^1 \phi^2(u)\mathrm{d}u - \left(\int_0^1 \phi(u)\mathrm{d}u\right)^2, \tag{6.2.3}$$

and

$$\phi(u,f) = -\frac{f'(F(u))}{f(F^{-1}(u))}, \quad 0 < u < 1.$$

The score is defined by

$$a_n(i) = \mathbb{E}\left[\phi(U_{n_i})\right] \quad U_{n_i} = \frac{i}{n+1}, \quad 1 \le i \le n \tag{6.2.4}$$

for $n \ge 1$, where $U_{n_1} \le \dots \le U_{n_n}$ are the order statistics from a sample of size n from $U(0,1)$.
(iii) Let

$$\boldsymbol{C}_n = \frac{1}{n}\sum_{i=1}^{n}(\boldsymbol{x}_i - \bar{\boldsymbol{x}})(\boldsymbol{x}_i - \bar{\boldsymbol{x}})^\top, \tag{6.2.5}$$

where $\boldsymbol{x}_i$ is the ith row of $\boldsymbol{X}$ and

$$\bar{\boldsymbol{x}} = \frac{1}{n}\sum_{i=1}^{n}\boldsymbol{x}_i.$$

Next, we assume that

$$\text{(a)} \lim_{n\to\infty} \boldsymbol{C}_n = \boldsymbol{C}, \quad \text{(b)} \lim_{n\to\infty} \max(\boldsymbol{x}_i - \bar{\boldsymbol{x}})^\top \boldsymbol{C}_n^{-1}(\boldsymbol{x}_i - \bar{\boldsymbol{x}}) = 0. \tag{6.2.6}$$

For R-estimation of $\boldsymbol{\beta} \in \mathbb{R}^p$, the rank of $y_i - \boldsymbol{x}_i^\top\boldsymbol{\beta}$ among $y_1 - \boldsymbol{x}_1^\top\boldsymbol{\beta}, \dots, y_n - \boldsymbol{x}_n^\top\boldsymbol{\beta}$ is denoted by $R_{n_i}(\boldsymbol{\beta})$. Note that $R_{n_i}(\boldsymbol{\beta})$ is translation invariant.

Now, for each $n \ge 1$, consider the set of scores $a_n(1) \le \dots \le a_n(n)$ and define the dispersion function as

$$D_n(\boldsymbol{\beta}) = \sum_{i=1}^{n}(y_i - (\boldsymbol{x}_i - \bar{\boldsymbol{x}})^\top\boldsymbol{\beta})a_n(R_{n_i}(\boldsymbol{\beta})). \tag{6.2.7}$$

The unrestricted R-estimator (URE) of $\boldsymbol{\beta}$ is defined by any central point of the set

$$\hat{\boldsymbol{\beta}}_n^{\mathrm{UR}} = \underset{\boldsymbol{\beta}\in\mathbb{R}^p}{\operatorname{argmin}}\{D_n(\boldsymbol{\beta})\}. \tag{6.2.8}$$

Equivalently, the $\hat{\boldsymbol{\beta}}^{\text{UR}}$ can be defined as

$$\hat{\boldsymbol{\beta}}_n^{\text{UR}} = \{\boldsymbol{\beta} | \boldsymbol{L}_n(\boldsymbol{\beta}) = \boldsymbol{0}\} \tag{6.2.9}$$

where $\boldsymbol{L}_n(\boldsymbol{\beta})$ is obtained from the derivative of $D_n(\boldsymbol{\beta})$, i.e. Eq. (6.2.7), and given by

$$\boldsymbol{L}_n(\boldsymbol{\beta}) = \frac{1}{\sqrt{n}} \sum_{i=1}^{n} (\boldsymbol{x}_i - \bar{\boldsymbol{x}}) a_n(R_{n_i}(\boldsymbol{\beta})). \tag{6.2.10}$$

Let us consider the linear rank statistics (LRS) as

$$\mathbf{L}_{n_i}(\boldsymbol{\beta}) = \frac{1}{\sqrt{n}} (\boldsymbol{x}_i - \bar{\boldsymbol{x}}) a_n(R_{n_i}(\boldsymbol{\beta})).$$

Using the properties of derivative functions, we can rewrite Eq. (6.2.9) as

$$\hat{\boldsymbol{\beta}}_n^{\text{UR}} = \left\{ \boldsymbol{\beta} | \sum_{i=1}^{p} L_{n_i}(\boldsymbol{\beta}) = 0 \right\}$$

If we set $||\boldsymbol{a}|| = \sum_{j=1}^{p} |a_j|$ for $\boldsymbol{a} = (a_1, \dots, a_p)^{\top}$, then it is easier to find the URE of $\boldsymbol{\beta}$ by

$$\hat{\boldsymbol{\beta}}_n^{\text{UR}} = \underset{\boldsymbol{\beta} \in \mathbb{R}^p}{\operatorname{argmin}} \{ ||\boldsymbol{L}_n(\boldsymbol{\beta})|| \}. \tag{6.2.11}$$

Then, by the asymptotic uniform linearity (AUL) results of Jurečková (1971), for any $\kappa > 0$ and $\varepsilon > 0$,

$$\lim_{n \to \infty} P \left(\sup_{||\boldsymbol{w}|| \le \kappa} \left\| \boldsymbol{L}_n(\boldsymbol{\beta} + \frac{\boldsymbol{w}}{\sqrt{n}}) - \boldsymbol{L}_n(\boldsymbol{\beta}) + \gamma(\phi, f) \boldsymbol{C} \boldsymbol{w} \right\| \ge \varepsilon \right) = 0 \tag{6.2.12}$$

where

$$\gamma(\phi, u) = \int_0^1 \phi(u) \phi(u, f) \mathrm{d}u.$$

Then, it is well-known that

$$\sqrt{n} \left(\hat{\boldsymbol{\beta}}_n^{\text{UR}} - \boldsymbol{\beta} \right) \xrightarrow{\mathcal{D}} \mathcal{N}_p(\boldsymbol{0}, \eta^2 \boldsymbol{C}^{-1}), \tag{6.2.13}$$

where

$$\eta^2 = \frac{A_\phi^2}{\gamma^2(\phi, f)}.$$

Also,

$$n^{\frac{1}{2}} \left\| \boldsymbol{L}_n(\hat{\boldsymbol{\beta}}_n^{\text{UR}}) \right\| \xrightarrow{P} 0. \tag{6.2.14}$$

As described in Chapter 3, we consider the asymptotic distributional risk of each estimator under the L_2 loss denoted by

$$\text{ADL}_2\text{-risk}(\hat{\boldsymbol{\beta}}_n) = \text{ADV}(\hat{\boldsymbol{\beta}}_n) + \text{ADB}^2(\hat{\boldsymbol{\beta}}_n)$$

where $\text{ADV}(\hat{\boldsymbol{\beta}}_n)$ is the asymptotic distributional variance and $\text{ADB}^2(\hat{\boldsymbol{\beta}}_n)$ is the square of the asymptotic distributional bias. Hence, the ADL_2-risk of $\hat{\boldsymbol{\beta}}^{\text{UR}}$ is given by

$$\text{ADL}_2\text{-risk}(\hat{\boldsymbol{\beta}}_n^{\text{UR}}) = \eta^2 \, \text{tr}(\boldsymbol{C}^{-1}). \tag{6.2.15}$$

An estimator performs better if it has a lower ADL_2-risk. We will also be comparing the asymptotic distributional L_2-risk efficiency (ADRE) of the various estimators. As described in Chapter 3, the relative efficiency of two estimators is given by

$$\text{ADRE}(\hat{\boldsymbol{\beta}}_n : \hat{\boldsymbol{\beta}}_n^*) = \frac{\text{ADL}_2\text{-risk}(\hat{\boldsymbol{\beta}}_n^*)}{\text{ADL}_2\text{-risk}(\hat{\boldsymbol{\beta}}_n)}.$$

6.3 Model Sparsity and Detection

There are many applications where the data scientist expects sparsity in the parameter set. That is, some β are non-zero while other β are either zero or very close to zero. The non-zero set is considered to be significant while the rest can be neglected in order to create a compact model. In such cases, it is convenient to rewrite the multiple regression model in Eq. (6.2.1) as

$$\boldsymbol{y} = \boldsymbol{X}_1\boldsymbol{\beta}_1 + \boldsymbol{X}_2\boldsymbol{\beta}_2 + \boldsymbol{\varepsilon} \tag{6.3.1}$$

where $\boldsymbol{\beta}_1$ is a p_1-vector while $\boldsymbol{\beta}_2$ is a p_2-vector and $\boldsymbol{X} = (\boldsymbol{X}_1, \boldsymbol{X}_2)$ is an $n \times p$ matrix of real numbers, where the dimension of $\boldsymbol{X}_1$ and $\boldsymbol{X}_2$ are $n \times p_1$, and $n \times p_2$, respectively. The question becomes one of figuring out how to find the sizes of these two $\boldsymbol{\beta}$-vectors.

One approach is to use the LASSO R-estimator as follows

$$\hat{\boldsymbol{\beta}}_n^{\text{LassoR}}(\lambda) = \underset{\boldsymbol{\beta} \in \mathbb{R}^p}{\text{argmin}} \left\{ D_n(\boldsymbol{\beta}) + \lambda \sum_{j=1}^{p} |\beta_j| \right\}$$

where

$$D_n(\boldsymbol{\beta}) = \sum_{i=1}^{n} (y_i - (\boldsymbol{x}_i - \bar{\boldsymbol{x}})^\top \boldsymbol{\beta}) a_n(R_{n_i}(\boldsymbol{\beta})). \tag{6.3.2}$$

Subset selection is performed by the proper choice of λ such that if $\boldsymbol{\beta} = (\boldsymbol{\beta}_1^\top, \boldsymbol{\beta}_2^\top)^\top$ is well-defined then, correspondingly, the LASSO R-estimator produces a partitioning of the parameters as $\boldsymbol{\beta}_n = (\boldsymbol{\beta}_{1n}^\top, \boldsymbol{\beta}_{2n}^\top)^\top$.

When $\boldsymbol{\beta}_2 = \mathbf{0}$, the model in Eq. (6.3.1) reduces to the restricted model

$$\boldsymbol{y} = \boldsymbol{X}_1\boldsymbol{\beta}_1 + \varepsilon, \tag{6.3.3}$$

where

$$\boldsymbol{X}_1 = (\boldsymbol{x}_1^{(1)}, \dots, \boldsymbol{x}_n^{(1)}),\ \boldsymbol{x}_i^{(1)} \in \mathbb{R}^{p_1}.$$

Accordingly, the restricted R-estimator (RRE) of $\boldsymbol{\beta}_1$ is given by

$$\hat{\boldsymbol{\beta}}_{1n}^{\mathrm{RR}} = \underset{\boldsymbol{\beta}_1 \in \mathbb{R}^{p_1}}{\operatorname{argmin}} \{D_n(\boldsymbol{\beta}_1)\}$$

where

$$D_n(\boldsymbol{\beta}_1) = \sum_{i=1}^{n} (y_i - (\boldsymbol{x}_i^{(1)} - \bar{\boldsymbol{x}}^{(1)})^\top \boldsymbol{\beta}_1) a_n(R_{n_i}(\boldsymbol{\beta}_1))$$

with $\boldsymbol{x}_i^{(1)}$ being the ith row of $\boldsymbol{X}_1$ and $\bar{\boldsymbol{x}}^{(1)} = n^{-1}\sum_{i=1}^{n} \boldsymbol{x}_i^{(1)}$. Also, the rank of $y_i - (\boldsymbol{x}_i^{(1)})^\top \boldsymbol{\beta}_1$ among $y_1 - (\boldsymbol{x}_1^{(1)})^\top \boldsymbol{\beta}_1, \dots, y_n - (\boldsymbol{x}_n^{(1)})^\top \boldsymbol{\beta}_1$ is denoted by $R_{n_i}(\boldsymbol{\beta}_1)$.

Equivalently, the RRE can be obtained by

$$\hat{\boldsymbol{\beta}}_{1n}^{\mathrm{RR}} = \underset{\boldsymbol{\beta}_1 \in \mathbb{R}^{p_1}}{\operatorname{argmin}} \{\|\boldsymbol{L}_n(\boldsymbol{\beta}_1)\|\} \tag{6.3.4}$$

where

$$\boldsymbol{L}_n(\boldsymbol{\beta}_1) = \left(L_{n_1}(\boldsymbol{\beta}_1), \dots, L_{n_{p_1}}(\boldsymbol{\beta}_1)\right)^\top.$$

Here

$$L_{n_i}(\boldsymbol{\beta}_1) = \frac{1}{\sqrt{n}} (\boldsymbol{x}_i^{(1)} - \bar{\boldsymbol{x}}^{(1)}) a_n(R_{n_i}(\boldsymbol{\beta}_1)).$$

Then, using AUL of Jurečková (1971), we have that

$$\sqrt{n}\left(\hat{\boldsymbol{\beta}}_{1n}^{\mathrm{RR}} - \boldsymbol{\beta}_1\right) \xrightarrow{\mathcal{D}} \mathcal{N}_{p_1}(\mathbf{0}, \eta^2 \boldsymbol{C}_{11}^{-1}) \tag{6.3.5}$$

where

$$\boldsymbol{C} = \begin{pmatrix} \boldsymbol{C}_{11} & \boldsymbol{C}_{12} \\ \boldsymbol{C}_{21} & \boldsymbol{C}_{22} \end{pmatrix}.$$

Let $\boldsymbol{C}^{-1} = (C^{ij})$ and C^{jj} be the jth block diagonal of $\boldsymbol{C}^{-1}$. Then

$$\boldsymbol{C}^{-1} = \begin{pmatrix} \boldsymbol{C}^{11} & \boldsymbol{C}^{12} \\ \boldsymbol{C}^{21} & \boldsymbol{C}^{22} \end{pmatrix}.$$

We are basically interested in the R-estimation of $\boldsymbol{\beta} = (\boldsymbol{\beta}_1^\top, \boldsymbol{\beta}_2^\top)^\top$ when we suspect that $\boldsymbol{\beta}_2$, a p_2-vector of coefficients, is zero and $\boldsymbol{\beta}_1$, a p_1-vector of coefficients, is non-zero. Then, we may write, the URE and RRE of $\boldsymbol{\beta}$, respectively, as

$$\hat{\boldsymbol{\beta}}_n^{\text{UR}} = \begin{pmatrix} \hat{\boldsymbol{\beta}}_{1n}^{\text{UR}} \\ \\ \hat{\boldsymbol{\beta}}_{2n}^{\text{UR}} \end{pmatrix} \quad \text{and} \quad \hat{\boldsymbol{\beta}}_n^{\text{RR}} = \begin{pmatrix} \hat{\boldsymbol{\beta}}_{1n}^{\text{RR}} \\ \\ \boldsymbol{0} \end{pmatrix}. \tag{6.3.6}$$

From the marginal distribution of $\hat{\boldsymbol{\beta}}_{1n}^{\text{UR}}$ and $\hat{\boldsymbol{\beta}}_{2n}^{\text{UR}}$, we have

$$\sqrt{n}\left(\hat{\boldsymbol{\beta}}_{1n}^{\text{UR}} - \boldsymbol{\beta}_1\right) \xrightarrow{\mathcal{D}} \mathcal{N}_{p_1}\left(\boldsymbol{0}, \eta^2 \boldsymbol{C}_{11\cdot2}^{-1}\right)$$

$$\sqrt{n}\left(\hat{\boldsymbol{\beta}}_{2n}^{\text{UR}} - \boldsymbol{\beta}_2\right) \xrightarrow{\mathcal{D}} \mathcal{N}_{p_2}\left(\boldsymbol{0}, \eta^2 \boldsymbol{C}_{22\cdot1}^{-1}\right),$$

where

$$\boldsymbol{C}_{11\cdot2} = \boldsymbol{C}_{11} - \boldsymbol{C}_{12}\boldsymbol{C}_{22}^{-1}\boldsymbol{C}_{12}$$

$$\boldsymbol{C}_{22\cdot1} = \boldsymbol{C}_{22} - \boldsymbol{C}_{21}\boldsymbol{C}_{11}^{-1}\boldsymbol{C}_{21}.$$

The ADL$_2$-risk of $\hat{\boldsymbol{\beta}}_{1n}^{\text{UR}}$ is given by

$$\text{ADL}_2\text{-risk}(\hat{\boldsymbol{\beta}}_{1n}^{\text{UR}}) = \eta^2\left(\text{tr}(\boldsymbol{C}_{11\cdot2}^{-1}) + \text{tr}(\boldsymbol{C}_{22\cdot1}^{-1})\right). \tag{6.3.7}$$

Next, consider the local alternative as

$$\mathcal{H}_A : \boldsymbol{\beta}_{2(n)} = \frac{1}{\sqrt{n}}\boldsymbol{\delta}_2, \qquad \boldsymbol{\delta}_2 = (\delta_{2_1}, \dots, \delta_{2_{p_2}})^\top. \tag{6.3.8}$$

Under the local alternatives, ADL$_2$-risk$(\hat{\boldsymbol{\beta}}_n^{\text{RR}})$ is given by

$$\text{ADL}_2\text{-risk}(\hat{\boldsymbol{\beta}}_n^{\text{RR}}) = \eta^2\left(\text{tr}(\boldsymbol{C}_{11}^{-1}) + \frac{1}{\eta^2}\boldsymbol{\delta}_2^\top\boldsymbol{\delta}_2\right). \tag{6.3.9}$$

In what follows, based on the definition of Δ^2, we use the Courant theorem (see Theorem A.2.4 of Anderson, 2003) to obtain

$$\text{Ch}_{\min}(\boldsymbol{C}_{22\cdot1}^{-1}) \leq \frac{\boldsymbol{\delta}_2^\top\boldsymbol{\delta}_2}{\boldsymbol{\delta}_2^\top\boldsymbol{C}_{22\cdot1}\boldsymbol{\delta}_2} \leq \text{Ch}_{\max}(\boldsymbol{C}_{22\cdot1}^{-1}),$$

where $\mathrm{Ch}_{\max(\min)}(\boldsymbol{A})$ is the largest (smallest) eigenvalue of $\boldsymbol{A}$. We obtain

$$\eta^2\Delta^2\mathrm{Ch}_{\min}(\boldsymbol{C}_{22\cdot1}^{-1}) \le \boldsymbol{\delta}_2^\top\boldsymbol{\delta}_2 \le \eta^2\Delta^2\mathrm{Ch}_{\max}(\boldsymbol{C}_{22\cdot1}^{-1}). \tag{6.3.10}$$

Thus, the lower-bound of ADL$_2$-risk($\hat{\boldsymbol{\beta}}_n^{\mathrm{RR}}$), denoted using '$*$', is given by

$$\mathrm{ADL}_2^*\text{-risk}(\hat{\boldsymbol{\beta}}_n^{\mathrm{RR}}) = \eta^2\,\mathrm{tr}(\boldsymbol{C}_{11}^{-1}) + \Delta^2\mathrm{Ch}_{\min}(\boldsymbol{C}_{22\cdot1}^{-1}).$$

The ADRE of $\hat{\boldsymbol{\beta}}_n^{\mathrm{RR}}$ is given by

$$\begin{aligned}\mathrm{ADRE}(\hat{\boldsymbol{\beta}}_n^{\mathrm{RR}} : \hat{\boldsymbol{\beta}}_n^{\mathrm{UR}}) &= \frac{\mathrm{tr}(\boldsymbol{C}_{11\cdot2}^{-1}) + \mathrm{tr}(\boldsymbol{C}_{22\cdot1}^{-1})}{\mathrm{tr}(\boldsymbol{C}_{11}^{-1}) + \Delta^2\,\mathrm{Ch}_{\min}(\boldsymbol{C}_{22\cdot1}^{-1})},\\ &= \left(1 + \frac{\mathrm{tr}(\boldsymbol{C}_{22\cdot1}^{-1})}{\mathrm{tr}(\boldsymbol{C}_{11\cdot2}^{-1})}\right)\left(\frac{\mathrm{tr}(\boldsymbol{C}_{11\cdot2}^{-1})}{\mathrm{tr}(\boldsymbol{C}_{11}^{-1})}\right)\\ &\quad\times\left(1 + \frac{\Delta^2\mathrm{Ch}_{\min}(\boldsymbol{C}_{22\cdot1})}{\mathrm{tr}(\boldsymbol{C}_{11}^{-1})}\right)^{-1}\end{aligned} \tag{6.3.11}$$

where $\mathrm{Ch}_{\min}(\boldsymbol{A})$ is the minimum eigenvalue of matrix $\boldsymbol{A}$.

If we use weights $\boldsymbol{C}_{11\cdot2}$ and $\boldsymbol{C}_{22\cdot1}$, then we obtain

$$\mathrm{ADRE}(\hat{\boldsymbol{\beta}}_n^{\mathrm{RR}} : \hat{\boldsymbol{\beta}}_n^{\mathrm{UR}}) = \frac{p_1 + p_2}{\mathrm{tr}(\boldsymbol{C}_{11}^{-1}\boldsymbol{C}_{11\cdot2}) + \Delta^2}, \text{ where } \Delta^2 = \frac{\boldsymbol{\delta}_2^\top\boldsymbol{C}_{11\cdot2}\boldsymbol{\delta}_2}{\eta^2}. \tag{6.3.12}$$

6.4 General Shrinkage R-estimator of β

In this section, we assume sparsity of $\boldsymbol{\beta} = (\boldsymbol{\beta}_1^\top, \boldsymbol{\beta}_2^\top)^\top$, i.e. $\boldsymbol{\beta}_2$ may be at or near $\mathbf{0}$. In general, the shrinkage R-estimator of $\boldsymbol{\beta}$ is defined as

$$\hat{\boldsymbol{\beta}}_n^{\mathrm{ShrinkageR}}(c) = (\hat{\boldsymbol{\beta}}_{1n}^{\mathrm{URT}}, c\hat{\boldsymbol{\beta}}_{2n}^{\mathrm{URT}})^\top, \quad 0 < c < 1$$

where c may be constant or random. Thus, under local alternatives, $\boldsymbol{\beta}_{2(n)}$: $n^{-\frac{1}{2}}\boldsymbol{\delta}_2$, we have

$$\mathrm{ADB}(\hat{\boldsymbol{\beta}}_n^{\mathrm{ShrinkageR}}(c)) = (\mathbf{0}, -(1-c)\boldsymbol{\delta}_2)^\top,$$

and

$$\mathrm{ADL}_2\text{-risk}(\hat{\boldsymbol{\beta}}_n^{\mathrm{ShrinkageR}}(c)) = \eta^2\,\mathrm{tr}(\boldsymbol{C}_{11\cdot2}^{-1}) + c^2\eta^2\,\mathrm{tr}(\boldsymbol{C}_{22\cdot1}^{-1}) + (1-c)^2\boldsymbol{\delta}_2^\top\boldsymbol{\delta}_2. \tag{6.4.1}$$

For the asymptotic risk functions, whenever we use the lower bound of $\boldsymbol{\delta}_2^\top\boldsymbol{\delta}_2$ from Eq. (6.3.10), we write ADL$_2^*$-risk, instead of ADL$_2$-risk. Hence,

$$\begin{aligned}\text{ADL}_2^*\text{-risk}(\hat{\boldsymbol{\beta}}_n^{\text{ShrinkageR}}(c)) &= \eta^2 \operatorname{tr}(\boldsymbol{C}_{11\cdot2}^{-1}) + c^2\eta^2 \operatorname{tr}(\boldsymbol{C}_{22\cdot1}^{-1}) \\ &\quad +(1-c)^2\eta^2\Delta^2\text{Ch}_{\min}(\boldsymbol{C}_{22\cdot1}^{-1}).\end{aligned} \tag{6.4.2}$$

Minimizing the R.H.S. w.r.t. c, we have

$$c^* = \frac{\Delta^2\text{Ch}_{\min}(\boldsymbol{C}_{22\cdot1}^{-1})}{\operatorname{tr}(\boldsymbol{C}_{22\cdot1}^{-1}) + \Delta^2\text{Ch}_{\min}(\boldsymbol{C}_{22\cdot1}^{-1})}.$$

Thus, the lower bound of the ADL$_2$-risk of $\hat{\boldsymbol{\beta}}_n^{\text{ShrinkageR}}(c^*)$ is given by

$$\text{ADL}_2^*\text{-risk}(\hat{\boldsymbol{\beta}}_n^{\text{ShrinkageR}}(c^*)) = \eta^2\left(\operatorname{tr}(\boldsymbol{C}_{11\cdot2}^{-1}) + \frac{\operatorname{tr}(\boldsymbol{C}_{22\cdot1}^{-1})\Delta^2\text{Ch}_{\min}(\boldsymbol{C}_{22\cdot1}^{-1})}{\operatorname{tr}(\boldsymbol{C}_{22\cdot1}^{-1}) + \Delta^2\text{Ch}_{\min}(\boldsymbol{C}_{22\cdot1}^{-1})}\right). \tag{6.4.3}$$

The ADRE is then given by

$$\begin{aligned}\text{ADRE}(\hat{\boldsymbol{\beta}}_n^{\text{ShrinkageR}}(c^*) : \hat{\boldsymbol{\beta}}_n^{\text{UR}}) &= \left(1 + \frac{\operatorname{tr}(\boldsymbol{C}_{22\cdot1}^{-1})}{\operatorname{tr}(\boldsymbol{C}_{11\cdot2}^{-1})}\right) \\ &\times\left(1 + \frac{\operatorname{tr}(\boldsymbol{C}_{22\cdot1}^{-1})}{\operatorname{tr}(\boldsymbol{C}_{11\cdot2}^{-1})}\left(\frac{\Delta^2\text{Ch}_{\min}(\boldsymbol{C}_{22\cdot1}^{-1})}{\operatorname{tr}(\boldsymbol{C}_{22\cdot1}^{-1}) + \Delta^2\text{Ch}_{\min}(\boldsymbol{C}_{22\cdot1}^{-1})}\right)\right)^{-1} \geq 1.\end{aligned}$$

6.4.1 Preliminary Test R-estimators

The preliminary test R-estimator of $\boldsymbol{\beta}$ is given by

$$\hat{\boldsymbol{\beta}}_n^{\text{PTR}}(\lambda) = \begin{pmatrix}\hat{\boldsymbol{\beta}}_{1n}^{\text{UR}} \\ \hat{\boldsymbol{\beta}}_{2n}^{\text{UR}}\, I(\mathcal{L}_n > \lambda^2)\end{pmatrix}. \tag{6.4.4}$$

For the test of the null hypothesis $\mathcal{H}_o : \boldsymbol{\beta}_2 = \boldsymbol{0}$ vs. $\mathcal{H}_A : \boldsymbol{\beta}_2 \neq \boldsymbol{0}$, we consider the partitioning

$$\boldsymbol{L}_n(\boldsymbol{\beta}_1, \boldsymbol{\beta}_2) = (\boldsymbol{L}_{n(1)}^\top(\boldsymbol{\beta}_1, \boldsymbol{\beta}_2), \boldsymbol{L}_{n(2)}^\top(\boldsymbol{\beta}_1, \boldsymbol{\beta}_2))^\top, \tag{6.4.5}$$

where

$$\boldsymbol{L}_{n(1)}(\boldsymbol{\beta}_1, \boldsymbol{\beta}_2) = (\boldsymbol{L}_{n_1}(\boldsymbol{\beta}_1, \boldsymbol{\beta}_2), \dots, \boldsymbol{L}_{n_{p_1}}(\boldsymbol{\beta}_1, \boldsymbol{\beta}_2))^\top,$$

$$\boldsymbol{L}_{n(2)}(\boldsymbol{\beta}_1, \boldsymbol{\beta}_2) = (\boldsymbol{L}_{n_{p+1}}(\boldsymbol{\beta}_1, \boldsymbol{\beta}_2), \dots, \boldsymbol{L}_{n_p}(\boldsymbol{\beta}_1, \boldsymbol{\beta}_2))^\top.$$

Let

$$\hat{\boldsymbol{L}}_{n(2)}(\hat{\boldsymbol{\beta}}_{1n}^{\text{UR}}, \mathbf{0}) = (\hat{\boldsymbol{L}}_{n_{p+1}}(\hat{\boldsymbol{\beta}}_{1n}^{\text{UR}}, \mathbf{0}), \dots, \hat{\boldsymbol{L}}_{n_p}(\hat{\boldsymbol{\beta}}_{1n}^{\text{UR}}, \mathbf{0}))^\top. \tag{6.4.6}$$

Then, we propose the following test statistic

$$\mathcal{L}_n = A_n^{-2} \hat{\boldsymbol{L}}_{n(2)}^\top (\hat{\boldsymbol{\beta}}_{1n}^{\text{UR}}, \mathbf{0}) \boldsymbol{C}_{22\cdot 1}^{-1} \hat{\boldsymbol{L}}_{n(2)}(\hat{\boldsymbol{\beta}}_{1n}^{\text{UR}}, \mathbf{0}), \tag{6.4.7}$$

where

$$A_n^2 = \frac{1}{n-1} \sum_{i=1}^n (a_n(i) - \bar{a}_n)^2, \quad \bar{a}_n = \frac{1}{n} \sum_{i=1}^n a_n(i).$$

Note that $\mathcal{L}_n$ may be approximated as

$$\mathcal{L}_n \stackrel{\mathcal{D}}{=} \frac{\gamma^2(\phi, f)}{A_\phi^2} \eta \hat{\boldsymbol{\beta}}_{1n}^{\text{UR}^\top} \boldsymbol{C}_{22\cdot 1} \hat{\boldsymbol{\beta}}_{1n}^{\text{UR}} + o_p(1)$$

Under the assumed regularity conditions of Eq. (6.2.6) as $n \to \infty$, $\mathcal{L}_n$ follows the χ^2 distribution with p_2 d.f. under $\mathcal{H}_o$.

We will find that

$$\text{ADB}(\hat{\boldsymbol{\beta}}_n^{\text{PTR}}(\lambda)) = \begin{pmatrix} \mathbf{0}_{p_1} \\ -\boldsymbol{\delta}_2 \mathcal{H}_{p_2+2}(\lambda^2, \Delta^2) \end{pmatrix} \tag{6.4.8}$$

and

$$\begin{aligned} \text{ADL}_2\text{-risk}(\hat{\boldsymbol{\beta}}_n^{\text{PTR}}(\lambda)) &= \eta^2\big(\operatorname{tr}(\boldsymbol{C}_{11\cdot 2}^{-1}) + \operatorname{tr}(\boldsymbol{C}_{22\cdot 1}^{-1}) \left(1 - \mathcal{H}_{p_2+2}(\lambda^2, \Delta^2)\right) \\ &\quad + \eta^{-2} \boldsymbol{\delta}_2^\top \boldsymbol{\delta}_2 Z(\lambda, \Delta)\big) \end{aligned} \tag{6.4.9}$$

where

$$Z(\lambda, \Delta) = 2\mathcal{H}_{p_2+2}(\lambda^2, \Delta^2) - \mathcal{H}_{p_2+4}(\lambda^2, \Delta^2). \tag{6.4.10}$$

Hence, the lower bound is given by

$$\begin{aligned} \text{ADL}_2^*\text{-risk}(\hat{\boldsymbol{\beta}}_n^{\text{PTR}}(\lambda)) &= \eta^2\big(\operatorname{tr}(\boldsymbol{C}_{11\cdot 2}^{-1}) + \operatorname{tr}(\boldsymbol{C}_{22\cdot 1}^{-1}) \left(1 - \mathcal{H}_{p_2+2}(\lambda^2, \Delta^2)\right) \\ &\quad + \Delta^2 \text{Ch}_{\min}(\boldsymbol{C}_{22\cdot 1}^{-1}) Z(\lambda, \Delta)\big) \end{aligned}$$

$$= \eta^2 \operatorname{tr}(\boldsymbol{C}_{11\cdot2}^{-1})\Bigg(1 + \frac{\operatorname{tr}(\boldsymbol{C}_{22\cdot1}^{-1})}{\operatorname{tr}(\boldsymbol{C}_{11\cdot2}^{-1})}(1 - \mathcal{H}_{p_2+2}(\lambda^2, \Delta^2))$$

$$+ \frac{\Delta^2 \operatorname{Ch}_{\min}(\boldsymbol{C}_{22\cdot1}^{-1})}{\operatorname{tr}(\boldsymbol{C}_{11\cdot2}^{-1})} Z(\lambda, \Delta)\Bigg). \tag{6.4.11}$$

Then, the lower bound ADRE of $\hat{\boldsymbol{\beta}}_n^{\text{PTR}}(\lambda)$ is given by

$$\text{ADRE}(\hat{\boldsymbol{\beta}}_n^{\text{PTR}}(\lambda) : \hat{\boldsymbol{\beta}}_n^{\text{UR}}) = \left(1 + \frac{\operatorname{tr}(\boldsymbol{C}_{22\cdot1}^{-1})}{\operatorname{tr}(\boldsymbol{C}_{11\cdot2}^{-1})}\right)$$

$$\times\left(1 + \frac{\operatorname{tr}(\boldsymbol{C}_{22\cdot1}^{-1})}{\operatorname{tr}(\boldsymbol{C}_{11\cdot2}^{-1})}(1 - \mathcal{H}_{p_2+2}(\lambda^2, \Delta^2)) + \frac{\Delta^2 \operatorname{Ch}_{\min}(\boldsymbol{C}_{22\cdot1}^{-1})}{\operatorname{tr}(\boldsymbol{C}_{11\cdot2}^{-1})} Z(\lambda, \Delta)\right)^{-1}. \tag{6.4.12}$$

This estimator is a discrete process and unstable such that a small change in data may result in different models being selected.

One may note further that, for the case when

$$0 < \Delta^2 \leq \frac{\operatorname{tr}(\boldsymbol{C}_{22\cdot1}^{-1})}{\operatorname{Ch}_{\min}(\boldsymbol{C}_{22\cdot1}^{-1})} \frac{\mathcal{H}_{p_2+2}(\lambda^2; \Delta^2)}{[2\mathcal{H}_{p_2+2}(\lambda^2; \Delta^2) - \mathcal{H}_5(\lambda^2; \Delta^2)]} = \Delta^2(\lambda),$$

we find that $\hat{\boldsymbol{\beta}}_{2n}^{\text{PTR}}(\lambda)$ is better than $\hat{\boldsymbol{\beta}}_{2n}^{\text{UR}}$. Otherwise, $\hat{\boldsymbol{\beta}}_{2n}^{\text{UR}}$ is better than $\hat{\boldsymbol{\beta}}_{2n}^{\text{PTR}}(\lambda)$ if $\Delta^2 > \Delta^2(\lambda)$. Hence, none of them dominate the others uniformly.

6.4.2 Stein–Saleh-type R-estimator

The Stein–Saleh-type R-estimator of $\boldsymbol{\beta} = (\boldsymbol{\beta}_1^\top, \boldsymbol{\beta}_2^\top)^\top$ is given by

$$\hat{\boldsymbol{\beta}}_n^{\text{SSR}}(\lambda) = \begin{pmatrix} \hat{\boldsymbol{\beta}}_{1n}^{\text{UR}} \\ \cdots\cdots\cdots\cdots\cdots\cdots \\ (1 - \lambda\mathcal{L}_n^{-1})\hat{\boldsymbol{\beta}}_{2n}^{\text{UR}} \end{pmatrix} \tag{6.4.13}$$

where λ is a shrinkage factor and usually taken to be $\lambda = p_2 - 2$.

The ADB and ADL$_2$-risk of $\hat{\boldsymbol{\beta}}_n^{SSR}(\lambda)$ are then given by

$$\text{ADB}(\hat{\boldsymbol{\beta}}_n^{\text{SSR}}(\lambda)) = \begin{pmatrix} \mathbf{0}_{p_1} \\ -\boldsymbol{\delta}_2\lambda\mathbb{E}\left[\chi_{p_2+2}^{-2}(\Delta^2)\right] \end{pmatrix} \tag{6.4.14}$$

and

$$\text{ADL}_2\text{-risk}(\hat{\boldsymbol{\beta}}_n^{\text{SSR}}(\lambda)) = \eta^2\Big(\text{tr}(\boldsymbol{C}_{11\cdot2}^{-1}) + \text{tr}(\boldsymbol{C}_{22\cdot1}^{-1}) - \lambda\,\text{tr}(\boldsymbol{C}_{22\cdot1}^{-1})X(\lambda,\Delta^2) + \lambda(\lambda+4)\boldsymbol{\delta}_2^{\top}\boldsymbol{\delta}_2\mathbb{E}\left[\chi_{p_2+4}^{-4}(\Delta^2)\right]\Big) \tag{6.4.15}$$

where

$$X(\lambda,\Delta^2) = 2\mathbb{E}\left[\chi_{p_2+2}^{-2}(\Delta^2)\right] - \lambda\mathbb{E}\left[\chi_{p_2+2}^{-4}(\Delta^2)\right] \tag{6.4.16}$$

and the lower bound is

$$\text{ADL}_2^*\text{-risk}(\hat{\boldsymbol{\beta}}_n^{\text{SSR}}(\lambda)) = \eta^2\Big[\text{tr}(\boldsymbol{C}_{11\cdot2}^{-1}) + \text{tr}(\boldsymbol{C}_{22\cdot1}^{-1}) - \lambda\,\text{tr}(\boldsymbol{C}_{22\cdot1}^{-1})X(\lambda,\Delta^2) + \lambda(\lambda+4)\Delta^2\text{Ch}_{\min}(\boldsymbol{C}_{22\cdot1}^{-1})\mathbb{E}\left[\chi_{p_2+4}^{-4}(\Delta^2)\right]\Big]. \tag{6.4.17}$$

Hence, the ADRE expression is given by

$$\text{ADRE}(\hat{\boldsymbol{\beta}}_n^{\text{SSR}}(\lambda) : \hat{\boldsymbol{\beta}}_n^{\text{UR}}) = \left(1 + \frac{\text{tr}(\boldsymbol{C}_{22\cdot1}^{-1})}{\text{tr}(\boldsymbol{C}_{11\cdot2}^{-1})}\right)\Bigg[1 + \frac{\text{tr}(\boldsymbol{C}_{22\cdot1}^{-1})}{\text{tr}(\boldsymbol{C}_{11\cdot2}^{-1})}(1 - \lambda X(\lambda,\Delta^2)) + \lambda(\lambda+4)\frac{\Delta^2\text{Ch}_{\min}(\mathcal{C}_{22\cdot1}^{-1})}{\text{tr}(\boldsymbol{C}_{11\cdot2}^{-1})}\mathbb{E}\left[\chi_{p_2+4}^{-4}(\Delta^2)\right]\Bigg]^{-1}. \tag{6.4.18}$$

6.4.3 Positive-rule Stein–Saleh-type R-estimator

The positive-rule Stein–Saleh-type R-estimator (PRSSRE) is defined by

$$\hat{\boldsymbol{\beta}}_n^{\text{SSR+}}(\lambda) = \begin{pmatrix} \hat{\boldsymbol{\beta}}_{1n}^{\text{UR}} \\ \cdots\cdots\cdots\cdots\cdots \\ (1 - \lambda\mathcal{L}_n^{-1})I(\mathcal{L}_n > \lambda)\hat{\boldsymbol{\beta}}_{2n}^{\text{UR}} \end{pmatrix} \tag{6.4.19}$$

where $\lambda = p_2 - 2$ is the shrinkage factor. Consequently, one may obtain the ADB and ADL$_2$-risk of $\hat{\boldsymbol{\beta}}^{\text{SSR+}}(\lambda)$ given by

$$\text{ADB}(\hat{\boldsymbol{\beta}}_n^{\text{SSR+}}(\lambda)) = \begin{pmatrix} \mathbf{0}_{p_1} \\ -\boldsymbol{\delta}_2\mathbb{E}\left[\left(1 - \lambda\chi_{p_2+2}^{-2}(\Delta^2)\right)I(\chi_{p_2+2}^{2}(\Delta^2) \le \lambda)\right] \end{pmatrix}$$

and

$$\text{ADL}_2\text{-risk}(\hat{\boldsymbol{\beta}}_n^{\text{SSR+}}(\lambda)) = \text{ADL}_2\text{-risk}(\hat{\boldsymbol{\beta}}_n^{\text{SSR}}(\lambda)) + \boldsymbol{\delta}_2^{\top}\boldsymbol{\delta}_2 Y(\lambda,\Delta^2)$$

$$-\eta^2 \operatorname{tr}(\boldsymbol{C}_{22\cdot1}^{-1})\mathbb{E}\left[\left(1-\lambda\chi_{p_2+2}^{2}(\Delta^2)\right)^2 I(\chi_{p_2+2}^{2}(\Delta^2)\le\lambda)\right] \tag{6.4.20}$$

where

$$\begin{aligned} Y(\lambda,\Delta^2) &= 2\mathbb{E}\left[(1-\lambda\chi_{p_2+2}^{-2}(\Delta^2))I(\chi_{p_2+2}^{2}(\Delta^2)\le\lambda)\right] \\ &\quad -\mathbb{E}\left[\left(1-\lambda\chi_{p_2+4}^{-2}(\Delta^2)\right)^2 I(\chi_{p_2+4}^{2}(\Delta^2)\le\lambda)\right]. \end{aligned} \tag{6.4.21}$$

Thus, the lower-bound of the ADL$_2$-risk of $\hat{\boldsymbol{\beta}}_n^{\mathrm{SSR+}}(\lambda)$ is given by

$$\begin{aligned} \mathrm{ADL}_2^*\text{-risk}(\hat{\boldsymbol{\beta}}_n^{\mathrm{SSR+}}(\lambda)) &= \mathrm{ADL}_2^*\text{-risk}(\hat{\boldsymbol{\beta}}_n^{\mathrm{SSR}}(\lambda)) + \Delta^2\mathrm{Ch}_{\min}(\boldsymbol{C}_{22\cdot1}^{-1})Y(\lambda,\Delta^2) \\ &\quad -\eta^2 \operatorname{tr}(\boldsymbol{C}_{22\cdot1}^{-1})\mathbb{E}\left[\left(1-\lambda\chi_{p_2+2}^{2}(\Delta^2)\right)^2 I(\chi_{p_2+2}^{2}(\Delta^2)\le\lambda)\right]. \end{aligned} \tag{6.4.22}$$

Hence,

$$\mathrm{ADRE}(\hat{\boldsymbol{\beta}}_n^{\mathrm{SSR+}}(\lambda):\hat{\boldsymbol{\beta}}_n^{\mathrm{UR}}) = \left(1+\frac{\operatorname{tr}(\boldsymbol{C}_{22\cdot1}^{-1})}{\operatorname{tr}(\boldsymbol{C}_{11\cdot1}^{-1})}\right)\left[\frac{\mathrm{ADL}_2\text{-risk}(\hat{\boldsymbol{\beta}}_n^{\mathrm{SSR}}(\lambda))}{\eta^2\operatorname{tr}(\boldsymbol{C}_{11\cdot2}^{-1})}+A\right]^{-1} \tag{6.4.23}$$

where

$$\begin{aligned} A &= \frac{\Delta^2\mathrm{Ch}_{\min}(\boldsymbol{C}_{22\cdot1}^{-1})}{\operatorname{tr}(\boldsymbol{C}_{11\cdot2}^{-1})}Y(\lambda,\Delta^2) \\ &\quad -\frac{\operatorname{tr}(\boldsymbol{C}_{22\cdot1}^{-1})}{\operatorname{tr}(\boldsymbol{C}_{11\cdot2}^{-1})}\mathbb{E}\left[\left(1-\lambda\chi_{p_2+2}^{2}(\Delta^2)\right)^2 I(\chi_{p_2+2}^{2}(\Delta^2)\le\lambda)\right]. \end{aligned} \tag{6.4.24}$$

6.5 Subset Selectors

6.5.1 Preliminary Test Subset Selector R-estimator

We want to define the preliminary test R-estimator for our considered restriction, $\boldsymbol{\beta}_2 = \mathbf{0}$. In this regard,

$$\hat{\boldsymbol{\beta}}_n^{\mathrm{PTSSR}}(\lambda) = \left(\hat{\beta}_j^{\mathrm{UR}} I(|\hat{\beta}_j^{\mathrm{UR}}| > \lambda\eta\sqrt{C^{jj}}),\ j=1,\ldots,p\right)^{\mathsf{T}} \tag{6.5.1}$$

where $\hat{\beta}_j^{\text{UR}}$ is the jth component of the vector $\hat{\boldsymbol{\beta}}_n^{\text{PTSSR}}(\lambda)$ and C^{jj} is the jth diagonal element of matrix $\boldsymbol{C}_{22\cdot 1}^{-1}$. Let us consider

$$Z_j = \frac{\sqrt{n}\hat{\beta}_j^{\text{UR}}}{\eta\sqrt{C^{jj}}}, j = 1, \dots, p.$$

Under the local alternative of Eq. (6.3.8), $K_{(n)} : \boldsymbol{\beta}_2 = n^{-\frac{1}{2}}\boldsymbol{\delta}_2, \boldsymbol{\delta}_2 = (\delta_1, \dots, \delta_p)^\top$,

$$\sqrt{n}\left(\hat{\beta}_j^{\text{UR}} - \beta_j\right) \xrightarrow{\mathcal{D}} \mathcal{N}(\delta_j, \eta^2 C^{jj}), \quad j = 1, \dots, p.$$

Therefore,

$$Z_j \xrightarrow{\mathcal{D}} \mathcal{N}(\Delta_j, 1), \quad \text{where} \quad \Delta_j = \frac{\delta_{2_j}}{\eta\sqrt{C^{jj}}}.$$

This definition is due to Donoho and Johnstone (1994) and called a hard threshold R-estimator (HTE). We can rewrite $\hat{\boldsymbol{\beta}}_n^{\text{PTSSR}}(\lambda)$ based on Z_j as

$$\sqrt{n}\hat{\boldsymbol{\beta}}_n^{\text{PTSSR}}(\lambda) = \left(\eta\sqrt{C^{jj}} Z_j I(|Z_j| > \lambda),\ j = 1, \dots, p\right)^\top. \tag{6.5.2}$$

Thus, the ADB and ADL_2-risk of this estimator are readily found to be

$$\text{ADB}(\hat{\boldsymbol{\beta}}_n^{\text{PTSSR}}(\lambda)) = -\eta(\sqrt{C^{jj}}\Delta_j \mathcal{H}_3(\lambda^2, \Delta^2), j = 1, \dots, p)^\top$$

and

$$\text{ADL}_2\text{-risk}(\hat{\boldsymbol{\beta}}_n^{\text{PTSSR}}(\lambda)) = \eta^2\Bigg[\text{tr}(\boldsymbol{C}_{11\cdot 2}^{-1}) + \sum_{j=p_1+1}^{p} C^{jj}\left(1 - \mathcal{H}_3(\lambda^2, \Delta_j^2)\right)$$
$$+ \sum_{j=p_1+1}^{p} \Delta_j^2\left(2\mathcal{H}_3(\lambda^2, \Delta_j^2) - \mathcal{H}_5(\lambda^2, \Delta_j^2)\right)\Bigg], j = p_1 + 1, \dots, p. \tag{6.5.3}$$

Therefore, the ADRE, for $j = 1, \dots, p$, is given by

$$\text{ADRE}(\hat{\boldsymbol{\beta}}_n^{\text{PTSSR}}(\lambda) : \hat{\boldsymbol{\beta}}_n^{\text{UR}}) = \left(1 + \frac{\text{tr}(\boldsymbol{C}_{22\cdot 1}^{-1})}{\text{tr}(\boldsymbol{C}_{11\cdot 2}^{-1})}\right)$$

$$\Bigg[1 + \frac{1}{\text{tr}(\boldsymbol{C}_{11\cdot 2}^{-1})}\Bigg(\sum_{j=p_1+1}^{p} C^{jj}\left(1 - \mathcal{H}_3(\lambda^2, \Delta_j^2)\right)$$

$$+ \sum_{j=p_1+1}^{p} \Delta_j^2\left(2\mathcal{H}_3(\lambda^2, \Delta_j^2) - \mathcal{H}_5(\lambda^2, \Delta_j^2)\right)\Bigg)\Bigg]^{-1}.$$

6.5.2 Stein–Saleh-type R-estimator

We want to define the Stein–Saleh-type R-estimator for our considered restriction, $\boldsymbol{\beta}_2 = \mathbf{0}$. For this purpose, we use

$$\hat{\boldsymbol{\beta}}_n^{\mathrm{SR}}(\lambda) = \left(\hat{\beta}_j^{\mathrm{UR}}\left(1 - \frac{\lambda\eta\sqrt{C^{jj}}}{\sqrt{n}|\hat{\beta}_j^{\mathrm{UR}}|}\right), \mid j = 1, \ldots, p\right)^{\mathrm{T}}, \tag{6.5.4}$$

which is equivalent to

$$\hat{\beta}_j^{\mathrm{SR}}(\lambda) = \hat{\beta}_j^{\mathrm{UR}} - \lambda\eta\sqrt{C^{jj}}\mathrm{sgn}(\hat{\beta}_j^{\mathrm{UR}})/\sqrt{n}, \quad j = 1, \ldots, p. \tag{6.5.5}$$

Thus, we can rewrite $\hat{\boldsymbol{\beta}}^{\mathrm{SR}}(\lambda)$ w.r.t. Z_j as

$$\sqrt{n}\hat{\beta}_j^{\mathrm{SR}}(\lambda) = \sqrt{n}\hat{\beta}_j^{\mathrm{UR}}\eta\sqrt{C^{jj}}\left(Z_j - \lambda\mathrm{sgn}(Z_j)\right), \; j = 1, \ldots, p. \tag{6.5.6}$$

Then, finally, we obtain $\hat{\boldsymbol{\beta}}_n^{\mathrm{SR}}(\lambda)$ as

$$\hat{\beta}_j^{\mathrm{SR}}(\lambda) = \eta\sqrt{C^{jj}}\mathrm{sgn}(Z_j)(|Z_j| - \lambda), \quad j = 1, \ldots, p.$$

Next, the ADB and ADL$_2$-risk of this estimator are given by

$$\mathrm{ADB}(\hat{\boldsymbol{\beta}}_n^{\mathrm{SR}}(\lambda)) = \left(-\lambda\eta\sqrt{C^{jj}}\left(2\Phi(\Delta_j) - 1\right) | j = 1, \ldots, p\right)^{\mathrm{T}}$$

and

$$\mathrm{ADL}_2\text{-risk}(\hat{\boldsymbol{\beta}}_n^{\mathrm{SR}}(\lambda)) = \eta^2\left[\mathrm{tr}(\boldsymbol{C}_{11\cdot 2}^{-1}) + \sum_{j=p_1+1}^{p} C^{jj}\left(1 + \lambda^2 - 2\lambda\sqrt{\frac{2}{\pi}}\exp\left\{-\frac{\Delta_j^2}{2}\right\}\right)\right]. \tag{6.5.7}$$

The optimum value of λ is

$$\lambda_{\mathrm{opt}} = \sqrt{\frac{2}{\pi}}\exp\left\{-\frac{\Delta_j^2}{2}\right\},$$

with a maximum

$$\lambda^* = \sqrt{\frac{2}{\pi}} \quad \text{when } \Delta_j^2 = 0 \;\; \forall j$$

and at $\Delta_j^2 = 0$, the optimum value is

$$\text{ADL}_2\text{-risk}(\hat{\boldsymbol{\beta}}_n^{\text{SR}}(\lambda^*)) = \eta^2 \left[\text{tr}(\boldsymbol{C}_{11\cdot2}^{-1}) \right.$$

$$\left. + \sum_{j=p_1+1}^{p} C^{jj} \left[1 - \frac{2}{\pi} \left(2\exp\left\{ -\frac{\Delta_j^2}{2} \right\} - 1 \right) \right] \right].$$

Finally,

$$\text{ADRE}(\hat{\boldsymbol{\beta}}_n^{\text{SR}}(\lambda^*) : \hat{\boldsymbol{\beta}}_n^{\text{UR}}) = \left(1 + \frac{\text{tr}(\boldsymbol{C}_{22\cdot1}^{-1})}{\text{tr}(\boldsymbol{C}_{11\cdot2}^{-1})} \right)$$

$$\times \left[1 + \frac{1}{\text{tr}(\boldsymbol{C}_{11\cdot2}^{-1})} \sum_{j=p_1+1}^{p} C^{jj} \left[1 - \frac{2}{\pi} \left(2\exp\left\{ -\frac{\Delta_j^2}{2} \right\} - 1 \right) \right] \right]^{-1} \geq 1.$$

Notice that this estimator changes sign as $|Z_j| \to \infty$. As such, we need to define a related R-estimator that removes this anomaly.

6.5.3 Positive-rule Stein–Saleh-type R-estimator (LASSO-type)

We noted that the Stein–Saleh-type R-estimator will eventually change signs. A simple solution is to keep its sign until it reaches zero, at which point it remains at zero. This version is the positive-rule Stein–Saleh R-estimator (PRSRE) which is also called LASSO-type R-estimator, a reduced version of LASSO (Tibshirani, 1996), and the soft-threshold estimator (Donoho and Johnstone, 1994).

The positive-rule Stein–Saleh-type subset R-estimator (PRSRE) is defined as

$$\hat{\boldsymbol{\beta}}_n^{\text{PRSR}}(\lambda) = \left(\eta\sqrt{C^{jj}}\,\text{sgn}(Z_j) \left(|Z_j| - \lambda \right)^+ \mid j = 1, \ldots, p \right)^{\top} \tag{6.5.8}$$

where $a^+ = \max\{0, a\}$.

The $\hat{\boldsymbol{\beta}}_n^{\text{PRSR}}(\lambda)$ is a perturbed LASSO. For the orthonormal case, the positive-rule Stein–Saleh-type R-estimator is the rank version LASSO of Tibshirani (1996). It must be noted that for the non-orthonormal case, the LASSO R-estimator does not have a closed form.

It may be shown that

$$\mathrm{ADB}(\hat{\boldsymbol{\beta}}_n^{\mathrm{PRSR}}(\lambda)) = -\eta\Big(\big[\Delta_j \mathcal{H}_3(\lambda^2, \Delta_j^2) - \lambda(\Phi(\lambda - \Delta_j)$$

$$-\Phi(\lambda + \Delta_j))\big] \mid j = 1, \dots, p\Big)^{\top} \tag{6.5.9}$$

and

$$\mathrm{ADL}_2\text{-risk}(\hat{\boldsymbol{\beta}}_n^{\mathrm{PRSR}}(\lambda)) = \eta^2\Bigg(\mathrm{tr}(\boldsymbol{C}_{22\cdot1}^{-1}) + \sum_{j=1}^{p}\Big\{C^{jj}\left(1 - \mathcal{H}_3(\lambda^2, \Delta_j^2)\right)$$

$$+C^{jj}\Delta_j^2\left(2\mathcal{H}_3(\lambda^2, \Delta_j^2) - \mathcal{H}_5(\lambda^2, \Delta_j^2)\right)$$

$$+C^{jj}\lambda^2\left(1 - \mathcal{H}_1(\lambda^2, \Delta_j^2)\right) - 2\lambda\left(\phi(\lambda - \Delta_j) + \phi(\lambda + \Delta_j)\right)\Big\}\Bigg).$$

Thus, the ADRE is obtained by

$$\mathrm{ADRE}(\hat{\boldsymbol{\beta}}_n^{\mathrm{PRSR}}(\lambda) : \hat{\boldsymbol{\beta}}_n^{\mathrm{UR}}) = \left(1 + \frac{\mathrm{tr}(\boldsymbol{C}_{22\cdot1}^{-1})}{\mathrm{tr}(\boldsymbol{C}_{11\cdot2}^{-1})}\right)\left(1 + \frac{1}{\mathrm{tr}(\boldsymbol{C}_{11\cdot2}^{-1})}\right.$$

$$\times \sum_{j=p_1+1}^{p}\Bigg[C^{jj}\Big\{\left(1 - \mathcal{H}_3(\lambda^2, \Delta_j^2)\right) + \Delta_j^2\left(2\mathcal{H}_3(\lambda^2, \Delta_j^2) - \mathcal{H}_5(\lambda^2, \Delta_j^2)\right)$$

$$\left.+\lambda^2\left(1 - \mathcal{H}_1(\lambda^2, \Delta_j^2) - 2\lambda\left(\phi(\lambda - \Delta_j) + \phi(\lambda + \Delta_j)\right)\right)\Big\}\Bigg]^{-1}. \tag{6.5.10}$$

Note that we may write that

$$\mathrm{ADL}_2\text{-risk}(\hat{\boldsymbol{\beta}}_n^{\mathrm{PRSR}}(\lambda)) = \mathrm{ADL}_2\text{-risk}(\hat{\boldsymbol{\beta}}_n^{\mathrm{SR}}(\lambda))$$

$$+ \sum_{j=p_1+1}^{p} C^{jj}\left[\lambda^2\left(1 - \mathcal{H}_1(\lambda^2, \Delta_j^2) - 2\lambda\left(\phi(\lambda - \Delta_j) + \phi(\lambda + \Delta_j)\right)\right)\right]$$

and

$$\mathrm{ADL}_2\text{-risk}(\hat{\boldsymbol{\beta}}_n^{\mathrm{PRSR}}(\lambda)) \le \mathrm{ADL}_2\text{-risk}(\hat{\boldsymbol{\beta}}_n^{\mathrm{SSR}}(\lambda)).$$

6.5.4 Ridge-type Subset Selector

We define ridge-type subset selector as

$$\hat{\boldsymbol{\beta}}_n^{\text{ridgeR}}(\lambda) = \left(\frac{\hat{\boldsymbol{\beta}}_{jn}^{\text{UR}}}{1 + \lambda C^{jj}} | j = 1, \ldots, p \right)^{\top}.$$

Thus, we may derive the two quantities of interest as

$$\text{ADB}(\hat{\boldsymbol{\beta}}_n^{\text{ridgeR}}(\lambda)) = \left(\frac{-\lambda C^{jj}}{1 + \lambda C^{jj}} \Delta | j = 1, \ldots, p \right)^{\top}$$

and

$$\text{ADL}_2\text{-risk}(\hat{\boldsymbol{\beta}}_n^{\text{ridgeR}}(\lambda)) = \eta^2 \Big(\big[\operatorname{tr}[(\boldsymbol{I} + \lambda \boldsymbol{C}^{-1})^{-2} \boldsymbol{C}^{-1}] + \lambda^2 \boldsymbol{\Delta}^{\top} (\boldsymbol{C} + \lambda \boldsymbol{I})^{-1} \boldsymbol{\Delta} \big] \Big),$$

where $\boldsymbol{\Delta} = (\Delta_1, \ldots, \Delta_p)^{\top}$.

6.5.5 Elastic Net-type R-estimator

In this case, we define the naive elastic-net R-estimator (nEnetRE) as

$$\begin{aligned}
\sqrt{n}\hat{\boldsymbol{\beta}}_n^{\text{nEnetR}}(\lambda, \alpha) &= \left(\frac{\operatorname{sgn}(\sqrt{n}\hat{\beta}_{jn}^{\text{UR}})}{[1 + \lambda(1-\alpha)C^{jj}]} \left(|\sqrt{n}\hat{\beta}_{jn}^{\text{UR}}| - \lambda\alpha\eta\sqrt{C^{jj}} \right)^{+} \middle| j = 1, \cdots, p \right)^{\top} \\
&= \underset{\boldsymbol{\beta} \in \mathbb{R}^p}{\operatorname{argmin}} \left\{ D_n(\boldsymbol{\beta}) + \lambda \left[\frac{1}{2}\alpha\eta \boldsymbol{D}|\boldsymbol{\beta}| + \frac{1}{2}(1-\alpha)\boldsymbol{\beta}^{\top}\boldsymbol{C}\boldsymbol{\beta} \right] \right\} \\
&= \left(\eta\sqrt{C^{jj}} \frac{\operatorname{sgn}(Z_j)}{[1 + \lambda(1-\alpha)C^{jj}]} (|Z_j| - \lambda\alpha)^{+} \middle| j = 1, \cdots, p \right)^{\top}
\end{aligned} \tag{6.5.11}$$

where $\boldsymbol{D} = \operatorname{Diag}\left(\frac{1}{\sqrt{C^{jj}}}, j = 1, \cdots, p \right)$ and $Z_j = \frac{\sqrt{n}}{\eta\sqrt{C^{jj}}} \hat{\beta}_{jn}^{\text{UR}}$.

This estimator ties in with the positive-rule Stein–Saleh-type R-estimator perfectly. Indeed, we have

$$\hat{\boldsymbol{\beta}}_n^{\text{nEnetR}}(\lambda, \alpha) = [\boldsymbol{I} + \lambda(1-\alpha)C^{-1}]^{-1} \hat{\boldsymbol{\beta}}_n^{\text{LassoR}}(\lambda\alpha). \tag{6.5.12}$$

Then,

$$\text{ADB}(\hat{\boldsymbol{\beta}}_n^{\text{nEnetR}}(\lambda, \alpha)) = [\boldsymbol{I}_p + \lambda(1-\alpha)C^{-1}]^{-1} [\boldsymbol{B} - \lambda(1-\alpha)\boldsymbol{C}^{-1})\Delta]$$

where

$$\boldsymbol{B} = \left(\sqrt{C^{jj}}\left[\Delta_j \mathcal{H}_3(\lambda^2\alpha^2, \Delta_j^2) - \lambda\alpha\left(\Phi(\lambda\alpha - \Delta_j) - \Phi(\lambda\alpha + \Delta_j)\right)\right]\Big| j = 1, \ldots, p\right)^{\top}$$

and

$$\begin{aligned}\text{ADL}_2\text{-risk}(\hat{\boldsymbol{\beta}}_n^{\text{nEnetR}}(\lambda, \alpha)) = \eta^2\Big[& \operatorname{tr}\{(\boldsymbol{I} + \lambda(1-\alpha)\boldsymbol{C}^{-1})^{-2}\boldsymbol{C}^{-1}\} \\ & + \lambda^2(1-\alpha)^2\boldsymbol{\Delta}^{\top}(\boldsymbol{C} + \lambda(1-\alpha)\boldsymbol{I})^{-2}\boldsymbol{\Delta} \\ & + 2\lambda(1-\alpha)\boldsymbol{B}^{\top}(\boldsymbol{C} + \lambda(1-\alpha)\boldsymbol{I})^{-2}\boldsymbol{\Delta}\Big].\end{aligned}$$

6.6 Adaptive LASSO

6.6.1 Introduction

LASSO is a popular technique for simultaneous estimation and variable selection (Tibshirani, 1996). The Lasso R-estimator is defined as

$$\hat{\boldsymbol{\beta}}_n^{\text{LassoR}}(\lambda_n) = \underset{\boldsymbol{\beta} \in \mathbb{R}^p}{\operatorname{argmin}}\left\{D_n(\boldsymbol{\beta}) + \lambda_n \sum_{j=1}^{n} |\beta_j|\right\} \tag{6.6.1}$$

where λ_n is the tuning parameter for the "L_1-penalty" which the key to its success. Alternatively, we call it positive-rule Stein–Saleh-type estimator as it is related to the Stein-type estimator with one degree of freedom. The usual best subset selection rule is named "preliminary test subset selector" which is a discrete process. It is also extremely variable whereas LASSO continuously shrinks the coefficients towards 0 as λ_n increases and some shrink to 0 exactly if λ_n is sufficiently large. One may notice that the penalized $\boldsymbol{\beta}$ are of equal weight, which does not offer any flexibility to seek better solutions. As an improvement, one may define a weighted "L_1-penalty" as $\sum_{j=1}^{p} w_j|\beta_j|$ given by

$$\hat{\boldsymbol{\beta}}_n^{\text{aLassoR}}(\lambda_n) = \operatorname{argmin}\left\{D_n(\boldsymbol{\beta}) + \lambda_n \sum_{j=1}^{p} w_j|\beta_j|\right\}. \tag{6.6.2}$$

This weighting factor opens the door to the notion of an "oracle property". Accordingly, we define a procedure to be "oracle" if estimates $\hat{\boldsymbol{\beta}}_n$ asymptotically have the following "oracle features" (Fan and Li, 2001).

(i) Identifies the right-subset model, where $A = \{j : \beta_j \neq 0\}$ and $|A| = p_0 < p$.
(ii) Has the optimal rate. $\sqrt{n}(\hat{\boldsymbol{\beta}}_S - \boldsymbol{\beta}_S) \xrightarrow{\mathcal{D}} \mathcal{N}_{p_0}(\mathbf{0}, \boldsymbol{\Sigma}^*)$, where $\boldsymbol{\Sigma}^*$ is the covariance matrix of the true sub-model.

Let $A_n = \{j : \hat{\beta}_j \neq 0\}$. The LASSO selection is consistent if $P(A_n = A) = 1$.

Now, we consider Eq. (6.6.2) with a specific (data-dependent) weight function, $\hat{w}_j$, to produce the "adaptive LASSO" (aLASSO) to obtain $A_n = \{j : \hat{\beta}_{jn}^{\mathrm{UR}} \neq 0\}$ where

$$\begin{aligned}
\hat{\boldsymbol{\beta}}_n^{\mathrm{aLassoR}}(\lambda_n) &= \underset{\boldsymbol{\beta}\in\mathbb{R}^p}{\operatorname{argmin}} \left\{ D_n(\boldsymbol{\beta}) + \lambda_n \sum_{j=1}^{p} \hat{w}_j |\beta_j| \right\} \\
&= \left(\operatorname{sgn}(\hat{\beta}_j^{\mathrm{UR}}) \left(|\hat{\beta}_j^{\mathrm{UR}}| - \lambda_n \hat{w}_j \eta \sqrt{C^{jj}} \right)^+ \Big| j = 1, \ldots, p \right)^\top \\
&= \left(\eta \sqrt{C^{jj}} \operatorname{sgn}(Z_j) \left(|Z_j| - \lambda_n \hat{w}_j \right)^+ \Big| j = 1, \ldots, p \right)^\top
\end{aligned} \tag{6.6.3}$$

where $\hat{\beta}_j^{\mathrm{UR}}$ is the usual unrestricted R-estimator discussed in Section 6.2 and $\hat{w}_j$ is selected to be

$$\hat{w}_j = \frac{1}{|\hat{\beta}_j^{\mathrm{UR}}|^{\gamma^*}}. \quad \gamma^* > 1. \tag{6.6.4}$$

6.6.2 Asymptotics for LASSO-type R-estimator

We estimate $\boldsymbol{\beta}$ by minimizing a penalized Jaeckel's rank dispersion function, $Q_n(\boldsymbol{\phi}) + \lambda_n \sum_{j=1}^{p} |\phi_j|$ for a given λ_n. Hence, the objective function given by

$$Q_n(\boldsymbol{\phi}) = \sum_{i=1}^{n} (y_i - \boldsymbol{x}_i^\top \boldsymbol{\phi}) a_n(R_{n_i}(\boldsymbol{\phi})) + \lambda_n \sum_{j=1}^{p} |\phi_j| \tag{6.6.5}$$

is minimized at $\boldsymbol{\phi} = \hat{\boldsymbol{\beta}}_n^{\mathrm{LassoR}}$. The following shows that $\hat{\boldsymbol{\beta}}_n^{\mathrm{LassoR}}$ is consistent provided that $\lambda_n = o(n)$.

Lemma 6.1 If $\boldsymbol{C}$ is non-singular and $\lambda_n/n \to \lambda_0 \geq 0$, then

$$\hat{\boldsymbol{\beta}}_n^{\text{LassoR}} - \text{argmin}\{Q(\boldsymbol{\phi})\} \xrightarrow{P} 0,$$

where

$$Q(\boldsymbol{\phi}) = (\boldsymbol{\phi} - \boldsymbol{\beta})^\top \boldsymbol{C}(\boldsymbol{\phi} - \boldsymbol{\beta}) + \lambda_0 \sum_{j=1}^{p} |\phi_j|. \tag{6.6.6}$$

Thus, if $\lambda_0 = o(n)$, $\text{argmin}(Q(\boldsymbol{\phi})) = \boldsymbol{\beta}$. So, $\hat{\boldsymbol{\beta}}_n^{\text{LassoR}}$ is consistent.

Proof: Consider Eq. (6.6.5). We need to show that

$$\sup_{\boldsymbol{\phi} \in \mathcal{K}} |Q_n(\boldsymbol{\phi}) - Q(\boldsymbol{\phi}) - \eta^2| \xrightarrow{P} 0 \tag{6.6.7}$$

for any compact set $\mathcal{K}$ and $\hat{\boldsymbol{\beta}}_n^{\text{LassoR}} = O_p(1)$. Under Eqs. (6.6.6) and (6.6.7),

$$\left| \underset{\boldsymbol{\phi} \in \mathcal{K}}{\text{argmin}}\{Q_n(\boldsymbol{\phi})\} - \underset{\boldsymbol{\phi} \in \mathcal{K}}{\text{argmin}}\{Q(\boldsymbol{\phi})\} \right| \xrightarrow{P} 0. \tag{6.6.8}$$

Here, $Q_n(\boldsymbol{\phi})$ is convex. Then Eqs. (6.6.6) and (6.6.7) follow from the point-wise convergence in probability of $Q_n(\boldsymbol{\phi})$ to $Q(\boldsymbol{\phi}) + \eta^2$ by standard results (Anderson and Gill, 1982; Pollard, 1991). □

Lemma 6.2 If $\lambda_n/\sqrt{n} \to \lambda_0 \geq 0$ and $\boldsymbol{C}$ is non-singular, then

$$\sqrt{n}\left(\hat{\boldsymbol{\beta}}_n^{\text{LassoR}} - \boldsymbol{\beta}\right) \xrightarrow{P} \underset{\boldsymbol{u} \in \mathbb{R}^p}{\text{argmin}}(V(\boldsymbol{u})), \tag{6.6.9}$$

where

$$\begin{aligned} V(\boldsymbol{u}) &= -2\boldsymbol{u}^\top \boldsymbol{W} + \gamma(\phi, f)\boldsymbol{u}^\top \boldsymbol{C}\boldsymbol{u} \\ &\quad + \lambda_0 \sum_{j=1}^{p} [u_j \text{sgn}(\beta_j) I(\beta_j \neq 0) + |u_j| I(\beta_j = 0)] \end{aligned} \tag{6.6.10}$$

where $\boldsymbol{W}$ has a $\mathcal{N}_p(\boldsymbol{0}, A_\phi^2 \boldsymbol{C})$ distribution.

Proof: Define $V_n(\boldsymbol{u})$ as

$$\begin{aligned} V_n(\boldsymbol{u}) &= \sup_{\|U\| \leq Mn^{-1/2}} \Bigg| D_n(\boldsymbol{\beta}^* + \frac{\boldsymbol{u}}{\sqrt{n}}) - D_n(\boldsymbol{\beta}^*) \\ &\quad + n^{-1/2} \boldsymbol{L}_n^\top(\boldsymbol{\beta}^*)\boldsymbol{u} - \frac{1}{2}\gamma(\phi, f) n^{-1/2} \boldsymbol{u}^\top \boldsymbol{C}\boldsymbol{u} \Bigg| \xrightarrow{P} 0 \end{aligned}$$

$$= \quad -2\boldsymbol{u}^\top \boldsymbol{W} + \frac{1}{2}\gamma(\phi, f)\boldsymbol{u}^\top \boldsymbol{C}\boldsymbol{u} + \lambda_n \sum_{j=1}^{n} \left(|\beta_j^* + \frac{u_j}{\sqrt{n}}| - |\beta_j^*| \right) \quad (6.6.11)$$

where we have

$$\lambda_n \sum_{j=1}^{p} \left[\left|\beta_j^* + \frac{u_j}{\sqrt{n}}\right| - \left|\beta_j^*\right| \right] \to \lambda_0 \sum_{j=1}^{p} \left[u_j \mathrm{sgn}(\beta_j^*) I(\beta_j^* \neq 0) + |u_j| I(\beta_j^* = 0) \right]. \tag{6.6.12}$$

Thus, $V_n(\boldsymbol{u}) \overset{\mathcal{D}}{\to} V(\boldsymbol{u})$ with finite dimensional convergence. Since $V(\boldsymbol{u})$ is convex and $V(\boldsymbol{u})$ has a unique minimum, it follows from Geyer (1996); Knight and Fu (2000) that

$$\mathrm{argmin}\{V_n(\boldsymbol{u})\} = \sqrt{n}(\hat{\boldsymbol{\beta}}_n^{\mathrm{LassoR}} - \boldsymbol{\beta}) \overset{\mathcal{D}}{\to} \mathrm{argmin}\{V(\boldsymbol{u})\}. \tag{6.6.13}$$

If $\lambda_n = 0$, then $\mathrm{argmin}\{V(\boldsymbol{u})\} = \boldsymbol{C}^{-1}\boldsymbol{W}$, $\boldsymbol{W} \sim \mathcal{N}_p(\boldsymbol{0}, \eta^2 \boldsymbol{C}^{-1})$. □

This lemma implies that when $\lambda_n = O(n^{\frac{1}{2}})$, essentially $\mathcal{A}_n$ cannot actually be $\mathcal{A}$ with a positive probability.

Lemma 6.3 If $\lambda_n/n \to 0$ and $\lambda_n/\sqrt{n} \to \infty$ and $\lambda_n \to \infty$, then

$$\frac{n}{\lambda_n}(\hat{\boldsymbol{\beta}}_n^{\mathrm{LassoR}} - \boldsymbol{\beta}) \overset{P}{\to} \mathrm{argmin}\{V_3(\boldsymbol{u})\} \tag{6.6.14}$$

where

$$V_3(\boldsymbol{u}) = \frac{1}{2}\gamma(\phi, f)\boldsymbol{u}^\top \boldsymbol{C}\boldsymbol{u} + \sum_{j=1}^{p}[u_j \mathrm{sgn}(\beta_j^*) I(\beta_j^* \neq 0) + |u_j| I(\beta_j^* = 0)]. \tag{6.6.15}$$

6.6.3 Oracle Property of aLASSO

Now, we present the theorem regarding the key "oracle" property of the aLASSO R-estimator.

Theorem 6.1 Suppose that $\lambda_n/\sqrt{n} \to 0$ and $\lambda_n n^{(\gamma-1)/2} \to \infty$. Then, the aLASSO R-estimator satisfies asymptotic normality, $\sqrt{n}(\hat{\boldsymbol{\beta}}_{\mathcal{A}}^{\mathrm{aLassoR}} - \boldsymbol{\beta}_{\mathcal{A}}) \overset{\mathcal{D}}{\to} \mathcal{N}_{p_0}(\boldsymbol{0}, \eta^2 \boldsymbol{C}_{11}^{-1})$.

Proof: First, we prove asymptotic normality. Consider

$$V_n(\boldsymbol{u}) = \frac{1}{2}\gamma(\phi, f)\boldsymbol{u}^\top \boldsymbol{C}\boldsymbol{u} - 2\boldsymbol{u}^\top \boldsymbol{W} + \frac{\lambda_n}{\sqrt{n}}\sum_{j=1}^{p} \hat{\omega}_j \sqrt{n}\left(\left|\beta_j^* + \frac{u_j}{\sqrt{n}}\right| - \left|\beta_j^*\right|\right). \tag{6.6.16}$$

If $\beta_j^* \neq 0$, then $\hat{\omega}_j \xrightarrow{P} |\beta_j^*|^{-\gamma^*}$ and $\sqrt{n}\left(\left|\beta_j^* + \frac{u_j}{\sqrt{n}}\right| - \left|\beta_j^*\right|\right) \to u_j \text{sgn}(\beta_j^*)$. By Slutsky's theorem, we have

$$\frac{\lambda_n}{\sqrt{n}}\hat{\omega}_j \sqrt{n}\left(\left|\beta_j^* + \frac{u_j}{\sqrt{n}}\right| - |\beta_j^*|\right) \xrightarrow{P} 0. \tag{6.6.17}$$

If $\beta_j^* = 0$, then $\sqrt{n}\left(\left|\beta_j^* + \frac{u_j}{\sqrt{n}}\right| - \left|\beta_j^*\right|\right) = |u_j|$ and

$$\frac{\lambda_n}{\sqrt{n}}\hat{\omega}_j = \frac{\lambda_n}{n} n^{\frac{\gamma^*}{2}} \left(|\sqrt{n}\hat{\beta}_j^{\text{UR}}|\right)^{-\gamma^*} \xrightarrow{P} \infty,$$

where $\sqrt{n}\hat{\beta}_j^{\text{UR}} = O_p(1)$.

Then, by Slutsky's theorem, we see that

$$V_n(\boldsymbol{u}) \xrightarrow{\mathcal{D}} V_0(\boldsymbol{u}) \quad \text{for every } \boldsymbol{u} \tag{6.6.18}$$

where

$$V_n(\boldsymbol{u}) = \begin{cases} \boldsymbol{u}_{\mathcal{A}}^\top \boldsymbol{C}_{11}\boldsymbol{u}_{\mathcal{A}} - 2\boldsymbol{u}_{\mathcal{A}}^\top \boldsymbol{W}_{\mathcal{A}} & \text{if} \quad u_j = 0 \quad \forall j \notin \mathcal{A} \\ \infty & \text{otherwise} \end{cases} \tag{6.6.19}$$

and $V_0(\boldsymbol{u})$ is convex with a unique minimum given by $(\boldsymbol{C}_{11}^{-1}\boldsymbol{W}_{\mathcal{A}}, \boldsymbol{0})^\top$. Using the epi-convergence results of Geyer (1994) and Knight and Fu (2000), we have

$$\hat{u}_{\mathcal{A}}^{(n)} \xrightarrow{\mathcal{D}} \boldsymbol{C}_{11}^{-1}\boldsymbol{W}_{\mathcal{A}} \quad \text{and} \quad \hat{u}_{\mathcal{A}^c}^{(n)} \xrightarrow{\mathcal{D}} 0. \tag{6.6.20}$$

Hence, we have proved that $\sqrt{n}(\hat{\boldsymbol{\beta}}_{\mathcal{A}}^{\text{aLassoR}} - \boldsymbol{\beta}_{\mathcal{A}}) \xrightarrow{\mathcal{D}} \mathcal{N}_{p_0}(\boldsymbol{0}, \eta^2 \boldsymbol{C}_{11}^{-1})$.

□

Theorem 6.2 (Johnson and Peng, 2008) Let $\hat{\boldsymbol{\beta}}_n^* = \underset{\boldsymbol{\beta}^* \in \mathbb{R}^p}{\text{argmin}}\{Q + n(\boldsymbol{\beta})\}$, where

$$Q_n(\boldsymbol{\beta}^*) \quad = \quad D_n(\boldsymbol{\beta}^*) + \lambda_n \sum_{j=1}^{p} \frac{|\beta_j^*|}{|\hat{\beta}_j^{\text{UR}}|},$$

$$D_n(\boldsymbol{\beta}^*) \quad = \quad \sum_{i=1}^{p}(y_i - \boldsymbol{x}_i^\top \boldsymbol{\beta}^*)a_n(R_{n_i}(\boldsymbol{\beta}^*)). \tag{6.6.21}$$

Then (a) if $\sqrt{n}\lambda_n = O_p(1)$, and we have that $\left\|\hat{\boldsymbol{\beta}}_n^* - \boldsymbol{\beta}^*\right\| = O_p(n^{-\frac{1}{2}})$, (b) if $\sqrt{n}\lambda_n \to \lambda_0 \in (0,\infty)$ and $n\lambda_n \to \infty$, then $P(\boldsymbol{\beta}_2 = \mathbf{0}) = 1$ where $\boldsymbol{\beta}_2 = \left\{\beta_j^* = 0, j = p_0 + 1, \dots, p\right\}$ and

$$n^{\frac{1}{2}}\gamma(\phi, f)\boldsymbol{C}_{11}\left\{(\hat{\boldsymbol{\beta}}_{1n}^* - \boldsymbol{\beta}_1) + n^{-\frac{1}{2}}A_\phi^{-1}\boldsymbol{C}_{11}^{-1}\lambda_0\boldsymbol{b}_1\right\} \xrightarrow{\mathcal{D}} \mathcal{N}_{p_0}(\mathbf{0}, \eta^2\boldsymbol{C}_{11}^{-1})$$

where

$$\boldsymbol{C} = \begin{pmatrix} \boldsymbol{C}_{11} & \boldsymbol{C}_{12} \\ \boldsymbol{C}_{21} & \boldsymbol{C}_{22} \end{pmatrix} \quad \text{and} \quad \boldsymbol{b}_1 = \left(\frac{\operatorname{sgn}(\boldsymbol{\beta}_1^*)}{|\boldsymbol{\beta}_1^*|}, \dots, \frac{\operatorname{sgn}(\boldsymbol{\beta}_{p_0}}{|\boldsymbol{\beta}_{p_0}^*|}\right).$$

Proof: Let

$$\begin{aligned} \mathcal{W}_n(\boldsymbol{u}) &= -\left\{\boldsymbol{u}^\top \frac{1}{\sqrt{n}} \boldsymbol{L}_n(\boldsymbol{\beta}^*)\right\} + \frac{1}{2}\gamma(\phi, f)\boldsymbol{u}^\top \boldsymbol{C}\boldsymbol{u} \\ &\quad + n\lambda_n \sum_{j=1}^{p} \left(\frac{|\hat{\beta}_j^* + n^{-\frac{1}{2}}u_j| - |\beta_j|}{|\hat{\beta}_j^{\mathrm{UR}}|}\right) \\ &\geq -O_p(1)\sum_{j=1}^{p}|u_j| + \frac{1}{2}\gamma(\phi, f)\boldsymbol{u}^\top\boldsymbol{C}\boldsymbol{u} - \sqrt{n}\lambda_n \sum_{j=1}^{p}\frac{u_j}{|\beta_j^*|} \\ &\geq -O_p(1)M + \frac{1}{2}\gamma(\phi, f)a_o \left\|\boldsymbol{U}\right\|^2 - O_p(1)\sum_{j=1}^{p_0}\frac{|u_j|}{|\beta_j|}. \end{aligned}$$

We then have $\mathcal{W}_n(u) > 0$ for all $\|\boldsymbol{U}\| = M$, where M is chosen sufficiently large. By convexity of $D(\boldsymbol{\beta}^*)$, it follows that $\mathcal{W}_n > 0, \forall \|\boldsymbol{U}\| \geq M$. This indicates that all minimizer of $Q_n(\boldsymbol{\beta}^*)$ must lie in the M ball around $\boldsymbol{\beta}^*$. Therefore, $\left\|\hat{\boldsymbol{\beta}}_n^* - \boldsymbol{\beta}^*\right\| = O_p(n^{-\frac{1}{2}})$. To show $P(\boldsymbol{\beta}_2 = \mathbf{0}) = 1$, we note from asymptotic linearity of $L_n(\boldsymbol{\beta}^*)$,

$$\begin{aligned} n^{-\frac{1}{2}} \left.\frac{\partial Q_n(\boldsymbol{\beta})}{\partial \beta_j}\right|_{\boldsymbol{\beta}=\boldsymbol{\beta}^*} &= -n^{-\frac{1}{2}}L_{n_j}(\boldsymbol{\beta}^*) + A_\phi \boldsymbol{C}_{(j)}\left\{\sqrt{n}\left(\hat{\boldsymbol{\beta}}^* - \boldsymbol{\beta}^*\right)\right\} \\ &\quad + n\lambda_n \frac{\operatorname{sgn}(\boldsymbol{\beta}^*)}{|\sqrt{n}\hat{\beta}_j^{\mathrm{UR}}|} + o(1), \end{aligned}$$

where $\boldsymbol{C}_{(j)}$ is the jth row of $\boldsymbol{C}$.

Because $n^{\frac{1}{2}}\boldsymbol{L}_n(\boldsymbol{\beta}^*) = O_p(1)$ and $n^{\frac{1}{2}}|\hat{\beta}_j^{\mathrm{UR}}| = O_p(1)$ for $j = p_0 + 1, \ldots, p$, we have

$$n^{-\frac{1}{2}} \frac{\partial Q_n(\boldsymbol{\beta})}{\partial \beta_j}\bigg|_{\boldsymbol{\beta}=\boldsymbol{\beta}^*} = O_p(1) + n\lambda_n \frac{\mathrm{sgn}(\beta_j^*)}{O_p(1)}, \quad j = p_0 + 1, \ldots, p. \qquad (6.6.22)$$

Further, $n\lambda_n \to \infty$, the sign of $\frac{\partial Q_n(\boldsymbol{\beta})}{\partial \beta_j}\Big|_{\boldsymbol{\beta}=\boldsymbol{\beta}^*}$ is the same as that of $\beta_j^* \neq 0$, $j = p_0 + 1, \ldots, p$ as n is large enough. This implies that $P(\boldsymbol{\beta}_2 = 0) \to 1$. Coupled with the definition $\hat{\boldsymbol{\beta}}_n^*$, we can show that

$$\begin{aligned}
\frac{\partial Q_n(\boldsymbol{\beta})}{\partial \beta_j}\bigg|_{\boldsymbol{\beta}=\boldsymbol{\beta}^*} &= n^{-\frac{1}{2}}\boldsymbol{L}_{n_1}\begin{pmatrix}\hat{\boldsymbol{\beta}}_{1n}^* \\ \boldsymbol{0}\end{pmatrix} + \sqrt{n}\lambda_n \left(\frac{\mathrm{sgn}((\hat{\boldsymbol{\beta}})_1^{\mathrm{UR}})}{|\hat{\beta}_1^{\mathrm{UR}}|}, \ldots, \frac{\mathrm{sgn}((\hat{\boldsymbol{\beta}})_{p_0}^{\mathrm{UR}})}{|\hat{\beta}_{p_0}^{\mathrm{UR}}|}\right) \\
&= n^{-\frac{1}{2}}\boldsymbol{L}_{n_1}(\hat{\boldsymbol{\beta}}_{1n}) + \gamma(\phi, f)\boldsymbol{C}_{11}(\hat{\boldsymbol{\beta}}_{1n}^* - \boldsymbol{\beta}_1) \\
&\quad + n^{\frac{1}{2}}\lambda_n \left(\frac{\mathrm{sgn}((\hat{\boldsymbol{\beta}})_1^{\mathrm{UR}})}{|\hat{\beta}_1^{\mathrm{UR}}|}, \ldots, \frac{\mathrm{sgn}((\hat{\boldsymbol{\beta}})_{p_0}^{\mathrm{UR}})}{|\hat{\beta}_{p_0}^{\mathrm{UR}}|}\right).
\end{aligned}$$

Now, $\sqrt{n}\lambda_n \to \lambda_0$ and $\hat{\beta}_j^{\mathrm{UR}} \xrightarrow{P} \beta_j^* \neq 0$ for $j = 1, \ldots, p_0$ and

$$n^{\frac{1}{2}}\gamma(\phi, f)\boldsymbol{C}_{11}((\hat{\boldsymbol{\beta}}_{1n}^* - \boldsymbol{\beta}_1^*) + n^{-\frac{1}{2}}\gamma^{-1}(\phi, f)\boldsymbol{C}_{11}^{-1}\lambda_0 \boldsymbol{b}_1) = n^{-\frac{1}{2}}\boldsymbol{L}_{n_1}(\boldsymbol{\beta}^*) + o_p(1).$$

Given that we have $n^{\frac{1}{2}}\boldsymbol{L}_{n_1}(\boldsymbol{\beta}^*) \overset{\mathcal{D}}{\sim} \mathcal{N}_{p_0}\left(\boldsymbol{0}, A_\phi^2 \boldsymbol{C}_{11}\right)$, this leads to the result that $\sqrt{n}(\hat{\boldsymbol{\beta}}_{1n}^* - \boldsymbol{\beta}_1) \xrightarrow{\mathcal{D}} \mathcal{N}_{p_0}\left(\boldsymbol{0}, \eta^2 \boldsymbol{C}_{11}^{-1}\right)$. □

6.7 Summary

In this chapter, we considered the non-orthogonal multiple regression model. We provided comparative performance characteristics of the primary penalty estimators, namely, ridge, LASSO, elastic net and aLASSO together with the unrestricted, restricted, preliminary test and Stein–Saleh type R-estimators. We used the marginal distribution theory and asymptotic distributional L_2-risk function to evaluate the estimators. We also detailed the oracle properties of aLASSO.

6.8 Problems

6.1. For the general shrinkage R-estimators of Section 6.4, derive the formulas for ADB($\hat{\boldsymbol{\beta}}_n^{\text{ShrinkageR}}(c)$) and ADL$_2$-risk($\hat{\boldsymbol{\beta}}_n^{\text{ShrinkageR}}(c)$).

6.2. For the preliminary test R-estimator in Section 6.4.1, prove the ADB($\hat{\boldsymbol{\beta}}_n^{\text{PTR}}(\lambda)$), ADL$_2$-risk($\hat{\boldsymbol{\beta}}_n^{\text{PTR}}(\lambda)$) and ADRE($\hat{\boldsymbol{\beta}}_n^{\text{PTR}}(\lambda) : \hat{\boldsymbol{\beta}}_n^{\text{UR}}$) equations.

6.3. For the Stein–Saleh R-estimator of Section 6.4.2, derive the ADL-risk($\hat{\boldsymbol{\beta}}_n^{\text{STR}}(\lambda)$) and ADRE($\hat{\boldsymbol{\beta}}_n^{\text{STR}}(\lambda) : \hat{\boldsymbol{\beta}}_n^{\text{UR}}$) expressions and show that ADRE($\hat{\boldsymbol{\beta}}_n^{\text{STR}}(\lambda) : \hat{\boldsymbol{\beta}}_n^{\text{UR}}$) ≥ 1.

6.4. For the positive-rule Stein–Saleh R-estimator of Section 6.4.3 (LASSO-type), prove the expression ADL$_2$-risk($\hat{\boldsymbol{\beta}}_n^{\text{STR+}}(\lambda)$) and show, after deriving the expression, that ADRE($\hat{\boldsymbol{\beta}}_n^{\text{STR+}}(\lambda) : \hat{\boldsymbol{\beta}}_n^{\text{UR}}$) ≥ 1.

6.5. For the preliminary test subset selector of Section 6.5.1, derive the equations for ADL$_2$-risk($\hat{\boldsymbol{\beta}}_n^{\text{PTSSR}}(\lambda)$) and its lower bound. Also, derive the full expression for ADRE($\hat{\boldsymbol{\beta}}_n^{\text{PTSSR}}(\lambda) : \hat{\boldsymbol{\beta}}_n^{\text{UR}}$).

6.6. For the Stein–Saleh subset selector estimator of Section 6.5.2, find ADB($\hat{\boldsymbol{\beta}}_n^{\text{SR}}(\lambda)$), ADL$_2$-risk($\hat{\boldsymbol{\beta}}_n^{\text{SR}}(\lambda)$) and ADRE($\hat{\boldsymbol{\beta}}_n^{\text{SR}}(\lambda) : \hat{\boldsymbol{\beta}}_n^{\text{UR}}$).

6.7. For the positive-rule Stein–Saleh R-estimator (LASSO-type) of Section 6.5.3, determine the ADB($\hat{\boldsymbol{\beta}}_n^{\text{PRSR}}(\lambda)$), ADL$_2$-risk($\hat{\boldsymbol{\beta}}_n^{\text{PRSR}}(\lambda)$) and the lower bound of the ADL$_2$-risk($\hat{\boldsymbol{\beta}}_n^{\text{PRSR}}(\lambda)$).

6.8. For the ridge-type selector of Section 6.5.4, find the expression for the selector and the lower bound of ADL$_2$-risk.

6.9. Derive the naive elastic net selector following Section 6.5.5 and the lower bound of the ADL$_2$-risk expression.

6.10. Show that the adaptive LASSO (aLASSO) R-estimator has the oracle property.

7

Partially Linear Multiple Regression Model

7.1 Introduction

In this chapter, we consider the usual partially linear model (PLM) of the form

$$y_i = \boldsymbol{x}_i^\top \boldsymbol{\beta} + g(t_i) + \varepsilon_i, \quad i = 1, \dots, n \tag{7.1.1}$$

where y_i is the response, $\boldsymbol{x}_i = (x_{i1}, \dots, x_{ip})^\top$ is a vector of explanatory variables, $\boldsymbol{\beta} = (\beta_1, \dots, \beta_p)^\top$ is an unknown p-dimensional parameter vector, t_i is known and non-random in some bounded domain $D \in \mathbb{R}$, and $g(t_i)$ is an unknown smooth function. The ε_i's are i.i.d. random errors with absolutely continuous c.d.f. $F(\cdot)$ and p.d.f. $f(\cdot) > 0$ having the first derivative $f'(\cdot)$ satisfying finite Fisher information

$$I(f) = \int_{-\infty}^{\infty} \left(-\frac{f'(x)}{f(x)} \right)^2 f(x) \mathrm{d}x < \infty, \tag{7.1.2}$$

and $\mathbb{E}(\epsilon_i) = 0, \forall i$, which is independent of $(\boldsymbol{x}_i, t_i)$. Here, we assume that the response y_i linearly depends on $\boldsymbol{x}_i$, but that it is nonlinearly related to t. Further,

$$\gamma(\varphi, \psi) = \int_0^1 \varphi(u)\psi(u)\mathrm{d}u, \quad \psi(u) = -\frac{f'(F^{-1}(u))}{f(F^{-1}(u))}, \tag{7.1.3}$$

and $\varphi(u)$ is a non-decreasing score generating function, $0 \le u \le 1$.

Surveys regarding the estimation and application of the model of Eq. (7.1.1) can be found in the monograph of Hardle et al. (2000). For more studies about the PLM, we refer the reader to Roozbeh and Arashi (2016) and Saleh et al. (2019) among others. For the estimation of $\boldsymbol{\beta}$, we first need to estimate the nonparametric component $g(\cdot)$, where we use the kernel smoothing method.

Rank-Based Methods for Shrinkage and Selection: With Application to Machine Learning.
First Edition. A. K. Md. Ehsanes Saleh, Mohammad Arashi, Resve A. Saleh, and
Mina Norouzirad.

To estimate $g(\cdot)$, assume that $(y_i, \boldsymbol{x}_i, t_i)$, $i = 1, \dots, n$ satisfy the model of Eq. (7.1.1). Since $\mathbb{E}(\varepsilon_i) = 0$, we have $g(t_i) = \mathbb{E}(y_i - \boldsymbol{x}_i^\top \boldsymbol{\beta})$ for $i = 1, \dots, n$. Hence, if we know $\boldsymbol{\beta}$, a natural nonparametric estimator of $g(\cdot)$ is given by

$$\hat{g}(t, \boldsymbol{\beta}) = \sum_{i=1}^{n} W_{ni}(t)(y_i - \boldsymbol{x}_i^\top \boldsymbol{\beta}), \tag{7.1.4}$$

where the probability weight function, $W_{ni}(\cdot)$, satisfies the following three regularity conditions

(i) $\max_{1 \le i \le n} \sum_{j=1}^{n} W_{ni}(t_j) = O(1)$,

(ii) $\max_{1 \le i,j \le n} W_{ni}(t_j) = O(n^{-\frac{2}{3}})$,

(iii) $\max_{1 \le i \le n} \sum_{j=1}^{n} W_{ni}(t_j) I(|t_i - t_j| > c_n) = O(d_n)$, where $I(A)$ is the indicator function of the set A,

$$\limsup_{n \to \infty} n c_n^3 < \infty,$$
$$\limsup_{n \to \infty} n d_n^3 < \infty.$$

The above assumptions guarantee the existence of $\hat{g}(t, \boldsymbol{\beta})$ with the optimal convergence rate of $n^{-4/5}$ in the PLM with probability one. See Muller and Ronz (2000) for more details. Throughout the rest of this chapter, we do not discuss further the estimation of $g(\cdot)$ in the PLMs, since the main concern is the estimation of $\boldsymbol{\beta}$.

7.2 Rank Estimation in the PLM

Consider the PLM defined by

$$y_i = \boldsymbol{x}_{i(1)}^\top \boldsymbol{\beta}_1 + \boldsymbol{x}_{i(2)}^\top \boldsymbol{\beta}_2 + g(t_i) + \varepsilon_i, \quad i = 1, \dots, n \tag{7.2.1}$$

where y_i are responses, $\boldsymbol{x}_{i(1)} = (x_{i1}, \dots, x_{ip_1})^\top$ and $\boldsymbol{x}_{i(2)} = (x_{i(p_1+1)}, \dots, x_{ip})^\top$ are covariates of the unknown vectors, namely $\boldsymbol{\beta}_1 = (\beta_1, \dots, \beta_{p_1})^\top$ and $\boldsymbol{\beta}_2 = (\beta_{p_1+1}, \dots, \beta_p)^\top$, respectively, $t_i \in [0, 1]$ are design points, $g(\cdot)$ is a real valued function defined on $[0, 1]$, and $\varepsilon_1, \dots, \varepsilon_n$ are i.i.d. errors with zero mean and constant $I(f)$.

Our main objective is to estimate $(\boldsymbol{\beta}_1^\top, \boldsymbol{\beta}_2^\top)^\top$ when one suspects that the p_2-dimensional sparsity condition $\boldsymbol{\beta}_2 = \mathbf{0}$ may hold. In this situation, we first estimate the parameter $\boldsymbol{\beta} = (\boldsymbol{\beta}_1^\top, \boldsymbol{\beta}_2^\top)^\top$ using Speckman's (1988) approach involving a partial kernel smoothing method, which attains the usual parametric convergence rate $n^{-1/2}$ without under-smoothing the nonparametric component $g(\cdot)$.

Assume that $\left\{(\boldsymbol{x}_{j(1)}^{\top}, \boldsymbol{x}_{j(2)}^{\top}), t_j, y_j | j = 1, 2, \ldots, n\right\}$ satisfies the model of Eq. (7.2.1). If $\boldsymbol{\beta} = (\boldsymbol{\beta}_1^{\top}, \boldsymbol{\beta}_2^{\top})^{\top}$ is the true parameter, then for $j = 1, \ldots, n$, we have

$$g(t_i) = \mathbb{E}(y_i - \boldsymbol{x}_{i(1)}^{\top}\boldsymbol{\beta}_1 - \boldsymbol{x}_{i(2)}^{\top}\boldsymbol{\beta}_2).$$

Thus, a natural nonparametric estimator of $g(\cdot)$ given $\boldsymbol{\beta} = (\boldsymbol{\beta}_1^{\top}, \boldsymbol{\beta}_2^{\top})^{\top}$ is

$$\tilde{g}(\boldsymbol{t}, \boldsymbol{\beta}_1, \boldsymbol{\beta}_2) = \sum_{i=1}^{n} W_{ni}(t_i)\left(Y_i - \boldsymbol{x}_{i(1)}^{\top}\boldsymbol{\beta}_1 - \boldsymbol{x}_{i(2)}^{\top}\boldsymbol{\beta}_2\right), \quad \boldsymbol{t} = (t_1, \ldots, t_n) \tag{7.2.2}$$

with the probability weight function $W_{ni}(\cdot)$ that satisfies the three regularity conditions (i)–(iii), given before Section 7.2.

Suppose that we partition the design matrix as follows

$$\boldsymbol{Z} = (\underset{n\times p_1}{\boldsymbol{Z}_1} \;\vdots\; \underset{n\times p_2}{\boldsymbol{Z}_2})^{\top}$$

and define others as

$$\hat{\boldsymbol{y}} = (\hat{y}_1, \ldots, \hat{y}_n)^{\top}, \quad \boldsymbol{Z} = (\boldsymbol{z}_1, \ldots, \boldsymbol{z}_n)^{\top},$$

where

$$\hat{y}_i = y_i - \sum_{k=1}^{n} W_{nk}(t_i)y_k, \quad \boldsymbol{z}_i = \boldsymbol{x}_i - \sum_{k=1}^{n} W_{nk}(t_i)\boldsymbol{x}_k, \quad j = 1, \ldots, n.$$

Further, assume that $(\boldsymbol{x}_i, t_i)$ are i.i.d. random variables and let

$$\mathbb{E}(x_{is}|t_i) = h_s(t_i), \qquad s = 1, \ldots, p.$$

Then, the conditional variance of x_{is} given t_i is $\mathbb{E}(\boldsymbol{u}_i\boldsymbol{u}_i^{\top}) = \boldsymbol{C}$, where $\boldsymbol{u}_i = (u_{i1}, \ldots, u_{ip})^{\top}$ with $u_{is} = x_{is} - h_s(t_i)$. Then, by L.L.N.,

$$\lim_{n\to\infty} \frac{1}{n}\sum_{k=1}^{n} u_{ik}u_{kj} = C_{ij} \qquad i, j = 1, \ldots, p,$$

and the matrix $\boldsymbol{C} = (C_{ij})$ (see Eq. (7.2.11)) is non-singular with probability one. Readers are referred to Shi and Lau (2000), Gao (1995, 1997) and Liang and Hardle (1999) among others. Moreover, for any permutation $(j_1, \ldots, j_k)$ of $(1, \ldots, n)$ as $n \to \infty$,

$$\left\|\max \sum_{i=1}^{n} W_{n_i}(t_j)u_i\right\| = O\left(n^{-\frac{1}{6}}\right). \qquad (7.2.3)$$

We also assume the functions $g(\cdot)$ and $h_s(\cdot)$ satisfy the Lipschitz condition of order 1 on $[0, 1]$ for $s = 1, \ldots, p$.

To estimate $\boldsymbol{\beta} = (\boldsymbol{\beta}_1^\top, \boldsymbol{\beta}_2^\top)^\top$, we minimize the Jaeckel's rank dispersion function

$$\begin{aligned} D_n(\boldsymbol{\beta}_1, \boldsymbol{\beta}_2) &= \sum_{i=1}^{n}(y_i - \boldsymbol{x}_{i(1)}^\top\boldsymbol{\beta}_1 - \boldsymbol{x}_{i(2)}^\top\boldsymbol{\beta}_2 - \tilde{g}(t_i, \boldsymbol{\beta}_1, \boldsymbol{\beta}_2))a_n(R_{ni}(\boldsymbol{\beta}_1, \boldsymbol{\beta}_2)) \\ &= \sum_{i=1}^{n}(\hat{y}_i - \boldsymbol{z}_{i(1)}^\top\boldsymbol{\beta}_1 - \boldsymbol{z}_{i(2)}^\top\boldsymbol{\beta}_2)a_n(R_{ni}(\boldsymbol{\beta}_1, \boldsymbol{\beta}_2)), \qquad (7.2.4) \end{aligned}$$

with respect to $(\boldsymbol{\beta}_1, \boldsymbol{\beta}_2)$ where

$$a_n(R_{ni}(\boldsymbol{\beta}_1, \boldsymbol{\beta}_2)) = \varphi\left(\frac{R_{ni}(\boldsymbol{\beta}_1, \boldsymbol{\beta}_2)}{n+1}\right), \quad i = 1, 2, \ldots, n \qquad (7.2.5)$$

are the scores based on the ranks of $\hat{y}_i - z_{i(1)}\boldsymbol{\beta}_1 - z_{i(2)}\boldsymbol{\beta}_2$ among

$$(\hat{y}_1 - z_{1(1)}\boldsymbol{\beta}_1 - z_{1(2)}\boldsymbol{\beta}_2), \ldots, (\hat{y}_n - z_{n(1)}\boldsymbol{\beta}_1 - z_{n(2)}\boldsymbol{\beta}_2)$$

with $\int_0^1 \varphi(u)\mathrm{d}u = 0$ and

$$A_\varphi^2 = \int_0^1 |\varphi(u)|\{u(1-u)\}^{\frac{1}{2}}\mathrm{d}u < \infty. \qquad (7.2.6)$$

Recall that $\varphi : (0, 1) \mapsto \mathbb{R}$ is the score generating function which is assumed to be non-constant, non-decreasing and square integrable on $(0, 1)$.

In addition to Eq. (7.2.5), the scores can also be defined the following way:

$$a_n(i) = \mathbb{E}\left[\varphi(U_{i:n})\right], \quad i = 1, \ldots, n,$$

where $U_{1:n} \leq \ldots \leq U_{n:n}$ are order statistics from a sample of size n from the uniform distribution $\mathcal{U}(0, 1)$.

Consider that

$$\begin{aligned} \frac{\partial D_n(\boldsymbol{\beta}_1, \boldsymbol{\beta}_2)}{\partial(\boldsymbol{\beta}_1, \boldsymbol{\beta}_2)} &= -\boldsymbol{L}_n(\boldsymbol{\beta}_1, \boldsymbol{\beta}_2) \\ &= -(L_{n1}(\boldsymbol{\beta}_1, \boldsymbol{\beta}_2), \ldots, L_{nn}(\boldsymbol{\beta}_1, \boldsymbol{\beta}_2))^\top, \qquad (7.2.7) \end{aligned}$$

where

$$\boldsymbol{L}_n(\boldsymbol{\beta}_1, \boldsymbol{\beta}_2) = \sum_{i=1}^{n}(z_i - \bar{z}_n)a_n(R_{ni}(\boldsymbol{\beta}_1, \boldsymbol{\beta}_2)) \qquad (7.2.8a)$$

$$L_{nj}(\boldsymbol{\beta}_1, \boldsymbol{\beta}_2) = \sum_{i=1}^{n}(z_{ij} - \bar{z}_{nj})a_n(R_{nj}(\boldsymbol{\beta}_1, \boldsymbol{\beta}_2)),\ j = 1, \ldots, p. \tag{7.2.8b}$$

Then, define the URE of $(\boldsymbol{\beta}_1^\top, \boldsymbol{\beta}_2^\top)^\top$ as

$$\hat{\boldsymbol{\beta}}_n^{\mathrm{UR}} = (\hat{\boldsymbol{\beta}}_{1n}^{\mathrm{UR}^\top}, \hat{\boldsymbol{\beta}}_{2n}^{\mathrm{UR}^\top})^\top = \underset{\boldsymbol{\beta}_1 \in \mathbb{R}^{p_1}, \boldsymbol{\beta}_2 \in \mathbb{R}^{p_2}}{\operatorname{argmin}} \{D_n(\boldsymbol{\beta}_1, \boldsymbol{\beta}_2)\}. \tag{7.2.9}$$

Note that if $\boldsymbol{\beta}_2 = \mathbf{0}$ is true, then the RRE of $\boldsymbol{\beta}_1$ is

$$\hat{\boldsymbol{\beta}}_n^{\mathrm{RR}} = (\hat{\boldsymbol{\beta}}_{1n}^{\mathrm{UR}^\top}, \mathbf{0}^\top)^\top, \quad \text{where} \quad \hat{\boldsymbol{\beta}}_{1n}^{\mathrm{UR}} = \underset{\boldsymbol{\beta}_1 \in \mathbb{R}^{p_1}}{\operatorname{argmin}} \left\{D_n(\boldsymbol{\beta}_1, \mathbf{0})\right\}. \tag{7.2.10}$$

It can be verified that

$$\hat{\boldsymbol{\beta}}_{1n}^{\mathrm{RR}} = \underset{\boldsymbol{\beta}_1 \in \mathbb{R}^{p_1}}{\operatorname{argmin}} \sum_{j=1}^{p_1} |L_{n_j}(\boldsymbol{\beta}_1, \mathbf{0})|.$$

Let us define

$$\boldsymbol{C}_n = \frac{1}{n}\sum_{i=1}^{n}(\boldsymbol{z}_i - \bar{\boldsymbol{z}}_n)(\boldsymbol{z}_i - \bar{\boldsymbol{z}}_n)^\top, \tag{7.2.11}$$

where $\boldsymbol{z}_i$ is the ith row of $\boldsymbol{Z}$ and $\bar{\boldsymbol{z}}_n = \frac{1}{n}\sum_{i=1}^{n} \boldsymbol{z}_i$. We assume $\boldsymbol{C}_n$ is invertible, and

$$\text{(a)} \lim_{n\to\infty} \boldsymbol{C}_n = \boldsymbol{C} \quad \text{and} \quad \text{(b)} \lim_{n\to\infty} \max_{1\le i\le n} (\boldsymbol{z}_i - \bar{\boldsymbol{z}}_n)^\top \boldsymbol{C}_n^{-1}(\boldsymbol{z}_i - \bar{\boldsymbol{z}}_n) = 0, \tag{7.2.12}$$

where $\boldsymbol{C}$ is an invertible matrix of dimension p.

In the sequel, we shall use the following partitioning where needed:

$$\boldsymbol{C}_n = \begin{bmatrix} \boldsymbol{C}_{n11} & \boldsymbol{C}_{n12} \\ \boldsymbol{C}_{n21} & \boldsymbol{C}_{n22} \end{bmatrix}, \quad \boldsymbol{C} = \begin{bmatrix} \boldsymbol{C}_{11} & \boldsymbol{C}_{12} \\ \boldsymbol{C}_{21} & \boldsymbol{C}_{22} \end{bmatrix}$$

$$\boldsymbol{C}_n^{-1} = \begin{bmatrix} \boldsymbol{C}_{n11\cdot2}^{-1} & -\boldsymbol{C}_{n11\cdot2}^{-1}\boldsymbol{C}_{n12}\boldsymbol{C}_{n22}^{-1} \\ -\boldsymbol{C}_{n22}^{-1}\boldsymbol{C}_{n21}\boldsymbol{C}_{n11\cdot2}^{-1} & \boldsymbol{C}_{n22\cdot1}^{-1} \end{bmatrix},$$

where $\boldsymbol{C}_{n22\cdot1}$ and $\boldsymbol{C}_{n11\cdot2}$ are the Schur complements of $\boldsymbol{C}_{n11}$ and $\boldsymbol{C}_{n22}$, respectively, given by

$$\boldsymbol{C}_{n22\cdot1} = \boldsymbol{C}_{n22} - \boldsymbol{C}_{n21}\boldsymbol{C}_{n11}^{-1}\boldsymbol{C}_{n12},$$

$$\boldsymbol{C}_{n11\cdot2} = \boldsymbol{C}_{n11} - \boldsymbol{C}_{n12}\boldsymbol{C}_{n22}^{-1}\boldsymbol{C}_{n21}.$$

Theorem 7.1 Under the assumed regularity conditions, as $n \to \infty$,

$$\sqrt{n}(\hat{\boldsymbol{\beta}}_n^{\mathrm{UR}} - \boldsymbol{\beta}) \xrightarrow{\mathcal{D}} \mathcal{N}_p(\mathbf{0}, \eta^2 \mathbf{C}^{-1}), \quad \eta^2 = \frac{A_\varphi^2}{\gamma^2(\varphi, \psi)},$$

where $\gamma(\varphi, \psi)$ and A_φ are given by Eqs. (7.1.3) and (7.2.6), respectively.

As a result of Theorem 7.1, we have the following limiting cases as $n \to \infty$,

(i) $\sqrt{n}(\hat{\boldsymbol{\beta}}_{1n}^{\mathrm{UR}} - \boldsymbol{\beta}_1) \xrightarrow{\mathcal{D}} \mathcal{N}_{p_1}(\mathbf{0}, \eta^2 \mathbf{C}_{11\cdot 2}^{-1})$,

(ii) $\sqrt{n}(\hat{\boldsymbol{\beta}}_{2n}^{\mathrm{UR}} - \boldsymbol{\beta}_2) \xrightarrow{\mathcal{D}} \mathcal{N}_{p_2}(\mathbf{0}, \eta^2 \mathbf{C}_{22\cdot 1}^{-1})$,

(iii) $\sqrt{n}(\hat{\boldsymbol{\beta}}_{1n}^{\mathrm{RR}} - \boldsymbol{\beta}_1) \xrightarrow{\mathcal{D}} \mathcal{N}_{p_1}(\mathbf{0}, \eta^2 \mathbf{C}_{11}^{-1})$,

(iv) $\sqrt{n}(\hat{\beta}_{jn}^{\mathrm{UR}} - \beta_j) \xrightarrow{\mathcal{D}} \mathcal{N}(0, \eta^2 C^{jj})$, where C^{jj} is the jth diagonal element of $\mathbf{C}^{-1}$.

7.2.1 Penalty R-estimators

We define the LASSO R-estimator as

$$\begin{aligned}
\sqrt{n}\hat{\boldsymbol{\beta}}_n^{\mathrm{LassoR}}(\lambda) &= \underset{(\boldsymbol{\beta}_1, \boldsymbol{\beta}_2) \in \mathbb{R}^p}{\operatorname{argmin}} \left\{ D_n(\boldsymbol{\beta}_1, \boldsymbol{\beta}_2) + \lambda \eta \sqrt{C^{jj}} \mathbf{1}_p^\top |\boldsymbol{\beta}| \right\} \\
&= \left(\operatorname{sgn}(\sqrt{n}\hat{\beta}_{jn}^{\mathrm{UR}}) \left(|\sqrt{n}\hat{\beta}_{jn}^{\mathrm{UR}}| - \lambda \eta \sqrt{C^{jj}} \right)^+ \Big| \; j = 1, \dots, p \right)^\top \\
&= \left(\eta \sqrt{C^{jj}} \operatorname{sgn}(Z_j)(|Z_j| - \lambda)^+ \Big| \; j = 1, \dots, p \right)^\top, \qquad (7.2.13)
\end{aligned}$$

where $Z_j = \frac{\sqrt{n}}{\eta\sqrt{C^{jj}}} \hat{\beta}_{jn}^{\mathrm{UR}}$.

Note that

$$\sqrt{n}\hat{\beta}_{jn}^{\mathrm{LassoR}}(\lambda) = \begin{cases} \eta\sqrt{C^{jj}}(Z_j - \lambda \operatorname{sgn}(Z_j)), & \text{if} \quad |Z_j| > \lambda, \\ 0, & \text{otherwise.} \end{cases}$$

Let $|Z_1| > \lambda, \dots, |Z_{p_1}| > \lambda$ and the rest of the p_2 components be 0. Then

$$\sqrt{n}\hat{\boldsymbol{\beta}}_n^{\mathrm{LassoR}}(\lambda) = (\eta\sqrt{C^{jj}}(Z_j - \lambda \operatorname{sgn}(Z_j)), j = 1, \dots, p_1, \mathbf{0}^\top)^\top$$

is the vector of the LASSO R-estimator.

Next, we consider the ridge R-estimator of $\boldsymbol{\beta}$. Due to Saleh et al. (2019), it is given by

$$\hat{\boldsymbol{\beta}}_n^{\text{ridgeR}}(k) = (\boldsymbol{I}_p + k\boldsymbol{C}^{-1})^{-1}\hat{\boldsymbol{\beta}}_n^{\text{UR}}, \tag{7.2.14}$$

for some ridge parameter $k > 0$.

Using the result of Theorem 7.1 marginal theory, we can define the subsequent ridge R-estimators as

$$\hat{\boldsymbol{\beta}}_{1n}^{\text{ridgeR}}(k) = (\boldsymbol{I}_{p_1} + k\boldsymbol{C}_{11\cdot2}^{-1})^{-1}\hat{\boldsymbol{\beta}}_{1n}^{\text{UR}} \tag{7.2.15}$$

$$\hat{\boldsymbol{\beta}}_{2n}^{\text{ridgeR}}(k) = (\boldsymbol{I}_{p_2} + k\boldsymbol{C}_{22\cdot1}^{-1})^{-1}\hat{\boldsymbol{\beta}}_{2n}^{\text{UR}}, \tag{7.2.16}$$

respectively.

If one suspects that $\boldsymbol{\beta}_2 = \mathbf{0}$ may hold, then the ridge R-estimator has the form

$$\hat{\boldsymbol{\beta}}_n^{\text{ridgeR}}(k) = \left(\hat{\boldsymbol{\beta}}_{1n}^{\text{UR}^\top}, \hat{\boldsymbol{\beta}}_{2n}^{\text{ridgeR}}(k)^\top\right)^\top. \tag{7.2.17}$$

Finally, as a special case, when $\boldsymbol{\beta}_2 = \mathbf{0}$, we have the following simple form

$$\begin{aligned}\hat{\boldsymbol{\beta}}_n^{\text{ridgeR}}(k) &= \underset{(\boldsymbol{\beta}_1,\boldsymbol{\beta}_2)\in\mathbb{R}^p}{\operatorname{argmin}} \{D_n(\boldsymbol{\beta}_1, \boldsymbol{\beta}_2) + k\|\boldsymbol{\beta}_2\|^2\} \\ &= \left(\hat{\boldsymbol{\beta}}_{1n}^{\text{UR}^\top}, \frac{1}{1+k}\hat{\boldsymbol{\beta}}_{2n}^{\text{UR}^\top}\right)^\top.\end{aligned} \tag{7.2.18}$$

We can also define the elastic net (Enet) R-estimator as

$$\begin{aligned}\hat{\boldsymbol{\beta}}_n^{\text{EnetR}}(\lambda, \alpha) &= \underset{\boldsymbol{\beta}\in\mathbb{R}^p}{\operatorname{argmin}}\left\{D_n(\boldsymbol{\beta}_1, \boldsymbol{\beta}_2) + \lambda\left[\alpha\eta\mathbf{1}_p^\top\boldsymbol{C}^{\frac{1}{2}}|\boldsymbol{\beta}| + \frac{1}{2}(1-\alpha)\boldsymbol{\beta}^\top\boldsymbol{C}\boldsymbol{\beta}\right]\right\} \\ &= \left[\boldsymbol{I}_p + \lambda(1-\alpha)\boldsymbol{C}^{-1}\right]^{-1}\hat{\boldsymbol{\beta}}_n^{\text{LassoR}}(\lambda\alpha),\end{aligned}$$

where

$$\hat{\boldsymbol{\beta}}_n^{\text{LassoR}}(\lambda\alpha) = \left(\operatorname{sgn}(\hat{\beta}_{jn}^{\text{UR}})\left(|\hat{\beta}_{jn}^{\text{UR}}| - \lambda\alpha\eta\sqrt{C^{jj}}\right)^+ \mid j = 1, \ldots, p\right)^\top.$$

For the orthonormal case, where $\boldsymbol{C} = \boldsymbol{I}_p$, the Enet R-estimator has the form

$$\begin{aligned}\hat{\boldsymbol{\beta}}_n^{\text{EnetR}}(\lambda, \alpha) &= [1 + \lambda(1-\alpha)]^{-1}\hat{\boldsymbol{\beta}}_n^{\text{LassoR}}(\lambda\alpha) \\ &= [1 + \lambda(1-\alpha)]^{-1}\left(\operatorname{sgn}(\hat{\beta}_{jn}^{\text{UR}})\left(|\hat{\beta}_{jn}^{\text{UR}}| - \lambda\alpha\eta\right)^+;\ j = 1, \ldots, p\right)^\top.\end{aligned}$$

As $\alpha \to 0$, the Enet R-estimator tends to $\hat{\boldsymbol{\beta}}_n^{\text{ridgeR}}(\lambda)$. As $\alpha \to 1$, the Enet R-estimator tends to the LASSO R-estimator.

The next section is devoted to the definition of shrinkage R-estimators.

7.2.2 Preliminary Test and Stein–Saleh-type R-estimator

If we suspect a sparsity condition that $\boldsymbol{\beta}_2 = \mathbf{0}$, we need to test it before making use of the RRE, $\hat{\boldsymbol{\beta}}_n^{\text{RR}} = (\hat{\boldsymbol{\beta}}_{1n}^{\text{RR}^\top}, \mathbf{0}^\top)^\top$. For the test of the null hypothesis $\mathcal{H}_o : \boldsymbol{\beta}_2 = \mathbf{0}$ vs. $\mathcal{H}_A : \boldsymbol{\beta}_2 \neq \mathbf{0}$, we consider the partitioning

$$\boldsymbol{L}_n(\boldsymbol{\beta}_1, \boldsymbol{\beta}_2) = (\boldsymbol{L}_{n(1)}^\top(\boldsymbol{\beta}_1, \boldsymbol{\beta}_2), \boldsymbol{L}_{n(2)}^\top(\boldsymbol{\beta}_1, \boldsymbol{\beta}_2))^\top, \tag{7.2.19}$$

where

$$\begin{aligned} \boldsymbol{L}_{n(1)}(\boldsymbol{\beta}_1, \boldsymbol{\beta}_2) &= (L_{n1}(\boldsymbol{\beta}_1, \boldsymbol{\beta}_2), \dots, L_{np_1}(\boldsymbol{\beta}_1, \boldsymbol{\beta}_2))^\top, \\ \boldsymbol{L}_{n(2)}(\boldsymbol{\beta}_1, \boldsymbol{\beta}_2) &= (L_{np_1+1}(\boldsymbol{\beta}_1, \boldsymbol{\beta}_2), \dots, L_{np}(\boldsymbol{\beta}_1, \boldsymbol{\beta}_2))^\top. \end{aligned}$$

Let

$$\hat{\boldsymbol{L}}_{n(2)}(\hat{\boldsymbol{\beta}}_{1n}^{\text{RR}}, \mathbf{0}) = (\hat{\boldsymbol{L}}_{np_1+1}(\hat{\boldsymbol{\beta}}_{1n}^{\text{RR}}, \mathbf{0}), \dots, \hat{\boldsymbol{L}}_{np}(\hat{\boldsymbol{\beta}}_{1n}^{\text{RR}}, \mathbf{0}))^\top. \tag{7.2.20}$$

Then, we propose the following test statistic

$$\mathcal{L}_n = A_n^{-2} \hat{\boldsymbol{L}}_{n(2)}^\top(\hat{\boldsymbol{\beta}}_{1n}^{\text{RR}}, \mathbf{0}) \boldsymbol{C}_{22\cdot1}^{-1} \hat{\boldsymbol{L}}_{n(2)}(\hat{\boldsymbol{\beta}}_{1n}^{\text{RR}}, \mathbf{0}), \tag{7.2.21}$$

where

$$A_n^2 = \frac{1}{n-1} \sum_{i=1}^{n} (a_n(i) - \bar{a}_n)^2, \quad \bar{a}_n = \frac{1}{n} \sum_{i=1}^{n} a_n(i).$$

Under the assumed regularity conditions of Eq. (7.2.12), as $n \to \infty$, $\mathcal{L}_n$ follows the χ^2 distribution with p_2 d.f. under $\mathcal{H}_o$.

Then, we define the PTRE of $(\boldsymbol{\beta}_1^\top, \boldsymbol{\beta}_2^\top)^\top$ as

$$\hat{\boldsymbol{\beta}}_n^{\text{PTR}}(\alpha) = (\hat{\boldsymbol{\beta}}_{1n}^{\text{UR}^\top}, \hat{\boldsymbol{\beta}}_{2n}^{\text{UR}^\top} - \hat{\boldsymbol{\beta}}_{2n}^{\text{UR}^\top} I(\mathcal{L}_n \leq \chi_{p_2}^2(\alpha)))^\top, \tag{7.2.22}$$

where $I(A)$ is the indicator function of the set A and $\chi_{p_2}^2(\alpha)$ is the α-level critical value of χ^2 distribution with p_2 degrees of freedom.

Similarly, we define the Stein–Saleh-type shrinkage R-estimator as

$$\hat{\boldsymbol{\beta}}_n^{\text{SSR}} = (\hat{\boldsymbol{\beta}}_{1n}^{\text{UR}^\top}, \hat{\boldsymbol{\beta}}_{2n}^{\text{UR}^\top}(1 - (p_2 - 2)\mathcal{L}_n^{-1}))^\top. \tag{7.2.23}$$

Finally, the positive-rule Stein–Saleh-type shrinkage R-estimator is given by

$$\hat{\boldsymbol{\beta}}_n^{\mathrm{SSR+}} = (\hat{\boldsymbol{\beta}}_{1n}^{\mathrm{UR}^\top}, \hat{\boldsymbol{\beta}}_{2n}^{\mathrm{UR}^\top}(1-(p_2-2)\mathcal{L}_n^{-1})I(\mathcal{L}_n > p_2-2))^\top. \tag{7.2.24}$$

7.3 ADB and ADL$_2$-risk

Since $\boldsymbol{\beta}_2 = \mathbf{0}$ is uncertain, we test the null hypothesis $\mathcal{H}_o : \boldsymbol{\beta}_2 = \mathbf{0}$ vs. $\mathcal{H}_A : \boldsymbol{\beta}_2 \neq \mathbf{0}$ based on the rank statistic $\mathcal{L}_n$ of Eq. (7.2.21). This test is consistent and its power tends to unity as $n \to \infty$ for fixed alternatives. Thus, we consider a sequence of Pitman alternatives, $K_{(n)}$, defined by

$$K_{(n)} : \boldsymbol{\beta}_n = n^{-1/2}\boldsymbol{\delta} = n^{-1/2}(\boldsymbol{\delta}_1^\top, \boldsymbol{\delta}_2^\top)^\top. \tag{7.3.1}$$

If $\boldsymbol{\delta}_2 = \mathbf{0}$, $K_{(n)} = \mathcal{H}_o$. Then, under $K_{(n)}$, the marginal asymptotic distribution of the term given by $\sqrt{n}(\hat{\boldsymbol{\beta}}_{jn}^{\mathrm{UR}} - \boldsymbol{\beta}_j)$ is $\mathcal{N}_{p_j}(\mathbf{0}, \eta^2\boldsymbol{C}_{jj.i}^{-1})$, for $j(i) = 1(2), 2(1)$. Also, under the class of local alternatives, $K_{(n)}$, we have

$$\hat{\boldsymbol{\beta}}_{1n}^{\mathrm{RR}} = \hat{\boldsymbol{\beta}}_{1n}^{\mathrm{UR}} + \boldsymbol{C}_{11}^{-1}\boldsymbol{C}_{12}\hat{\boldsymbol{\beta}}_{2n}^{\mathrm{UR}} + o_p(n^{-\frac{1}{2}}) \tag{7.3.2}$$

$$\mathcal{L}_n = \eta^{-2}\hat{\boldsymbol{\beta}}_{2n}^{\mathrm{UR}^\top}\boldsymbol{C}_{22\cdot 1}\hat{\boldsymbol{\beta}}_{2n}^{\mathrm{UR}} + o_p(1). \tag{7.3.3}$$

In the sequel, we compute the asymptotic distributional bias and weighted L$_2$-risk expressions of the estimators. We study the comparative performance of the R-estimators based on the weighted L$_2$-risks defined by

$$\begin{aligned}\text{ADL}_2\text{-risk}(\hat{\boldsymbol{\beta}}_n^* : \boldsymbol{W}_1, \boldsymbol{W}_2) &= \lim_{n\to\infty} \mathbb{E}[(\hat{\boldsymbol{\beta}}_{1n}^* - \boldsymbol{\beta}_1)^\top \boldsymbol{W}_1(\hat{\boldsymbol{\beta}}_{1n}^* - \boldsymbol{\beta}_1)] \\ &\quad + \lim_{n\to\infty} \mathbb{E}[(\hat{\boldsymbol{\beta}}_{2n}^* - \boldsymbol{\beta}_2)^\top \boldsymbol{W}_2(\hat{\boldsymbol{\beta}}_{2n}^* - \boldsymbol{\beta}_2)]\end{aligned}$$

where $\hat{\boldsymbol{\beta}}_n^* = (\hat{\boldsymbol{\beta}}_{1n}^{*\top}, \hat{\boldsymbol{\beta}}_{2n}^{*\top})^\top$ is any estimator of $\boldsymbol{\beta} = (\boldsymbol{\beta}_1^\top, \boldsymbol{\beta}_2^\top)^\top$, and $\boldsymbol{W}_1$ and $\boldsymbol{W}_2$ are weight matrices. For convenience, when $\boldsymbol{W}_1 = \boldsymbol{I}_{p_1}$ and $\boldsymbol{W}_2 = \boldsymbol{I}_{p_2}$, we get the ADL$_2$-risk and write ADL$_2$-risk$(\hat{\boldsymbol{\beta}}_n^*) = \lim_{n\to\infty} \mathbb{E}[\|\hat{\boldsymbol{\beta}}_n^* - \boldsymbol{\beta}\|_2^2]$.

Let

$$\Delta^2 = \frac{1}{\eta^2}\boldsymbol{\delta}_2^\top \boldsymbol{C}_{22\cdot 1}\boldsymbol{\delta}_2. \tag{7.3.4}$$

Clearly, if $\boldsymbol{\delta}_2 = \mathbf{0}$, i.e. under the null-hypothesis, then $\Delta^2 = 0$.
In what follows, we use the Courant theorem (see Theorem A.2.4 Anderson, 2003) to obtain

$$\text{Ch}_{\min}(\boldsymbol{C}_{22\cdot1}^{-1}) \leq \frac{\boldsymbol{\delta}_2^{\top}\boldsymbol{\delta}_2}{\boldsymbol{\delta}_2^{\top}\boldsymbol{C}_{22\cdot1}\boldsymbol{\delta}_2} \leq \text{Ch}_{\max}(\boldsymbol{C}_{22\cdot1}^{-1}),$$

where $\text{Ch}_{\max(\min)}(\boldsymbol{A})$ is the largest (smallest) eigenvalue of $\boldsymbol{A}$.

Using Eq. (7.3.4), we obtain

$$\eta^2\Delta^2\text{Ch}_{\min}(\boldsymbol{C}_{22\cdot1}^{-1}) \leq \boldsymbol{\delta}_2^{\top}\boldsymbol{\delta}_2 \leq \eta^2\Delta^2\text{Ch}_{\max}(\boldsymbol{C}_{22\cdot1}^{-1}). \tag{7.3.5}$$

For the asymptotic risk functions, whenever we use the lower bound of $\boldsymbol{\delta}_2^{\top}\boldsymbol{\delta}_2$ from Eq. (7.3.5), we write ADL_2^*-risk, instead of ADL_2-risk.

Theorem 7.2 The ADB and ADL_2-risk of the R-estimators are given by

(i) Unrestricted R-estimator

$$\begin{aligned}\text{ADB}(\hat{\boldsymbol{\beta}}_n^{\text{UR}}) &= \begin{pmatrix}\mathbf{0}_{p_1}\\ \\ \mathbf{0}_{p_2}\end{pmatrix},\\ \text{ADL}_2^*\text{-risk}(\hat{\boldsymbol{\beta}}_n^{\text{UR}}) &= \eta^2\left[\text{tr}(\boldsymbol{C}_{11\cdot2}^{-1}) + \text{tr}(\boldsymbol{C}_{22\cdot1}^{-1})\right].\end{aligned}$$

Thus, we conclude

$$\text{ADL}_2\text{-risk}(\hat{\boldsymbol{\beta}}_n^{\text{UR}} : \boldsymbol{C}_{11\cdot2}, \boldsymbol{C}_{22\cdot1}) = \eta^2(p_1 + p_2).$$

(ii) Restricted R-estimator

$$\begin{aligned}\text{ADB}(\hat{\boldsymbol{\beta}}_n^{\text{RR}}) &= \begin{pmatrix}\mathbf{0}_{p_1}\\ \\ -\boldsymbol{\delta}_2\end{pmatrix},\\ \text{ADL}_2^*\text{-risk}(\hat{\boldsymbol{\beta}}_n^{\text{RR}}) &= \eta^2\left[\text{tr}(\boldsymbol{C}_{11}^{-1}) + \Delta^2\text{Ch}_{\min}(\boldsymbol{C}_{22\cdot1}^{-1})\right].\end{aligned}$$

Furthermore,

$$\text{ADL}_2\text{-risk}(\hat{\boldsymbol{\beta}}_n^{\text{RR}} : \boldsymbol{C}_{11\cdot2}, \boldsymbol{C}_{22\cdot1}) = \eta^2\left(\text{tr}(\boldsymbol{C}_{11}^{-1}\boldsymbol{C}_{11\cdot2}) + \Delta^2\right).$$

(iii) Preliminary test R-estimator

$$\begin{aligned}\text{ADB}\left(\hat{\boldsymbol{\beta}}_n^{\text{PTR}}(\alpha)\right) &= \begin{pmatrix}\mathbf{0}_{p_1}\\ \\ -\boldsymbol{\delta}_2\mathcal{H}_{p_2+2}(\chi_{p_2}^2(\alpha); \Delta^2)\end{pmatrix},\\ \text{ADL}_2^*\text{-risk}\left(\hat{\boldsymbol{\beta}}_n^{\text{PTR}}(\alpha)\right) &= \eta^2\Big[\text{tr}(\boldsymbol{C}_{11\cdot2}^{-1})\\ &\quad + \text{tr}\,\boldsymbol{C}_{22\cdot1}^{-1}(1 - \mathcal{H}_{p_2+2}(\chi_{p_2}^2(\alpha); \Delta^2)) + \Delta^2\text{Ch}_{\min}(\boldsymbol{C}_{22\cdot1}^{-1})\end{aligned}$$

$$\times(2\mathcal{H}_{p_2+2}(\chi^2_{p_2}(\alpha);\Delta^2) - \mathcal{H}_{p_2+4}(\chi^2_{p_2}(\alpha);\Delta^2))\Big]$$

where $\mathcal{H}_\nu(c;\Delta^2)$ is the c.d.f. of χ^2-distribution with ν d.f. and non-centrality parameter Δ^2 evaluated at c. Further,

$$\begin{aligned}\text{ADL}_2\text{-risk}\left(\hat{\boldsymbol{\beta}}_n^{\text{PTR}}(\alpha) : \boldsymbol{C}_{11\cdot2}, \boldsymbol{C}_{22\cdot1}\right) &= \eta^2 p_1 \\ &+\eta^2 p_2(1 - \mathcal{H}_{p_2+2}(\chi^2_{p_2}(\alpha);\Delta^2)) \\ &+\Delta^2(2\mathcal{H}_{p_2+2}(\chi^2_{p_2}(\alpha);\Delta^2) - \mathcal{H}_{p_2+4}(\chi^2_{p_2}(\alpha);\Delta^2)).\end{aligned}$$

(iv) Stein–Saleh-type Shrinkage R-estimator

$$\begin{aligned}\text{ADB}(\hat{\boldsymbol{\beta}}_n^{\text{SSR}}) &= \begin{pmatrix} \mathbf{0}_{p_1} \\ -\boldsymbol{\delta}_2(p_2-2)\mathbb{E}[\chi^{-2}_{p_2+2}(\Delta^2)] \end{pmatrix}, \\ \text{ADL}_2^*\text{-risk}\left(\hat{\boldsymbol{\beta}}_n^{\text{SSR}}\right) &= \eta^2\Big[\operatorname{tr}(\boldsymbol{C}_{11\cdot2}^{-1}) + \operatorname{tr}(\boldsymbol{C}_{22\cdot1}^{-1}) \\ -\Big\{(p_2-2)\operatorname{tr}(\boldsymbol{C}_{22\cdot1}^{-1})&\left(2\mathbb{E}[\chi^{-2}_{p_2+2}(\Delta^2)] - (p_2-2)\mathbb{E}[\chi^{-4}_{p_2+2}(\Delta^2)]\right)\Big\} \\ &+(p_2^2-4)\Delta^2\text{Ch}_{\min}(\boldsymbol{C}_{22\cdot1}^{-1})\mathbb{E}[\chi^{-4}_{p_2+4}(\Delta^2)]\Big]\end{aligned}$$

where

$$\mathbb{E}[\chi^{-2\nu}_{p_2}(\Delta^2)] = \int_0^\infty x^{-2\nu} d\mathcal{H}_{p_2}(x;\Delta^2).$$

Furthermore,

$$\text{ADL}_2\text{-risk}\left(\hat{\boldsymbol{\beta}}_n^{\text{SSR}} : \boldsymbol{C}_{11\cdot2}, \boldsymbol{C}_{22\cdot1}\right) = \eta^2 p_1 + \eta^2\left(p_2 - (p_2-2)^2\mathbb{E}[\chi^{-2}_{p_2+2}(\Delta^2)]\right).$$

(v) Positive-rule Stein–Saleh-type Shrinkage R-estimator

$$\text{ADB}(\hat{\boldsymbol{\beta}}_n^{\text{SSR+}}) = \begin{pmatrix} \mathbf{0}_{p_1} \\ \text{ADB}(\hat{\boldsymbol{\beta}}_{2n}^{\text{SSR}}) - \boldsymbol{\delta}_2\mathbb{E}[(1-(p_2-2)\chi^{-2}_{p_2+2}(\Delta^2))I(\chi^2_{p_2+2}(\Delta^2) < p_2-2)] \end{pmatrix},$$

$$\begin{aligned}\text{ADL}_2^*\text{-risk}\left(\hat{\boldsymbol{\beta}}_n^{\text{SSR+}}\right) &= \text{ADL}_2^*\text{-risk}\left(\hat{\boldsymbol{\beta}}_n^{\text{SSR}}\right) \\ -\eta^2\operatorname{tr}(\boldsymbol{C}_{22\cdot1}^{-1})&\mathbb{E}[(1-(p_2-2)\chi^{-2}_{p_2+2}(\Delta^2))^2 I(\chi^2_{p_2+2}(\Delta^2) < p_2-2)] \\ -\Delta^2\text{Ch}_{\min}(\boldsymbol{C}_{22\cdot1}^{-1})&\{2\mathbb{E}[(1-(p_2-2)\chi^{-2}_{p_2+2}(\Delta^2))I(\chi^2_{p_2+2}(\Delta^2) < p_2-2)]\end{aligned}$$

$$-\mathbb{E}[(1-(p_2-2)\chi^{-2}_{p_2+4}(\Delta^2))I(\chi^2_{p_2+4}(\Delta^2)<p_2-2)]\}.$$

Hence, the ADL$_2$-risk of $\hat{\boldsymbol{\beta}}_n^{\text{SSR+}}$, in the orthonormal case, is given by

$$\begin{aligned}
&\eta^2[p-(p_2-2)^2\mathbb{E}[\chi^{-2}_{p_2}(\Delta^2)]]\\
&-\eta^2[p_2\mathbb{E}[(1-(p_2-2)\chi^{-2}_{p_2+2}(\Delta^2))^2I(\chi^2_{p_2+2}(\Delta^2)<p_2-2)]\\
&-\Delta^2\{2\mathbb{E}[(1-(p_2-2)\chi^{-2}_{p_2+2}(\Delta^2))I(\chi^2_{p_2+2}(\Delta^2)<p_2-2)]\\
&-\mathbb{E}[(1-(p_2-2)\chi^{-2}_{p_2+4}(\Delta^2))I(\chi^2_{p_2+4}(\Delta^2)<p_2-2)]\}]\\
=\;&\eta^2[(p_1+p_2)-(p_2-2)^2\mathbb{E}[\chi^{-2}_{p_2}(\Delta^2)]-R^*],
\end{aligned}$$

where

$$\begin{aligned}
R^* =\;& p_2\mathbb{E}[(1-(p_2-2)\chi^{-2}_{p_2+2}(\Delta^2))^2I(\chi^2_{p_2+2}(\Delta^2)<p_2-2)\\
&-\Delta^2\Big\{2\mathbb{E}\left[\left((p_2-2)\chi^{-2}_{p_2+2}(\Delta^2)-1\right)I(\chi^2_{p_2+2}(\Delta^2)<p_2-2)\right]\\
&-\mathbb{E}\left[\left(1-(p_2-2)\chi^{-2}_{p_2+4}(\Delta^2)\right)^2I(\chi^2_{p_2+4}(\Delta^2)<p_2-2)\right]\Big\}.
\end{aligned}$$

(vi) Ridge-type R-estimator

$$\begin{aligned}
\text{ADB}(\hat{\boldsymbol{\beta}}_n^{\text{ridgeR}}(k)) &= \begin{pmatrix}\mathbf{0}_{p_1}\\ -k\boldsymbol{\delta}_2[\boldsymbol{I}_{p_2}+k\boldsymbol{C}^{-1}_{22\cdot1}]^{-1}\end{pmatrix}\\
\text{ADL}_2^*\text{-risk}\left(\hat{\boldsymbol{\beta}}_n^{\text{ridgeR}}(k)\right) &= k^2\boldsymbol{\delta}_2^\top(\boldsymbol{C}_{22\cdot1}+k\boldsymbol{I}_{p_2})^{-2}\boldsymbol{\delta}_2\\
&\quad -\eta^2\left[\text{tr}(\boldsymbol{C}^{-1}_{11\cdot2})+\text{tr}\left((\boldsymbol{I}_{p_2}+k\boldsymbol{C}^{-1}_{22\cdot1})^{-2}\boldsymbol{C}^{-1}_{22\cdot1}\right)\right].
\end{aligned}$$

Note by Theorem 8.4.1 of Saleh et al. (2014), there exists a $k\in(0,k^*)$, with $k^*=\eta^2/\delta_{2\max}$, where $\delta_{2\max}=\max\{\delta_{21},\ldots,\delta_{2p_2})\}$, such that $\hat{\boldsymbol{\beta}}_{2n}^{\text{ridgeR}}(k)$ has smaller risk than $\hat{\boldsymbol{\beta}}_{2n}^{\text{UR}}$.

Also

$$\text{ADL}_2\text{-risk}\left(\hat{\boldsymbol{\beta}}_n^{\text{ridgeR}}(k):\boldsymbol{C}_{11\cdot2},\boldsymbol{C}_{22\cdot1}\right)=\eta^2p_1+\frac{\eta^2}{(k+1)^2}(p_2+k^2\Delta^2).$$

(vii) LASSO-type R-estimator

$$\text{ADB}\left(\hat{\beta}_n^{\text{LassoR}}(\lambda)\right) = \begin{pmatrix} -\lambda[\boldsymbol{I}_{p_1} + \lambda \boldsymbol{C}_{11\cdot2}^{-1}]^{-1}\boldsymbol{\delta}_1 \\ -\boldsymbol{\delta}_2 \end{pmatrix}$$

using Donoho and Johnstone (1994) and Tibshirani (1996).

We obtain the lower bound risk

$$\text{ADL}_2^*\text{-risk}\left(\hat{\boldsymbol{\beta}}_n^{\text{LassoR}}(\lambda)\right) = \eta^2\left[\text{tr}(\boldsymbol{C}_{11\cdot2}^{-1}) + \Delta^2\text{Ch}_{\min}(\boldsymbol{C}_{22\cdot1}^{-1})\right].$$

(viii) Enet R-estimator

$$\begin{aligned}\text{ADB}\left(\hat{\beta}_{jn}^{\text{EnetR}}(\lambda,\alpha)\right) &= \Big(\Delta_j\mathcal{H}_3(\lambda\alpha,\Delta_j) \\ &\quad -\lambda\alpha(\Phi(\lambda\alpha-\Delta_j)-\Phi(\lambda\alpha+\Delta_j))\Big|j=1,\dots,p\Big)^{\top} = \boldsymbol{B}.\end{aligned}$$

Also,

$$\begin{aligned}\text{ADL}_2^*\text{-risk}\left(\hat{\boldsymbol{\beta}}_n^{\text{EnetR}}(\lambda,\alpha)\right) &= \eta^2\,\text{tr}\left([\boldsymbol{I}+\lambda(1-\alpha)\boldsymbol{C}^{-1}]^{-2}\boldsymbol{C}^{-1}\right) \\ &\quad +\lambda^2(1-\alpha)^2\boldsymbol{\Delta}^{\top}[\boldsymbol{C}_{11\cdot2}+\lambda(1-\alpha)\boldsymbol{I}_{p_1}]^{-2}\boldsymbol{\Delta} \\ &\quad +2\lambda(1-\alpha)\boldsymbol{B}^{\top}(\boldsymbol{C}+\lambda(1-\alpha)\boldsymbol{I})^{-2}\boldsymbol{\Delta}),\end{aligned}$$

where $\boldsymbol{\Delta} = (\Delta_1,\dots,\Delta_p)^{\top}$.

Next, we will compare the performance of the R-estimators analytically using the ADL_2-risk.

7.4 ADL_2-risk Comparisons

In this section, we compare the R-estimators with respect to their ADL_2-risk functions.

(i) RRE vs. URE

$$\text{ADL}_2^*\text{-risk}\left(\hat{\boldsymbol{\beta}}_n^{\text{RR}}\right) - \text{ADL}_2^*\text{-risk}\left(\hat{\boldsymbol{\beta}}_n^{\text{UR}}\right) =$$

$$\eta^2\left[\mathrm{tr}(\boldsymbol{C}_{11}^{-1})+\Delta^2\mathrm{Ch}_{\min}(\boldsymbol{C}_{22\cdot1}^{-1})\right]-\eta^2\left[\mathrm{tr}(\boldsymbol{C}_{11\cdot2}^{-1})+\mathrm{tr}(\boldsymbol{C}_{22\cdot1}^{-1})\right].$$

Hence, the URE outperforms RRE for

$$\Delta^2>\frac{\mathrm{tr}(\boldsymbol{C}_{11\cdot2}^{-1})+\mathrm{tr}(\boldsymbol{C}_{22\cdot1}^{-1})-\mathrm{tr}(\boldsymbol{C}_{11}^{-1})}{\mathrm{Ch}_{\min}(\boldsymbol{C}_{22\cdot1}^{-1})}.$$

Otherwise, RRE outperforms URE. Neither RRE nor URE dominates the other uniformly.

(ii) PTRE vs. URE

$$\begin{aligned}
&\mathrm{ADL}_2^*\text{-risk}\left(\hat{\boldsymbol{\beta}}_n^{\mathrm{PTR}}(\alpha)\right)-\mathrm{ADL}_2^*\text{-risk}\left(\hat{\boldsymbol{\beta}}_n^{\mathrm{UR}}\right)\\
&=\eta^2\Big[\mathrm{tr}(\boldsymbol{C}_{11\cdot2}^{-1})+\mathrm{tr}(\boldsymbol{C}_{22\cdot1}^{-1})(1-\mathcal{H}_{p_2+2}(\chi^2_{p_2}(\alpha);\Delta^2))\\
&\quad+\Delta^2\mathrm{Ch}_{\min}(\boldsymbol{C}_{22\cdot1}^{-1})(2\mathcal{H}_{p_2+2}(\chi^2_{p_2}(\alpha);\Delta^2)-\mathcal{H}_{p_2+4}(\chi^2_{p_2}(\alpha);\Delta^2))\Big]\\
&\quad-\eta^2\left[\mathrm{tr}\,\boldsymbol{C}_{11\cdot2}^{-1}+\mathrm{tr}\,\boldsymbol{C}_{22\cdot1}^{-1}\right].
\end{aligned}$$

Hence, if

$$0<\Delta^2\leq\frac{\mathrm{tr}(\boldsymbol{C}_{22\cdot1}^{-1})\mathcal{H}_{p_2+2}(\chi^2_{p_2}(\alpha);\Delta^2)}{\mathrm{Ch}_{\min}(\boldsymbol{C}_{22\cdot1}^{-1})\left[2\mathcal{H}_{p_2+2}(\chi^2_{p_2}(\alpha);\Delta^2)-\mathcal{H}_{p_2+4}(\chi^2_{p_2}(\alpha);\Delta^2)\right]}$$

then PTRE dominates URE. If

$$\Delta^2>\frac{\mathrm{tr}(\boldsymbol{C}_{22\cdot1}^{-1})\mathcal{H}_{p_2+2}(\chi^2_{p_2}(\alpha);\Delta^2)}{\mathrm{Ch}_{\min}(\boldsymbol{C}_{22\cdot1}^{-1})\left[2\mathcal{H}_{p_2+2}(\chi^2_{p_2}(\alpha);\Delta^2)-\mathcal{H}_{p_2+4}(\chi^2_{p_2}(\alpha);\Delta^2)\right]},$$

then URE dominates PTRE. Thus, neither PTRE nor URE dominates the other uniformly.

(iii) Stein–Saleh-type vs. URE

$$\begin{aligned}
&\mathrm{ADL}_2^*\text{-risk}\left(\hat{\boldsymbol{\beta}}_n^{\mathrm{SSR}}\right)-\mathrm{ADL}_2\text{-risk}\left(\hat{\boldsymbol{\beta}}_n^{\mathrm{UR}}\right)\\
&=\eta^2\Big[\mathrm{tr}(\boldsymbol{C}_{11\cdot2}^{-1})+\mathrm{tr}(\boldsymbol{C}_{22\cdot1}^{-1})\\
&\quad-\left\{(p_2-2)\,\mathrm{tr}(\boldsymbol{C}_{22\cdot1}^{-1})\left(2\mathbb{E}[\chi^{-2}_{p_2+2}(\Delta^2)]-(p_2-2)\mathbb{E}[\chi^{-4}_{p_2+4}(\Delta^2)]\right)-1\right\}\\
&\quad+(p_2^2-4)\Delta^2\mathrm{Ch}_{\min}(\boldsymbol{C}_{22\cdot1}^{-1})\mathbb{E}[\chi^{-4}_{p_2+4}(\Delta^2)]\Big]
\end{aligned}$$

for all $\Delta^2\geq 0$. Hence, Stein–Saleh-type dominates URE uniformly.

(iv) Stein–Saleh-type vs. PRSRE

$$\begin{aligned}
& \text{ADL}_2^*\text{-risk}\left(\hat{\boldsymbol{\beta}}_n^{\text{SSR}}\right) - \text{ADL}_2\text{-risk}\left(\hat{\boldsymbol{\beta}}_n^{\text{SSR+}}\right) \\
= \;& \eta^2 \operatorname{tr}(\boldsymbol{C}_{22\cdot1}^{-1})\mathbb{E}[(1-(p_2-2)\chi_{p_2+2}^{-2}(\Delta^2))^2 I(\chi_{p_2+2}^2(\Delta^2) < p_2-2)] \\
& +\Delta^2 \text{Ch}_{\min}(\boldsymbol{C}_{22\cdot1}^{-1})\Big\{2\mathbb{E}[(1-(p_2-2)\chi_{p_2+2}^{-2}(\Delta^2))I(\chi_{p_2+2}^2(\Delta^2) < p_2-2)] \\
& +\mathbb{E}[(1-(p_2-2)\chi_{p_2+4}^{-2}(\Delta^2))I(\chi_{p_2+4}^2(\Delta^2) < p_2-2)]\Big\} \geq 0
\end{aligned}$$

for all $\Delta^2 \geq 0$, since

$$0 < \chi_{p_2+2}^2(\Delta^2) < p_2 - 2 \Rightarrow \left((p_2-2)\chi_{p_2+2}^2(\Delta^2) - 1\right) \geq 0.$$

Then, combining (iii) and (iv) we find that

$$\text{ADL}_2^*\text{-risk}\left(\hat{\boldsymbol{\beta}}_n^{\text{SSR+}}\right) \leq \text{ADL}_2^*\text{-risk}\left(\hat{\boldsymbol{\beta}}_n^{\text{SSR}}\right) \leq \text{ADL}_2^*\text{-risk}\left(\hat{\boldsymbol{\beta}}_n^{\text{UR}}\right).$$

(v) LASSO R estimator vs. URE

$$\text{ADL}_2^*\text{-risk}\left(\hat{\boldsymbol{\beta}}_n^{\text{LassoR}}(\lambda)\right) - \text{ADL}_2^*\text{-risk}\left(\hat{\boldsymbol{\beta}}_n^{\text{UR}}\right) = \eta^2\left[\Delta^2 \text{Ch}_{\min}(\boldsymbol{C}_{22\cdot1}^{-1}) - \operatorname{tr}(\boldsymbol{C}_{22\cdot1}^{-1})\right].$$

LASSO R-estimator performs better than the URE whenever

$$\Delta^2 \leq \frac{\operatorname{tr}(\boldsymbol{C}_{22\cdot1}^{-1})}{\text{Ch}_{\min}(\boldsymbol{C}_{22\cdot1}^{-1})}.$$

Otherwise, the URE is superior. Hence, none of the estimators outperforms the other uniformly.

(vi) LASSO R estimator vs. PTRE

$$\begin{aligned}
& \text{ADL}_2^*\text{-risk}\left(\hat{\boldsymbol{\beta}}_n^{\text{PTR}}(\alpha)\right) - \text{ADL}_2^*\text{-risk}\left(\hat{\boldsymbol{\beta}}_n^{\text{LassoR}}(\lambda)\right) \\
= \;& \eta^2\Big[\operatorname{tr}(\boldsymbol{C}_{22\cdot1}^{-1})(1-\mathcal{H}_{p_2+2}(\chi_{p_2}^2(\alpha);\Delta^2))
\end{aligned}$$

$$+\Delta^2 \mathrm{Ch}_{\min}(\boldsymbol{C}_{22\cdot1}^{-1})(2\mathcal{H}_{p_2+2}(\chi^2_{p_2}(\alpha);\Delta^2) - \mathcal{H}_{p_2+4}(\chi^2_{p_2}(\alpha);\Delta^2) - 1)\Big].$$

Then, for

$$0 \le \Delta^2 \le \frac{\mathrm{tr}(\boldsymbol{C}_{22\cdot1}^{-1})[1 - \mathcal{H}_{p_2+2}(\chi^2_{p_2}(\alpha);\Delta^2)]}{\mathrm{Ch}_{\min}(\boldsymbol{C}_{22\cdot1}^{-1})\left[1 - 2\mathcal{H}_{p_2+2}(\chi^2_{p_2}(\alpha);\Delta^2) + \mathcal{H}_{p_2+4}(\chi^2_{p_2}(\alpha);\Delta^2)\right]}$$

LASSO outperforms PTRE. On the other hand, outside this interval, PTRE outperforms LASSO. Hence, neither LASSO nor PTRE dominates the other uniformly.

(vii) LASSO vs. Stein–Saleh-type

$$\begin{aligned}
&\mathrm{ADL}_2^*\text{-risk}\left(\hat{\boldsymbol{\beta}}_n^{\mathrm{SSR}}\right) - \mathrm{ADL}_2^*\text{-risk}\left(\hat{\boldsymbol{\beta}}_n^{\mathrm{LassoR}}(\lambda)\right) = \eta^2\Big[\mathrm{tr}(\boldsymbol{C}_{22\cdot1}^{-1})\\
&-\Big\{(p_2-2)\,\mathrm{tr}(\boldsymbol{C}_{22\cdot1}^{-1})\left(2\mathbb{E}[\chi^{-2}_{p_2+2}(\Delta^2)] - (p_2-2)\mathbb{E}[\chi^{-4}_{p_2+4}(\Delta^2)]\right)\Big\}\\
&+(p_2^2-4)\Delta^2 \mathrm{Ch}_{\min}(\boldsymbol{C}_{22\cdot1}^{-1})\left(\mathbb{E}[\chi^{-4}_{p_2+4}(\Delta^2)] - 1\right)\Big].
\end{aligned}$$

Hence, for

$$0 \le \Delta^2 \le \frac{\mathrm{tr}(\boldsymbol{C}_{22\cdot1}^{-1}) - \Big\{(p_2-2)\,\mathrm{tr}(\boldsymbol{C}_{22\cdot1}^{-1})\left(2\mathbb{E}[\chi^{-2}_{p_2+2}(\Delta^2)] - (p_2-2)\mathbb{E}[\chi^{-4}_{p_2+4}(\Delta^2)]\right)\Big\}}{(p_2^2-4)Ch_{\min}(\boldsymbol{C}_{22\cdot1}^{-1})\left(1 - \mathbb{E}[\chi^{-4}_{p_2+4}(\Delta^2)]\right)}$$

LASSO outperforms Stein–Saleh-type, otherwise Stein–Saleh-type outperforms LASSO. Hence, neither LASSO nor Stein–Saleh-type outperforms the other uniformly.

(viii) LASSO R estimator vs. PRSRE

For the last comparison in this section, we proceed with the orthonormal case.

$$\begin{aligned}
&\mathrm{ADL}_2\text{-risk}\left(\hat{\boldsymbol{\beta}}_n^{\mathrm{SSR+}} : \boldsymbol{C}_{11\cdot2}, \boldsymbol{C}_{22\cdot1}\right) - \mathrm{ADL}_2^*\text{-risk}\left(\hat{\boldsymbol{\beta}}_n^{\mathrm{LassoR}}(\lambda)\right)\\
&= \eta^2\left[\mathrm{ADL}_2\text{-risk}\left(\hat{\boldsymbol{\beta}}_n^{\mathrm{SSR}}\right) - R^* - (p_1+\Delta^2)\right]\\
&= \eta^2\left[p_2 - (p_2-2)^2\mathbb{E}[\chi^{-2}_{p_2}(\Delta^2)] - R^* - \Delta^2\right]\\
&= \eta^2(A - \Delta^2 B),
\end{aligned}$$

where

$$A = p_2 - (p_2-2)^2\mathbb{E}[\chi^{-2}_{p_2}(\Delta^2)]$$

$$-p_2\mathbb{E}[(1-(p_2-2)\chi^{-2}_{p_2+2}(\Delta^2))^2 I(\chi^2_{p_2+2}(\Delta^2) < p_2 - 2)]$$

and

$$\begin{aligned} B &= 1 + 2\mathbb{E}\left[\left((p_2-2)\chi^{-2}_{p_2+2}(\Delta^2) - 1\right) I(\chi^2_{p_2+2}(\Delta^2) < p_2 - 2)\right] \\ &+ \mathbb{E}\left[\left(1-(p_2-2)\chi^{-2}_{p_2+4}(\Delta^2)\right)^2 I(\chi^2_{p_2+4}(\Delta^2) < p_2 - 2)\right]. \end{aligned}$$

Again, for $0 \leq \Delta^2 \leq A/B$, LASSO R estimator outperforms PRSRE, otherwise PRSE outperforms LASSO. Neither LASSO nor PRSE dominates the other.

7.4.0.1 Ridge vs. others

In this section, we specifically compare the ridge R-estimator with other rank estimators for both orthonormal and non-orthonormal cases. It will be shown that neither PTRE, Stein–Saleh-type R-estimator nor PRSRE dominate the ridge R-estimator. However, for $\boldsymbol{C} = \boldsymbol{I}_p$, ridge dominates them uniformly.

Ridge vs. URE:

$$\begin{aligned} &\text{ADL}_2\text{-risk}\left(\hat{\boldsymbol{\beta}}_n^{\text{UR}}\right) - \text{ADL}_2^*\text{-risk}\left(\hat{\boldsymbol{\beta}}_n^{\text{ridgeR}}(k)\right) \\ &= \eta^2\left[\text{tr}(\boldsymbol{C}^{-1}_{22\cdot1}) - \text{tr}\left((\boldsymbol{I}_{p_2} + k\boldsymbol{C}^{-1}_{22\cdot1})^{-2}\boldsymbol{C}^{-1}_{22\cdot1}\right)\right] - k^2\boldsymbol{\delta}_2^{\top}(\boldsymbol{C}_{22\cdot1} + k\boldsymbol{I}_{p_2})^{-2}\boldsymbol{\delta}_2 \geq 0. \end{aligned}$$

Hence, the ridge R-estimator dominates uniformly over URE, using theorem 8.4.1 of Saleh et al. (2014).

RRE vs. Ridge:

$$\begin{aligned} &\text{ADL}_2^*\text{-risk}\left(\hat{\boldsymbol{\beta}}_n^{\text{RR}}\right) - \text{ADL}_2^*\text{-risk}\left(\hat{\boldsymbol{\beta}}_n^{\text{ridgeR}}(k)\right) = \eta^2\left[\text{tr}(\boldsymbol{C}^{-1}_{11}) + \Delta^2\text{Ch}_{\min}(\boldsymbol{C}^{-1}_{22\cdot1})\right] \\ &- \eta^2\left[\text{tr}(\boldsymbol{C}^{-1}_{11\cdot2}) + \text{tr}\left((\boldsymbol{I}_{p_2} + k\boldsymbol{C}^{-1}_{22\cdot1})^{-2}\boldsymbol{C}^{-1}_{22\cdot1}\right)\right] - k^2\boldsymbol{\delta}_2^{\top}(\boldsymbol{C}_{22\cdot1} + k\boldsymbol{I}_{p_2})^{-2}\boldsymbol{\delta}_2. \end{aligned}$$

Hence, the ridge R-estimator dominates the RRE and LASSO R-estimator whenever

$$\Delta^2 \geq \frac{\text{ADL}_2^*\text{-risk}\left(\hat{\boldsymbol{\beta}}_n^{\text{ridgeR}}(k)\right) - \eta^2\,\text{tr}\,\boldsymbol{C}^{-1}_{11}}{\text{Ch}_{\min}(\boldsymbol{C}^{-1}_{22\cdot1})}.$$

PTRE vs. Ridge:

$$\text{ADL}_2^*\text{-risk}\left(\hat{\boldsymbol{\beta}}_n^{\text{PTR}}(\alpha)\right) - \text{ADL}_2^*\text{-risk}\left(\hat{\boldsymbol{\beta}}_n^{\text{ridgeR}}(k_{\text{opt}})\right)$$

$$
\begin{aligned}
&= \eta^2\Big[\operatorname{tr}(\boldsymbol{C}_{22\cdot1}^{-1})(1-\mathcal{H}_{p_2+2}(\chi^2_{p_2}(\alpha);\Delta^2)) \\
&\quad +\Delta^2 \operatorname{Ch}_{\min}(\boldsymbol{C}_{22\cdot1}^{-1})(2\mathcal{H}_{p_2+2}(\chi^2_{p_2}(\alpha);\Delta^2)-\mathcal{H}_{p_2+4}(\chi^2_{p_2}(\alpha);\Delta^2))\Big] \\
&\quad -\eta^2\left[\operatorname{tr}(\boldsymbol{I}_{p_2}+k\boldsymbol{C}_{22\cdot1}^{-1})^{-2}\boldsymbol{C}_{22\cdot1}^{-1}\right] \\
&\quad -k^2\boldsymbol{\delta}_2^{\top}(\boldsymbol{C}_{22\cdot1}+k\boldsymbol{I}_{p_2})^{-2}\boldsymbol{\delta}_2 \\
&= \eta^2\Big[\operatorname{tr}(\boldsymbol{C}_{22\cdot1}^{-1})-\Big\{\operatorname{tr}\left((\boldsymbol{I}_{p_2}+k\boldsymbol{C}_{22\cdot1}^{-1})^{-2}\boldsymbol{C}_{22\cdot1}^{-1}\right) \\
&\quad +k^2\boldsymbol{\delta}_2^{\top}(\boldsymbol{C}_{22\cdot1}^{-1}+k\boldsymbol{I}_{p_2})^{-2}\boldsymbol{\delta}_2\Big\}-\Big\{\operatorname{tr}(\boldsymbol{C}_{22\cdot1}^{-1})\mathcal{H}_{p_2+2}(\chi_{p_2}(\alpha);\Delta^2) \\
&\quad -\Delta^2\operatorname{Ch}_{\min}(\boldsymbol{C}_{22\cdot1}^{-1})\left(2\mathcal{H}_{p_2+2}(\chi^2_{p_2}(\alpha);\Delta^2)-\mathcal{H}_5(\chi^2_{p_2}(\alpha);\Delta^2)\right)\Big\}\Big] \geq 0.
\end{aligned}
$$

Let $\Delta^2 = 0$. Then the difference equals

$$
\operatorname{tr}(\boldsymbol{C}_{22\cdot1}^{-1})-\operatorname{tr}((\boldsymbol{I}_{p_2}+k\boldsymbol{C}_{22\cdot1})^{-2}\boldsymbol{C}_{22\cdot1}^{-1})-\operatorname{tr}(\boldsymbol{C}_{22\cdot1}^{-1})\mathcal{H}_{p_2+2}(\chi^2_{p_2}(\alpha);0)\geq 0
$$

and for $\Delta^2 \to \infty$, the difference is

$$
\eta^2\left[\operatorname{tr}(\boldsymbol{C}_{22\cdot1}^{-1})-\operatorname{tr}\left((\boldsymbol{I}_{p_2}+k\boldsymbol{C}_{22\cdot1}^{-1})^{-2}\boldsymbol{C}_{22\cdot1}^{-1}\right)\right]\geq 0.
$$

Further, a graph of the expression in the second bracket would be below the graph of the first expression. Hence, the ADL$_2$-risk-difference is non-zero and ridge is uniformly better than $\hat{\boldsymbol{\beta}}_n^{\mathrm{PTR}}(\alpha)$.

Now, consider

$$
\begin{aligned}
&\text{ADL}_2\text{-risk}\left(\hat{\boldsymbol{\beta}}_n^{\mathrm{PTR}}(\alpha):\boldsymbol{C}_{11\cdot2},\boldsymbol{C}_{22\cdot1}\right)-\text{ADL}_2\text{-risk}\left(\hat{\boldsymbol{\beta}}_n^{\mathrm{ridgeR}}(k_{\mathrm{opt}}):\boldsymbol{C}_{11\cdot2},\boldsymbol{C}_{22\cdot1}\right) \\
&\quad =\eta^2\Big[p_2(1-\mathcal{H}_{p_2+2}(\chi^2_{p_2}(\alpha);\Delta^2))+\Delta^2(2\mathcal{H}_{p_2+2}(\chi^2_{p_2}(\alpha);\Delta^2) \\
&\qquad -\mathcal{H}_{p_2+4}(\chi^2_{p_2}(\alpha);\Delta^2))-\frac{p_2\Delta^2}{p_2+\Delta^2}\Big].
\end{aligned}
$$

See Saleh et al. (2019).

Note that the risk of PTRE is an increasing function of Δ^2, crossing the p_2-line to a maximum and then it drops monotonically towards the p_2-line as $\Delta^2 \to \infty$. At $\Delta^2 = 0$, the value of this risk is $p_2[1-\mathcal{H}_{p_2+2}(\chi^2_{p_2}(\alpha);0)] < p_2$. On the other hand, for the orthogonal case, the risk of the ridge R-estimator is equal to $\frac{p_2\Delta^2}{p_2+\Delta^2}$ which is an increasing function of Δ^2 below the p_2-line with a minimum value 0 at $\Delta^2 = 0$ and it converges to p_2 as $\Delta^2 \to \infty$. Hence, the risk difference is non-negative for all $\Delta^2 \geq 0$ and the ridge R-estimator outperforms the PTRE uniformly.

Stein–Saleh-type vs. Ridge:
From the risk functions, we have

$$\text{ADL}_2^*\text{-risk}\left(\hat{\boldsymbol{\beta}}_n^{\text{SSR}}\right) - \text{ADL}_2^*\text{-risk}\left(\hat{\boldsymbol{\beta}}_n^{\text{ridgeR}}(k)\right) = \eta^2\Big[\operatorname{tr}\boldsymbol{C}_{11\cdot2}^{-1} + \operatorname{tr}(\boldsymbol{C}_{22\cdot1}^{-1})$$
$$-\Big\{(p_2-2)\operatorname{tr}\boldsymbol{C}_{22\cdot1}^{-1}\left(2\mathbb{E}[\chi_{p_2+2}^{-2}(\Delta^2)] - (p_2-2)\mathbb{E}[\chi_{p_2+2}^{-4}(\Delta^2)]\right)\Big\}$$
$$+(p_2^2-4)\Delta^2\text{Ch}_{\min}(\boldsymbol{C}_{22\cdot1}^{-1})\mathbb{E}[\chi_{p_2+4}^{-4}(\Delta^2)]\Big]$$
$$-\eta^2\left[\operatorname{tr}(\boldsymbol{I}_{p_2} + k\boldsymbol{C}_{22\cdot1}^{-1})^{-2}\boldsymbol{C}_{22\cdot1}^{-1}\right]$$
$$-k^2\boldsymbol{\delta}_2^\top(\boldsymbol{C}_{22\cdot1} + k\boldsymbol{I}_{p_2})^{-2}\boldsymbol{\delta}_2$$
$$=\eta^2\Big[\operatorname{tr}(\boldsymbol{C}_{22\cdot1}^{-1})\Big\{(p_2-2)\operatorname{tr}\boldsymbol{C}_{22\cdot1}^{-1}\Big(2\mathbb{E}[\chi_{p_2+2}^{-2}(\Delta^2)]$$
$$-(p_2-2)\mathbb{E}[\chi_{p_2+2}^{-4}(\Delta^2)]\Big)\Big\} - \eta^2\left[\operatorname{tr}(\boldsymbol{I}_{p_2} + k\boldsymbol{C}_{22\cdot1}^{-1})^{-2}\boldsymbol{C}_{22\cdot1}^{-1}\right]$$
$$-k^2\boldsymbol{\delta}_2^\top(\boldsymbol{C}_{22\cdot1} + k\boldsymbol{I}_{p_2})^{-2}\boldsymbol{\delta}_2 + (p_2^2-4)\Delta^2\text{Ch}_{\min}(\boldsymbol{C}_{22\cdot1}^{-1})\mathbb{E}\left[\chi_{p_2+4}^{-4}(\Delta^2)\right]\Big] \geq 0.$$

On the other hand, we can obtain

$$\text{ADL}_2\text{-risk}\left(\hat{\boldsymbol{\beta}}_n^{\text{SSR}} : \boldsymbol{C}_{11\cdot2}, \boldsymbol{C}_{22\cdot1}\right) - \text{ADL}_2\text{-risk}\left(\hat{\boldsymbol{\beta}}_n^{\text{ridgeR}}(k_{\text{opt}}) : \boldsymbol{C}_{11\cdot2}, \boldsymbol{C}_{22\cdot1}\right)$$
$$= \eta^2\left\{p_2 - (p_2-2)^2\mathbb{E}[\chi_{p_2}^{-2}(\Delta^2)] - \frac{p_2\Delta^2}{p_2+\Delta^2}\right\}.$$

See Saleh et al. (2019).

Note that the first function is increasing in Δ^2 with its minimum at $\Delta^2 = 0$ and as $\Delta^2 \to \infty$, it tends to p_2. The second function is also increasing in Δ^2 with a value 0 at Δ^2 and approaches the value p_2 as $\Delta^2 \to \infty$. Hence, the risk difference is non-negative for all $\Delta^2 \geq 0$. Therefore, the ridge-type uniformly outperforms Stein–Saleh-type.

PRSRE vs. Ridge-type:
From the risk functions we obtain

$$\text{ADL}_2^*\text{-risk}\left(\hat{\boldsymbol{\beta}}_n^{\text{SSR+}}\right) - \text{ADL}_2^*\text{-risk}\left(\hat{\boldsymbol{\beta}}_n^{\text{ridgeR}}(k)\right)$$
$$= \text{ADL}_2\text{-risk}(\hat{\boldsymbol{\beta}}_n^{\text{SSR}}) - \text{ADL}_2\text{-risk}(\hat{\boldsymbol{\beta}}_n^{\text{ridgeR}}(k))$$
$$-\eta^2\operatorname{tr}(\boldsymbol{C}_{11\cdot2})\mathbb{E}[(1-(p_2-2)\chi_{p_2+2}^{-2}(\Delta^2))I(\chi_{p_2+2}^2(\Delta^2) < p_2-2)]$$
$$+\Delta^2\text{Ch}_{\min}(\boldsymbol{C}_{22\cdot1}^{-1})\Big\{2\mathbb{E}[(1-(p_2-2)\chi_{p_2+2}^{-2}(\Delta^2))I(\chi_{p_2+2}^2(\Delta^2) < p_2-2)]$$

$$-\mathbb{E}[(1-(p_2-2)\chi^{-2}_{p_2+4}(\Delta^2))I(\chi^2_{p_2+4}(\Delta^2)<p_2-2)]\Big\}\geq 0 \quad (7.4.1)$$

since

$$\text{ADL}_2\text{-risk}(\hat{\boldsymbol{\beta}}_n^{\text{SSR}})\geq \eta^2\,\text{tr}(\boldsymbol{C}_{11\cdot 2})\mathbb{E}[(1-(p_2-2)\chi^{-2}_{p_2+2}(\Delta^2))I(\chi^2_{p_2+2}(\Delta^2)<p_2-2)]$$
$$-\Delta^2\text{Ch}_{\min}(\boldsymbol{C}^{-1}_{22\cdot 1})\Big\{2\mathbb{E}[(1-(p_2-2)\chi^{-2}_{p_2+2}(\Delta^2))I(\chi^2_{p_2+2}(\Delta^2)<p_2-2)]$$
$$-\mathbb{E}[(1-(p_2-2)\chi^{-2}_{p_2+4}(\Delta^2))I(\chi^2_{p_2+4}(\Delta^2)<p_2-2)]\Big\}$$

and

$$\text{ADL}_2\text{-risk}(\hat{\boldsymbol{\beta}}_n^{\text{SSR}})\geq \text{ADL}_2\text{-risk}(\hat{\boldsymbol{\beta}}_n^{\text{ridgeR}}(k)). \quad (7.4.2)$$

On the other hand, we have

$$\text{ADL}_2\text{-risk}\left(\hat{\boldsymbol{\beta}}_n^{\text{SSR+}}:\boldsymbol{C}_{11\cdot 2},\boldsymbol{C}_{22\cdot 1}\right)-\text{ADL}_2\text{-risk}\left(\hat{\boldsymbol{\beta}}_n^{\text{ridgeR}}(k_{\text{opt}}):\boldsymbol{C}_{11\cdot 2},\boldsymbol{C}_{22\cdot 1}\right)$$
$$=\eta^2\left\{p_2-(p_2-2)^2\mathbb{E}[\chi_{p_2}^{-2}(\Delta^2)]-\frac{p_2\Delta^2}{p_2+\Delta^2}-R^*\right\}\geq 0,\quad \forall\,\Delta^2\geq 0,$$

where

$$R^* = p_2\mathbb{E}[(1-(p_2-2)\chi^{-2}_{p_2+2}(\Delta^2))^2 I(\chi^2_{p_2+2}(\Delta^2)<p_2-2)]$$
$$-\Delta^2\{2\mathbb{E}[(1-(p_2-2)\chi^{-2}_{p_2+2}(\Delta^2))I(\chi^2_{p_2+2}(\Delta^2)<p_2-2)]$$
$$-\mathbb{E}[(1-(p_2-2)\chi^{-2}_{p_2+4}(\Delta^2))I(\chi^2_{p_2+4}(\Delta^2)<p_2-2)]\}.$$

Consider the risk of PRSRE. It is an increasing function of Δ^2. At $\Delta^2=0$, its value is

$$(p_1+2)-p_2\mathbb{E}[(1-(p_2-2)\chi^{-2}_{p_2+2}(0))I(\chi^2_{p_2+2}(0)<p_2-2)]\geq 0$$

and as $\Delta^2\to\infty$, it tends to p_1+p_2. For the ridge-type, at $\Delta^2=0$, the value is p_1 and as $\Delta^2\to\infty$, it tends to p_1+p_2. Hence, the risk difference is non-negative and therefore the ridge-type outperforms PRSRE uniformly. See Saleh et al. (2018).

7.5 Summary: L_2-risk Efficiencies

In this section, we provide the final expressions of the asymptotic distributional L_2-risk efficiencies (ADRE) of the proposed R-estimators. For our purposes, we use the lower bound of Eq. (7.3.5) of the L_2-risk functions for all estimators except the unrestricted R-estimator. This result will enhance the provided comparisons.

For any estimator $\hat{\boldsymbol{\beta}}_n$, we define

$$\text{ADRE}(\hat{\boldsymbol{\beta}}_n : \hat{\boldsymbol{\beta}}_n^{\text{UR}}) = \frac{\text{ADL}_2^*\text{-risk}\left(\hat{\boldsymbol{\beta}}_n^{\text{UR}}\right)}{\text{ADL}_2^*\text{-risk}\left(\hat{\boldsymbol{\beta}}_n\right)}.$$

Theorem 7.3 The ADRE of the R-estimators are given by

(i) $$\text{ADRE}(\hat{\boldsymbol{\beta}}_n^{\text{RR}} : \hat{\boldsymbol{\beta}}_n^{\text{UR}}) = \frac{\text{tr}(\boldsymbol{C}_{11\cdot2}^{-1}) + \text{tr}(\boldsymbol{C}_{22\cdot1}^{-1})}{\text{tr}(\boldsymbol{C}_{11}^{-1}) + \Delta^2\text{Ch}_{\min}(\boldsymbol{C}_{22\cdot1}^{-1})},$$

(ii) $$\text{ADRE}(\hat{\boldsymbol{\beta}}_n^{\text{LassoR}}(\lambda_{\text{opt}}) : \hat{\boldsymbol{\beta}}_n^{\text{UR}}) = \frac{\text{tr}(\boldsymbol{C}_{11\cdot2}^{-1}) + \text{tr}(\boldsymbol{C}_{22\cdot1}^{-1})}{\text{tr}(\boldsymbol{C}_{11\cdot2}^{-1}) + \Delta^2\text{Ch}_{\min}(\boldsymbol{C}_{22\cdot1}^{-1})}$$

$$= \left(1 + \frac{\text{tr}(\boldsymbol{C}_{22\cdot1}^{-1})}{\text{tr}(\boldsymbol{C}_{11\cdot2}^{-1})}\right)\left(1 + \frac{\Delta^2\text{Ch}_{\min}(\boldsymbol{C}_{22\cdot1}^{-1})}{\text{tr}(\boldsymbol{C}_{11\cdot2}^{-1})}\right)^{-1},$$

(iii) $$\text{ADRE}(\hat{\boldsymbol{\beta}}_n^{\text{ridgeR}}(k) : \hat{\boldsymbol{\beta}}_n^{\text{UR}})$$

$$= \frac{\text{tr}(\boldsymbol{C}_{11\cdot2}^{-1}) + \text{tr}(\boldsymbol{C}_{22\cdot1}^{-1})}{\text{tr}(\boldsymbol{C}_{11\cdot2}^{-1}) + \text{tr}(\boldsymbol{I}_{p_2} + k\boldsymbol{C}_{22\cdot1}^{-1})^{-2}\boldsymbol{C}_{22\cdot1} + \Delta^2\text{Ch}_{\min}(\boldsymbol{C}_{22\cdot1}^{-1})}$$

$$= \left(1 + \frac{\text{tr}(\boldsymbol{C}_{22\cdot1}^{-1})}{\text{tr}(\boldsymbol{C}_{11\cdot2}^{-1})}\right)\left(1 + \frac{\text{tr}(\boldsymbol{I}_{p_2} + k\boldsymbol{C}_{22\cdot1}^{-1})^{-2}\boldsymbol{C}_{22\cdot1}}{\text{tr}(\boldsymbol{C}_{11\cdot2}^{-1})} + \frac{\Delta^2\text{Ch}_{\min}(\boldsymbol{C}_{22\cdot1}^{-1})}{\text{tr}(\boldsymbol{C}_{11\cdot2}^{-1})}\right)^{-1},$$

(iv) $$\text{ADRE}(\hat{\boldsymbol{\beta}}_n^{\text{PTR}} : \hat{\boldsymbol{\beta}}_n^{\text{UR}}) = \left(1 + \frac{\text{tr}(\boldsymbol{C}_{22\cdot1}^{-1})}{\text{tr}(\boldsymbol{C}_{11\cdot2}^{-1})}\right)$$

$$\times\left\{1 + \frac{\text{tr}(\boldsymbol{C}_{22\cdot1}^{-1})}{\text{tr}(\boldsymbol{C}_{11\cdot2}^{-1})}(1 - \mathcal{H}_{p_2+2}(\chi^2_{p_2}(\alpha); \Delta^2))\right.$$

$$\left. + \frac{\Delta^2\text{Ch}_{\min}(\boldsymbol{C}_{22\cdot1}^{-1})}{\text{tr}(\boldsymbol{C}_{11\cdot2}^{-1})}(2\mathcal{H}_{p_2+2}(\chi^2_{p_2}(\alpha); \Delta^2) - \mathcal{H}_{p_2+4}(\chi^2_{p_2}(\alpha); \Delta^2))\right\}^{-1},$$

(v) $$\text{ADRE}(\hat{\boldsymbol{\beta}}_n^{\text{SSR}} : \hat{\boldsymbol{\beta}}_n^{\text{UR}}) = \left(1 + \frac{\text{tr}(\boldsymbol{C}_{22\cdot1}^{-1})}{\text{tr}(\boldsymbol{C}_{11\cdot2}^{-1})}\right)$$

$$\times\left\{1 + \frac{\text{tr}(\boldsymbol{C}_{22\cdot1}^{-1})}{\text{tr}(\boldsymbol{C}_{11\cdot2}^{-1})}\right.$$

$$\times\left\{(p_2 - 2)\,\text{tr}(\boldsymbol{C}_{22\cdot1}^{-1})\left(2\mathbb{E}\left[\chi^{-2}_{p_2+2}(\Delta^2)\right] - (p_2 - 2)\mathbb{E}\left[\chi^{-4}_{p_2+4}(\Delta^2)\right]\right)\right\}$$

$$+\frac{(p_2^2-4)\Delta^2(Ch)_{\min}(\boldsymbol{C}_{22\cdot1}^{-1})}{\operatorname{tr}(\boldsymbol{C}_{11\cdot2}^{-1})}\mathbb{E}\left[\chi_{p_2+4}^{-4}(\Delta^2)\right]\Bigg\}^{-1},$$

(vi)
$$\operatorname{ADRE}(\hat{\boldsymbol{\beta}}_n^{\mathrm{SSR+}}:\hat{\boldsymbol{\beta}}_n^{\mathrm{UR}})=\left(1+\frac{\operatorname{tr}(\boldsymbol{C}_{22\cdot1}^{-1})}{\operatorname{tr}(\boldsymbol{C}_{11\cdot2}^{-1})}\right)$$

$$\times\Bigg\{1+\frac{\operatorname{tr}(\boldsymbol{C}_{22\cdot1}^{-1})}{\operatorname{tr}(\boldsymbol{C}_{11\cdot2}^{-1})}$$

$$\times\left\{(p_2-2)\operatorname{tr}(\boldsymbol{C}_{22\cdot1}^{-1})\left(2\mathbb{E}\left[\chi_{p_2+2}^{-2}(\Delta^2)\right]-(p_2-2)\mathbb{E}\left[\chi_{p_2+4}^{-4}(\Delta^2)\right]\right)\right\}$$

$$+\frac{(p_2^2-4)\Delta^2\mathrm{Ch}_{\min}(\boldsymbol{C}_{22\cdot1}^{-1})}{\operatorname{tr}(\boldsymbol{C}_{11\cdot2}^{-1})}\mathbb{E}\left[\chi_{p_2+4}^{-4}(\Delta^2)\right]$$

$$-\frac{\operatorname{tr}(\boldsymbol{C}_{22\cdot1}^{-1})}{\operatorname{tr}(\boldsymbol{C}_{11\cdot2}^{-1})}\mathbb{E}\left[\left(1-(p_2-2)\chi_{p_2+2}^{-2}(\Delta^2)\right)^2 I\left(\chi_{p_2+2}^{2}(\Delta^2)<(p_2-2)\right)\right]$$

$$+\frac{\Delta^2\mathrm{Ch}_{\min}(\boldsymbol{C}_{22\cdot1}^{-1})}{\operatorname{tr}(\boldsymbol{C}_{11\cdot2}^{-1})}\Bigg[2\mathbb{E}\left[\left(1-(p_2-2)\chi_{p_2+2}^{-2}(\Delta^2)\right)I\left(\chi_{p_2+2}^{2}(\Delta^2)<(p_2-2)\right)\right]$$

$$-\mathbb{E}\left[\left(1-(p_2-2)\chi_{p_2+4}^{-2}(\Delta^2)\right)^2 I\left(\chi_{p_2+4}^{2}(\Delta^2)<(p_2-2)\right)\right]\Bigg]\Bigg\}^{-1},$$

(vii)
$$\operatorname{ADRE}(\hat{\boldsymbol{\beta}}_n^{\mathrm{EnetR}}:\hat{\boldsymbol{\beta}}_n^{\mathrm{UR}})=\frac{\operatorname{tr}(\boldsymbol{C}_{11\cdot2}^{-1})+\operatorname{tr}(\boldsymbol{C}_{22\cdot1}^{-1})}{J(\boldsymbol{\Delta})}$$

where

$$\begin{aligned}J(\boldsymbol{\Delta}) &= \operatorname{tr}\left((\boldsymbol{I}_{p_1}+\lambda(1-\alpha)\boldsymbol{C}_{11\cdot2}^{-1})^{-2}\boldsymbol{C}_{11\cdot2}\right)\\ &\quad+\lambda^2(1-\alpha)^2\boldsymbol{\Delta}^\top\left(\boldsymbol{C}_{22\cdot1}+\lambda(1-\alpha)\boldsymbol{I}_{p_2}\right)^{-2}\boldsymbol{\Delta}\\ &\quad+2\lambda(1-\alpha)\boldsymbol{B}^\top\left(\boldsymbol{C}_{22\cdot1}+\lambda(1-\alpha)\boldsymbol{I}_{p_2}\right)^{-2}\boldsymbol{\Delta}.\end{aligned} \tag{7.5.1}$$

7.6 Problems

7.1. Prove Theorem 7.2.
7.2. Verify the ADL$_2$-risk of $\hat{\boldsymbol{\beta}}_n^{\mathrm{LassoR}}(\lambda)$.
7.3. Does the LASSO-type R-estimator satisfy the oracle properties?
7.4. Compare the R-estimators using the lower bound of ADL$_2$-risk.
7.5. Define the adaptive LASSO-type R-estimator of $(\boldsymbol{\beta}_1^\top,\boldsymbol{\beta}_2^\top)^\top$.
7.6. Verify the expressions for ADL$_2$-risk of (a) $\hat{\boldsymbol{\beta}}_n^{\mathrm{PTR}}(\lambda)$, (b) $\hat{\boldsymbol{\beta}}_n^{\mathrm{RR}}$, (c) $\hat{\boldsymbol{\beta}}_n^{\mathrm{SSR}}$, and (d) $\hat{\boldsymbol{\beta}}_n^{\mathrm{SSR+}}$.

8

Liu Regression Models

8.1 Introduction

When multicollinearity is present among regressors, it causes inaccurate estimates of the regression coefficients. According to the ridge notion (see Saleh et al., 2019, section 1.2), to avoid explosion of regression coefficients, we penalize the squared norm and the resultant is the ridge estimator of Hoerl and Kennard (1970). This estimator has a nonlinear nature with respect to the ridge/tuning/biasing parameter. As a linear alternative, Liu (1993) proposed a multicollinear resistant estimator as a linear combination of the ridge and Stein-type estimators. We shall interpret this estimator as the linear unified estimator.

In this chapter, we borrow the analysis of Arashi et al. (2018), and concentrate only on the Liu estimator in the linear regression model for the whole components set $\boldsymbol{\beta}$. Therefore, we briefly review the rationale behind the Liu estimator in the linear regression model and then consider its rank-estimation extension with application.

8.2 Linear Unified (Liu) Estimator

Consider the multiple regression model

$$\boldsymbol{y} = \boldsymbol{X}\boldsymbol{\beta} + \boldsymbol{\varepsilon}, \tag{8.2.1}$$

where $\boldsymbol{X}$ is the design matrix of full column rank p, $\boldsymbol{\beta}$ is the p-dimensional vector of regression coefficients and $\boldsymbol{\varepsilon} = (\varepsilon_1, \dots, \varepsilon_n)^\top$ is the error term. The OLS estimator is

$$\hat{\boldsymbol{\beta}}_n^{\text{LS}} = (\boldsymbol{X}^\top \boldsymbol{X})^{-1} \boldsymbol{X}^\top \boldsymbol{y}.$$

Rank-Based Methods for Shrinkage and Selection: With Application to Machine Learning. First Edition. A. K. Md. Ehsanes Saleh, Mohammad Arashi, Resve A. Saleh, and Mina Norouzirad.

The ridge regression estimator can be obtained by minimizing the squared error loss estimation with an additional penalty term, to combat multicollinearity in regression. It has the form

$$\hat{\boldsymbol{\beta}}_n^{\text{ridge}}(k) = (\boldsymbol{X}^\top\boldsymbol{X} + k\boldsymbol{I}_n)^{-1}\boldsymbol{X}^\top\boldsymbol{y}.$$

We refer to Saleh et al. (2019) for an extensive study on this estimator from both theory and application. Although this estimator is very useful and practical in multicollinear and high-dimensional situations (see Shao and Deng, 2012), because of its nonlinearity with respect to the ridge parameter, k, many endeavors have been made to refine or propose a multicollinear resistant alternative estimators. One such alternative is the linear unified estimator proposed by Liu (1993). He proposed the following estimator, which combines the benefits of Hoerl and Kennard (1970) and Stein (1956).

$$\hat{\boldsymbol{\beta}}_n^{\text{Liu}}(d) = (\boldsymbol{X}^\top\boldsymbol{X} + \boldsymbol{I}_p)^{-1}(\boldsymbol{X}^\top\boldsymbol{y} + d\hat{\boldsymbol{\beta}}_n^{\text{LS}}),$$

where $0 < d < 1$ is the biasing/Liu parameter.

The logic behind this construction methodology is to find a way of reducing the distance between $\boldsymbol{\beta}$ and $\hat{\boldsymbol{\beta}}_n^{\text{LS}}$. In this regard, Mayer and Willke (1973) suggested minimizing the sum of squares error subject to making $||d\hat{\boldsymbol{\beta}}^{\text{LS}} - \boldsymbol{\beta}||^2$ small. With this idea in hand, the Liu estimator $\hat{\boldsymbol{\beta}}_n^{\text{Liu}}$ is the solution to the following optimization problem

$$\min_{\boldsymbol{\beta}\in\mathbb{R}^p} ||\boldsymbol{y} - \boldsymbol{X}\boldsymbol{\beta}||^2 \quad \text{such that} \quad ||d\hat{\boldsymbol{\beta}}^{\text{LS}} - \boldsymbol{\beta}||^2 < \lambda,$$

for some $\lambda > 0$.

Among different variations and extensions, we address some important forms here and refer the reader to Akdeniz and Kaciranlar (1995); Kaciranlar et al. (1999); Liu (2003); Gruber (2012); Kibria (2012); Xu and Yang (2012); Arashi et al. (2014); Akdeniz Duran et al. (2015); Kurtoglu and Ozkale (2016); Asar and Genc (2016); Arashi et al. (2017b); Asar et al. (2017); Wu et al. (2018); Filzmoser and Kurnaz (2018); Karbalaee et al. (2019), for further studies.

The generalized Liu estimator (Liu, 1993) is

$$\hat{\boldsymbol{\beta}}_n^{\text{GeneralizedLiu}}(\boldsymbol{D}) = (\boldsymbol{X}^\top\boldsymbol{X} + \boldsymbol{I}_p)^{-1}(\boldsymbol{X}^\top\boldsymbol{y} + \boldsymbol{D}\hat{\boldsymbol{\beta}}_n^{\text{LS}}),$$

where $\boldsymbol{D} = \text{Diag}(d_1, \cdots, d_p)$, $0 < d_i < 1$, $i = 1, \cdots, p$.

The estimator of Ozkale and Kaciranlar (2007) is given by

$$\hat{\boldsymbol{\beta}}_n^{\text{OK}}(k, d) = (\boldsymbol{X}^\top\boldsymbol{X} + k\boldsymbol{I}_p)^{-1}(\boldsymbol{X}^\top\boldsymbol{y} + kd\hat{\boldsymbol{\beta}}_n^{\text{LS}}),\ k > 0,\ d \in (0, 1). \tag{8.2.2}$$

The estimator of Yang and Chang (2010) is

$$\hat{\boldsymbol{\beta}}_n^{\text{YC}}(k,d) = (\boldsymbol{X}^\top\boldsymbol{X} + \boldsymbol{I}_p)^{-1}(\boldsymbol{X}^\top\boldsymbol{X} + d\boldsymbol{I}_p)(\boldsymbol{X}^\top\boldsymbol{X} + k\boldsymbol{I}_p)^{-1}\boldsymbol{X}^\top\boldsymbol{y},$$

$$k > 0,\ d \in (0,1).$$

The estimator of Kurnaz and Akay (2015) is given by:

$$\hat{\boldsymbol{\beta}}_n^{\text{KA}}(k) = (\boldsymbol{X}^\top\boldsymbol{X} + k\boldsymbol{I}_p)^{-1}(\boldsymbol{X}^\top\boldsymbol{y} + g(k)\hat{\boldsymbol{\beta}}^*), \quad k > 0,$$

where $\hat{\boldsymbol{\beta}}_n^*$ can be any estimator and $g(\cdot)$ is a continuous function.

Regarding simultaneous optimization of both tuning parameters k and d in Eq. (8.2.2) using cross validation, we refer to Roozbeh et al. (2020). Knowing the fact that $\boldsymbol{X}^\top\boldsymbol{X}$ is invertible in the full rank model of Eq. (8.2.1), we may write

$$\begin{aligned}\hat{\boldsymbol{\beta}}_n^{\text{ridge}}(k) &= (\boldsymbol{X}^\top\boldsymbol{X} + k\boldsymbol{I}_n)^{-1}\boldsymbol{X}^\top\boldsymbol{X}(\boldsymbol{X}^\top\boldsymbol{X})^{-1}\boldsymbol{X}^\top\boldsymbol{y} \\ &= \boldsymbol{R}_n(k)\hat{\boldsymbol{\beta}}_n^{\text{LS}}\end{aligned}$$

where

$$\boldsymbol{R}_n(k) = (\boldsymbol{X}^\top\boldsymbol{X} + k\boldsymbol{I}_n)^{-1}\boldsymbol{X}^\top\boldsymbol{X} = (\boldsymbol{I}_n + k(\boldsymbol{X}^\top\boldsymbol{X})^{-1})^{-1}.$$

Hence, one way of constructing methodology for ridge-type estimators is to pre-multiply the expression $\boldsymbol{R}_n(k)$ to the OLS estimator or unrestricted estimator.

In a similar manner, we may represent

$$\begin{aligned}\hat{\boldsymbol{\beta}}_n^{\text{Liu}}(d) &= (\boldsymbol{X}^\top\boldsymbol{X} + \boldsymbol{I}_p)^{-1}(\boldsymbol{X}^\top\boldsymbol{X} + d\boldsymbol{I}_p)\hat{\boldsymbol{\beta}}_n^{\text{LS}} \\ &= \boldsymbol{F}_n(d)\hat{\boldsymbol{\beta}}_n^{\text{LS}},\end{aligned}$$

where

$$\boldsymbol{F}_n(d) = (\boldsymbol{X}^\top\boldsymbol{X} + \boldsymbol{I}_p)^{-1}(\boldsymbol{X}^\top\boldsymbol{X} + d\boldsymbol{I}_p).$$

We must point out that, in the high-dimensional case where $p > n$, the Liu estimator and some of its variants are not applicable due to the fact that the OLS estimator does not exist. This is a substantial shortcoming of the Liu estimator dealing with high-dimensional data. As a remedy, borrowing the structure of $\hat{\boldsymbol{\beta}}_n^{\text{Liu}}$, one may replace $\hat{\boldsymbol{\beta}}_n^{\text{LS}}$ with a selector estimator such as LASSO, aLASSO, Enet. However, we kill the variable selection property of the selectors. To fulfil both the selection and multicollinearity resistance, Arashi et al. (2017a) proposed the following optimization problem

$$\min_{\boldsymbol{\beta}\in\mathbb{R}^p} \|\boldsymbol{y} - \boldsymbol{X}\boldsymbol{\beta}\|^2 \quad \text{such that} \quad \|d\hat{\boldsymbol{\beta}}^{\text{LS}} - \boldsymbol{\beta}\|^2 < \lambda_1,\ \|\boldsymbol{\beta}\| < \lambda_2, \tag{8.2.3}$$

where, in this case, $\|\boldsymbol{\beta}\| = \sum_{j=1}^{p} |\beta_j|$. The authors argue that we can use

$$\hat{\boldsymbol{\beta}}_n^{\text{LassoLiu}}(d) = \boldsymbol{F}_n(d)\hat{\boldsymbol{\beta}}^{\text{LassoLS}} \tag{8.2.4}$$

as the solution to Eq. (8.2.3). Recently, Genc and Ozkale (2020) solved the optimization problem of Eq. (8.2.3) using gradient descent. Although they did not obtain a closed-form estimator, they proved the grouping effect of the solution of Eq. (8.2.3) following the approach in Zou and Hastie (2005). Very recently, Arashi et al. (2021) discussed the low-dimensional theoretical properties and derived the asymptotic distribution of Eq. (8.2.4).

8.2.1 Liu-type R-estimator

Recall in the multiple regression model of Eq. (8.2.1), under the named regularity conditions, following the rank-based objective function of Jaeckel (1972) defined by

$$D_n(\boldsymbol{\beta}) = \sum_{i=1}^{n}(y_i - \boldsymbol{x}_i^{\top}\boldsymbol{\beta})a_n(R_{ni}(\boldsymbol{\beta})),$$

the unrestricted R-estimator (URE) is obtained by $\hat{\boldsymbol{\beta}}_n^{\text{UR}} = \operatorname{argmin}_{\boldsymbol{\beta}\in\mathbb{R}^p}\{D_n(\boldsymbol{\beta})\}$. In 2014, Turkmen and Ozturk considered the penalized objective function and defined the rank-based ridge estimator as

$$\hat{\boldsymbol{\beta}}_n^{\text{ridgeR}} = \operatorname{argmin}_{\boldsymbol{\beta}\in\mathbb{R}^p}\{D_n(\boldsymbol{\beta}) + n\lambda_n\|\boldsymbol{\beta}\|^2\},$$

where $\lambda_n \geq 0$ is the regularization parameter. Due to the form of the objective function, the solution does not have a closed form. In this regard, we have the following important result due to Turkmen and Ozturk (2014).

Theorem 8.1 Under the model (8.2.1), in addition to the named regularity conditions, assume $\lim_{n\to\infty} \sqrt{n}\lambda_n = \lambda_o < \infty$ and $\lim_{n\to\infty} \lambda_n = 0$. Then, we have

$$\sqrt{n}(\hat{\boldsymbol{\beta}}_n^{\text{ridgeR}} - \boldsymbol{\beta}_o) \xrightarrow{\mathcal{D}} \mathcal{N}(-2\lambda_o\eta\boldsymbol{\beta}_o, \eta^2\boldsymbol{I}_p)$$

where $\boldsymbol{\beta}_o$ is the true regression parameter and $\eta^2 = \frac{A_\phi^2}{\gamma^2(\phi,f)}$ where $A_\phi^2 = \int_0^1 \phi^2(u)du - \left(\int_0^1 \phi(u)du\right)^2$, and $\gamma(\phi, f) = \int_0^1 \phi(u)\psi(u, f)du$ where $\psi(u, f) = \frac{f'(F^{-1}(u))}{f(F^{-1}(u))}, 0 < u < 1$.

Following Turkmen and Ozturk (2014) one may define a rank-based Liu estimator as

$$\hat{\boldsymbol{\beta}}_n^{\text{LiuR}} = \underset{\boldsymbol{\beta}\in\mathbb{R}^p}{\text{argmin}}\{D_n(\boldsymbol{\beta}) + \|d_n\hat{\boldsymbol{\beta}}_n^{\text{UR}} - \boldsymbol{\beta}\|^2\},$$

where $d_n \in (0,1)$ is the regularization parameter.

In order to find a closed form ridge-type rank estimator based on $\hat{\boldsymbol{\beta}}_n^{\text{UR}}$, Saleh et al. (2019) suggested to minimize the distance between $\boldsymbol{\beta}$ and the R-estimator $\hat{\boldsymbol{\beta}}_n^{\text{UR}}$ to combat multicollinearity in robust regression. Using the marginal theory, they derived a closed-form estimator given by

$$\hat{\boldsymbol{\beta}}_n^{\text{ridgeR}}(k) = \left(\hat{\boldsymbol{\beta}}_{1n}^{\text{UR}^\top}, \frac{1}{1+k}\hat{\boldsymbol{\beta}}_{2n}^{\text{UR}^\top}\right)^\top.$$

Indeed, the form $\frac{1}{1+k}\hat{\boldsymbol{\beta}}_{2n}^{\text{UR}^\top}$ is reminiscent of the contraction estimator defined by Mayer and Willke (1973) in classical theory.

The marginal theory due to Saleh et al. (2019) can also be used to define a closed form Liu-type R-estimator. In this case, one may consider it as

$$\begin{aligned}\hat{\boldsymbol{\beta}}_n^{\text{LiuR}}(d) &= \underset{(\boldsymbol{\beta}_1,\boldsymbol{\beta}_2)\in\mathbb{R}^p}{\text{argmin}}\{D_n(\boldsymbol{\beta}_1,\boldsymbol{\beta}_2) + \|d\hat{\boldsymbol{\beta}}_{2n}^{\text{UR}^\top} - \boldsymbol{\beta}_2\|^2\}\\ &= \left(\hat{\boldsymbol{\beta}}_{1n}^{\text{UR}^\top}, \frac{1+d}{2}\hat{\boldsymbol{\beta}}_{2n}^{\text{UR}^\top}\right)^\top\end{aligned}$$

where one suspects that $\boldsymbol{\beta}_2 = \mathbf{0}$ may hold. Hence, the performance analysis of $\hat{\boldsymbol{\beta}}_n^{\text{LiuR}}(d)$ is similar to $\hat{\boldsymbol{\beta}}_n^{\text{ridgeR}}(k)$. In order to add some fresh ideas, we follow the construction methodology that was initially used by Akdeniz and Kaciranlar (2001), i.e. we pre-multiply the R-estimator by a factor. For this and the rest of our study regarding shrinkage Liu-type R-estimators, we shall assume in the regression model (8.2.1).

(1) Errors $\boldsymbol{\varepsilon} = (\varepsilon_1, \dots, \varepsilon_n)^\top$ are i.i.d. random variables with (unknown) c.d.f. F having an absolutely continuous p.d.f. f with finite and non-zero Fisher information,

$$0 < I(f) = \int_{-\infty}^{\infty}\left(-\frac{f'(x)}{f(x)}\right)^2 f(x)\mathrm{d}x < \infty. \tag{8.2.5}$$

(2) For the definition of linear rank statistics, we consider the score generating function $\varphi : (0,1) \mapsto \mathbb{R}$ which is assumed to be non-constant, non-decreasing and square integrable on $(0,1)$ so that

$$A_{\varphi}^{2} = \int_{0}^{1} \varphi^{2}(u)\mathrm{d}u - \left(\int_{0}^{1} \varphi(u)\mathrm{d}u\right)^{2}. \tag{8.2.6}$$

The scores are defined in either of the following ways:

$$a_n(i) = \mathbb{E}\left[\varphi(U_{i:n})\right], \quad \text{or} \quad a_n(i) = \varphi\left(\frac{i}{n+1}\right), \quad i = 1, \dots, n,$$

where $U_{1:n} \leq \dots \leq U_{n:n}$ are order statistics from a sample of size n from the uniform distribution $\mathcal{U}(0,1)$.

(3) Let us define

$$\boldsymbol{C}_n = \frac{1}{n}\sum_{i=1}^{n} \boldsymbol{x}_i \boldsymbol{x}_i^{\top}, \tag{8.2.7}$$

where $\boldsymbol{x}_i$ is the ith row of $\boldsymbol{X}$. We assume that

$$\text{(a)} \ \lim_{n\to\infty} \boldsymbol{C}_n = \boldsymbol{C} \quad \text{and (b)} \ \lim_{n\to\infty} \max_{1\leq i\leq n} \boldsymbol{x}_i^{\top}\boldsymbol{C}_n^{-1}\boldsymbol{x}_i = 0 \tag{8.2.8}$$

where $\boldsymbol{C}$ is a non-negative definite matrix.

(4)

$$\lim_{n\to\infty} \boldsymbol{F}_n(d) = \boldsymbol{F}_d, \quad \boldsymbol{F}_d = (\boldsymbol{C} + \boldsymbol{I}_p)^{-1}(\boldsymbol{C} + d\boldsymbol{I}_p),$$

where

$$\boldsymbol{F}_n(d) = (\boldsymbol{C}_n + \boldsymbol{I}_p)^{-1}(\boldsymbol{C}_n + d\boldsymbol{I}_p).$$

Then, we define the unrestricted Liu-type R-estimator (ULRE) as

$$\hat{\boldsymbol{\beta}}_n^{\text{ULiuR}}(d) = \boldsymbol{F}_n(d)\hat{\boldsymbol{\beta}}_n^{\text{UR}}, \quad 0 < d < 1. \tag{8.2.9}$$

8.3 Shrinkage Liu-type R-estimators

Suppose we are provided with the sub-space restriction, incorporated as the prior information, $\boldsymbol{H}\boldsymbol{\beta} = \boldsymbol{h}$, where $\boldsymbol{H}$ is a $q{\times}p$ $(q < p)$ matrix and $\boldsymbol{h}$ is a q-vector of known constants. Under such prior information, we define the restricted Liu-type R-estimator (RLRE) as

$$\hat{\boldsymbol{\beta}}_n^{\text{RLiuR}}(d) = \boldsymbol{F}_n(d)\hat{\boldsymbol{\beta}}_n^{\text{RR}}, \tag{8.3.1}$$

where $\hat{\boldsymbol{\beta}}_n^{\text{RR}}$ is the restricted R-estimator (RRE) given by

$$\hat{\boldsymbol{\beta}}_n^{\text{RR}} = \hat{\boldsymbol{\beta}}_n^{\text{UR}} - \boldsymbol{C}_n^{-1}\boldsymbol{H}^\top\boldsymbol{G}_n^{-1}(\boldsymbol{H}\hat{\boldsymbol{\beta}}_n^{\text{UR}} - \boldsymbol{h}), \quad \boldsymbol{G}_n = \boldsymbol{H}\boldsymbol{C}_n^{-1}\boldsymbol{H}^\top. \tag{8.3.2}$$

Then, the preliminary test Liu-type R-estimator (PTLRE) is

$$\hat{\boldsymbol{\beta}}_n^{\text{PTLiuR}}(d,\alpha) = \boldsymbol{F}_n(d)\hat{\boldsymbol{\beta}}_n^{\text{PTR}}(\alpha), \tag{8.3.3}$$

where

$$\hat{\boldsymbol{\beta}}_n^{\text{PT}}(\alpha) = \hat{\boldsymbol{\beta}}_n^{\text{UR}} - (\hat{\boldsymbol{\beta}}_n^{\text{UR}} - \hat{\boldsymbol{\beta}}_n^{\text{RR}})I(\mathcal{L}_n \le \chi_q^2(\alpha)), \tag{8.3.4}$$

with

$$\mathcal{L}_n = \frac{n(\boldsymbol{H}\hat{\boldsymbol{\beta}}_n^{\text{UR}} - \boldsymbol{h})^\top\boldsymbol{G}_n^{-1}(\boldsymbol{H}\hat{\boldsymbol{\beta}}_n^{\text{UR}} - \boldsymbol{h})}{\eta^2}. \tag{8.3.5}$$

Then, the Stein-type shrinkage Liu-type R-estimator (SSLRE) has the form

$$\hat{\boldsymbol{\beta}}_n^{\text{SSLiuR}}(d) = \boldsymbol{F}_n(d)\hat{\boldsymbol{\beta}}_n^{\text{SSR}}, \tag{8.3.6}$$

with

$$\hat{\boldsymbol{\beta}}_n^{\text{SSR}} = \hat{\boldsymbol{\beta}}_n^{\text{UR}} - (q-2)(\hat{\boldsymbol{\beta}}_n^{\text{UR}} - \hat{\boldsymbol{\beta}}_n^{\text{RR}})\mathcal{L}_n^{-1}. \tag{8.3.7}$$

Finally, the positive-rule shrinkage Liu-type R-estimator (PRSLRE) is given by

$$\hat{\boldsymbol{\beta}}_n^{\text{SSLiuR+}}(d) = \boldsymbol{F}_n(d)\hat{\boldsymbol{\beta}}_n^{\text{SSR+}} \tag{8.3.8}$$

where

$$\begin{aligned}\hat{\boldsymbol{\beta}}_n^{\text{SSR+}} &= \hat{\boldsymbol{\beta}}_n^{\text{SSR}} - (1-(q-2)\mathcal{L}_n^{-1})I(\mathcal{L}_n \le q-2)(\hat{\boldsymbol{\beta}}_n^{\text{UR}} - \hat{\boldsymbol{\beta}}_n^{\text{RR}}) \\ &= \hat{\boldsymbol{\beta}}_n^{\text{SSR}} - (\hat{\boldsymbol{\beta}}_n^{\text{UR}} - \hat{\boldsymbol{\beta}}_n^{\text{RR}})I(\mathcal{L}_n \le q-2). \end{aligned} \tag{8.3.9}$$

8.4 Asymptotic Distributional Risk

For the asymptotic distributional risk of the shrinkage estimators, we have the following results.

Theorem 8.2 Under the local alternatives $K_n : \boldsymbol{H\beta} = \boldsymbol{h} + \frac{\boldsymbol{\gamma}}{\sqrt{n}}$, where $\boldsymbol{\gamma} = (\gamma_1, \dots, \gamma_q)^\top \neq \boldsymbol{0}$,

$$\begin{pmatrix} \sqrt{n}(\hat{\boldsymbol{\beta}}_n^{\text{ULiuR}}(d) - \boldsymbol{\beta}) \\ \sqrt{n}(\hat{\boldsymbol{\beta}}_n^{\text{RLiuR}}(d) - \boldsymbol{\beta}) \\ \sqrt{n}(\hat{\boldsymbol{\beta}}_n^{\text{ULiuR}}(d) - \hat{\boldsymbol{\beta}}_n^{\text{RLiuR}}(d)) \end{pmatrix} \xrightarrow{\mathcal{D}} \mathcal{N}_{3p}\left(\begin{pmatrix} \boldsymbol{b}_1 \\ \boldsymbol{b}_2 \\ \boldsymbol{b}_1 - \boldsymbol{b}_2 \end{pmatrix}, \eta^2 \boldsymbol{V} \right)$$

where

$$\begin{aligned} \boldsymbol{b}_1 &= -(1-d)(\boldsymbol{C} + \boldsymbol{I}_p)^{-1}\boldsymbol{\beta}, \\ \boldsymbol{b}_2 &= \boldsymbol{b}_1 - \boldsymbol{F}_d\boldsymbol{\eta}, \quad \boldsymbol{\eta} = \boldsymbol{C}^{-1}\boldsymbol{H}^\top\boldsymbol{G}^{-1}\boldsymbol{\gamma}, \end{aligned}$$

$$\boldsymbol{V} = \begin{pmatrix} \boldsymbol{F}_d^\top\boldsymbol{C}^{-1}\boldsymbol{F}_d & \boldsymbol{F}_d^\top(\boldsymbol{C}^{-1} - \boldsymbol{A})\boldsymbol{F}_d & \boldsymbol{F}_d^\top\boldsymbol{A}\boldsymbol{F}_d \\ \boldsymbol{F}_d^\top(\boldsymbol{C}^{-1} - \boldsymbol{A})\boldsymbol{F}_d & \boldsymbol{F}_d^\top(\boldsymbol{C}^{-1} - \boldsymbol{A})\boldsymbol{F}_d & \boldsymbol{0} \\ \boldsymbol{F}_d^\top\boldsymbol{A}\boldsymbol{F}_d & \boldsymbol{0} & \boldsymbol{F}_d^\top\boldsymbol{A}\boldsymbol{F}_d \end{pmatrix}$$

and $\boldsymbol{A} = \boldsymbol{C}^{-1}\boldsymbol{H}^\top\boldsymbol{G}^{-1}\boldsymbol{H}\boldsymbol{C}^{-1}$. where $\boldsymbol{G} = \boldsymbol{H}\boldsymbol{C}^{-1}\boldsymbol{H}^\top$

Theorem 8.3 The ADL$_2$-risk of R-estimators are given by

$$\begin{aligned} \text{ADL}_2\text{-risk}(\hat{\boldsymbol{\beta}}_n^{\text{ULiuR}}(d)) &= \eta^2 \operatorname{tr}(\boldsymbol{F}_d^\top\boldsymbol{C}^{-1}\boldsymbol{F}_d) + (1-d)^2\boldsymbol{\beta}^\top(\boldsymbol{C} + \boldsymbol{I}_p)^{-2}\boldsymbol{\beta}, \\ \text{ADL}_2\text{-risk}(\hat{\boldsymbol{\beta}}_n^{\text{RLiuR}}(d)) &= \eta^2 \operatorname{tr}(\boldsymbol{F}_d^\top\boldsymbol{C}^{-1}\boldsymbol{F}_d) - \eta^2 \operatorname{tr}(\boldsymbol{F}_d^\top\boldsymbol{A}\boldsymbol{F}_d) \\ &\quad + \boldsymbol{\eta}^\top\boldsymbol{F}_d^\top\boldsymbol{F}_d\boldsymbol{\eta} + 2(1-d)\boldsymbol{\eta}^\top\boldsymbol{F}_d^\top(\boldsymbol{C} + \boldsymbol{I}_p)^{-1}\boldsymbol{\beta} \\ &\quad + (1-d)^2\boldsymbol{\beta}^\top(\boldsymbol{C} + \boldsymbol{I}_p)^{-2}\boldsymbol{\beta}, \\ \text{ADL}_2\text{-risk}(\hat{\boldsymbol{\beta}}_n^{\text{PTLiuR}}(d, \alpha)) &= \eta^2 \operatorname{tr}(\boldsymbol{F}_d^\top\boldsymbol{C}^{-1}\boldsymbol{F}_d) + \boldsymbol{\eta}^\top\boldsymbol{F}_d^\top\boldsymbol{F}_d\boldsymbol{\eta}Z(\alpha, \Delta^2) \\ &\quad - \eta^2 \operatorname{tr}(\boldsymbol{F}_d^\top\boldsymbol{A}\boldsymbol{F}_d)\mathcal{H}_{q+2}(\chi^2_{q,\alpha}; \Delta^2) \\ &\quad + 2(1-d)\boldsymbol{\eta}^\top\boldsymbol{F}_d^\top(\boldsymbol{C} + \boldsymbol{I}_p)^{-1}\boldsymbol{\beta}\mathcal{H}_{q+2}(\chi^2_{q,\alpha}; \Delta^2) \\ &\quad + (1-d)^2\boldsymbol{\beta}^\top(\boldsymbol{C} + \boldsymbol{I}_p)^{-2}\boldsymbol{\beta}, \end{aligned}$$

$$
\begin{aligned}
\text{ADL}_2\text{-risk}(\hat{\boldsymbol{\beta}}_n^{\text{SSLiuR}}(d);\boldsymbol{\beta}) &= \eta^2\,\text{tr}(\boldsymbol{F}_d^\top \boldsymbol{C}^{-1}\boldsymbol{F}_d)\\
&\quad -\eta^2(q-2)\,\text{tr}(\boldsymbol{F}_d^\top \boldsymbol{A}\boldsymbol{F}_d)X(\Delta^2)\\
&\quad +(q-2)\boldsymbol{\eta}^\top \boldsymbol{F}_d^\top \boldsymbol{F}_d\boldsymbol{\eta}Y(\Delta^2)\\
&\quad +2(q-2)(1-d)\boldsymbol{\eta}^\top \boldsymbol{F}_d^\top[\boldsymbol{C}+\boldsymbol{I}_p]^{-1}\boldsymbol{\beta}\mathbb{E}\left[\chi_{q+2}^{-2}(\Delta^2)\right]\\
&\quad +(1-d)^2\boldsymbol{\beta}^\top[\boldsymbol{C}+\boldsymbol{I}_p]^{-2}\boldsymbol{\beta},
\end{aligned}
$$

$$
\begin{aligned}
\text{ADL}_2\text{-risk}(\hat{\boldsymbol{\beta}}_n^{\text{SSLiuR+}}(d)) &= \text{ADL}_2\text{-risk}(\hat{\boldsymbol{\beta}}_n^{\text{SSLiuR}}(d))\\
&-\eta^2\left\{\text{tr}(\boldsymbol{F}_d^\top \boldsymbol{A}\boldsymbol{F}_d)\mathbb{E}\left[(1-(q-2)\chi_{q+2}^{-2}(\Delta^2))^2 I(\chi_{q+2}^{2}(\Delta^2)\le q-2)\right]\right.\\
&\left.+\,\eta^2\left(\boldsymbol{\eta}^\top \boldsymbol{F}_d^\top \boldsymbol{F}_d\boldsymbol{\eta}\right)\mathbb{E}\left[(1-(q-4)\chi_{q+4}^{-2}(\Delta^2))^2 I(\chi_{q+4}^{2}(\Delta^2)\le q-4)\right]\right\}\\
&-2(\boldsymbol{\eta}^\top \boldsymbol{F}_d^\top \boldsymbol{F}_d\boldsymbol{\eta})\mathbb{E}\left[((q-2)\chi_{q+2}^{-2}(\Delta^2)-1)I(\chi_{q+2}^{2}(\Delta^2)\le q-2)\right]\\
&-2(1-d)\boldsymbol{\eta}^\top \boldsymbol{F}_d^\top[\boldsymbol{C}+\boldsymbol{I}_p]^{-1}\boldsymbol{\beta}\mathbb{E}\left[((q-2)\chi_{q+2}^{-2}(\Delta^2)-1)I(\chi_{q+2}^{2}(\Delta^2)\le q-2)\right],
\end{aligned}
$$

where $\Delta^2 = \eta^2(\boldsymbol{H}\boldsymbol{\beta}-\boldsymbol{h})^\top \boldsymbol{G}^{-1}(\boldsymbol{H}\boldsymbol{\beta}-\boldsymbol{h})$, and

$$
\begin{aligned}
X(\Delta^2) &= 2\mathbb{E}\left[\chi_{q+2}^{-2}(\Delta^2)\right]-(q-2)\mathbb{E}\left[\chi_{q+2}^{-4}(\Delta^2)\right]\\
Y(\Delta^2) &= 2\mathbb{E}\left[\chi_{q+2}^{-2}(\Delta^2))\right]-2\mathbb{E}\left[\chi_{q+4}^{-2}(\Delta^2))\right]+(q-2)\mathbb{E}\left[\chi_{q+4}^{-4}(\Delta^2)\right]\\
Z(\alpha,\Delta^2) &= 2\mathcal{H}_{q+2}(\chi_{q,\alpha}^2;\Delta^2)-\mathcal{H}_{q+4}(\chi_{q,\alpha}^2;\Delta^2).
\end{aligned}
$$

8.5 Asymptotic Distributional Risk Comparisons

In this section, we compare the R-estimators with respect to their ADL$_2$-risk functions. The study is focused on the shrinkage Liu-type R-estimators. In this section, we compare the R-estimators with respect to their ADL$_2$-risk functions.

Using the spectral decomposition of $\boldsymbol{C}$,

$$\boldsymbol{\Gamma}^\top \boldsymbol{C}\boldsymbol{\Gamma} = \boldsymbol{\Lambda} = \text{Diag}(\lambda_1,\lambda_2,\dots,\lambda_p),$$

with $\lambda_1 \ge \lambda_2 \ge \dots \ge \lambda_p > 0$, the eigenvalues of $\boldsymbol{F}_d = (\boldsymbol{C}+\boldsymbol{I}_p)^{-1}(\boldsymbol{C}+d\boldsymbol{I}_p)$ are

$$(\frac{\lambda_1+d}{\lambda_1+1},\frac{\lambda_2+d}{\lambda_2+1}\dots,\frac{\lambda_p+d}{\lambda_p+1})$$

while the eigenvalues of $(\boldsymbol{C}+\boldsymbol{I}_p)$ are $(\lambda_1+1, \lambda_2+1 \dots, \lambda_p+1)$. Then, we obtain the following identities

$$\begin{aligned}
\operatorname{tr}(\boldsymbol{F}_d^\top \boldsymbol{C}^{-1}\boldsymbol{F}_d) &= \sum_{i=1}^{p} \frac{(\lambda_i+d)^2}{\lambda_i(\lambda_i+1)^2}, \\
\boldsymbol{\beta}^\top(\boldsymbol{C}+\boldsymbol{I}_p)^{-2}\boldsymbol{\beta} &= \sum_{i=1}^{p} \frac{\theta_i^2}{(\lambda_i+1)^2}; \quad \theta = \boldsymbol{\Gamma}^\top\boldsymbol{\beta}, \\
\operatorname{tr}(\boldsymbol{F}_d^\top \boldsymbol{A}\boldsymbol{F}_d) &= \sum_{i=1}^{p} \frac{q_{ii}(\lambda_i+d)^2}{(\lambda_i+1)^2},
\end{aligned}$$

where $q_{ii} \geq 0$ is the ith diagonal element of the matrix $\boldsymbol{Q} = \boldsymbol{\Gamma}^\top\boldsymbol{A}\boldsymbol{\Gamma}$. Also,

$$\boldsymbol{\eta}^\top \boldsymbol{F}_d^\top \boldsymbol{F}_d \boldsymbol{\eta} = \sum_{i=1}^{p} \frac{\eta_i^{\star 2}(\lambda_i+d)^2}{(\lambda_i+1)^2},$$

where $\eta_i^\star$ is the ith element of vector $\boldsymbol{\eta}^\star = \boldsymbol{\eta}^\top\boldsymbol{\Gamma}$. Similarly,

$$\boldsymbol{\eta}^\top \boldsymbol{F}_d^\top (\boldsymbol{C}+\boldsymbol{I}_p)^{-1}\boldsymbol{\beta} = \sum_{i=1}^{p} \frac{\theta_i \eta_i^\star(\lambda_i+d)}{(\lambda_i+1)^2}.$$

8.5.1 Comparison of SSLRE and PTLRE

We will be using the risk difference to compare different estimators. In this case, $\text{ADL}_2\text{-risk}(\hat{\boldsymbol{\beta}}_n^{\text{SSLiuR}}(d)) - \text{ADL}_2\text{-risk}(\hat{\boldsymbol{\beta}}_n^{\text{PTLiuR}}(d))$ will be non-positive if

$$\boldsymbol{\eta}^\top \boldsymbol{F}_d^\top \boldsymbol{F}_d \boldsymbol{\eta} \geq \frac{f_1(\Delta^2, d, \alpha)}{Z(\alpha, \Delta^2) - (q-2)Y(\Delta^2)}, \tag{8.5.1}$$

where

$$\begin{aligned}
f_1(\Delta^2, d, \alpha) &= \operatorname{tr}(\boldsymbol{F}_d^\top \boldsymbol{A}\boldsymbol{F}_d)\{\mathcal{H}_{q+2}(\chi^2_{q,\alpha}; \Delta^2) - (q-2)X(\Delta^2)\} \\
&+ 2(1-d)\eta^2\boldsymbol{\eta}^\top \boldsymbol{F}_d^\top(\boldsymbol{C}+\boldsymbol{I}_p)^{-1}\boldsymbol{\beta}\left\{(q-2)\mathbb{E}(\chi^{-2}_{q+2}(\Delta^2)) - \mathcal{H}_{q+2}(\chi^2_{q,\alpha}; \Delta^2)\right\}.
\end{aligned}$$

Since the departure parameter $\Delta^2 > 0$, we assume that both the numerator and the denominator of Eq. (8.5.1) are positive or negative, respectively. Then $\hat{\boldsymbol{\beta}}_n^{\text{SSLiuR}}(d)$ dominates $\hat{\boldsymbol{\beta}}_n^{\text{PTLiuR}}(d, \alpha)$ when

$$\Delta^2 \leq \Delta_1^2(\Delta^2, d, \alpha) = \frac{f_1(\Delta^2, d, \alpha)}{\text{Ch}_{\max}[(\boldsymbol{F}_d^\top \boldsymbol{F}_d)\boldsymbol{C}^{-1}]\{Z(\alpha, \Delta^2) - (q-2)Y(\Delta^2)\}},$$

and $\hat{\boldsymbol{\beta}}_n^{\text{PTLiuR}}(d,\alpha)$ dominates $\hat{\boldsymbol{\beta}}_n^{\text{SSLiuR}}(d)$ when

$$\Delta^2 > \Delta_2^2(\Delta^2, d, \alpha) = \frac{f_1(\Delta^2, d, \alpha)}{\text{Ch}_{\min}[(\boldsymbol{F}_d^\top \boldsymbol{F}_d)\boldsymbol{C}^{-1}]\{Z(\alpha, \Delta^2) - (q-2)Y(\Delta^2)\}},$$

where $\text{Ch}_{\min}(\boldsymbol{A})$ and $\text{Ch}_{\max}(\boldsymbol{A})$ are the minimum and maximum characteristic roots of matrix $\boldsymbol{A}$.

Now, we consider the risk difference

$$\text{ADL}_2\text{-risk}(\hat{\boldsymbol{\beta}}_n^{\text{SSLiuR}}(d)) - \text{ADL}_2\text{-risk}(\hat{\boldsymbol{\beta}}_n^{\text{PTLiuR}}(d))$$

as a function of eigenvalues and define

$$d_1(\Delta^2, \alpha) = \frac{f_2(\Delta^2, \alpha)}{g_1(\Delta^2, \alpha)},$$

where

$$\begin{aligned} f_2(\Delta^2, \alpha) \quad &= \quad \max_i \Big\{ \lambda_i \eta_i^{\star^2} \{Z(\alpha, \Delta^2) - (q-2)Y(\Delta^2)\} \\ &\quad - \eta^2 q_{ii} \lambda_i \{\mathcal{H}_{q+2}(\chi^2_{q,\alpha}; \Delta^2) - (q-2)X(\Delta^2)\} \\ &\quad + 2\theta_i \eta_i^{\star^2} \{\mathcal{H}_{q+2}(\chi^2_{q,\alpha}; \Delta^2) - (q-2)\mathbb{E}(\chi^{-2}_{q+2}(\Delta^2))\} \Big\} \end{aligned}$$

and

$$\begin{aligned} g_1(\Delta^2, \alpha) \quad &= \quad \min_i \Big\{ \tau_\psi^2 q_{ii} \{\mathcal{H}_{q+2}(\chi^2_{q,\alpha}; \Delta^2) - (q-2)X(\Delta^2)\} \\ &\quad - \eta_i^{\star^2} \{Z(\alpha, \Delta^2) - (q-2)Y(\Delta^2)\} \\ &\quad + 2\theta_i \eta_i^{\star^2} \{\mathcal{H}_{q+2}(\chi^2_{q,\alpha}; \Delta^2) - (q-2)\mathbb{E}(\chi^{-2}_{q+2}(\Delta^2))\} \Big\}. \end{aligned}$$

Suppose that $d > 0$. Then, we have the following:

(1) If $g_1(\Delta^2, \alpha) > 0$ and $f_2(\Delta^2, \alpha) > 0$, it follows that for each positive d with $d < d_1(\Delta^2, \alpha)$, $\hat{\boldsymbol{\beta}}_n^{\text{SSLiuR}}(d)$ has risk value less than that of $\hat{\boldsymbol{\beta}}_n^{\text{PTLiuR}}(d, \alpha)$.
(2) If $g_1(\Delta^2, \alpha) < 0$ and $f_2(\Delta^2, \alpha) < 0$, it follows that for each positive d with $d > d_1(\Delta^2, \alpha)$, $\hat{\boldsymbol{\beta}}_n^{\text{SSLiuR}}(d)$ has risk value less than that of $\hat{\boldsymbol{\beta}}_n^{\text{PTLiuR}}(d, \alpha)$.

Under $\mathcal{H}_o : \boldsymbol{H\beta} = \boldsymbol{h}$, the risk difference reduces to

$$\eta^2 \sum_{i=1}^{p} \frac{q_{ii}(\lambda_i + d)^2}{(\lambda_i + 1)^2} \{\mathcal{H}_{q+2}(\chi^2_{q,\alpha}; 0) - (q-2)X(0)\},$$

where

$$\mathcal{H}_{q+2}(\chi^2_{q,\alpha}; 0) = \chi^2_{q+2}(\frac{q\chi^2_{q,\alpha}}{q+2}; 0) = 1 - \alpha.$$

Thus, the risk of $\hat{\boldsymbol{\beta}}_n^{\text{SSLiuR}}(d)$ is smaller than that of $\hat{\boldsymbol{\beta}}_n^{\text{PTLiuR}}(d, \alpha)$ when the critical value $\chi^2_{q,\alpha}$ satisfies

$$\chi^2_{q,\alpha} \leq \frac{q+2}{q} \chi^{-2}_{q+2}((q-2)X(0)).$$

Otherwise, the risk of $\hat{\boldsymbol{\beta}}_n^{\text{PTLiuR}}(d, \alpha)$ is smaller than the risk of $\hat{\boldsymbol{\beta}}_n^{\text{SSLiuR}}(d)$.

Remark 8.1 For $\alpha = 1$, we will have the comparison between SSLRE and ULRE and, for $\alpha = 0$, we will have comparison between SSLRE and RLRE.

8.5.2 Comparison of PRSLRE and PTLRE

Case 1. Under the null hypothesis, $\mathcal{H}_o : \boldsymbol{H\beta} = \boldsymbol{h}$.
The risk difference is

$$\text{ADL}_2\text{-risk}(\hat{\boldsymbol{\beta}}_n^{\text{SSLiuR+}}(d)) - \text{ADL}_2\text{-risk}(\hat{\boldsymbol{\beta}}_n^{\text{PTLiuR}}(d)) = \sum_{i=1}^{p} \frac{q_{ii}(\lambda_i + d)^2\eta^2}{(\lambda_i + 1)^2}$$
$$\times \left\{\mathcal{H}_{q+2}(\chi^2_{q,\alpha}; 0) - (q-2)X(0) - \mathbb{E}\left[(1 - (q-2)\chi^{-2}_{q+2}(0))^2 I(\chi^2_{q,\alpha} \leq q-2)\right]\right\} \geq 0$$

for all α satisfying the condition

$$\left\{\alpha : \chi^2_{q,\alpha} \geq \mathcal{H}_{q+2}\left((q-2)X(0) + \mathbb{E}\left[(1-(q-2)\chi^{-2}_{q+2}(0))^2 I(\chi^2_{q+2}(0) \leq q-2\right]; 0\right)\right\}. \tag{8.5.2}$$

Thus, the risk of PTLRE is smaller than that of PRLRE when the critical value $\chi^2_{q,\alpha}$ satisfies the relation in Eq. (8.5.2). However, the risk of PRLRE is smaller than that of PTLRE when the critical value $\chi^2_{q,\alpha}$ satisfies the opposite relation to Eq. (8.5.2).

Case 2. When the null hypothesis does not hold.
The risk difference ADL$_2$-risk($\hat{\boldsymbol{\beta}}_n^{\text{SSLiuR+}}(d)$)−ADL$_2$-risk($\hat{\boldsymbol{\beta}}_n^{\text{PTLiuR}}(d)$) will be non-positive when

$$\boldsymbol{\eta}^\top \boldsymbol{F}_d^\top \boldsymbol{F}_d \boldsymbol{\eta} \geq \frac{f_3(\Delta^2, d, \alpha)}{g_2(\Delta^2, \alpha)} \tag{8.5.3}$$

where

$$f_3(\Delta^2, d, \alpha) = \eta^2 \operatorname{tr}(\boldsymbol{F}_d \boldsymbol{A} \boldsymbol{F}_d^\top)\{\mathcal{H}_{q+2}(\chi^2_{q,\alpha}; \Delta^2) - (q-2)X(\Delta^2) + a_1\}$$
$$-2(1-d)\boldsymbol{\eta}^\top \boldsymbol{F}_d^\top (\boldsymbol{C} + \boldsymbol{I}_p)^{-1}\boldsymbol{\beta}\left\{(q-2)\mathbb{E}(\chi^{-2}_{q+2}(\Delta^2)) - \mathcal{H}_{q+2}(\chi^2_{q,\alpha}; \Delta^2) - a_3\right\}$$

and

$$g_2(\Delta^2, \alpha) = Z(\alpha, \Delta^2) - (q-2)Y(\Delta^2) + a_2 + 2a_3.$$

Also,

$$a_1 = \mathbb{E}\left[\left(1-(q-2)\chi^{-2}_{q+2}(\Delta^2)\right)^2 I\left(\chi^2_{q+2}(\Delta^2) \leq q-2\right)\right]$$
$$a_2 = \mathbb{E}\left[\left(1-(q-4)\chi^{-2}_{q+4}(\Delta^2)\right)^2 I\left(\chi^2_{q+4}(\Delta^2) \leq q-4\right)\right]$$
$$a_3 = \mathbb{E}\left[\left((q-2)\chi^{-2}_{q+2}(\Delta^2) - 1\right) I\left(\chi^2_{q+2}(\Delta^2) \leq q-2\right)\right].$$

Since $\Delta^2 > 0$, assume that both the numerator and the denominator of Eq. (8.5.3) are positive or negative, respectively. Then $\hat{\boldsymbol{\beta}}_n^{\text{SSLiuR+}}(d)$ dominates $\hat{\boldsymbol{\beta}}_n^{\text{PTLiuR}}(d, \alpha)$ when

$$\Delta^2 \leq \Delta_3^2(\Delta^2, d, \alpha) = \frac{f_3(\Delta^2, d, \alpha)}{\text{Ch}_{\max}[(\boldsymbol{F}_d^\top \boldsymbol{F}_d)\boldsymbol{C}^{-1}] \times g_2(\Delta^2, \alpha)},$$

and $\hat{\boldsymbol{\beta}}_n^{\text{PTLiuR}}(d, \alpha)$ dominates $\hat{\boldsymbol{\beta}}_n^{\text{SSLiuR+}}(d)$ when

$$\Delta^2 > \Delta_4^2(\Delta^2, d, \alpha) = \frac{f_3(\Delta^2, d, \alpha)}{\text{Ch}_{\min}[(\boldsymbol{F}_d^\top \boldsymbol{F}_d)\boldsymbol{C}^{-1}] \times g_2(\Delta^2, \alpha)}.$$

Now, we consider the risk difference of PRLRE and PTLRE as a function of the eigenvalues and define

$$d_2(\Delta^2, \alpha) = \frac{f_4(\Delta^2, \alpha)}{g_3(\Delta^2, \alpha)},$$

where

$$f_4(\Delta^2, \alpha) = \max_i \left\{\lambda_i q_{ii} P_1 + \lambda_i \eta_i^{\star 2} P_2 + \theta_i \eta_i^\star P_3\right\}$$

and

$$g_3(\Delta^2,\alpha) = \min_i \left\{2 + \theta_i\lambda_i\eta_i^{\star}P_3 - \lambda_i q_{ii}P_1 - \lambda_i\eta_i^{\star^2}P_2\right\}.$$

Also,

$$\begin{aligned}
P_1 &= \mathcal{H}_{q+2}(\chi^2_{q,\alpha};\Delta^2) - (q-2)X(\Delta^2) + a_1, \\
P_2 &= (q-2)Y(\Delta^2) - a_2 - 2a_3 - Z(\alpha,\Delta^2), \\
P_3 &= (q-2)\mathbb{E}\left[\chi^{-2}_{q+2}(\Delta^2)\right] - a_3 - \mathcal{H}_{q+2}(\chi^2_{q,\alpha};\Delta^2) - (q-2)X(\Delta^2).
\end{aligned}$$

Suppose that $d > 0$. Then, we have the following statements.

(1) If $g_3(\Delta^2,\alpha) > 0$ and $f_4(\Delta^2,\alpha) > 0$, it follows that for each positive d with $d > d_2(\Delta^2,\alpha)$, $\hat{\boldsymbol{\beta}}_n^{\text{SSLiuR+}}(d)$ has risk value less than that of $\hat{\boldsymbol{\beta}}_n^{\text{PTLiuR}}(d,\alpha)$.
(2) If $g_3(\Delta^2,\alpha) < 0$ and $f_4(\Delta^2,\alpha) < 0$, it follows that for each positive d with $d < d_2(\Delta^2,\alpha)$, $\hat{\boldsymbol{\beta}}_n^{\text{SSLiuR+}}(d)$ has risk value less than that of $\hat{\boldsymbol{\beta}}_n^{\text{PTLiuR}}(d,\alpha)$.

Remark 8.2 For $\alpha = 0$, we obtain the superiority condition of $\hat{\boldsymbol{\beta}}_n^{\text{SSLiuR+}}(d)$ over $\hat{\boldsymbol{\beta}}_n^{\text{RLiuR}}(d)$ and, for $\alpha = 1$, we obtain the superiority condition of $\hat{\boldsymbol{\beta}}_n^{\text{SSLiuR+}}(d)$ over $\hat{\boldsymbol{\beta}}_n^{\text{ULiuR}}(d)$.

8.5.3 Comparison of PRLRE and SSLRE

In this case, we have

$$\begin{aligned}
&\text{ADL}_2\text{-risk}(\hat{\boldsymbol{\beta}}_n^{\text{SSLiuR+}}(d)) - \text{ADL}_2\text{-risk}(\hat{\boldsymbol{\beta}}_n^{\text{SSLiuR}}(d)) \ = \\
&-\eta^2\Big\{\text{tr}(\boldsymbol{F}_d^\top \boldsymbol{A}\boldsymbol{F}_d)\mathbb{E}\left[\left(1-(q-2)\chi^{-2}_{q+2}(\Delta^2)\right)^2 I(\chi^2_{q+2}(\Delta^2)\le q-2)\right] \\
&+\eta^2(\boldsymbol{\eta}^\top\boldsymbol{F}_d^\top\boldsymbol{F}_d\boldsymbol{\eta})\mathbb{E}\left[\left(1-(q-2)\chi^{-2}_{q+4}(\Delta^2)\right)^2 I(\chi^2_{q+4}(\Delta^2)\le q-2)\right]\Big\} \\
&-2(\boldsymbol{\eta}^\top\boldsymbol{F}_d^\top\boldsymbol{F}_d\boldsymbol{\eta})\mathbb{E}\left[\left((q-2)\chi^{-2}_{q+2}(\Delta^2)-1\right)I(\chi^2_{q+2}(\Delta^2)\le q-2)\right] \\
&-2(1-d)\boldsymbol{\eta}^\top\boldsymbol{F}_d^\top(\boldsymbol{C}+\boldsymbol{I}_p)^{-1}\boldsymbol{\beta} \\
&\times\mathbb{E}\left[\left((q-2)\chi^{-2}_{q+2}(\Delta^2)-1\right)I(\chi^2_{q+2}(\Delta^2)\le q-2)\right].
\end{aligned} \tag{8.5.4}$$

Case 1: Suppose $\boldsymbol{\eta}^\top \boldsymbol{F}_d^\top(\boldsymbol{C} + \boldsymbol{I}_p)^{-1}\boldsymbol{\beta}$ is positive, then the R.H.S. of Eq. (8.5.4) is negative, since the expectation of a positive random variable is positive. Thus, for all Δ^2 and d

$$\text{ADL}_2\text{-risk}(\hat{\boldsymbol{\beta}}_n^{\text{SSLiuR+}}(d)) \leq \text{ADL}_2\text{-risk}(\hat{\boldsymbol{\beta}}_n^{\text{SSLiuR}}(d)).$$

Therefore, under this condition, the PRLRE not only confirms the inadmissibility of SSLRE but also provides a simple superior estimator for the ill-conditioned data.

Case 2: Suppose $\boldsymbol{\eta}^\top \boldsymbol{F}_d^\top(\boldsymbol{C} + \boldsymbol{I}_p)^{-1}\boldsymbol{\beta}$ is negative, then the difference in Eq. (8.5.4) will be positive when

$$\boldsymbol{\eta}^\top \boldsymbol{F}_d^\top \boldsymbol{F}_d \boldsymbol{\eta} \geq \frac{f_5(\Delta^2, d)}{g_4(\Delta^2)}, \tag{8.5.5}$$

where

$$f_5(\Delta^2, d) = 2(1-d)\boldsymbol{\eta}^\top \boldsymbol{F}_d^\top(\boldsymbol{C} + \boldsymbol{I}_p)^{-1}\boldsymbol{\beta}\, a_3 - \eta^2 \operatorname{tr}(\boldsymbol{F}_d^\top \boldsymbol{A} \boldsymbol{F}_d)\, a_1$$

and

$$g_4(\Delta^2) = -(a_2 + 2a_3).$$

Since $\Delta^2 > 0$, assume that both the numerator and the denominator of Eq. (8.5.6) are positive or negative, respectively. Then $\hat{\boldsymbol{\beta}}_n^{\text{SSLiuR+}}(d)$ dominates $\hat{\boldsymbol{\beta}}_n^{\text{SSLiuR}}(d)$, when

$$\Delta^2 \leq \Delta_5^2(\Delta^2, d) = \frac{f_9(\Delta^2, d)}{\text{Ch}_{\max}[(\boldsymbol{F}_d^\top \boldsymbol{F}_d)\boldsymbol{C}^{-1}] \times g_7(\Delta^2)}$$

and $\hat{\boldsymbol{\beta}}_n^{\text{SSLiuR}}(d)$ dominates $\hat{\boldsymbol{\beta}}_n^{\text{SSLiuR+}}(d)$, when

$$\Delta^2 > \Delta_6^2(\Delta^2, d) = \frac{f_9(\Delta^2, d)}{\text{Ch}_{\min}[(\boldsymbol{F}_d^\top \boldsymbol{F}_d)\boldsymbol{C}^{-1}] \times g_7(\Delta^2)}.$$

8.5.4 Comparison of Liu-Type Rank Estimators With Counterparts

Here, we only consider the case when the null hypothesis does not hold, which is more general. Further we only compare SSLRE with SSRE. For more details and similar comparisons, we refer the reader to Arashi et al. (2014).

The risk difference between SSRE and SSLRE is

$$\begin{aligned}
&\text{ADL}_2\text{-risk}(\hat{\boldsymbol{\beta}}_n^{\text{SSR}}) - \text{ADL}_2\text{-risk}(\hat{\boldsymbol{\beta}}_n^{\text{SSLiuR}}(d)) \quad = \\
&\eta^2 \operatorname{tr}(\boldsymbol{c}^{-1} - \boldsymbol{F}_d^\top \boldsymbol{C}^{-1}\boldsymbol{F}_d) - \eta^2(q-2)\operatorname{tr}(\boldsymbol{A} - \boldsymbol{F}_d^\top \boldsymbol{A}\boldsymbol{F}_d)X(\Delta^2) \\
&+(q-2)\boldsymbol{\eta}^\top(\boldsymbol{I} - \boldsymbol{F}_d^\top \boldsymbol{F}_d)\boldsymbol{\eta} Y(\Delta^2) \\
&+2(q-2)(1-d)\boldsymbol{\eta}^\top \boldsymbol{F}_d^\top[\boldsymbol{C} + \boldsymbol{I}_p]^{-1}\boldsymbol{\beta}\mathbb{E}\left[\chi_{q+2}^{-2}(\Delta^2)\right]
\end{aligned}$$

$$+(1-d)^2\boldsymbol{\beta}^\top[\boldsymbol{C}+\boldsymbol{I}_p]^{-2}\boldsymbol{\beta}.$$

The above difference will be greater than or equal to 0, when

$$\boldsymbol{\eta}^\top(\boldsymbol{I}-\boldsymbol{F}_d^\top\boldsymbol{F}_d)\boldsymbol{\eta} \geq \frac{h_1(\Delta^2,d)}{Y(\Delta^2)} \tag{8.5.6}$$

where

$$\begin{aligned} h_1(\Delta^2,d) &= -\eta^2\,\mathrm{tr}(\boldsymbol{c}^{-1}-\boldsymbol{F}_d^\top\boldsymbol{C}^{-1}\boldsymbol{F}_d)+\eta^2(q-2)\,\mathrm{tr}(\boldsymbol{A}-\boldsymbol{F}_d^\top\boldsymbol{A}\boldsymbol{F}_d)X(\Delta^2) \\ &\quad -2(q-2)(1-d)\boldsymbol{\eta}^\top\boldsymbol{F}_d^\top[\boldsymbol{C}+\boldsymbol{I}_p]^{-1}\boldsymbol{\beta}\mathbb{E}\left[\chi_{q+2}^{-2}(\Delta^2)\right] \\ &\quad -(1-d)^2\boldsymbol{\beta}^\top[\boldsymbol{C}+\boldsymbol{I}_p]^{-2}\boldsymbol{\beta}. \end{aligned}$$

Since $\Delta^2 > 0$, we assume that the numerator of Eq. (8.5.6) is positive. Then, the SSLRE will dominate SSRE whenever $\Delta^2 \leq \Delta_7^2(\Delta^2,d)$, where

$$\Delta_7^2(\Delta^2,d) = \frac{h_1(\Delta^2,d)}{(q-2)\mathrm{Ch}_{\max}[(\boldsymbol{I}-\boldsymbol{F}_d^\top\boldsymbol{F}_d)\boldsymbol{C}^{-1}]Y(\Delta^2)}.$$

However, SSRE dominates SSLRE when

$$\Delta^2 < \Delta_8^2(\Delta^2,d) = \frac{h_1(\Delta^2,d)}{(q-2)\mathrm{Ch}_{\min}[(\boldsymbol{I}-\boldsymbol{F}_d^\top\boldsymbol{F}_d)\boldsymbol{C}^{-1}]Y(\Delta^2)}.$$

Then, we compare these two estimators based on the biasing parameter, d. For our purposes, we have

$$\begin{aligned} \mathrm{ADL}_2\text{-risk}(\hat{\boldsymbol{\beta}}_n^{\mathrm{SSLiuR}}(d)) &= \eta^2\sum_{i=1}^p\frac{(\lambda_i+d)^2}{\lambda_i(\lambda_i+1)^2} \\ &\quad -\eta^2\sum_{i=1}^p\frac{(q-2)q_{ii}(\lambda_i+d)^2}{(\lambda_i+1)^2}X(\Delta^2) \\ &\quad +2(q-2)(1-d)\sum_{i=1}^p\frac{\theta_i\eta_i^\star(\lambda_i+d)}{(\lambda_i+1)^2}\mathbb{E}\left[\chi_{q+2}^{-2}(\Delta)\right] \\ &\quad +\sum_{i=1}^p\frac{(q-2)\eta_i^{\star 2}(\lambda_i+d)^2}{(\lambda_i+1)^2}Y(\Delta^2) \\ &\quad +(1-d)^2\sum_{i=1}^p\frac{\theta_i^2}{(\lambda_i+1)^2}. \end{aligned} \tag{8.5.7}$$

Now differentiating Eq. (8.5.7) with respect to the biasing parameter d, we get

$$\frac{\partial\mathrm{ADL}_2\text{-risk}(\hat{\boldsymbol{\beta}}_n^{\mathrm{SSLiuR}}(d))}{\partial d} = \eta^2\sum_{i=1}^p\frac{2(q-2)}{\lambda_i(\lambda_i+1)^2}$$

$$\times\{(\lambda_i + d) - \lambda_i q_{ii}(\lambda_i + d)X(\Delta^2) + \eta^2\lambda_i\eta_i^{\star 2}(\lambda_i + d)Y(\Delta^2)$$

$$+ \eta^2\lambda_i(\theta_i\eta_i^\star - \theta_i\eta_i^\star\lambda_i - 2\theta_i\eta_i^\star d)\mathbb{E}\left[\chi_{q+2}^{-2}(\Delta^2)\right] - (q-2)^{-1}\eta^2\lambda_i(1-d)\theta_i^2\right\}$$

$$= \eta^2 \sum_{i=1}^{p} \frac{2}{(\lambda_i + 1)^2}(h_2(\Delta^2, \alpha)d - h_3(\Delta^2, \alpha))$$

where

$$\begin{aligned} h_2(\Delta^2, \alpha) &= \min_i \Big\{1 - \lambda_i q_{ii}X(\Delta^2) + \eta^{-2}\lambda_i\eta_i^{\star 2}Y(\Delta^2) \\ &\quad -2\eta^{-2}\lambda_i\theta_i\eta_i^\star\mathbb{E}\left[\chi_{q+2}^{-2}(\Delta^2)\right] + (q-2)^{-1}\eta^{-2}\lambda_i\theta_i^2\Big\} \end{aligned}$$

and

$$\begin{aligned} h_3(\Delta^2, \alpha) &= \max_i \Big\{\lambda_i\Big(\lambda_i q_{ii}X(\Delta^2) - 1 - \eta^{-2}\lambda_i\eta_i^{\star 2}Y(\Delta^2) \\ &\quad -\tau_\psi^{-2}\theta_i\eta_i^\star(1-\lambda_i) + (q-2)^{-1}\eta^2\theta_i^2\Big)\Big\}. \end{aligned}$$

Now we define

$$d_3(\Delta^2, \alpha) = \frac{h_2(\Delta^2, \alpha)}{h_3(\Delta^2, \alpha)}.$$

Suppose $d > 0$, then we have the following:

1. If $h_i(\Delta^2, \alpha) > 0, i = 2, 3$, it follows that for each positive d with $d > d_3(\Delta^2, \alpha)$, $\hat{\boldsymbol{\beta}}_n^{\text{SSLiuR}}(d)$ has risk value less than that of $\hat{\boldsymbol{\beta}}_n^{\text{SSR}}$.
2. If $h_i(\Delta^2, \alpha) < 0, i = 2, 3$, it follows that for each positive d with $d < d_3(\Delta^2, \alpha)$, $\hat{\boldsymbol{\beta}}_n^{\text{SSLiuR}}(d)$ has risk value less than that of $\hat{\boldsymbol{\beta}}_n^{\text{SSR}}$.

8.6 Estimation of *d*

Consider the ADL$_2$-risk of the ULRE is given by

$$\begin{aligned} f(d) = \text{ADL}_2\text{-risk}(\hat{\boldsymbol{\beta}}_n^{\text{ULiuR}}(d)) &= \eta^2 \operatorname{tr}(\boldsymbol{F}_d^\top \boldsymbol{C}^{-1}\boldsymbol{F}_d) \\ &\quad +(1-d)^2\boldsymbol{\beta}^\top(\boldsymbol{C} + \boldsymbol{I}_p)^{-2}\boldsymbol{\beta}, \\ &= \eta^2 \sum_{i=1}^{p} \frac{(\lambda_i + d)^2}{\lambda_i(\lambda_i + 1)^2} + (1-d)^2 \sum_{i=1}^{p} \frac{\theta_i^2}{(\lambda_i + 1)^2} \\ &= \delta_1(d) + \delta_2(d). \end{aligned} \tag{8.6.1}$$

For the Liu estimator, one would like to find a value of d such that the decrease in variance ($\delta_1(d)$) is greater than the increase in squared bias ($\delta_1(d)$). In order to show that such a value less than 1 exists so that

$$\text{ADL}_2\text{-risk}(\hat{\boldsymbol{\beta}}_n^{\text{ULiuR}}(d)) < \text{ADL}_2\text{-risk}(\hat{\boldsymbol{\beta}}_n^{\text{UR}}),$$

we will take derivative of Eq. (8.6.1) with respect to d as follows,

$$\frac{\partial f(d)}{\partial d} = 2\eta^2 \sum_{i=1}^{p} \frac{(\lambda_i + d)}{\lambda_i(\lambda_i + 1)^2} - 2(1-d) \sum_{i=1}^{p} \frac{\theta_i^2}{(\lambda_i + 1)^2}. \tag{8.6.2}$$

For $d = 1$ in Eq. (8.6.2), we obtain

$$\frac{\partial f(d)}{\partial d} = 2\sum_{i=1}^{p} \frac{1}{\lambda_i(\lambda_i + 1)} > 0, \quad \text{since} \quad \lambda_i \geq 0. \tag{8.6.3}$$

Therefore, there exists a value of d $(0 < d < 1)$, such that

$$\text{ADL}_2\text{-risk}(\hat{\boldsymbol{\beta}}_n^{\text{ULiuR}}(d)) < \text{ADL}_2\text{-risk}(\hat{\boldsymbol{\beta}}_n^{\text{UR}}).$$

Now from Eq. (8.6.2), the risk of $\hat{\boldsymbol{\beta}}_n^{\text{ULiuR}}(d)$ is minimized at

$$d = \frac{\sum_{i=1}^{p} \frac{\theta_i^2 - \tau_\psi^2}{(\lambda_i+1)^2}}{\sum_{i=1}^{p} \frac{(\eta^2 + \lambda_i \theta_i^2)}{\lambda_i(\lambda_i+1)^2}},$$

where θ_i is the ith component of the vector, $\boldsymbol{\theta} = \boldsymbol{\Gamma}^\top \boldsymbol{\beta}$. Replacing θ_i^2 and η^2 by their corresponding unbiased estimate $\left(\hat{\theta}_i^{\text{UR}}\right)^2 - \frac{\hat{\eta}^2}{\lambda_i}$ and $\hat{\eta}^2$, we get the optimum value of d as

$$d_{\text{opt}} = 1 - \hat{\eta}^2 \left\{ \frac{\sum_{i=1}^{p} \frac{1}{\lambda_i(\lambda_i+1)}}{\sum_{i=1}^{p} \frac{\left(\hat{\theta}_i^{\text{UR}}\right)^2}{(\lambda_i+1)^2}} \right\}.$$

Refer to Akdeniz and Ozturk (2005) and Arashi et al. (2017b) for further information on the estimation of d, among others.

8.7 Diabetes Data Analysis

In this section, we conduct performance analysis of the Liu estimators by adopting a real example from Arashi et al. (2018). The data set consists of 10 covariates: age, sex, body mass index (BMI), blood pressure (BP), and six blood

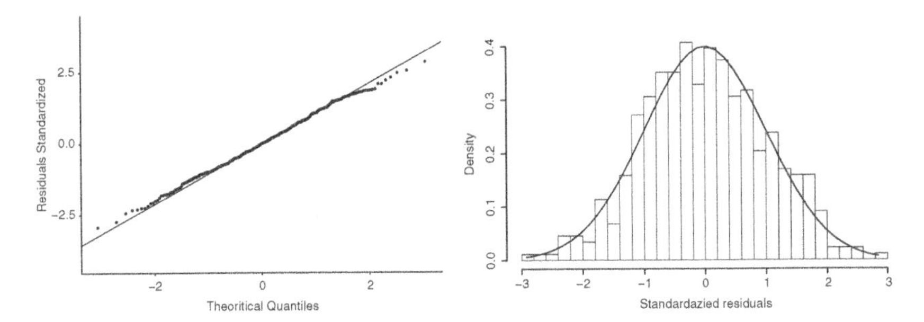

Figure 8.1 Left: the Q-Q plot for the diabetes data sets; Right: the distribution of the residuals.

Table 8.1 The VIF values of the diabetes data set.

Variable	Age	Sex	BMI	BP	S1	S2	S3	S4	S5	S6
VIF	1.22	1.28	1.51	1.46	59.20	39.19	15.40	8.89	10.07	1.48

serum measurements, S1, S2, S3, S4, S5, and S6 for each of $n = 442$ patients. The response variable is a quantitative measure of disease progression after 1 year of follow-up. The objective of this study is to examine the disease progression associated with important prognostic variables. Note that the predictors are scaled to have a mean of zero and unit variance before we start the analysis.

First, we verify the characteristics of this data set: the Bonferonni test reports the 57th observation is an outlier; the left plot of Figure 8.1 confirms it. The distribution of the residuals is depicted on the right of Figure 8.1 that shows it follows a normal model. To verify the multicollinearity, the VIF values are reported in Table 8.1. The values show that there is multicollinearity in the variables, S1-S5. Hence, this data set is suitably chosen for our proposed usage.

8.7.1 Penalty Estimators

For the purpose of comparison and adaptivity with other parts of the book, we give the penalty estimators again according to the model setup of Eq. (8.2.1). The key idea in penalized regression methods is minimizing an objection function $J_{\rho,\lambda}(\boldsymbol{\beta})$ in the form of

$$J_{\rho,\lambda}(\boldsymbol{\beta}) = (\boldsymbol{y} - \boldsymbol{X}\boldsymbol{\beta})^{\top}(\boldsymbol{y} - \boldsymbol{X}\boldsymbol{\beta}) + \lambda P_{\rho}(\boldsymbol{\beta}) \tag{8.7.1}$$

to obtain the estimates of the parameter. The first term in the objective function is the sum of the squared error loss, the second term of Eq. (8.7.1) is a penalty function, and λ is a tuning parameter which controls the trade-off between two components of $J_{\rho,\lambda}(\boldsymbol{\beta})$.

The penalty function is usually chosen as a norm on $\mathbb{R}^p$, such as

$$P_\rho(\boldsymbol{\beta}) = \sum_{j=1}^{p} |\beta_j|^q, \; q > 0.$$

This class of estimator is called the bridge estimator, proposed by Frank and Freidman (1993).

Ridge A well-known member of the class of bridge estimators is the ridge estimator which is obtained by considering $q = 2$. Ridge regression minimizes the residual sum of squares subject to an L_2-penalty, $\sum_{j=1}^{p} \beta_j^2$. The ridge R-estimator is defined as

$$\hat{\boldsymbol{\beta}}_n^{\text{ridgeLiuR}} = \underset{\boldsymbol{\beta}}{\text{argmin}} \left\{ D(\boldsymbol{\beta}) + \lambda \sum_{j=1}^{p} \beta_j^2 \right\}.$$

LASSO The LASSO performs variable selection and parameter estimation simultaneously. The Lasso R-estimator is defined by

$$\hat{\boldsymbol{\beta}}_n^{\text{LassoR}} = \underset{\boldsymbol{\beta}}{\text{argmin}} \left\{ D(\boldsymbol{\beta}) + \lambda \sum_{j=1}^{p} |\beta_j| \right\}.$$

This estimator belongs to class of bridge estimators for $q = 1$.

aLASSO Zou (2006) introduced the adaptive LASSO (aLASSO) by modifying the LASSO penalty by using adaptive weights on L_1-penalty with the regression coefficients.

The aLasso R estimator, $\hat{\boldsymbol{\beta}}_n^{\text{aLassoR}}$ is obtained by

$$\hat{\boldsymbol{\beta}}_n^{\text{aLassoR}} = \underset{\boldsymbol{\beta}}{\text{argmin}} \left\{ D(\boldsymbol{\beta}) + \lambda \sum_{j=1}^{p} \hat{\omega}_j |\beta_j| \right\},$$

where the weight function is

$$\hat{\omega}_j = \frac{1}{|\hat{\beta}_j^*|^\gamma}; \quad \gamma > 0,$$

and $\hat{\beta}_j^*$ is a $\sqrt{n}$-consistent estimator of $\boldsymbol{\beta}$. The minimization procedure for aLASSO solution does not induce any computational difficulty and can be solved very efficiently.

SCAD Although the LASSO method does both shrinkage and variable selection due to the nature of the L_1-penalty by setting many coefficients identically to zero, it does not possess oracle properties, as discussed in Fan and Li (2001). To overcome the inefficiency of traditional variable selection procedures, Fan and Li (2001) proposed smoothly clipped absolute deviation (SCAD) to select variables and estimate the coefficients of variables automatically and simultaneously. The SCAD R-estimator is defined as

$$\hat{\boldsymbol{\beta}}_n^{\text{scadR}} = \underset{\boldsymbol{\beta}}{\operatorname{argmin}} \left\{ D(\boldsymbol{\beta}) + \lambda \sum_{j=1}^{p} p_{\alpha,\lambda}(\beta_j) \right\}.$$

Here $p_{\alpha,\lambda}(\cdot)$ is the SCAD penalty. It is a symmetric and quadratic spline on $[0, \infty)$ with knots at λ and $\alpha\lambda$, whose first order derivative is given by

$$p_{\alpha,\lambda}(\beta) = \begin{cases} \lambda\,|\beta|\,, & |\beta| \leq \lambda \\ -\left(\beta^2 - 2\alpha\lambda\,|\lambda| + \lambda^2\right) / \left[2(\alpha - 1)\right], & \lambda < |\beta| \leq \alpha\lambda\,. \\ (\alpha + 1)\lambda^2/2 & |\beta| > \alpha\lambda \end{cases}$$

Here $\lambda > 0$ and $\alpha > 2$ are the tuning parameters. For $\alpha = \infty$, the expression in this equation is equivalent to the L_1-penalty.

MCP Zhang (2010) introduced a new penalization method called minimax concave penalty (MCP) for variable selection. The MCP R-estimator is given by

$$\hat{\boldsymbol{\beta}}_n^{\text{mcpR}} = \underset{\boldsymbol{\beta}}{\operatorname{argmin}} \left\{ D(\boldsymbol{\beta}) + \sum_{j=1}^{p} \rho(|\beta_j|; \lambda) \right\},$$

where $\rho(\cdot; \lambda)$ is the MCP penalty given by

$$\rho(t; \lambda) = \lambda \int_0^{\mathrm{T}} \left(1 - \frac{x}{\gamma\lambda}\right)_+ \mathrm{d}x$$

and $\gamma > 0$ and λ are regularization and penalty parameters, respectively.

Enet The elastic net (Enet) was proposed by Zou and Hastie (2005) to overcome the limitations of the LASSO and ridge methods. The Enet R-estimator is defined as

$$\hat{\beta}_n^{\text{EnetR}} = \underset{\beta}{\operatorname{argmin}} \left\{ D(\beta) + \lambda_1 \sum_{j=1}^{p} |\beta_j| + \lambda_2 \sum_{j=1}^{p} \beta_j^2 \right\}, \tag{8.7.2}$$

where λ_1 is the LASSO penalty parameter for penalizing the sum of the absolute values of the regression coefficients and λ_2 is the ridge penalty parameter for penalizing the sum of the squared regression coefficients, respectively.

MNET Huang et al. (2016) introduced the MNET estimation which use MCP penalty term instead of the L_1-penalty in (8.7.2). The Mnet R-estimator is defined as

$$\hat{\beta}_n^{\text{MnetR}} = \underset{\beta}{\operatorname{argmin}} \left\{ D(\beta) + \sum_{j=1}^{p} \rho(|\beta_j|;\lambda_1) + \lambda_2 \sum_{j=1}^{p} \beta_j^2 \right\}.$$

Similar to the Enet of Zou and Hastie (2005) the MNET also tends to select or drop highly correlated predictors together. However, unlike the Enet, the MNET is selection consistent and equal to the oracle ridge estimator with high probability under reasonable conditions.

8.7.2 Performance Analysis

The performance of any of the methods described above is assessed by the bootstrapped prediction error criteria. The average prediction errors are calculated via five-fold CV for each bootstrap replicate.

In order to construct the suggested methods, we consider a two-step approach here since the prior information is not available as long as there is no one who knows which predictor explains significantly the response variable. In the first step, a candidate sub-model may be selected with the help of step-wise or subset selection procedures and a model selection criteria. Here, we used the forward selection method to pick important covariates. It shows that the variables of AGE, S3 and S6 may be ignored and the other variables can be used a sub model. Hence, in the second step, the suggested methods can be constructed by using this candidate sub-model.

We consider the unknown regression parameters, with the aid of the forward selection method, to be

$$\beta_{\text{true}} = (0, -11.3, 25.2, 15.9, -29.3, 16.8, 0, 6.4, 33.6, 0)^{\top},$$

as the true parameters in the bootstrap method. In the following, K-fold cross validation is used to obtain an estimate of the prediction errors of the model. In a K-fold cross validation, the data set is randomly divided into K subsets of roughly equal size. One subset is left aside, $\{(\boldsymbol{X}^{\text{test}}, \boldsymbol{y}^{\text{test}})\}$, termed the test set, while the remaining $K-1$ subsets, called the training set, are used to fit the model. The result estimator is called $\hat{\boldsymbol{\beta}}^{\text{train}}$. The fitted model is then used to predict the responses of the test data set. Finally, prediction errors are obtained by taking the squared deviation of the observed and predicted values in the test set, i.e.

$$\text{PE}^k = \left\| \boldsymbol{y}_k^{\text{test}} - \hat{\boldsymbol{y}}_k^{\text{test}} \right\|^2; \qquad k = 1, 2, \ldots, K,$$

where $\hat{\boldsymbol{y}}_k^{\text{test}} = \boldsymbol{X}_k^{\text{test}} \hat{\boldsymbol{\beta}}_k^{\text{train}}$. The process is repeated for all K subsets and the prediction errors are combined. To account for the random variation of the cross validation, the process is reiterated N times and the median prediction error (MPE) is estimated using

$$\text{MPE} = \text{median}\left\{ \frac{1}{K} \sum_{k=1}^{K} \text{PE}_1^k, \ldots, \frac{1}{K} \sum_{k=1}^{K} \text{PE}_N^k \right\},$$

where PE_i^k is the prediction error considering the kth test set in the ith iteration.

The performance of the arbitrary estimator, $\hat{\boldsymbol{\beta}}_n^{\star}$, with respect to the full model estimator, $\hat{\boldsymbol{\beta}}_n^{\text{ULiuR}}(d)$, is obtained by the efficiency (Eff) formula defined as

$$\text{RMPE}(\hat{\boldsymbol{\beta}}^{\star}; \hat{\boldsymbol{\beta}}_n^{\text{ULiuR}}(d)) = \frac{\text{MPE}(\hat{\boldsymbol{\beta}}_n^{\text{ULiuR}}(d))}{\text{MPE}(\hat{\boldsymbol{\beta}}^{\star})}.$$

If the value of RMPE is greater than 1, then $\hat{\boldsymbol{\beta}}^{\star}$ performs better than $\hat{\boldsymbol{\beta}}_n^{\text{ULiuR}}(d)$.

Our results are based on $N = 500$ resampled bootstrap samples. In Table 8.2, we report the estimations for the 10 parameters, and the MPE and RMPE of each method. Note, PTLRE1, PTLRE2, PTLRE3, and PTLRE4 stand for PTLRE with $\alpha = 0.01$, $\alpha = 0.05$, $\alpha = 0.15$, and $\alpha = 0.25$, respectively. From this table, the restricted Liu-type R-estimator (RLRE) performs better than other estimators, based on an MPE of 17.92 and an RMPE of 1.54. When comparing the other penalized estimators, the positive-rule Stein-type shrinkage Liu-type R-estimator (PRSLRE) has a lower median prediction error of only 19.66 while the rest are above 20.0. Among the remaining penalized estimators, aLASSO is the best at 20.11, while MNET stands in the second position with 21.94.

The results are somewhat counter to conventional wisdom whereby a sparse regression model should benefit greatly from the use of penalty estimators such as LASSO and aLASSO. But the results in the table indicate that PRSLRE (with RMPE of 1.40) is superior to LASSO (with RMPE of 1.17) and competitive with or slightly better than aLASSO (with RMPE of 1.37) on this data set. As seen earlier in Chapter 2, there is sparsity in this data set and PRSLRE was able to

Table 8.2 Estimations for the diabetes data*. (The numbers in parentheses are the corresponding standard deviations).

	ULRE	RLRE	PTLRE1	PTLRE2	PTLRE3	PTLRE4
Age	−0.80 (0.43)	0.04 (0.01)	0.04 (0.01)	0.04 (0.01)	0.04 (0.01)	0.04 (0.01)
Sex	−12.68 (0.52)	−12.65 (0.52)	−12.65 (0.52)	−12.65 (0.52)	−12.65 (0.52)	−12.65 (0.52)
BMI	25.25 (0.65)	25.42 (0.64)	25.42 (0.64)	25.42 (0.64)	25.42 (0.64)	25.42 (0.64)
BP	16.02 (0.53)	16.23 (0.56)	16.23 (0.56)	16.23 (0.56)	16.23 (0.56)	16.22 (0.56)
S1	−28.93 (2.60)	−22.86 (1.19)	−22.86 (1.19)	−22.86 (1.19)	−22.86 (1.19)	−22.88 (1.20)
S2	14.82 (2.14)	10.35 (1.39)	10.35 (1.40)	10.35 (1.40)	10.35 (1.40)	10.36 (1.41)
S3	0.53 (1.39)	−3.06 (0.16)	−3.06 (0.16)	−3.06 (0.16)	−3.06 (0.16)	−3.06 (0.16)
S4	7.53 (1.43)	5.80 (1.12)	5.81 (1.12)	5.81 (1.12)	5.81 (1.12)	5.81 (1.13)
S5	33.47 (1.22)	31.88 (0.92)	31.89 (0.92)	31.89 (0.92)	31.89 (0.92)	31.89 (0.93)
S6	2.43 (0.55)	0.06 (0.01)	0.07 (0.01)	0.07 (0.01)	0.07 (0.01)	0.07 (0.01)
MPE	27.62 (0.39)	17.92 (0.30)	17.92 (0.30)	17.92 (0.30)	17.92 (0.30)	18.08 (0.30)
RMPE	1.00	1.54	1.55	1.55	1.55	1.55

	SSLRE	PRSLRE	UR	LS	ridge R	LASSO R
Age	−0.01 (0.48)	0.02 (0.10)	−0.85 (0.43)	−0.42 (0.42)	−0.10 (0.39)	−0.24 (0.35)
Sex	−12.65 (0.52)	−12.67 (0.52)	−12.79 (0.53)	−11.35 (0.49)	−10.43 (0.45)	−11.09 (0.49)
BMI	25.44 (0.65)	25.37 (0.64)	25.20 (0.65)	24.84 (0.59)	24.05 (0.54)	24.89 (0.60)
BP	16.13 (0.61)	16.11 (0.56)	16.09 (0.53)	15.49 (0.50)	14.81 (0.45)	15.28 (0.50)

(Continued)

Table 8.2 *(Continued)*

S1	−22.65 (3.11)	−23.86 (1.61)	−37.95 (3.38)	−28.56 (2.25)	−5.67 (0.49)	−26.92 (1.31)
S2	10.29 (2.47)	11.32 (1.62)	21.94 (2.74)	15.67 (1.93)	−2.38 (0.52)	14.11 (1.16)
S3	−2.75 (1.42)	−2.75 (0.46)	4.46 (1.67)	0.80 (1.27)	−8.57 (0.49)	0.00 (0.91)
S4	6.06 (1.57)	6.30 (1.26)	8.63 (1.50)	7.47 (1.28)	5.46 (0.81)	6.30 (1.33)
S5	31.96 (1.39)	32.19 (1.02)	36.93 (1.45)	32.33 (1.10)	22.53 (0.55)	31.37 (1.10)
S6	0.19 (0.75)	0.12 (0.36)	2.37 (0.55)	3.31 (0.53)	3.86 (0.48)	3.17 (0.52)
MPE	25.01 (1.16)	19.66 (0.33)	27.03 (0.34)	29.98 (0.43)	41.38 (0.34)	23.51 (0.30)
RMPE	1.10	1.40	1.02	0.92	0.66	1.17

	aLASSO R	**Enet R**	**Mnet R**	**SCAD R**	**MCP R**
Age	0.00 (0.01)	−0.27(0.35)	0.00 (0.18)	0.00 (0.11)	0.00 (0.13)
Sex	−10.61 (0.49)	−11.10 (0.49)	−11.17 (0.52)	−11.21 (0.49)	−11.20 (0.51)
BMI	25.26 (0.62)	24.88 (0.59)	25.00 (0.62)	25.04 (0.62)	25.06 (0.64)
BP	15.21 (0.54)	15.31 (0.50)	15.48 (0.51)	15.52 (0.53)	15.53 (0.53)
S1	−25.10 (2.01)	−26.97 (1.37)	−25.10 (8.11)	−24.08 (9.31)	−24.55 (9.08)
S2	12.64 (2.12)	14.18 (1.12)	13.12 (6.02)	12.25 (6.92)	12.74 (6.84)
S3	0.00 (0.32)	0.00 (0.98)	0.00 (4.09)	0.00 (4.72)	0.00 (4.53)
S4	5.53 (1.80)	6.32 (1.32)	4.64 (2.61)	3.58 (2.56)	3.89 (2.62)
S5	31.58 (1.50)	31.35 (1.11)	30.21 (2.10)	30.11 (2.45)	30.11 (2.30)
S6	1.84 (0.67)	3.17 (0.52)	1.98 (0.91)	1.84 (0.83)	1.80 (0.96)
MPE	20.11 (0.31)	23.32 (0.30)	21.94 (0.28)	22.18 (0.30)	22.56 (0.30)
RMPE	1.37	1.18	1.26	1.24	1.22

produce a lower MPE. This result is not an anomaly. In fact, there are also some other studies confirming the superior performance of the shrinkage estimators in comparison with well-known penalty estimators, for example, in the work of Yuzbasi et al. (2020).

8.8 Summary

To summarize the contribution of this chapter, we give the following highlights:

1. We described the notion behind the construction of the linear unified (Liu) estimator for multicollinear situations and provided some recent variants.
2. We defined the Liu-type R-estimator using the marginal theory.
3. Shrinkage Liu-type R-estimators were reviewed and their asymptotic distributional performance was analyzed in detail.
4. Finally, numerical assessments were conducted to compare the performance of shrinkage Liu-type R-estimators with the well-known penalty estimators ridge, LASSO, aLASSO, Enet, MNet, MCP, and SCAD, in which the R-estimation methodology was not used for the penalty estimators.

To conclude this chapter, we turn the readers' attention to a surprising achievement. Commonly, we expect penalty estimators to provide better prediction in the sparse regression model. However, our numerical study has demonstrated that the positive-rule shrinkage Liu-type R-estimator is a strong competitor compared to the penalty estimators since it provides a smaller median prediction error, and is superior to the Enet and SCAD among all penalty estimators considered. This is surprising since the shrinkage Liu-type R-estimator does not select variables and therefore would be expected to give poor performance in sparse models. There are also some other studies confirming the superior performance of the shrinkage estimators in comparison with penalty estimators. Interested readers who wish to pursue further work in this area may refer to Yuzbasi et al. (2020), among others.

8.9 Problems

8.1. Prove Theorem 8.1.

8.2. Prove Theorem 8.2.

8.3. Prove Theorem 8.3.

8.4. Define the hard and soft threshold estimators based on $\hat{\boldsymbol{\beta}}_n^{\text{ULiuR}}(d)$ and derive the asymptotic properties.

8.5. Define the Liu-type R-estimator as

$$\hat{\boldsymbol{\beta}}_n^{\text{LiuR}}(\lambda) = \underset{\boldsymbol{\beta}_2 \in \mathbb{R}}{\operatorname{argmin}} \left\{ \sum_{i=1}^{p} \sum_{j=1}^{n_i} |y_{ij} - \beta_i| a_{n_i}^{+}(R_{n_i}^{+}(\theta_i)) + (\lambda \hat{\boldsymbol{\beta}}_{2n}^{\text{UR}} - \boldsymbol{\beta}_2)^{\top} (\lambda \hat{\boldsymbol{\beta}}_{2n}^{\text{UR}} - \boldsymbol{\beta}_2) \right\}. \tag{8.9.1}$$

Solve Eq. (8.9.1) and derive its ADB and ADL_2-risk.

8.6. Compare the Liu-type R-estimator given in Eq. (8.9.1) with the LASSO-type R-estimator w.r.t. the ADL_2-risk difference.

8.7. Write a gradient descent algorithm to find the following solution

$$\hat{\boldsymbol{\beta}}_n^{\text{aLassoR}}(d) = \underset{\boldsymbol{\beta} \in \mathbb{R}^p}{\operatorname{argmin}} \left\{ \mathrm{D}(\boldsymbol{\beta}) + \lambda \sum_{j=1}^{p} \hat{w}_j |d \hat{\beta}_j^{\text{UR}} - \theta_j| \right\},$$

where $\hat{\beta}_j^{\text{UR}}$ is the jth element of R estimator under the model of Eq. (8.7.1), the weight function is

$$\hat{w}_j = \frac{1}{|\hat{\beta}_j^*|^{\gamma}}; \quad \gamma > 0,$$

and $\hat{\beta}_j^*$ is a $\sqrt{n}$-consistent estimator of $\boldsymbol{\beta}$, and $0 < d < 1$.

9

Autoregressive Models

9.1 Introduction

An important model in the analysis of time-series data is the auto-regression (AR) model. In such a model, the output response is constructed as a linear function of up to p previous values of itself. The previous values are based on measurements made in uniform increments of time, for example, each day, month or year. Since the measurements are evenly spaced, the time series can be represented as occurring at different integer values of time. We use the notation AR(p) to refer to models that use p previous values in time to predict the next (i.e. most recent) value. Specifically, for a model of order p, we consider the random process $X_t, t = \dots, -1, 0, 1, \dots$ satisfying the relation

$$X_t = \rho_1 X_{t-1} + \dots + \rho_p X_{t-p} + \varepsilon_t, \quad 1 \leq t \leq n, p \geq 1 \tag{9.1.1}$$

where $\boldsymbol{\rho} = (\rho_1, \dots, \rho_p)^\top \in \mathbb{R}^p$ is a vector of unknown auto-regressive parameters and we assume that all roots of

$$x^p - \rho_1 x^{p-1} - \dots - \rho_p = 0 \tag{9.1.2}$$

are in $(-1, 1)$. The process $\{X_t\}$ satisfying Eqs. (9.1.1) and (9.1.2) is called the auto-regression process of dimension p (> 1).

Let $\boldsymbol{Y}_o = (X_0, X_{-1}, \dots, X_{1-p})^\top$ be an observable random vector independent of white noise given by $\varepsilon_1, \varepsilon_2, \dots, \varepsilon_n$. Now, let $\boldsymbol{\rho} = (\boldsymbol{\rho}_1^\top, \boldsymbol{\rho}_2^\top)^\top$, where $\boldsymbol{\rho}_1$ and $\boldsymbol{\rho}_2$ are p_1 and p_2 vectors. Then, Eq. (9.1.1) may be written as

$$X_t = \boldsymbol{\rho}_1^\top \boldsymbol{Y}_{t1} + \boldsymbol{\rho}_2^\top \boldsymbol{Y}_{t2} + \varepsilon_t, \quad 1 \leq t \leq n \tag{9.1.3}$$

where $\boldsymbol{Y}_{t1} = (X_{t-1}, \dots, X_{t-p_1})^\top$, and $\boldsymbol{Y}_{t2} = (X_{t-p_1-1}, \dots, X_{t-p})^\top$, $1 \leq t \leq n$.

We may refer to Eq. (9.1.3) as a sparse auto-regression model where $\boldsymbol{\rho}_2$ may contain elements which are small or negligible. If $\boldsymbol{\rho}_2 = \boldsymbol{0}$ exactly, then we define a restricted auto-regressive model of order p_1, namely,

Rank-Based Methods for Shrinkage and Selection: With Application to Machine Learning. First Edition. A. K. Md. Ehsanes Saleh, Mohammad Arashi, Resve A. Saleh, and Mina Norouzirad.

$$X_t = \rho_1 X_{t-1} + \dots + \rho_{p_1} X_{t-p_1} + \varepsilon_t, \quad 1 \le t \le n, \ p_1 \ge 1. \tag{9.1.4}$$

Our primary interest in this chapter is to develop the theory of R-estimation of various types, beginning with selection of subsets and ending with shrinkage R-estimation.

9.2 R-estimation of ρ for the AR(p)-Model

In this section, we introduce a class of R-estimators of $\boldsymbol{\rho}$ for the AR(p) model of Eq. (9.1.1) and discuss their large sample properties. Let $\boldsymbol{Y}_t = (X_t, \dots, X_{t-p+1})^\top$, $1 \le t \le n$, and $R_i(b)$ be the rank of $X_t - \boldsymbol{b}^\top \boldsymbol{Y}_{t-1}$ among $\{X_j - \boldsymbol{b}^\top Y_{j-1}, 1 \le j \le n\}$. Set $R_i(b) = 0$ for $i \le 0$. Let ϕ be a non-decreasing function from $[0,1]$ to the real line and define $\boldsymbol{L}_n(\boldsymbol{b}) = (L_1(\boldsymbol{b}), \dots, L_p(\boldsymbol{b}))^\top$, where

$$L_s(\boldsymbol{b}) = \frac{1}{\sqrt{n}} \sum_{t=s+1}^{n} X_{t-s} \phi\left(\frac{R_t(\boldsymbol{b})}{n+1}\right), \quad 1 \le t \le p, \ \boldsymbol{b} \in \mathbb{R}^p. \tag{9.2.1}$$

The class of rank statistics, $\boldsymbol{L}_n(\boldsymbol{b})$, one for each ϕ, is analogous to a class of similar rank statistics in linear regression models where one replaces the weights $\{X_{t-j}\}$ by appropriate design points.

Thus, for the R-estimation of $\boldsymbol{\rho}$, we apply Jaeckel (1972) to the AR(p)-model. Accordingly, we set

$$a_n(t) = \phi\left(\frac{t}{n+1}\right), \ 1 \le t \le n,$$

and $\boldsymbol{Z}_{(t)}(\boldsymbol{b})$ is the tth largest residual among $\{X_k - \boldsymbol{b}^\top \boldsymbol{Y}_{k-1}, 1 \le k \le n\}$, and

$$D_n(\boldsymbol{b}) = \sum_{t=1}^{n} a_n(t) \boldsymbol{Z}_{(t)}(\boldsymbol{b}), \quad \boldsymbol{b} \in \mathbb{R}^p. \tag{9.2.2}$$

If $\sum_{t=1}^{n} a_n(t) = 0$, then $D_n(\boldsymbol{b})$ may be shown to be convex on $\mathbb{R}^p$ with its derivative $-n^{-\frac{1}{2}} \boldsymbol{L}_n(\boldsymbol{b})$. Thus, a minimum of $D_n(\boldsymbol{b})$ exists.

Let us now define the R-estimator $\hat{\boldsymbol{\rho}}_n^{\mathrm{UR}}$ of $\boldsymbol{\rho}$ by the relation

$$\inf_{\boldsymbol{b} \in \mathbb{R}^p} \|\boldsymbol{L}_n(\boldsymbol{b})\| = \left\| \boldsymbol{L}_n(\hat{\boldsymbol{\rho}}_n^{\mathrm{UR}}) \right\|. \tag{9.2.3}$$

The following theorem gives the asymptotic uniform linearity (AUL) result similar to Jurečková (1971) for ordinary regression models.

Theorem 9.1 (Koul and Saleh, 1993) (AUL of Statistics). Assume Eqs. (9.1.1) and (9.1.2) hold. In addition, assume the following

(a) (i) $\mathbb{E}[\varepsilon] = 0, 0 < \mathbb{E}\left[\varepsilon^4\right] < \infty$,
  (ii) F has uniformly continuous density $f, f > 0$ a.e.
(b) ϕ is non-decreasing and differentiable with its derivative, ϕ', being uniformly continuous on $[0, 1]$. Then, for every $0 < M < \infty$,

$$\sup_{\|\boldsymbol{w}\| \leq M} \left\| \boldsymbol{L}_n(\boldsymbol{\rho} + n^{-\frac{1}{2}}\boldsymbol{w}) - \hat{\boldsymbol{L}}_n + \boldsymbol{w}^\top \boldsymbol{\Sigma}\gamma \right\| = o_p(1) \tag{9.2.4}$$

where $\hat{\boldsymbol{L}}_n = (\hat{L}_1, \dots, \hat{L}_p)^\top$ with

$$\hat{L}_j = \frac{1}{\sqrt{n}} \sum_{t=j+1}^{n} (X_{t-j} - \bar{X}_j)\left[\phi(F(\varepsilon_i)) - \bar{\phi}\right], \quad 1 \leq j \leq p \tag{9.2.5}$$

and

$$\begin{aligned}
\bar{\phi} &= \int_0^1 \phi(\omega)\mathrm{d}\omega \\
\bar{X}_t &= \frac{1}{n} \sum_{t=s+1}^{n} X_{t-s} \\
\gamma &= \int_{-\infty}^{+\infty} f(t)\mathrm{d}\phi(F(t)) \\
\boldsymbol{\Sigma} &= (\sigma(i-j)),\ i = 1, \dots, p;\ j = 1, \dots, p
\end{aligned}$$

where

$$\sigma(k) = \mathrm{Cov}(X_0, X_k), \qquad 1 \leq k \leq p.$$

The above theorem covers Wilcoxon-type scores but not the normal score. Using the AUL result of Eq. (9.2.4), one may obtain

$$\sqrt{n}(\hat{\boldsymbol{\rho}}_n^{\mathrm{UR}} - \boldsymbol{\rho}) = \gamma^{-1}\boldsymbol{\Sigma}^{-1}\hat{\boldsymbol{L}}_n + o_p(1),\ \gamma = \int_{-\infty}^{+\infty} f\mathrm{d}\phi(F). \tag{9.2.6}$$

Observe that $\hat{\boldsymbol{L}}_n$ is a vector of square integrable mean zero martingale with

$$\mathbb{E}\left[\hat{\boldsymbol{L}}_n\hat{\boldsymbol{L}}_n^\top\right] = A_\phi^2\boldsymbol{\Sigma}$$

where $A_\phi^2 = \text{Var}(\phi(\boldsymbol{u}))$. Thus, by a routine Cramér–Wold device and Corollary 3.1 of Hall and Heyde (1981), one obtains as $n \to \infty$,

$$\hat{\boldsymbol{L}}_n \xrightarrow{\mathcal{D}} \mathcal{N}_p(\boldsymbol{0}, A_\phi^2 \boldsymbol{\Sigma}), \quad A_\phi^2 = \int_0^1 \phi^2(u)du \tag{9.2.7}$$

and hence,

$$\sqrt{n}(\hat{\boldsymbol{\rho}}_n^{\text{UR}} - \boldsymbol{\rho}) \xrightarrow{\mathcal{D}} \mathcal{N}_p(\boldsymbol{0}, \eta^2 \boldsymbol{\Sigma}^{-1}), \quad \eta^2 = \frac{A_\phi^2}{\gamma^2}. \tag{9.2.8}$$

9.3 LASSO, Ridge, Preliminary Test and Stein–Saleh-type R-estimators

In this section, we present various R-estimators of $\boldsymbol{\rho}$. First, we consider the usual penalty R-estimators beginning with the hard threshold R-estimator (HTE). We shall refer to this estimator as the preliminary test subset selector (PTSS). It is defined by

$$\hat{\boldsymbol{\rho}}_n^{\text{PTSS}}(\lambda) = \left(\hat{\rho}_{jn}^{\text{UR}} I(|\hat{\rho}_{jn}^{\text{UR}} > \lambda\eta\sqrt{\sigma_n^{jj}}) \middle| j = 1, \dots, p \right)^{\top}, \tag{9.3.1}$$

where σ_n^{ii} is the (ii)-element of $\boldsymbol{\Sigma}_n^{-1}$. A continuous version of the PTSS is the Saleh-type selector

$$\hat{\boldsymbol{\rho}}_n^{\text{SS}}(\lambda) = \left(\hat{\rho}_{jn}^{\text{UR}} \left(1 - \frac{\lambda\eta}{|\hat{\rho}_{jn}^{\text{UR}}|} \sqrt{\sigma_n^{jj}} \right) \middle| j = 1, \dots, p \right)^{\top}. \tag{9.3.2}$$

One of the most important penalty R-estimators is the LASSO due to Tibshirani (1996) (alternatively referred to as soft-threshold or positive-rule Saleh-type) R-estimator which is given by

$$\hat{\boldsymbol{\rho}}_n^{\text{LassoR}}(\lambda) = \left(\text{sgn}(\hat{\rho}_{jn}^{\text{UR}})(|\hat{\rho}_{jn}^{\text{UR}}| - \lambda\eta\sqrt{\sigma_n^{jj}})^{+} \middle| j = 1, \dots, p \right)^{\top}, \tag{9.3.3}$$

where $\hat{\boldsymbol{\rho}}_n^{\text{UR}} = (\hat{\rho}_{1n}^{\text{UR}}, \dots, \hat{\rho}_{pn}^{\text{UR}})^{\top}$ and σ_n^{jj} is the jth diagonal element of the estimator of the matrix, $\boldsymbol{\Sigma}_n^{-1}$ defined by

$$\boldsymbol{\Sigma}_n = (\sigma_{nij}) \quad \text{with} \quad \sigma_{nij} = \sum_{k=1}^{n-\max\{i,j\}} (X_{k-i}-\bar{X}_i)(X_{k-j}-\bar{X}_j), \quad 1 \le i, j \le n \tag{9.3.4}$$

which is the estimator of $\boldsymbol{\Sigma}$ defined in Eq. (9.2.2). We partition $\boldsymbol{\Sigma}$ as

$$\boldsymbol{\Sigma} = \begin{pmatrix} \boldsymbol{\Sigma}_{11} & \boldsymbol{\Sigma}_{12} \\ \boldsymbol{\Sigma}_{21} & \boldsymbol{\Sigma}_{22} \end{pmatrix} \quad \text{and} \quad \boldsymbol{\Sigma}_n = \begin{pmatrix} \boldsymbol{\Sigma}_{n11} & \boldsymbol{\Sigma}_{n12} \\ \boldsymbol{\Sigma}_{n21} & \boldsymbol{\Sigma}_{n22} \end{pmatrix} \tag{9.3.5}$$

where $\boldsymbol{\Sigma}_{11}, \boldsymbol{\Sigma}_{12}$, and $\boldsymbol{\Sigma}_{22}$ are $p_1 \times p_1, p_1 \times p_2, p_2 \times p_2$ matrices, respectively, with $\boldsymbol{\Sigma}_{12} = \boldsymbol{\Sigma}_{21}$ and similarly for $\boldsymbol{\Sigma}_{n11}, \boldsymbol{\Sigma}_{n12}, \boldsymbol{\Sigma}_{n22}$ with $\boldsymbol{\Sigma}_{n12} = \boldsymbol{\Sigma}_{n21}$, respectively. We let

$$\boldsymbol{\Sigma}_{nrr.s} = \boldsymbol{\Sigma}_{nrr} - \boldsymbol{\Sigma}_{nrs}\boldsymbol{\Sigma}_{nss}^{-1}\boldsymbol{\Sigma}_{nsr}, r \neq s = 1, 2.$$

Define $\boldsymbol{\Sigma}_{rr.s}$ analogously $r \neq s = 1, 2$ using the above partition of $\boldsymbol{\Sigma}$ (or $\boldsymbol{\Sigma}_n$). Note that $E\left[\varepsilon_i^2\right] < \infty$ and the ergodic theorem implies $n^{-1}\boldsymbol{\Sigma}_n \to \boldsymbol{\Sigma}$ and $n^{-1}\boldsymbol{\Sigma}_{nrr.s} \to \boldsymbol{\Sigma}_{rr.s}$ as $n \to \infty$. Since $\boldsymbol{\Sigma}^{-1}$ exists, $\boldsymbol{\Sigma}_n^{-1}$ and $\boldsymbol{\Sigma}_{nrr.s}^{-1}$ also exist for a sufficiently large n.

Using subset selectors, namely, PTSS or LASSO, we may partition $\boldsymbol{\rho}$ into two subsets $\boldsymbol{\rho}^\top = (\boldsymbol{\rho}_1^\top, \boldsymbol{\rho}_2^\top)$ when $\boldsymbol{\rho}_i$ is a $p_i, i = 1, 2$ vector. We are primarily interested in the R-estimation of $\boldsymbol{\rho}$ when we suspect that $\boldsymbol{\rho}_2$ may be sparse. For this case, we partition $\boldsymbol{L}_n(\boldsymbol{u}) = \left(\boldsymbol{L}_{n(1)}^\top(\boldsymbol{u}), \boldsymbol{L}_{n(2)}^\top(\boldsymbol{u})\right)$ and $\boldsymbol{Y}_{i-1}^\top = (\boldsymbol{Y}_{i,1}^\top, \boldsymbol{Y}_{i,2}^\top)$ where

$$\boldsymbol{L}_{n(1)} = \left(L_1(u), \dots, L_{p_1}(u)\right)^\top, \quad \boldsymbol{L}_{n(2)} = \left(L_{p_1+1}(u), \dots, L_p(u)\right)^\top$$
$$\boldsymbol{Y}_{i,1} = \left(X_{i-1}, \dots, X_{i-p_1}\right)^\top, \quad \boldsymbol{Y}_{i,2} = \left(X_{i-p_1+1}, \dots, X_{i-p}\right)^\top, \ 1 \leq i \leq n.$$

Further, consider the estimator $\boldsymbol{\Sigma}_n$ of $\boldsymbol{\Sigma}$, when

$$\boldsymbol{\Sigma}_n = \left(\sigma_{nij}\right) \text{ with } \sigma_{nij} = \sum_{k=1}^{n-\max(i,j)} (X_{k-i} - \bar{X}_i)(X_{k-j} - \bar{X}_j), \quad 1 \leq i, j \leq n$$

as in Eq. (9.3.4).

Next, partition the R-estimation of $\boldsymbol{\rho}$ as $\hat{\boldsymbol{\rho}}_n^{\mathrm{UR}} = \left(\hat{\boldsymbol{\rho}}_{1n}^{\mathrm{UR}^\top}, \hat{\boldsymbol{\rho}}_{2n}^{\mathrm{UR}^\top}\right)^\top$ and call $\hat{\boldsymbol{\rho}}_{1n}^{\mathrm{UR}}$ the unrestricted rank estimator (URE) of $\boldsymbol{\rho}_1$. The restricted R-estimator of $\boldsymbol{\rho}_1$ is given by

$$\inf_{\boldsymbol{u} \in \mathbb{R}^{p_1}} \left\| \boldsymbol{L}_{n(1)}(\boldsymbol{u}, \boldsymbol{0}) \right\| = \left\| \boldsymbol{L}_{n(1)}(\hat{\boldsymbol{\rho}}_{1n}^{\mathrm{RR}}, \boldsymbol{0}) \right\|.$$

For the test of $\mathcal{H}_o : \boldsymbol{\rho}_2 = \boldsymbol{0}$, we consider the test-statistic as

$$\mathcal{L}_n = n\sigma^{-2}\hat{\boldsymbol{L}}_{n(2)}^\top \boldsymbol{\Sigma}_{n22\cdot1}^{-1} \hat{\boldsymbol{L}}_{n(2)}. \tag{9.3.6}$$

It can be shown that under $\mathcal{H}_o$, $\mathcal{L}_n$ asymptotically has the central chi-square distribution with p_2 d.f. Then, the preliminary test R-estimator (PTRE) is defined by

$$\hat{\boldsymbol{\rho}}_{1n}^{\mathrm{PTR}}(\lambda) = \hat{\boldsymbol{\rho}}_{1n}^{\mathrm{RR}} + (\hat{\boldsymbol{\rho}}_{1n}^{\mathrm{UR}} - \hat{\boldsymbol{\rho}}_{1n}^{\mathrm{RR}}) I(\mathcal{L}_n > \lambda^2). \tag{9.3.7}$$

The Stein–Saleh-type R-estimator (SSRE) of $\boldsymbol{\rho}$ is then defined by

$$\hat{\boldsymbol{\rho}}_{1n}^{\mathrm{SSR}}(\lambda) = \hat{\boldsymbol{\rho}}_{1n}^{\mathrm{RR}} + (\hat{\boldsymbol{\rho}}_{1n}^{\mathrm{UR}} - \hat{\boldsymbol{\rho}}_{1n}^{\mathrm{RR}})(1 - c\mathcal{L}_n^{-1}), \quad c = p_2 - 2 \tag{9.3.8}$$

and the positive-rule Stein–Saleh-type R-estimator (PRSSRE) is defined by

$$\hat{\boldsymbol{\rho}}_{n}^{\mathrm{SSR+}} = \hat{\boldsymbol{\rho}}_{1n}^{\mathrm{RR}} + (\hat{\boldsymbol{\rho}}_{1n}^{\mathrm{UR}} - \hat{\boldsymbol{\rho}}_{1n}^{\mathrm{RR}})(1 - c\mathcal{L}_n^{-1})\, I(\mathcal{L}_n > c), \quad c = p_2 - 2. \tag{9.3.9}$$

For the ridge regression R-estimator (RRRE), we may define

$$\hat{\boldsymbol{\rho}}_{n}^{\mathrm{ridgeR}}(k) = \left(\hat{\boldsymbol{\rho}}_{1n}^{\mathrm{UR}^{\top}}, \hat{\boldsymbol{\rho}}_{2n}^{\mathrm{UR}^{\top}}(\boldsymbol{I}_{p_2} + k\boldsymbol{\Sigma}_{n22.1}^{-1})^{-1}\right)^{\top}, \quad k > 0. \tag{9.3.10}$$

For the elastic net (Enet) R-estimator of $\boldsymbol{\rho}$, we define

$$\hat{\boldsymbol{\rho}}_{n}^{\mathrm{EnetR}}(\lambda, \alpha) = \left[\boldsymbol{I}_p + \lambda(1-\alpha)\boldsymbol{\Sigma}^{-1}\right]^{-1} \hat{\boldsymbol{\rho}}_{n}^{\mathrm{LassoR}}(\lambda\alpha). \tag{9.3.11}$$

If $\alpha = 0$, $\hat{\boldsymbol{\rho}}_{n}^{\mathrm{Enet}}(\lambda, 0) = \left[\boldsymbol{I}_p + \lambda\boldsymbol{\Sigma}^{-1}\right]^{-1} \hat{\boldsymbol{\rho}}_{n}^{\mathrm{UR}}$ and if $\alpha = 1$, $\hat{\boldsymbol{\rho}}_{n}^{\mathrm{Enet}}(\lambda, 1) = \hat{\boldsymbol{\rho}}_{n}^{\mathrm{LassoR}}(\lambda)$.

9.4 Asymptotic Distributional L_2-risk

In order to obtain the expressions for the ADL$_2$-risk of the above R-estimators, we follow Saleh et al. (2019) and use the sequence of local alternatives, $K_{(n)}$: $\boldsymbol{\rho}_{2(n)} = n^{-\frac{1}{2}}\boldsymbol{\delta}_2$, $\boldsymbol{\delta} \in \mathbb{R}^{p_2}$. In addition, we need the following assumption: the error distributional function F has absolutely continuous density, $f \geq 0$, with its derivative f' satisfying $I(f) = \int_{-\infty}^{+\infty} \left(\frac{f'}{f}\right)^2 \mathrm{d}F < \infty$.

In what follows, $\mathcal{H}_m(x; \Delta^2)$ stands for the c.d.f. of a non-central chi distribution with m d.f. and non-centrality parameter, Δ^2, and

$$\mathbb{E}\left[\chi_m^{-2}(\Delta^2)\right] = \int_{-\infty}^{+\infty} x^{-r} \mathrm{d}H_s(x, \Delta^2), \qquad s > 1.$$

Finally, for any R-estimator $\hat{\boldsymbol{\rho}}_{1n}^{*}$ of $\boldsymbol{\rho}_1$ for which $\sqrt{n}(\hat{\boldsymbol{\rho}}_{1n}^{*} - \boldsymbol{\rho}_1)$ converges in distribution under $\{K_{(n)}\}$ to a p_1-variate normal with mean $\mathbf{0}$ and positive-definite matrix, $\boldsymbol{V}^*$, define the ADL$_2$-risk with respect to the loss $n(\hat{\boldsymbol{\rho}}_{1n}^{*} - \boldsymbol{\rho}_1)^{\top}\boldsymbol{W}_1(\hat{\boldsymbol{\rho}}_{1n}^{*} - \boldsymbol{\rho}_1)$ to be

$$\mathrm{ADL_2\text{-}risk}(\boldsymbol{\rho}_{1n}^{*}) = \mathrm{tr}(\boldsymbol{W}_1\boldsymbol{V}^*) \tag{9.4.1}$$

where $\boldsymbol{W}_1$ is the positive-definite matrix and $\mathrm{tr}(\cdot)$ is the trace operator.

Akritas and Johnson (1982) have shown that under $a(i)$, $a(ii)$, and (b) of Theorem 9.1, $\{K_{(n)}\}$ is contiguous to $\mathcal{H}_o$. Using this argument, similar to the one appearing in the proof of theorem 3.1 of Sen and Saleh (1987), one obtains the following theorem.

Theorem 9.2 Under $K_{(n)}$ and the regularized conditions stated in Theorem 9.1, we have

(i) $\lim_{n\to\infty} P\left(\mathcal{L}_n \leq \lambda^2 \middle| K_{(n)}\right) = \mathcal{H}_{p_2}(\lambda^2, \Delta^2), \quad \Delta^2 = \sigma_\phi^2\left(\boldsymbol{\delta}_2^\top \boldsymbol{\Sigma}_{22\cdot1}\boldsymbol{\delta}_2\right)$

(ii) $\lim_{n\to\infty} P\left(\sqrt{n}\left(\hat{\boldsymbol{\rho}}_{1n}^{\mathrm{UR}} - \boldsymbol{\rho}_1\right) \leq \boldsymbol{x} \middle| K_{(n)}\right) = \Phi_{p_1}(\boldsymbol{x}; \mathbf{0}, \eta^2\boldsymbol{\Sigma}_{11\cdot2}^{-1})$

(iii) $\lim_{n\to\infty} P\left(\sqrt{n}\left(\hat{\boldsymbol{\rho}}_{1n}^{\mathrm{RR}} - \boldsymbol{\rho}_1\right) \leq \boldsymbol{x} \middle| K_{(n)}\right) = \Phi_{p_1}(\boldsymbol{x} + \boldsymbol{\Sigma}_{11}^{-1}\boldsymbol{\Sigma}_{12}\boldsymbol{\delta}_2; \mathbf{0}, \eta^2\boldsymbol{\Sigma}_{11}^{-1})$

(iv) $\lim_{n\to\infty} P\left(\sqrt{n}(\hat{\boldsymbol{\rho}}_{1n}^{\mathrm{PTR}}(\lambda) - \boldsymbol{\rho}_1) \leq \boldsymbol{x} \middle| K_{(n)}\right) =$

$$\mathcal{H}_{p_2}(\lambda^2, \Delta^2) G_{p_1}(\boldsymbol{x} + \boldsymbol{\Sigma}_{11}^{-1}\boldsymbol{\Sigma}_{12}\boldsymbol{\delta}_2, \mathbf{0}, \eta^2\boldsymbol{\Sigma}_{11}^{-1})$$
$$+ \int_{E(\boldsymbol{\delta}_2)} \Phi_{p_1}(\boldsymbol{x} - \boldsymbol{D}_{12}\boldsymbol{D}_{22}^{-1}\boldsymbol{z}; \mathbf{0}, \eta^2\boldsymbol{D}_{11\cdot2}) \mathrm{d}\Phi_{p_2}(\boldsymbol{z}, \mathbf{0}, \eta^2\boldsymbol{D}_{22})$$

where $\Phi_\nu(\cdot)$ is the multivariate normal c.d.f. $\boldsymbol{D} = \boldsymbol{\Sigma}^{-1}$ and the D_{ij} and $D_{ii\cdot j}$ are defined in the usual way. Further,

$$E(\boldsymbol{\delta}_2) = \{\boldsymbol{Z} : \eta^2(\boldsymbol{Z} + \boldsymbol{\delta}_2)^\top \boldsymbol{\Sigma}_{22\cdot1}(\boldsymbol{Z} + \boldsymbol{\delta}_2)\}.$$

Finally,

$$\text{(v)} \quad \sqrt{n}\left(\hat{\boldsymbol{\rho}}_{1n}^{\mathrm{SSR}} - \boldsymbol{\rho}_1\right) \xrightarrow{\mathcal{D}} \boldsymbol{D}_1\boldsymbol{U} + \frac{c\eta^2\boldsymbol{\Sigma}_{11}^{-1}\boldsymbol{\Sigma}_{12}(\boldsymbol{D}_2\boldsymbol{U} + \boldsymbol{\delta}_2)}{(\boldsymbol{D}_2\boldsymbol{U} + \boldsymbol{\delta}_2)^\top \boldsymbol{\Sigma}_{22\cdot1}(\boldsymbol{D}_2\boldsymbol{U} + \boldsymbol{\delta}_2)}$$

where $\boldsymbol{U} \sim \mathcal{N}_p(\mathbf{0}, \eta^2\boldsymbol{\Sigma})$ and $\boldsymbol{D} = \begin{pmatrix} \boldsymbol{D}_1 \\ \boldsymbol{D}_2 \end{pmatrix}$.

Using linearity results imply that

$$\hat{\boldsymbol{\rho}}_{1n}^{\mathrm{RR}} = \hat{\boldsymbol{\rho}}_{1n}^{\mathrm{UR}} + \boldsymbol{\Sigma}_{11}^{-1}\boldsymbol{\Sigma}_{12}\hat{\boldsymbol{\rho}}_{2n}^{\mathrm{UR}} + o_p(n^{-\frac{1}{2}}) \tag{9.4.2}$$

$$\mathcal{L}_n = \frac{1}{\eta^2}\left(\hat{\boldsymbol{\rho}}_{2n}^{\mathrm{UR}}\right)^\top \boldsymbol{\Sigma}_{22\cdot1}\hat{\boldsymbol{\rho}}_{2n}^{\mathrm{UR}} + o_p(1). \tag{9.4.3}$$

Therefore,

$$\sqrt{n}\left((\hat{\boldsymbol{\rho}}_{1n}^{\mathrm{UR}} - \boldsymbol{\rho}_1)^\top, (\hat{\boldsymbol{\rho}}_{2n}^{\mathrm{UR}} - n^{-\frac{1}{2}}\boldsymbol{\delta}_2)^\top\right)^\top \xrightarrow{\mathcal{D}} \mathcal{N}_p\left(\mathbf{0}, \eta^2\begin{pmatrix} \boldsymbol{\Sigma}_{11\cdot2}^{-1} & \mathbf{0} \\ \mathbf{0} & \boldsymbol{\Sigma}_{22\cdot1}^{-1} \end{pmatrix}\right). \tag{9.4.4}$$

Note that, by definition, $\boldsymbol{D}_{11} + \boldsymbol{\Sigma}_{11}^{-1}\boldsymbol{\Sigma}_{12}\boldsymbol{D}_{12} = \boldsymbol{\Sigma}_{11}^{-1}$ and $\boldsymbol{D}_{12} + \boldsymbol{\Sigma}_{11}^{-1}\boldsymbol{\Sigma}_{12}\boldsymbol{D}_{22} = \boldsymbol{0}$. Hence, using Eqs. (9.4.2) and (9.4.3), we may write

$$\hat{\boldsymbol{\rho}}_{1n}^{\mathrm{RR}} = \boldsymbol{T}_1\hat{\boldsymbol{\rho}}_{1n}^{\mathrm{UR}} + o_p(n^{\frac{1}{2}}) \quad \text{and} \quad \mathcal{L}_n = n\left(\hat{\boldsymbol{\rho}}_n^{\mathrm{UR}}\right)^{\top} \boldsymbol{T}_2\left(\hat{\boldsymbol{\rho}}_n^{\mathrm{UR}}\right) + o_p(1)$$

where $\boldsymbol{T}_1\boldsymbol{\Sigma}^{-1}\boldsymbol{T}_2 = \boldsymbol{0}$. Thus, using Eq. (9.4.4) we may conclude under $K_{(n)}$, $\sqrt{n}(\hat{\boldsymbol{\rho}}_{1n}^{\mathrm{RR}} - \boldsymbol{\rho}_1)$ and $\mathcal{L}_n$ are asymptotically independent. Further,

$$\sqrt{n}\begin{pmatrix} (\hat{\boldsymbol{\rho}}_{1n}^{\mathrm{UR}} - \boldsymbol{\rho}_1) \\ (\hat{\boldsymbol{\rho}}_{2n}^{\mathrm{UR}} - n^{-\frac{1}{2}}\boldsymbol{\delta}_2) \end{pmatrix} = \begin{pmatrix} \boldsymbol{\Sigma}_{11\cdot 2}^{-1} & \boldsymbol{0} \\ \boldsymbol{0} & \boldsymbol{\Sigma}_{22\cdot 1}^{-1} \end{pmatrix} \gamma^{-1} n^{-\frac{1}{2}} L_n(\boldsymbol{\rho}) + o_p(1) \tag{9.4.5}$$

where $n^{-\frac{1}{2}}\gamma^{-1}\boldsymbol{L}_n(\boldsymbol{\rho}) \xrightarrow{\mathcal{D}} \boldsymbol{U}$ and (iv) follows.

The next theorem presents the ADL$_2$-risk expressions of the R-estimators.

Theorem 9.3 Under $K_{(n)}$ and assumed regularity conditions of Theorem 9.2, we obtain the ADL$_2$-risk expressions:

$$\text{(i)}\quad \text{ADL}_2\text{-risk}(\hat{\boldsymbol{\rho}}_n^{\mathrm{UR}}) = \eta^2\left(\mathrm{tr}(\boldsymbol{\Sigma}_{11\cdot 2}^{-1}) + \mathrm{tr}(\boldsymbol{\Sigma}_{22\cdot 1}^{-1})\right), \tag{9.4.6}$$

$$\text{(ii)}\quad \text{ADL}_2^*\text{-risk}(\hat{\boldsymbol{\rho}}_n^{\mathrm{LassoR}}(\lambda)) = \eta^2\left(\mathrm{tr}(\boldsymbol{\Sigma}_{11\cdot 2}^{-1}) + \Delta^2\mathrm{Ch}_{\min}(\boldsymbol{\Sigma}_{22\cdot 1}^{-1})\right), \tag{9.4.7}$$

where ADL$_2^*$-risk stands for the lower bound of the risk function.

$$\text{(iii)}\quad \text{ADL}_2^*\text{-risk}(\hat{\boldsymbol{\rho}}_1^{\mathrm{PTSS}}(\lambda)) = \eta^2\left(\mathrm{tr}(\boldsymbol{\Sigma}_{11\cdot 2}^{-1}) + \Delta^2\mathrm{Ch}_{\min}(\boldsymbol{\Sigma}_{22\cdot 1}^{-1})\right), \tag{9.4.8}$$

$$\text{(iv)}\quad \text{ADL}_2^*\text{-risk}(\hat{\boldsymbol{\rho}}_{1n}^{\mathrm{RR}}, \boldsymbol{0}) = \eta^2\left(\mathrm{tr}(\boldsymbol{\Sigma}_{11}^{-1}) + \frac{1}{\eta^2}\boldsymbol{\delta}_2^{\top}\boldsymbol{M}\boldsymbol{\delta}_2\right),$$

$$\boldsymbol{M} = \boldsymbol{\Sigma}_{21}\boldsymbol{\Sigma}_{11}^{-2}\boldsymbol{\Sigma}_{12}, \tag{9.4.9}$$

$$\begin{aligned}\text{(v)}\quad \text{ADL}_2^*\text{-risk}(\hat{\boldsymbol{\rho}}_n^{\mathrm{PTR}}(\lambda)) = \eta^2\Big(& \mathrm{tr}(\boldsymbol{\Sigma}_{11\cdot 2}^{-1})\left(1 - \mathcal{H}_{p_2+2}(\lambda^2, \Delta^2)\right) \\ &+ \mathrm{tr}(\boldsymbol{\Sigma}_{11}^{-1})\mathcal{H}_{p_2+2}(\lambda^2, \Delta^2) + \boldsymbol{\delta}_2^{\top}\boldsymbol{M}\boldsymbol{\delta}_2 \\ &\times\left\{2\mathcal{H}_{p_2+2}(\lambda^2, \Delta^2) - \mathcal{H}_{p_2+4}(\lambda^2, \Delta^2)\right\}\Big),\end{aligned} \tag{9.4.10}$$

$$\text{(vi)}\quad \text{ADL}_2^*\text{-risk}(\hat{\boldsymbol{\rho}}_n^{\mathrm{SSR}}(\lambda)) = \eta^2\Big(\mathrm{tr}(\boldsymbol{\Sigma}_{11\cdot 2}^{-1}) - (p_2 - 2)\,\mathrm{tr}(\boldsymbol{M}\boldsymbol{\Sigma}_{22\cdot 1}^{-1})$$

$$\times\left(2\mathbb{E}\left[\chi_{p_2+2}^{-2}(\Delta^2)\right]-(p_2-2)\mathbb{E}\left[\chi_{p_2+2}^{-4}(\Delta^2)\right]\right)$$

$$+(p_2^2-4)(\boldsymbol{\delta}_2^\top \boldsymbol{M}\boldsymbol{\delta}_2)\mathbb{E}\left[\chi_{p_2+4}^{-4}(\Delta^2)\right]\Big), \tag{9.4.11}$$

$$\text{(vii)}\quad \text{ADL}_2^*\text{-risk}(\hat{\boldsymbol{\rho}}_n^{\text{SSR+}}(\lambda)) = \text{ADL}_2^*\text{-risk}(\hat{\boldsymbol{\rho}}_n^{\text{SSR}}(\lambda)) - \eta^2\Big[\operatorname{tr}(\boldsymbol{\Sigma}_{22\cdot1}^{-1})$$

$$\times\left\{\mathbb{E}\left[\left(1-(p_2-2)\chi_{p_{2n}+2}^{-2}(\Delta^2)\right)^2 I\left(\chi_{p_{2n}+2}^2(\Delta^2)<(p_2-2)\right)\right]\right\}$$

$$+\boldsymbol{\delta}_2{}^\top\boldsymbol{M}\boldsymbol{\delta}_2\Big\{2\mathbb{E}\left[\left(1-(p_2-2)\chi_{p_{2n}+2}^{-2}(\Delta^2)\right)I\left(\chi_{p_{2n}+2}^2(\Delta^2)<p_2-2\right)\right]$$

$$-\mathbb{E}\left[\left(1-(p_2-2)\chi_{p_{2n}+4}^{-2}(\Delta^2)\right)^2 I\left(\chi_{p_{2n}+4}^2(\Delta^2)<(p_2-2)\right)\right]\Big\}\Big],$$

$$\text{(viii)}\quad \text{ADL}_2^*\text{-risk}(\hat{\boldsymbol{\rho}}_n^{\text{ridgeR}}(k)) = \eta^2\Big[\operatorname{tr}(\boldsymbol{\Sigma}_{11\cdot2}^{-1})$$

$$+\operatorname{tr}\left(\left(\boldsymbol{I}_{p_2}+k\boldsymbol{\Sigma}_{22\cdot1}^{-1}\right)^{-2}\boldsymbol{\Sigma}_{22\cdot1}^{-1}\right)+k^2\boldsymbol{\delta}_2^\top(\boldsymbol{\Sigma}_{22\cdot1}+k\boldsymbol{I}_{p_2})^{-2}\boldsymbol{\delta}_2\Big],$$

$$\text{(ix)}\quad \text{ADL}_2^*\text{-risk}(\hat{\boldsymbol{\rho}}_n^{\text{EnetR}}(\lambda,\alpha)) = \eta^2\Big[\operatorname{tr}\left((\boldsymbol{I}_{p_2}+\lambda(1-\alpha)\boldsymbol{\Sigma}_{11.2}^{-1})^{-2}\boldsymbol{\Sigma}_{11\cdot2}^{-1}\right)$$

$$+\lambda^2(1-\alpha)^2\boldsymbol{\Delta}^\top(\boldsymbol{C}+\lambda(1-\alpha)\boldsymbol{I}_{p_2})^{-2}\boldsymbol{\Delta}$$

$$+2\lambda(1-\alpha)\boldsymbol{B}^\top\left(\boldsymbol{C}+\lambda(1-\alpha)\boldsymbol{I}\right)^{-2}\boldsymbol{\Delta}\Big], \tag{9.4.12}$$

where $\eta^2=\left(\int_{-\infty}^{+\infty}f(t)\mathrm{d}\phi(F(t))\right)^{-2}A_\phi^2$, $\boldsymbol{\Delta}=(\Delta_1,\dots,\Delta_p)^\top$, and

$$\boldsymbol{B}=\left(\Delta_j\mathcal{H}_3(\lambda^2,\Delta_j^2)-\lambda\alpha\left(\Phi(\lambda\alpha-\Delta_j)-\Phi(\lambda\alpha+\Delta_j)\right)\Big|\, j=1,\dots,p\right)^\top.$$

We proceed to discuss the asymptotic distributional risk efficiency (ADRE) of the R-estimators in the next section.

9.5 Asymptotic Distributional L_2-risk Analysis

In this section, we compare the R-estimators based on the ADL_2-risk criterion.

9.5.1 Comparison of Unrestricted vs. Restricted R-estimators

First, we compare $\hat{\boldsymbol{\rho}}_n^{\mathrm{RR}}$ with $\hat{\boldsymbol{\rho}}_n^{\mathrm{UR}}$ using their ADL$_2$-risk difference as follows,

$$\mathrm{ADL}_2\text{-risk}(\hat{\boldsymbol{\rho}}_{1n}^{\mathrm{UR}})-\mathrm{ADL}_2\text{-risk}(\hat{\boldsymbol{\rho}}_{1n}^{\mathrm{RR}}) = \eta^2\Big(\mathrm{tr}(\boldsymbol{\Sigma}_{11\cdot2}^{-1}-\mathrm{tr}(\boldsymbol{\Sigma}_{11}^{-1})-\Delta^2\mathrm{Ch}_{\min}(\boldsymbol{M}\boldsymbol{\Sigma}_{22\cdot1}^{-1})\Big).$$

Thus, the difference is non-negative whenever

$$0 < \Delta^2 \leq \frac{\mathrm{tr}(\boldsymbol{M}\boldsymbol{\Sigma}_{22\cdot1}^{-1})}{\mathrm{Ch}_{\max}(\boldsymbol{M}\boldsymbol{\Sigma}_{22\cdot1}^{-1})} \tag{9.5.1}$$

which means $\hat{\boldsymbol{\rho}}_{1n}^{\mathrm{RR}}$ outperforms $\hat{\boldsymbol{\rho}}_{1n}^{\mathrm{UR}}$ for Δ^2 in the above interval; otherwise, $\hat{\boldsymbol{\rho}}_{1n}^{\mathrm{UR}}$ outperforms $\hat{\boldsymbol{\rho}}_{1n}^{\mathrm{RR}}$. That is, neither $\hat{\boldsymbol{\rho}}_{1n}^{\mathrm{UR}}$ nor $\hat{\boldsymbol{\rho}}_{1n}^{\mathrm{RR}}$ outperforms the other uniformly.

9.5.2 Comparison of Unrestricted vs. Preliminary Test R-estimator

Here

$$\mathrm{ADL}_2\text{-risk}(\hat{\boldsymbol{\rho}}_n^{\mathrm{UR}}) - \mathrm{ADL}_2\text{-risk}(\hat{\boldsymbol{\rho}}_{1n}^{\mathrm{PTR}}(\lambda)) = \eta^2\Big(\mathrm{tr}(\boldsymbol{\Sigma}_{11\cdot2}^{-1} - \boldsymbol{\Sigma}_{11}^{-1})\mathcal{H}_{p_2+2}(\lambda^2,\Delta^2)$$
$$- \Delta^2\mathrm{Ch}_{\max}(\boldsymbol{M}\boldsymbol{\Sigma}_{22\cdot1}^{-1})\left(2\mathcal{H}_{p_2+2}(\lambda^2,\Delta^2) - \mathcal{H}_{p_2+4}(\lambda^2,\Delta^2)\right)\Big).$$

The expression is non-negative whenever

$$0 \leq \Delta^2 \leq \frac{\mathrm{tr}(\boldsymbol{M}\boldsymbol{\Sigma}_{22\cdot1}^{-1})\mathcal{H}_{p_2+2}(\lambda^2,\Delta^2)}{\mathrm{Ch}_{\max}(\boldsymbol{M}\boldsymbol{\Sigma}_{22\cdot1}^{-1})\left(2\mathcal{H}_{p_2+2}(\lambda^2,\Delta^2) - \mathcal{H}_{p_2+4}(\lambda^2,\Delta^2)\right)}. \tag{9.5.2}$$

Hence, $\hat{\boldsymbol{\rho}}_{1n}^{\mathrm{PTR}}$ outperforms $\hat{\boldsymbol{\rho}}_{1n}^{\mathrm{UR}}$ in this interval; otherwise, $\hat{\boldsymbol{\rho}}_{1n}^{\mathrm{UR}}$ outperforms $\hat{\boldsymbol{\rho}}_{1n}^{\mathrm{PTR}}$. It follows that PTRE fails to dominate the URE, $\hat{\boldsymbol{\rho}}_{1n}^{\mathrm{UR}}$.

9.5.3 Comparison of Unrestricted vs. Stein–Saleh-type R-estimators

We provide the comparison in a form of a theorem.

Theorem 9.4 Let

$$h = \frac{\mathrm{Ch}_{\max}(\boldsymbol{M})}{\mathrm{tr}(\boldsymbol{M})}, \quad 0 < h \leq 1.$$

Then, a sufficient condition for the asymptotic dominance of Stein–Saleh R-estimator, $\hat{\boldsymbol{\rho}}_{1n}^{\mathrm{SSR}}$, over URE, $\hat{\boldsymbol{\rho}}_{1n}^{\mathrm{UR}}$. That is ADL$_2$-risk$(\hat{\boldsymbol{\rho}}_{1n}^{\mathrm{UR}}) \geq$ ADL$_2$-risk$(\hat{\boldsymbol{\rho}}_{1n}^{\mathrm{SSR}})$ for all $\Delta^2 \in \mathbb{R}$ is that the shrinkage factor, c, is positive and it satisfies the inequality

$$2\mathbb{E}\left[\chi_{p_2+2}^{-2}(\Delta^2)\right]-c\mathbb{E}\left[\chi_{p_2+2}^{-4}(\Delta^2)\right]-(c+4)h\Delta^2\mathbb{E}\left[\chi_{p_2+4}^{-4}(\Delta^2)\right] \geq 0 \quad \forall\Delta^2 \in \mathbb{R}^+ \tag{9.5.3}$$

which requires that

$$p_2 \geq 3, \quad 0 < c < 2(p_2 - 2) \quad \text{and} \quad h(c+1) \leq 2. \tag{9.5.4}$$

Proof: We have

$$\begin{aligned}
&\text{ADL}_2\text{-risk}(\hat{\boldsymbol{\rho}}_{1n}^{\text{UR}}) - \text{ADL}_2\text{-risk}(\hat{\boldsymbol{\rho}}_{1n}^{\text{SSR}}) \\
&= c\,\text{tr}(\boldsymbol{M})\left(2\mathbb{E}\left[\chi^{-2}_{p_2+2}(\Delta^2)\right] - c\mathbb{E}\left[\chi^{-4}_{p_2+4}(\Delta^2)\right]\right) \\
&\quad -c(c+4)\Delta^2 \text{Ch}_{\max}(\boldsymbol{M}\boldsymbol{\Sigma}^{-1}_{22\cdot 1})\mathbb{E}\left[\chi^{-4}_{p_2+4}(\Delta^2)\right] \\
&= c\,\text{tr}(\boldsymbol{M})\left(2\mathbb{E}\left[\chi^{-2}_{p_2+2}(\Delta^2)\right] - c\mathbb{E}\left[\chi^{-4}_{p_2+4}(\Delta^2)\right]\right) \\
&\quad -c(c+4)h\,\text{tr}(\boldsymbol{M})\Delta^2\mathbb{E}\left[\chi^{-4}_{p_2+4}(\Delta^2)\right] \\
&= c\,\text{tr}(\boldsymbol{M})\left(\mathbb{E}\left[\chi^{-2}_{p_2+2}(\Delta^2)\right] + \Delta^2\mathbb{E}\left[\chi^{-4}_{p_2+4}(\Delta^2)\right]\right) \\
&\quad -c(c+4)h\,\text{tr}(\boldsymbol{M})\Delta^2\mathbb{E}\left[\chi^{-4}_{p_2+4}(\Delta^2)\right] \\
&= c\,\text{tr}(\boldsymbol{M})\mathbb{E}\left[\chi^{-2}_{p_2+2}(\Delta^2)\right] + c\,\text{tr}(\boldsymbol{M})\Delta^2\mathbb{E}\left[\chi^{-4}_{p_2+4}(\Delta^2)\right](1 + h(c+4)).
\end{aligned} \tag{9.5.5}$$

Using Eq. (9.5.3), we can rewrite Eq. (9.5.5) as

$$[2(p_2-2) - c]\,\mathbb{E}\left[\chi^{-4}_{p_2+2}(\Delta^2)\right] + [2 - h(c+4)]\,\Delta^2\mathbb{E}\left[\chi^{-4}_{p_2+4}(\Delta^2)\right] \geq 0. \tag{9.5.6}$$

From the above, at $\Delta^2 = 0$, we conclude that c has to be $\leq (p_2 - 2)$ (which requires that $p_2 \geq 3$). Furthermore, using a rough inequality that

$$\mathbb{E}\left[\chi^{-4}_{p_2+4}(\Delta^2)\right] \geq \frac{p_2 - 2}{p_2 + 2}\mathbb{E}\left[\chi^{-4}_{p_2+2}(\Delta^2)\right]$$

and allowing Δ^2 to be appropriately large, we conclude that for the above to be positive for all Δ^2, we must have $h(c+4) \leq 2$. This completes the proof of Eq. (9.5.4).

Consequently, by examining Eq. (9.5.4), we obtain

$$\begin{aligned}
\text{tr}(\boldsymbol{M}) &= \text{tr}(\boldsymbol{I}_{p_1} - \boldsymbol{\Sigma}^{-1}_{11}\boldsymbol{\Sigma}_{11\cdot 2}) \; (= \text{tr}(\boldsymbol{I}_{p_2} - \boldsymbol{\Sigma}^{-1}_{22}\boldsymbol{\Sigma}_{22\cdot 1}) \\
&= \min(p_1, p_2).
\end{aligned} \tag{9.5.7}$$

Also, for $h \geq \frac{1}{2}$, for any positive c, $h(c+4)$ cannot be $\leq \frac{1}{2}$. Thus, we redefine

$$p^* = \min(p_1, p_2) \geq 3, \quad h < \frac{1}{2}, \quad 0 < c \leq \min(2h^{-1} - 4, 2(p_2 - 2)). \tag{9.5.8}$$

Therefore, we need $h < \frac{1}{2}$ and $p^* \geq 3$. □

9.5.4 Comparison of the Preliminary Test vs. Stein–Saleh-type R-estimators

We capture the result in the following theorem.

Theorem 9.5 Under the regularity conditions of Theorem 9.4, the PTRE, $\hat{\boldsymbol{\rho}}_{1n}^{\text{PTR}}(\lambda)$, fails to dominate $\hat{\boldsymbol{\rho}}_{1n}^{\text{SS}}$. Also, if for λ^2, the level of significance of the preliminary test, we have

$$\mathcal{H}_{p_2+2}(\lambda^2, 0) \geq \frac{q(2(p_2-2)-q)}{p_2(p_2-2)}, \quad q = \max\{(p_2-2), (2h^{-1}-4)\}, \tag{9.5.9}$$

then $\hat{\boldsymbol{\rho}}_{1n}^{\text{SS}}$ fails to dominate $\hat{\boldsymbol{\rho}}_{1n}^{\text{PTR}}(\lambda)$.

Proof: Consider the ADL$_2$-risk difference of $\hat{\boldsymbol{\rho}}_{1n}^{\text{SS}}$ and $\hat{\boldsymbol{\rho}}_{1n}^{\text{PTR}}(\lambda)$ given by

$$\begin{aligned}
&\text{tr}(\boldsymbol{M}\boldsymbol{\Sigma}_{22\cdot1}^{-1})\left\{\mathcal{H}_{p_2+2}(\lambda^2, \Delta^2) - c\left[2\mathbb{E}\left[\chi_{p_2+2}^{-2}(\Delta^2)\right] - c\mathbb{E}\left[\chi_{p_2+2}^{-4}(\Delta^2)\right]\right]\right\} \\
&- \Delta^2 \text{Ch}_{\max}(\boldsymbol{M}\boldsymbol{\Sigma}_{22\cdot1}^{-1})\left\{2\mathcal{H}_{p_2+2}(\lambda^2, \Delta^2) - \mathcal{H}_{p_2+4}(\lambda^2, \Delta^2)\right. \\
&\left.- c(c+4)\mathbb{E}\left[\chi_{p_2+4}^{-4}(\Delta^2)\right]\right\}.
\end{aligned} \tag{9.5.10}$$

By Eqs. (9.5.3) and (9.5.4), the second term is always non-positive whereas the first term is negative whenever

$$\Delta^2 \text{Ch}_{\max}(\boldsymbol{M}\boldsymbol{\Sigma}_{22\cdot1}^{-1}) \geq \text{tr}(\boldsymbol{M}\boldsymbol{\Sigma}_{22\cdot1}^{-1})\left\{2 - \frac{\mathcal{H}_{p_2+4}(\lambda^2, \Delta^2)}{\mathcal{H}_{p_2+2}(\lambda^2, \Delta^2)}\right\}^{-1}. \tag{9.5.11}$$

Hence, the ADL$_2$-risk difference is negative outside a closed neighborhood of the pivot, so that PTRE fails to dominate $\hat{\boldsymbol{\rho}}_{1n}^{\text{SS}}$ when Eq. (9.5.6) or Eq. (9.5.8) holds. Next, we note under $\Delta^2 = 0$, the expression of Eq. (9.5.9) reduces to

$$\mathrm{tr}(\boldsymbol{M}\boldsymbol{\Sigma}_{22\cdot1}^{-1})\left\{\mathcal{H}_{p_2+2}(\lambda^2,0) - p_2^{-1}c(2-(p_2-2)^{-1}c\right\}. \tag{9.5.12}$$

Now, $cp_2^{-1}(2-c/(p_2-2))$ attains a maximum (over $c \in (0, 2(p_2-2))$ at $c = p_2-2$ while by Eq. (9.5.6), $c \leq 2h^{-1} - 4$. Hence, defining q as in Eq. (9.5.9), we note that Eq. (9.5.12) is bounded from below by

$$\mathrm{tr}(\boldsymbol{M}\boldsymbol{\Sigma}_{22\cdot1}^{-1})\left\{\mathcal{H}_{p_2+2}(\lambda^2,0) - q\frac{2(p_2-2)-q}{p_2(p_2-2)}\right\}. \tag{9.5.13}$$

Thus, ensuring positiveness of Eq. (9.5.12) implies that $\hat{\boldsymbol{\rho}}_{1n}^{\mathrm{SS}}$ fails to dominate $\hat{\boldsymbol{\rho}}_{1n}^{\mathrm{PTR}}(\lambda)$. □

9.6 Summary

In this chapter, we proposed shrinkage and penalty R-estimators for the vector parameter of the auto-regression model of order p. Given the detailed discussions and their usages for the R-estimators in previous chapters, we constructed this chapter concisely, where we only gave the form of the R-estimators along with their asymptotic distributional L_2-risk, through some theorems. The results of this chapter will be useful for time-series regression in the context of robust rank regression.

To end this chapter, in what follows, we provide the ADRE expressions. We first note that the ADL$_2$-risk($\hat{\boldsymbol{\rho}}_n^{\mathrm{UR}}$) is given by Eq. (9.4.6),

$$\mathrm{ADL_2\text{-}risk}(\hat{\boldsymbol{\rho}}_n^{\mathrm{UR}}) = \eta^2\left(\mathrm{tr}(\boldsymbol{\Sigma}_{11\cdot2}^{-1}) + \mathrm{tr}(\boldsymbol{\Sigma}_{22\cdot1}^{-1})\right). \tag{9.6.1}$$

Then, we have

$$\mathrm{ADRE}(\hat{\boldsymbol{\rho}}_{1n}^{\mathrm{PTSSR}}(\lambda) : \hat{\boldsymbol{\rho}}_{1n}^{\mathrm{UR}}) = \left(1 + \frac{\Delta^2\mathrm{Ch}_{\max}(\boldsymbol{M}\boldsymbol{\Sigma}_{22\cdot1}^{-1})}{\mathrm{tr}(\boldsymbol{\Sigma}_{11\cdot2}^{-1})}\right)^{-1} \tag{9.6.2}$$

$$\mathrm{ADRE}(\hat{\boldsymbol{\rho}}_{1n}^{\mathrm{LassoR}}(\lambda) : \hat{\boldsymbol{\rho}}_{1n}^{\mathrm{UR}}) = \left(1 + \frac{\Delta^2\mathrm{Ch}_{\max}(\boldsymbol{M}\boldsymbol{\Sigma}_{22\cdot1}^{-1})}{\mathrm{tr}(\boldsymbol{\Sigma}_{11\cdot2}^{-1})}\right)^{-1} \tag{9.6.3}$$

$$\mathrm{ADRE}(\hat{\boldsymbol{\rho}}_{1n}^{\mathrm{RR}} : \hat{\boldsymbol{\rho}}_{1n}^{\mathrm{UR}}) = \left(\frac{\mathrm{tr}(\boldsymbol{\Sigma}_{11\cdot2}^{-1})}{\mathrm{tr}(\boldsymbol{\Sigma}_{11}^{-1})}\right)\left(1 + \frac{\Delta^2\mathrm{Ch}_{\max}(\boldsymbol{M}\boldsymbol{\Sigma}_{22\cdot1}^{-1})}{\mathrm{tr}(\boldsymbol{\Sigma}_{11}^{1})}\right)^{-1} \tag{9.6.4}$$

$$\mathrm{ADRE}(\hat{\boldsymbol{\rho}}_{1n}^{\mathrm{PTR}}(\lambda) : \hat{\boldsymbol{\rho}}_{n}^{\mathrm{UR}}) = \left[1 + \frac{\mathrm{tr}(\boldsymbol{M}\boldsymbol{\Sigma}_{22\cdot1}^{-1})}{\mathrm{tr}(\boldsymbol{\Sigma}_{11\cdot2}^{-1})}\left(1 - \mathcal{H}_{p_2+2}(\lambda^2,\Delta^2)\right)\right.$$
$$\left. + \frac{\mathrm{tr}(\boldsymbol{\Sigma}_{11}^{-1})}{\mathrm{tr}(\boldsymbol{\Sigma}_{11\cdot2}^{-1})}\mathcal{H}_{p_2+2}(\lambda^2,\Delta^2)\right.$$

$$+\frac{\Delta^2 \text{Ch}_{\max}(\boldsymbol{M}\boldsymbol{\Sigma}_{22.1}^{-1})}{\text{tr}(\boldsymbol{\Sigma}_{11}^{-1})}\left(2\mathcal{H}_{p_2+2}(\lambda^2,\Delta^2)-\mathcal{H}_{p_2+4}(\lambda^2,\Delta^2)\right)\Bigg]^{-1} \tag{9.6.5}$$

$$\text{ADRE}(\hat{\boldsymbol{\rho}}_{1n}^{\text{SSR}}(\lambda):\hat{\boldsymbol{\rho}}_{1n}^{\text{UR}})=\Bigg[1+\frac{\text{tr}(\boldsymbol{M}\boldsymbol{\Sigma}_{22\cdot1}^{-1})}{\text{tr}(\boldsymbol{\Sigma}_{11\cdot2}^{-1})}\Big\{1-(p_2-2)$$

$$\times\left(2\mathbb{E}\left[\chi_{p_2+2}^{-2}(\Delta^2)\right]-(p_2-2)\mathbb{E}\left[\chi_{p_2+2}^{-4}(\Delta^2)\right]\right)\Big\}$$

$$+(p_2^2-4)\frac{\Delta^2\text{Ch}_{\max}(\boldsymbol{M}\boldsymbol{\Sigma}_{22\cdot1})}{\text{tr}(\boldsymbol{\Sigma}_{11\cdot2}^{-1})}\mathbb{E}\left[\chi_{p_2+4}^{-4}(\Delta^2)\right]\Bigg]^{-1} \tag{9.6.6}$$

$$\text{ADRE}(\hat{\boldsymbol{\rho}}_{1n}^{\text{SSR+}}(\lambda):\hat{\boldsymbol{\rho}}_{1n}^{\text{UR}})=\left[\text{ADRE}(\hat{\boldsymbol{\rho}}_{1n}^{\text{SSR}}(\lambda):\hat{\boldsymbol{\rho}}_{1n}^{\text{UR}})\right]^{-1}$$

$$-\frac{\text{tr}(\boldsymbol{M}\boldsymbol{\Sigma}_{22\cdot1}^{-1})}{\text{tr}(\boldsymbol{\Sigma}_{11\cdot2}^{-1})}\mathbb{E}\left[\left(1-(p_2-2)\chi_{p_2+2}^{-2}(\Delta^2)\right)^2 I(\chi_{p_2+2}^2(\Delta^2)<(p_2-2))\right]$$

$$+\frac{\Delta^2\text{Ch}_{\max}(\boldsymbol{M}\boldsymbol{\Sigma}_{22\cdot1}^{-1})}{\text{tr}(\boldsymbol{\Sigma}_{11\cdot2}^{-1})}\Big[2\mathbb{E}\left[\left(1-(p_2-2)\chi_{p_2+2}^{-2}(\Delta^2)\right)I(\chi_{p_2+2}^2(\Delta^2)<(p_2-2))\right]$$

$$-\mathbb{E}\left[\left(1-(p_2-2)\chi_{p_2+4}^{-2}(\Delta^2)\right)^2 I(\chi_{p_2+4}^2(\Delta^2)<(p_2-2))\right]\Big]^{-1} \tag{9.6.7}$$

$$\text{ADRE}(\hat{\boldsymbol{\rho}}_{1n}^{\text{ridgeR}}(\lambda):\hat{\boldsymbol{\rho}}_{n}^{\text{UR}})=\left(1+\frac{\text{tr}(\boldsymbol{M}\boldsymbol{\Sigma}_{22\cdot1}^{-1})}{\text{tr}(\boldsymbol{\Sigma}_{11\cdot2}^{-1})}\right)$$

$$\times\left\{1+\frac{\text{tr}\left(\left(\boldsymbol{I}_{p_1}+k\boldsymbol{\Sigma}_{22\cdot1}^{-1}\right)\boldsymbol{\Sigma}_{22\cdot1}^{-1}\right)}{\text{tr}(\boldsymbol{\Sigma}_{11\cdot2})}+k^2\frac{\boldsymbol{\delta}_2^\top(\boldsymbol{\Sigma}_{22\cdot1}+k\boldsymbol{I}_{p_2})\boldsymbol{\delta}_2}{\text{tr}(\boldsymbol{\Sigma}_{11\cdot2}^{-1})}\right\}^{-1} \tag{9.6.8}$$

$$\text{ADRE}(\hat{\boldsymbol{\rho}}_{1n}^{\text{EnetR}}(\lambda,\alpha):\hat{\boldsymbol{\rho}}_{1n}^{\text{UR}})=\frac{\text{tr}(\boldsymbol{\Sigma}_{11\cdot2}^{-1})+\text{tr}(\boldsymbol{\Sigma}_{22\cdot1}^{-1})}{G(\lambda,\alpha)} \tag{9.6.9}$$

where

$$G(\lambda,\alpha)=\text{tr}\left(\left(\boldsymbol{I}_{p_1}+\lambda(1-\alpha)\boldsymbol{\Sigma}_{11\cdot2}^{-1}\right)^{-2}\boldsymbol{\Sigma}_{11\cdot2}^{-1}\right)$$

$$+\lambda^2(1-\alpha)^2\boldsymbol{\Delta}^\top\left(\boldsymbol{\Sigma}_{22\cdot1}+\lambda(1-\alpha)\boldsymbol{I}_{p_2}\right)^{-2}\boldsymbol{\Delta}$$

$$+2\lambda(1-\alpha)\boldsymbol{B}^\top(\boldsymbol{C}+\lambda(1-\alpha)\boldsymbol{I}_{p_1})^{-1}\boldsymbol{\Delta}. \tag{9.6.10}$$

9.7 Problems

9.1. Prove Theorem 9.1.

9.2. Find the asymptotic null distribution of the test statistic $\mathcal{L}_n$ for testing $\mathcal{H}_o : \boldsymbol{\rho}_2 = \mathbf{0}$.

9.3. Prove Theorem 9.2.

9.4. Compare the asymptotic distribution L_2-risks of the ridge regression R-estimator with that of the Enet R-estimator of $\boldsymbol{\rho}$.

9.5. Show that ADL_2^*-risk$(\hat{\boldsymbol{\beta}}_n^{\text{EnetR}}(\lambda, \alpha))$ equals Theorem 9.3 (ix).

9.6. Verify Theorem 9.3.

9.7. Calculate the ADL_2-risk of $\hat{\boldsymbol{\rho}}_n^{\text{SSR}}$.

10

High-Dimensional Models

10.1 Introduction

In this chapter, we consider robust rank-based high-dimensional data analysis, where the data may arise from biological, medical, social and economical studies. High-dimensional problems are typically characterized by $p \gg n$, that is, the number of variables, p, greatly exceeds the number of observations, n. They also pertain to cases when $p \approx n$. In either of these domains, the nature of the problem changes significantly. We cannot simply apply the techniques described in the previous chapters without modification. Instead, we must rely on the fact that there is a high degree of sparsity in the parameter set. Under this assumption, we can perform subset selection to identify variables that are clearly zero, those that are clearly non-zero and those that lie between these two sets. Due to complexity and model prediction in the presence of outliers, statistical theory is very important using rank-based methodology. This chapter is an extension to chapter 11 of Saleh et al. (2019), where they studied the p_n-dimensional ridge regression from a post-selection viewpoint. The notation of p_n is used to show that the dimension of the coefficient set depends on n.

In the high-dimensional problem addressed in this chapter, we assume the model is sparse. As such we consider a multiple regression model for the ith response with covariate x_{ij} as follows,

$$y_i = \sum_{j=1}^{p_n} x_{ij}\beta_j^* + \varepsilon_i, \quad 1 \le i \le n. \tag{10.1.1}$$

In matrix form, the model is given by

$$\boldsymbol{y} = \boldsymbol{X}\boldsymbol{\beta}^* + \boldsymbol{\varepsilon}, \quad \boldsymbol{\beta}^* = \left(\beta_1^*, \dots, \beta_{p_n}^*\right)^\top, \tag{10.1.2}$$

Rank-Based Methods for Shrinkage and Selection: With Application to Machine Learning. First Edition. A. K. Md. Ehsanes Saleh, Mohammad Arashi, Resve A. Saleh, and Mina Norouzirad.

where $\boldsymbol{y} = (y_1, \dots, y_n)^\top$ is the response vector, $\boldsymbol{X} = (\boldsymbol{x}_1, \dots, \boldsymbol{x}_n)^\top$, $\boldsymbol{x}_i = (x_{i1}, \dots, x_{ip_n})^\top \in \mathbb{R}^{p_n}$, is an $n \times p_n$ fixed-design matrix, $\boldsymbol{\beta}^*$ is the true regression coefficient vector/set, and $\boldsymbol{\varepsilon} = (\varepsilon_1, \dots, \varepsilon_n)^\top$ is independently distributed with c.d.f. $F(x)$ having finite Fisher information,

$$I(f) = \int \left(\frac{f'(x)}{f(x)}\right)^2 \mathrm{d}F(x) < \infty. \tag{10.1.3}$$

The score generating function $\phi(u)$ satisfies

$$\begin{aligned}
&(i) \quad \int_0^1 |\phi(u)|\{n(1-u)\}^{-\frac{1}{2}} \mathrm{d}u < \infty, \\
&(ii) \quad A_\phi^2 = \int_0^1 \phi^2(u)\mathrm{d}u - \left(\int_0^1 \phi(u)\mathrm{d}u\right)^2, \\
&(iii) \quad \gamma(\phi, f) = \int_{\infty}^{+\infty} f(x)\mathrm{d}\phi(F(x)),\ \mathrm{d}\phi(F(x)) = \phi(u, f)\mathrm{d}u.
\end{aligned} \tag{10.1.4}$$

Also, let

$$\eta^2 = \frac{A_\phi^2}{\gamma^2(\phi, f)}. \tag{10.1.5}$$

For any subset $\mathcal{S} \subset \{1, \dots, p_n\}$, with cardinal value $|\mathcal{S}|$, denote $\boldsymbol{\beta}_{\mathcal{S}}^*$ as a subvector of $\boldsymbol{\beta}^*$ indexed by $\mathcal{S}$. Further, suppose we can divide the index set $\{1, \dots, p_n\}$ into three disjoint subsets: $\mathcal{S}_1$, $\mathcal{S}_2$, and $\mathcal{S}_3$. In particular, $\mathcal{S}_1$ includes indexes of non-zero $\boldsymbol{\beta}_i^*$ which are moderately large and easily detected (called effective coefficients); $\mathcal{S}_3$ includes indexes with only zero coefficients; $\mathcal{S}_2$, being the intermediate, includes indexes of those non-zero β_j^* with weak but non-zero effects. We further assume $\mathcal{S}_2 \cup \mathcal{S}_3$ are small effect coefficients and thus the complement of $\mathcal{S}_1$, i.e. $\mathcal{S}_1^c$ is negligible. In other words, $\sum_{j \in \mathcal{S}_1^c} |\beta_j^*|$ is small.

In this chapter, we shall consider the use of LASSO and ridge penalties for selection and estimation simultaneously, in order to keep calculations simple, under some model assumptions as follows:

A(i) There exist some β_j^*'s such that $|\beta_j^*| \geq c_1 \sqrt{\frac{\log(p_n)}{n}}$ for $\forall j \in \mathcal{S}_1$.
A(ii) Some β_j^* exist such that $\|\boldsymbol{\beta}_{\mathcal{S}_2}^*\| = O(n^\tau)$ for some $0 < \tau < 1$.
A(iii) $\beta_j^* = 0$ for $\forall j \in \mathcal{S}_3$.
A(iv) $\sum_{j \notin \mathcal{S}_1} |\beta_j^*| \leq \eta_1$ for some $\eta_1 \geq 0$.

10.2 Identifiability of β^* and Projection

Since we are dealing with a deterministic design matrix, $\boldsymbol{X}$, and high dimensions, $p_n > n$, we need $\boldsymbol{\beta}^*$ to be identifiable and it is so if and only if it lies in a set having a one-to-one correspondence with a linear space $\mathcal{R}(\boldsymbol{X})$ spanned by the rows of $\boldsymbol{X}$. Since $\dim\mathcal{R}(\boldsymbol{X}) = \text{rank}(\boldsymbol{X}) \leq \min(n, p_n) = n$, and hence $\boldsymbol{\beta}^*$ is not identifiable, it is not realistic to derive a consistent estimator of $\boldsymbol{\beta}^*$ or consistent variable selection. For statistical inference, we may not need to estimate vector $\boldsymbol{\beta}^*$ but only estimate $\boldsymbol{\theta}$, the projection of $\boldsymbol{\beta}^*$ onto $\mathcal{R}(\boldsymbol{X})$. We define $\boldsymbol{\beta}^*$ as identifiable if $\boldsymbol{\beta}_1^* \in B^*$ and $\boldsymbol{\beta}_2^* \in B^*$, then $\boldsymbol{X}\boldsymbol{\beta}_1^* = \boldsymbol{X}\boldsymbol{\beta}_2^*$ implies $\boldsymbol{\beta}_1^* = \boldsymbol{\beta}_2^*$, where B^* is the parameter space of $\boldsymbol{\beta}^*$. The following lemma due to Shao and Deng (2012) gives the necessary and sufficient conditions for identifiability of $\boldsymbol{\beta}^*$.

Lemma 10.1 Under the model of Eq. (10.1.2), with $p_n > k_n$, $\boldsymbol{\beta}^*$ is identifiable if and only if there exists a known function $\boldsymbol{\phi} : \mathbb{R}^{k_n} \to \mathbb{R}^{p_n-k_n}$ such that

$$B^* = \{\boldsymbol{\beta}^* : \boldsymbol{\beta}^* = \boldsymbol{Q}_1\boldsymbol{\xi} + \boldsymbol{Q}_2\boldsymbol{\phi}(\boldsymbol{\xi}), \quad \boldsymbol{\xi} \in \mathbb{R}^{k_n}\} \tag{10.2.1}$$

where $\boldsymbol{Q}_1$ is a $p_n \times k_n$ and $\boldsymbol{Q}_2$ is a $p_n \times p_n - k_n$ matrix such that $\boldsymbol{Q}_2^\top\boldsymbol{Q}_2 = \boldsymbol{I}_{p_n-k_n}$, $\boldsymbol{Q}_1^\top\boldsymbol{Q}_1 = \boldsymbol{I}_{k_n}$ and $\boldsymbol{Q}_1^\top\boldsymbol{Q}_2 = \boldsymbol{0}_{k_n}$.

Indeed, Lemma 10.1 shows that an identifiable $\boldsymbol{\beta}$ must be in a set having a one-to-one correspondence with $\mathcal{R}(\boldsymbol{X}) = \{\boldsymbol{\beta}^* = \boldsymbol{Q}_1\boldsymbol{\xi},\ \boldsymbol{\xi} \in \mathbb{R}^{k_n}\}$. Thus, we consider the projection of $\boldsymbol{\beta}$ onto $\mathcal{R}(\boldsymbol{X})$ given by

$$\boldsymbol{\theta} = \boldsymbol{X}^\top(\boldsymbol{X}\boldsymbol{X}^\top)^-\boldsymbol{X}\boldsymbol{\beta}^* = \boldsymbol{Q}_1\boldsymbol{Q}_1^\top\boldsymbol{\beta}^*. \tag{10.2.2}$$

Note that $\boldsymbol{\theta} \in \mathcal{R}(\boldsymbol{X})$ and $\boldsymbol{\theta} = \boldsymbol{\beta}^*$ if and only if $\boldsymbol{\beta}^* \in \mathcal{R}(\boldsymbol{X})$. For this, $\boldsymbol{X}\boldsymbol{\beta}^* = \boldsymbol{X}\boldsymbol{\theta}$ and the sparse model is then given by

$$\boldsymbol{y} = \boldsymbol{X}\boldsymbol{\theta} + \boldsymbol{\varepsilon} = \boldsymbol{X}_1\boldsymbol{\theta}_1 + \boldsymbol{X}_2\boldsymbol{\theta}_2 + \boldsymbol{\varepsilon}. \tag{10.2.3}$$

10.3 Parsimonious Model Selection

As indicated earlier, a penalized Jeackel rank dispersion (PRD) function is often used to select a parsimonious model for the high-dimensional regression model of Eq. (10.1.1),

$$\hat{\boldsymbol{\beta}}_n^{\text{PRD}} = \underset{\boldsymbol{\beta}^* \in \mathbb{R}^{p_n}}{\text{argmin}}\left\{D_n(\boldsymbol{\beta}^*) + \gamma_n\boldsymbol{1}_{p_n}^\top|\boldsymbol{\beta}^*|\right\}, \quad \boldsymbol{\beta}^* = (\beta_1^*, \dots, \beta_{p_n}^*)^\top, \tag{10.3.1}$$

where γ_n is the tuning parameter, $\mathbf{1}_{p_n}$ is a p_n-tuple of 1, and $|\boldsymbol{\beta}^*| = (|\beta_1^*|, \dots, |\beta_{p_n}^*|)^\top$,

$$D_n(\boldsymbol{\beta}) = \sum_{i=1}^{n}(y_i - \sum_{j=1}^{p_n} x_{ij}\beta_j)a_n(R_{n_i}(\boldsymbol{\beta})), \tag{10.3.2}$$

and $R_{n_i}(\boldsymbol{\beta})$ is the rank of $y_i - \sum_{j=1}^{p_n} x_{ij}\beta_j$ among $\{y_\alpha - \sum_{j=1}^{p_n} x_{\alpha j}\beta_j,\ 1 \leq \alpha \leq n\}$. As before, let $\hat{\mathcal{S}}_1 \subset \{1, \dots, p_n\}$ be the active subset from Eq. (10.3.1). Then, a data-adaptive candidate subset model is produced as

$$\hat{\beta}_{jn}^{\text{PRD}} = 0 \quad \text{if and only if} \quad j \notin \hat{\mathcal{S}}_1. \tag{10.3.3}$$

Without loss of generality, we rearrange the covariates as $\boldsymbol{X} = \left(\boldsymbol{X}_{\mathcal{S}_1}|\boldsymbol{X}_{\mathcal{S}_2}|\boldsymbol{X}_{\mathcal{S}_3}\right)$ where $\boldsymbol{X}_{\mathcal{S}_i}$ is a sub-matrix indexed by $\mathcal{S}_i \subset \{1, \dots, p_n\}$ and each $\boldsymbol{X}_{\mathcal{S}_i}$ is of dimension $n \times p_{in}$, $i = 1, 2, 3$. We designated $p_i = p_{in}$ to show the dependency on the sample size. In general, $\mathcal{S}_2$ cannot be separable from $\mathcal{S}_3$. Some post-selection R-estimations can be presented to improve prediction performance of the PRD estimator. Thus, we can collect together the variable $\boldsymbol{X}_{\hat{\mathcal{S}}_j}$ from the PRD subset selection and obtain the restricted R-estimator $\hat{\boldsymbol{\beta}}_{1n}^{\text{RR}}$ using the restricted model

$$y_i = \sum_{j=1}^{p_{1n}} x_{ij}\beta_j^* + \varepsilon_i, \quad 1 \leq i \leq n, \tag{10.3.4}$$

where

$$\hat{\boldsymbol{\beta}}_{1n}^{\text{RR}} = \underset{\boldsymbol{\beta}_{\hat{\mathcal{S}}_1}^* \in \mathbb{R}^{p_{1n}}}{\text{argmin}} \left\{ \sum_{j=1}^{p_{1n}} |L_{n_j}(\boldsymbol{\beta}_{\hat{\mathcal{S}}_1}^*, \mathbf{0})| \right\} \tag{10.3.5}$$

and

$$L_{n_j}(\boldsymbol{\beta}_{\hat{\mathcal{S}}_1}^*, \mathbf{0}) = \sum_{i=1}^{n} x_{ij}a_n\left(R_{n_i}(\boldsymbol{\beta}_{\hat{\mathcal{S}}_1}^*)\right) \tag{10.3.6}$$

with $R_{n_i}(\boldsymbol{\beta}_{\hat{\mathcal{S}}_1}^*)$ as the rank of $\left(y_i - \sum_{j=1}^{p_{1n}} x_{ij}\beta_j^*\right)$ among $y_i - \sum_{j=1}^{p_{1n}} x_{ij}\beta_j^*$, $i = 1, \dots, n$. Further, we obtain

$$\sqrt{n}\left(\hat{\boldsymbol{\beta}}_{1n}^{\text{RR}} - \boldsymbol{\beta}_{\hat{\mathcal{S}}_1}^*\right) \xrightarrow{\mathcal{D}} \mathcal{N}_{p_{1n}}(\mathbf{0}, \eta^2\boldsymbol{C}_{11}^{-1}), \tag{10.3.7}$$

where

$$\boldsymbol{C} = \begin{pmatrix} \boldsymbol{C}_{11} & \boldsymbol{C}_{12} \\ \boldsymbol{C}_{21} & \boldsymbol{C}_{22} \end{pmatrix}. \tag{10.3.8}$$

The next section delves deeper into the specifics of $\boldsymbol{C}$ and the partitions.

10.4 Some Notation and Separation

Based upon a subset partitioning of $\mathcal{S}_1$, $\mathcal{S}_2$, and $\mathcal{S}_3$, we can also partition the true parameters $\boldsymbol{\beta}^* = \left(\boldsymbol{\beta}_1^{*\top}, \boldsymbol{\beta}_2^{*\top}, \boldsymbol{\beta}_3^{*\top}\right)^\top$, without loss of generality. Some notations are shortened for convenience such as $\boldsymbol{\beta}^*_{\mathcal{S}_k} = \boldsymbol{\beta}^*_k$, $k = 1, 2, 3$. Similar notations are also adopted for other sub-vectors and matrices. For example, using the same partitions, the design matrix $\boldsymbol{X} = (\boldsymbol{x}_1, \dots, \boldsymbol{x}_n)^\top$, $\boldsymbol{x}_i \in \mathbb{R}^{p_n}$, $i = 1, \dots, n$, can be written as $\boldsymbol{X} = (\boldsymbol{X}_1, \boldsymbol{X}_2, \boldsymbol{X}_3)$. We also write $\boldsymbol{X} = (\boldsymbol{Z}, \boldsymbol{X}_3)$ with $\boldsymbol{Z} = (\boldsymbol{X}_1, \boldsymbol{X}_2)$. The row vector of $\boldsymbol{Z}$ is denoted as $\boldsymbol{z}_i = (z_{i1}, \dots, z_{i(p_{1n}+p_{2n})})$, for $1 \le i \le n$.

We denote $p_{kn} = |\mathcal{S}_k|$ for $k = 1, 2, 3$ and $p_n = p_{1n} + p_{2n} + p_{3n}$. In this chapter, we allow $p_n = \sum_{k=1}^{3} p_{kn}$ to be very large but restrict $p_{1n} + p_{2n} \le n$ such that $\boldsymbol{C}_n = n^{-1}\boldsymbol{Z}^\top\boldsymbol{Z}$ is non-singular. In the limit, we assume $\boldsymbol{C}_n \to \boldsymbol{C}$, where $\boldsymbol{C}$ is a positive definite matrix. If $\boldsymbol{C}_n$ is singular, then a generalized inverse matrix is adopted when needed in computations. Specifically, one can use the Moore–Penrose generalized inverse for uniqueness.

10.4.1 Special Matrices

For convenience and to avoid confusion with notation, we provide some special sub-matrices of $\boldsymbol{C}_n$ and their respective limiting cases as follows.

Let

$$
\begin{aligned}
\boldsymbol{C}_{nij} &= n^{-1}\boldsymbol{X}_{\mathcal{S}_i}^\top \boldsymbol{X}_{\mathcal{S}_j}, \quad i, j = 1, 2, \quad \boldsymbol{C}_{nii} = n^{-1}\boldsymbol{X}_{\mathcal{S}_i}^\top \boldsymbol{X}_{\mathcal{S}_i}, \; i = 1, 2 \\
\boldsymbol{C}_{n22\cdot1} &= n^{-1}\left(\boldsymbol{X}_{\mathcal{S}_2}^\top \boldsymbol{X}_{\mathcal{S}_2} - \boldsymbol{X}_{\mathcal{S}_2}^\top \boldsymbol{X}_{\mathcal{S}_1}(\boldsymbol{X}_{\mathcal{S}_1}^\top \boldsymbol{X}_{\mathcal{S}_1})^{-1}\boldsymbol{X}_{\mathcal{S}_1}^\top \boldsymbol{X}_{\mathcal{S}_2}\right) \\
\boldsymbol{C}_{n11\cdot2} &= n^{-1}\left(\boldsymbol{X}_{\mathcal{S}_1}^\top \boldsymbol{X}_{\mathcal{S}_1} - \boldsymbol{X}_{\mathcal{S}_1}^\top \boldsymbol{X}_{\mathcal{S}_2}(\boldsymbol{X}_{\mathcal{S}_2}^\top \boldsymbol{X}_{\mathcal{S}_2})^{-1}\boldsymbol{X}_{\mathcal{S}_2}^\top \boldsymbol{X}_{\mathcal{S}_1}\right) \\
\boldsymbol{M}_{\mathcal{S}_1^c} &= \boldsymbol{I}_n - \boldsymbol{X}_{\mathcal{S}_1^c}\left(\boldsymbol{X}_{\mathcal{S}_1^c}^\top \boldsymbol{X}_{\mathcal{S}_1^c} + \gamma_n \boldsymbol{I}_{q_n}\right)^{-1}\boldsymbol{X}_{\mathcal{S}_1^c}^\top.
\end{aligned}
\tag{10.4.1}
$$

Let $\boldsymbol{U} = (\boldsymbol{X}_{\mathcal{S}_2}, \boldsymbol{X}_{\mathcal{S}_3})$ be a $n \times (p_{2n} + p_{3n})$ sub-matrix. Let

$$
\boldsymbol{M}_{\mathcal{S}_1} = \boldsymbol{I}_n - \boldsymbol{X}_{\mathcal{S}_1}(\boldsymbol{X}_{\mathcal{S}_1}^\top \boldsymbol{X}_{\mathcal{S}_1})^{-1}\boldsymbol{X}_{\mathcal{S}_1}^\top.
$$

Then, $\boldsymbol{U}^\top \boldsymbol{M}_{\mathcal{S}_1}\boldsymbol{U}$ is a $(p_n - p_{1n}) \times (p_n - p_{1n})$ dimensional singular matrix with rank $k_n \ge 0$. We denote $\varrho_{1n} \le \dots \le \varrho_{k_n n}$ for all k_n positive eigenvalues of $\boldsymbol{U}^\top \boldsymbol{M}_{\mathcal{S}_1}\boldsymbol{U}$.

Here, we have the limits

$$
\lim_{n\to\infty} \boldsymbol{C}_{nij} = \boldsymbol{C}_{ij}, \quad i, j = 1, 2,
$$

$$\lim_{n\to\infty} \boldsymbol{C}_{n22\cdot1} = \boldsymbol{C}_{22\cdot1},$$

$$\lim_{n\to\infty} \boldsymbol{C}_{n11\cdot2} = \boldsymbol{C}_{11\cdot2}. \tag{10.4.2}$$

Our objective is now to estimate $\boldsymbol{\beta}^*_{\mathcal{S}_1}$. In the next section, we provide the steps of the estimation technique in preparation for the separation of disjoint coefficients sets.

10.4.2 Steps Towards Estimators

We are following three steps to organize the estimation of parameters:

1: Use Section 10.2 to obtain the LASSO solution to Eq. (10.3.1) for the model

$$y_i = \sum_{j=1}^{p_n} x_{ij}\beta_j^* + \varepsilon_i, \quad i = 1, \ldots, n, \tag{10.4.3}$$

and accordingly arrange the $n \times s_1$ matrix $\boldsymbol{X}_{\hat{\mathcal{S}}_1}$ and $n \times s_2$ matrix $\boldsymbol{X}_{\hat{\mathcal{S}}_1^c}$, where $s_1 = p_{1n} + o_p(1)$ and $s_2 = p_{2n} + p_{3n} + o_p(1)$. Then, using Eq. (10.3.5), obtain $\hat{\boldsymbol{\beta}}^{\text{RR}}_{1n}$.

2: Define the R-estimators of $\boldsymbol{\beta}^*_{\mathcal{S}_1}$ and $\boldsymbol{\beta}^*_{\mathcal{S}_1^c}$ for the full model of Eq. (10.1.1) using $\boldsymbol{X}_{\hat{\mathcal{S}}_1}$ and $\boldsymbol{X}_{\hat{\mathcal{S}}_1^c}$. Note that we need $p_{1n} + p_{2n} < n$ and $p_{3n} > 2$. These matrices are used to obtain the ridge regression estimators with penalty $\boldsymbol{\beta}^{*\top}_{\mathcal{S}_1^c}\boldsymbol{\beta}^*_{\mathcal{S}_1^c}$.

3: Use the post-selection ridge estimators to define the preliminary-test and Stein–Saleh-type R-estimators.

Remark 10.1 The reader must note $\boldsymbol{X}_{\mathcal{S}_1^c}$ is of dimension $n \times q_n$, where $q_n = p_{2n} + p_{3n}$.

10.4.3 Post-selection Ridge Estimation of $\boldsymbol{\beta}^*_{\mathcal{S}_1}$ and $\boldsymbol{\beta}^*_{\mathcal{S}_2}$

We suspect that $\boldsymbol{\beta}^*_{\mathcal{S}_1^c}$ may be negligible. Step 1 in Section 10.4.2 gives $\hat{\mathcal{S}}_1$. So we construct ridge R-estimators (ridgeREs) of $\boldsymbol{\beta}^*_{\mathcal{S}_1}$ and $\boldsymbol{\beta}^*_{\mathcal{S}_1^c}$ based upon minimizing the penalized PRD function

$$Q_n(\boldsymbol{\beta}^*|\hat{\mathcal{S}}_1) = D_n(\boldsymbol{\beta}^*|\hat{\mathcal{S}}_1) + \gamma_n\boldsymbol{\beta}^{*\top}_{\mathcal{S}_1^c}\boldsymbol{\beta}^*_{\mathcal{S}_1^c}, \quad |\hat{\mathcal{S}}_1^c| = q_n, \; q_n = p_{2n} + p_{3n}, \tag{10.4.4}$$

where

$$D_n(\boldsymbol{\beta}^*|\hat{\mathcal{S}}_1) = D_n(\boldsymbol{\beta}^*|\hat{\mathcal{S}}_1 = \mathcal{S}_1) = \sum_{i=1}^{n}\left(y_i - \boldsymbol{x}^\top_{i\mathcal{S}_1}\boldsymbol{\beta}^*_{\mathcal{S}_1} - \boldsymbol{x}^\top_{i\mathcal{S}_1^c}\boldsymbol{\beta}^*_{\mathcal{S}_1^c}\right)a_n(R_{n_i}(\boldsymbol{\beta})). \tag{10.4.5}$$

Let the solutions be given by

$$\left(\hat{\boldsymbol{\beta}}_{\hat{\mathcal{S}}_1}^{\text{ridgeR}}(\gamma_n)^\top, \hat{\boldsymbol{\beta}}_{\hat{\mathcal{S}}_1^c}^{\text{ridgeR}}(\gamma_n)^\top\right)^\top = \text{argmin}\left\{D_n(\boldsymbol{\beta}^* | \hat{\mathcal{S}}_1) + \gamma_n \|\boldsymbol{\beta}^*_{\hat{\mathcal{S}}_1^c}\|^2\right\}. \quad (10.4.6)$$

Next, we perform subset selection using the preliminary test subset selector (PTSS) as:

$$\hat{\beta}_{jn}^{\text{ridgeR}}(\gamma_n, a_n) = \begin{cases} \hat{\beta}_{jn}^{\text{ridgeR}}(\gamma_n), & j \in \hat{\mathcal{S}}_1 \\ \hat{\beta}_{jn}^{\text{ridgeR}}(\gamma_n) I\left(\hat{\beta}_{jn}^{\text{ridgeR}}(\gamma_n) > a_n\right), & j \in \hat{\mathcal{S}}_1^c \end{cases}, \quad (10.4.7)$$

where $I(\cdot)$ is the indicator function, and a_n is a threshold parameter given by $a_n = c_1 n^{-\alpha}, 0 < \alpha \leq 1/2$, for some $c_1 > 0$. Note here that

$$\hat{\mathcal{S}}_2 = \hat{\mathcal{S}}_2(\hat{\mathcal{S}}_1) = \left\{j \in \hat{\mathcal{S}}_1^c : \hat{\beta}_{jn}^{\text{ridgeR}}(\gamma_n, a_n) \neq 0\right\}. \quad (10.4.8)$$

We call $\hat{\boldsymbol{\beta}}^{\text{ridgeR}}(\gamma_n, a_n) = \left(\hat{\beta}_{jn}^{\text{ridgeR}}(\gamma_n, a_n) | j \in \hat{\mathcal{S}}_1^c\right)^\top$ the post-selection ridge R-estimator, where we used estimated values with the indexes in the selected set $\hat{\mathcal{S}}$.

As in Remark 10.1, we now use $\mathcal{S}_i$ instead of $\hat{\mathcal{S}}_i$ in what follows, unless there is a particular need to do otherwise.

10.4.4 Post-selection Ridge R-estimators for $\boldsymbol{\beta}^*_{\mathcal{S}_1}$ and $\boldsymbol{\beta}^*_{\mathcal{S}_2}$

Let $\boldsymbol{X}_{\mathcal{S}_3^c} = (\boldsymbol{z}_1, \dots, \boldsymbol{z}_n)^\top$. Then, solve the equation

$$\frac{\partial Q_n(\boldsymbol{\beta}^* | \mathcal{S}_1)}{\partial \boldsymbol{\beta}^*_{\mathcal{S}_3^c}} = \boldsymbol{0} \quad (10.4.9)$$

or

$$-\boldsymbol{L}_{\mathcal{S}_3^c}\left(\boldsymbol{\beta}^*_{\mathcal{S}_3^c}, \boldsymbol{0}\right) + \gamma_n \begin{pmatrix} \boldsymbol{0}_{p_{1n}} \\ \hat{\boldsymbol{\beta}}_{\mathcal{S}_2}^{\text{ridgeR}}(\gamma_n) \end{pmatrix} = O_{p_{1n}+p_{2n}}, \quad (10.4.10)$$

where

$$\boldsymbol{L}_{\mathcal{S}_3^c}\left(\boldsymbol{\beta}^*_{\mathcal{S}_3^c}, \boldsymbol{0}\right) = \sum_{i=1}^n \boldsymbol{z}_i a_n(R_{n_i}(\boldsymbol{\beta}^*_{\mathcal{S}_3^c})). \quad (10.4.11)$$

Also

$$\frac{\partial Q_n(\boldsymbol{\beta}|\mathcal{S}_1)}{\partial \boldsymbol{\beta}^*_{\mathcal{S}_3}} = \gamma_n \begin{pmatrix} \mathbf{0}_{p_{1n}+p_{2n}} \\ \hat{\boldsymbol{\beta}}^{\text{ridgeR}}_{\mathcal{S}_3}(\gamma_n) \end{pmatrix} = \begin{pmatrix} \mathbf{0}_{p_{1n}+p_{2n}} \\ \mathbf{0}_{p_n-p_{1n}-p_{2n}} \end{pmatrix}, \tag{10.4.12}$$

where $\hat{\boldsymbol{\beta}}^{\text{ridgeR}}_{\mathcal{S}_3}(\gamma_n) = \mathbf{0}$ with probability one by Theorem 10.1 (given later).

Next, using asymptotic linearity results, we have

$$\sup_{\|\boldsymbol{w}_{\mathcal{S}_3^c}\|<k} n^{-\frac{1}{2}} \left\| \boldsymbol{L}_{\mathcal{S}_3^c}\left(\boldsymbol{\beta}^*_{\mathcal{S}_3^c}, \mathbf{0}\right) - \boldsymbol{L}_{\mathcal{S}_3^c}(\mathbf{0},\mathbf{0}) + \gamma(\phi, f)\boldsymbol{C}\boldsymbol{w}_{\mathcal{S}_3^c} \right\| \xrightarrow{P} 0. \tag{10.4.13}$$

Thus, together with Eq. (10.4.10) we obtain

$$\begin{aligned}
\boldsymbol{w}_{\mathcal{S}_3^c} &= \sqrt{n}\left(\hat{\boldsymbol{\beta}}^{\text{RR}}_{\mathcal{S}_3^c}(\gamma_n) - \boldsymbol{\beta}^*_{\mathcal{S}_3^c}\right) \\
&= \gamma^{-1}(\phi, f)\boldsymbol{C}_n^{-1}\left[\boldsymbol{L}_{\mathcal{S}_3^c}(\mathbf{0},\mathbf{0}) - \gamma_n \begin{pmatrix} \mathbf{0} \\ \hat{\boldsymbol{\beta}}^{\text{ridgeR}}_{\mathcal{S}_2}(\gamma_n) \end{pmatrix}\right] + o_p(1) \\
&= \gamma^{-1}(\phi, f)\boldsymbol{C}_n^{-1}\boldsymbol{L}_{\mathcal{S}_3^c}(\mathbf{0},\mathbf{0}) + o_p(1).
\end{aligned} \tag{10.4.14}$$

To understand Eq. (10.4.14), using $\boldsymbol{d}_n = (\boldsymbol{d}_{1n}^\top, \boldsymbol{d}_{2n}^\top)^\top$, we can write

$$\begin{aligned}
\gamma^{-1}(\phi, f)n^{-\frac{1}{2}}\gamma_n\boldsymbol{d}_n^\top\boldsymbol{C}_n^{-1}\begin{pmatrix} \mathbf{0} \\ \hat{\boldsymbol{\beta}}^{\text{ridgeR}}_{\mathcal{S}_2}(\gamma_n) \end{pmatrix} &\leq \varrho_{1n}^{-1}n^{-\frac{1}{2}}\gamma_n\boldsymbol{d}_{2n}^\top\hat{\boldsymbol{\beta}}^{\text{ridgeR}}_{\mathcal{S}_2}(\gamma_n) \quad \text{from B(ii)} \\
&= O_p(\varrho_{1n}^{-1}n^{-\frac{1}{2}}\gamma_n\boldsymbol{d}_{2n}^\top\boldsymbol{\beta}^*_{\mathcal{S}_2}) \quad \text{from B(i)} \\
\leq O_p(\varrho_{1n}^{-1}n^{-\frac{1}{2}}\gamma_n\|\boldsymbol{d}_{2n}\|\|\boldsymbol{\beta}^*_{\mathcal{S}_2}\|) \quad &\text{Cauchy–Schwartz inequality} \\
&\leq O_p(\varrho_{1n}^{-1}n^{-(\frac{1}{2}-\tau)}\gamma_n) \quad \text{from A(ii)} \\
&= o_p(1),
\end{aligned} \tag{10.4.15}$$

since $\gamma_n = o(n^{(1/2)-\tau})$ for $\alpha < 1/4 - \tau/2$ for $0 < \tau < 1/2$. See Sections 10.1 and 10.6 for definitions of A(ii), B(i) and B(ii).

Now, from Eq. (10.4.14), we may exactly see that

$$\gamma^{-1}(\phi, f)\boldsymbol{C}_n^{-1}\boldsymbol{L}_{\mathcal{S}_3^c}(\mathbf{0},\mathbf{0}) \xrightarrow{\mathcal{D}} \mathcal{N}_{p_{1n}+p_{2n}}(\mathbf{0}, \eta^2\boldsymbol{C}^{-1}). \tag{10.4.16}$$

Next, we consider the test of hypothesis $\mathcal{H}_o : \boldsymbol{\beta}^*_{\mathcal{S}_2} = 0$. For this problem, we consider the statistic

$$\boldsymbol{L}_{n(2)}(\hat{\boldsymbol{\beta}}^{\text{RR}}_{1n}, \boldsymbol{0}) = \left(L_{n(2)1}(\hat{\boldsymbol{\beta}}^{\text{RR}}_{1n}, 0), \dots L_{n(2)p_{2n}}(\hat{\boldsymbol{\beta}}^{\text{RR}}_{1n}, 0)\right)^{\top} \tag{10.4.17}$$

where

$$\sup_{\|\boldsymbol{w}_{\mathcal{S}_2}\|<k} n^{-\frac{1}{2}} \left\| \boldsymbol{L}_{n(2)}\left(\hat{\boldsymbol{\beta}}^{\text{RR}}_{1n}, n^{-\frac{1}{2}}\boldsymbol{w}_{\mathcal{S}_2}\right) - \boldsymbol{L}_{n(2)}\left(\boldsymbol{\beta}^*_{\mathcal{S}_1}, \boldsymbol{0}\right) + \gamma(\phi, f)\boldsymbol{C}_{n22\cdot1}\boldsymbol{w}_{\mathcal{S}_2} \right\| \xrightarrow{P} 0. \tag{10.4.18}$$

Hence,

$$\begin{aligned}
\boldsymbol{w}_{\mathcal{S}_2} &= \sqrt{n}\left(\hat{\boldsymbol{\beta}}^{\text{ridgeR}}_{\mathcal{S}_2}(\gamma_n) - \boldsymbol{\beta}^*_{\mathcal{S}_2}\right) \\
&= \gamma^{-1}(\phi, f)\boldsymbol{C}^{-1}_{n22\cdot1}\boldsymbol{L}_{n(2)}(\boldsymbol{\beta}^*_{\mathcal{S}_1}, \boldsymbol{0}) + o_p(1) \\
&\xrightarrow{\mathcal{D}} \mathcal{N}_{p_{2n}}(\boldsymbol{0}, \eta^2\boldsymbol{C}^{-1}_{22\cdot1}).
\end{aligned} \tag{10.4.19}$$

For the test of $\mathcal{H}_o : \boldsymbol{\beta}^*_{\mathcal{S}_2} = 0$, we define the test statistic

$$\mathcal{L}_n = A_n^{-2}\left[\boldsymbol{L}_{n(2)}\left(\hat{\boldsymbol{\beta}}^{\text{RR}}_{1n}, \boldsymbol{0}\right)\right]^{\top} \boldsymbol{C}_{n22\cdot1}\left[\boldsymbol{L}_{n(2)}\left(\hat{\boldsymbol{\beta}}^{\text{RR}}_{1n}, \boldsymbol{0}\right)\right] \tag{10.4.20}$$

$$= \frac{1}{\eta^2}\left[\hat{\boldsymbol{\beta}}^{\text{ridgeR}}_{\mathcal{S}_2}(\gamma_n)\right]^{\top} \boldsymbol{C}_{n22\cdot1}\left[\hat{\boldsymbol{\beta}}^{\text{ridgeR}}_{\mathcal{S}_2}(\gamma_n)\right] + o_p(1). \tag{10.4.21}$$

If we use Eq. (10.4.21), then we can estimate η by using $\underset{1\le i\le n}{\text{median}}\, |y_\alpha - \sum_{j=1}^{p_{1n}} x_{\alpha j}\beta^*_j|$.

Further, we note that

$$\hat{\boldsymbol{\beta}}^{\text{RR}}_{1n} = \hat{\boldsymbol{\beta}}^{\text{ridgeR}}_{\mathcal{S}_1}(\gamma_n) + \boldsymbol{C}^{-1}_{n11}\boldsymbol{C}_{n12}\hat{\boldsymbol{\beta}}^{\text{ridgeR}}_{\mathcal{S}_2}(\gamma_n) + o_p(1). \tag{10.4.22}$$

10.5 Post-selection Shrinkage R-estimators

We now have two R-estimators, namely, $\hat{\boldsymbol{\beta}}^{\text{RR}}_{1n}$ and $\hat{\boldsymbol{\beta}}^{\text{ridgeR}}_{\mathcal{S}_1}(\gamma_n)$. From Sen and Saleh (1987), first we define the preliminary test estimator.

(a) Preliminary test R-estimator

$$\begin{aligned}
\hat{\boldsymbol{\beta}}^{\text{PTR}}_{\mathcal{S}_1}(\gamma_n) &= \hat{\boldsymbol{\beta}}^{\text{RR}}_{1n} I\left(\mathcal{L}_n \le \lambda^2\right) + \hat{\boldsymbol{\beta}}^{\text{ridgeR}}_{\mathcal{S}_1}(\gamma_n) I\left(\mathcal{L}_n > \lambda^2\right) \\
&= \hat{\boldsymbol{\beta}}^{\text{ridgeR}}_{\mathcal{S}_1}(\gamma_n) - \left(\hat{\boldsymbol{\beta}}^{\text{ridgeR}}_{\mathcal{S}_1}(\gamma_n) - \hat{\boldsymbol{\beta}}^{\text{RR}}_{1n}\right) I\left(\mathcal{L}_n \le \lambda^2\right)
\end{aligned}$$

$$= \hat{\boldsymbol{\beta}}_{\mathcal{S}_1}^{\text{ridgeR}}(\gamma_n) + \boldsymbol{C}_{n11}^{-1}\boldsymbol{C}_{n12}\hat{\boldsymbol{\beta}}_{\mathcal{S}_2}^{\text{ridgeR}}(\gamma_n) I(\mathcal{L}_n \le \lambda^2),$$

where the last equality follows from Eq. (10.4.22).

(b) Stein–Saleh-type R-estimator

Now, the continuous version of the $\hat{\boldsymbol{\beta}}_{\mathcal{S}_1}^{\text{PTR}}(\gamma_n)$ estimator is given by

$$\begin{aligned}
\hat{\boldsymbol{\beta}}_{\mathcal{S}_1}^{\text{SSR}}(\gamma_n) &= \hat{\boldsymbol{\beta}}_{1n}^{\text{RR}} + \left(\hat{\boldsymbol{\beta}}_{\mathcal{S}_1}^{\text{ridgeR}}(\gamma_n) - \hat{\boldsymbol{\beta}}_{1n}^{\text{RR}}\right)(1 - d\mathcal{L}_n^{-1}) \\
&= \hat{\boldsymbol{\beta}}_{\mathcal{S}_1}^{\text{ridgeR}}(\gamma_n) - \left(\hat{\boldsymbol{\beta}}_{\mathcal{S}_1}^{\text{ridgeR}}(\gamma_n) - \hat{\boldsymbol{\beta}}_{1n}^{\text{RR}}\right) d\mathcal{L}_n^{-1} \\
&= \hat{\boldsymbol{\beta}}_{\mathcal{S}_1}^{\text{ridgeR}}(\gamma_n) + \boldsymbol{C}_{n11}^{-1}\boldsymbol{C}_{n12}\hat{\boldsymbol{\beta}}_{\mathcal{S}_2}^{\text{ridgeR}}(\gamma_n) d\mathcal{L}_n^{-1},
\end{aligned}$$

where $d = s_2 - 2$, $s_2 = |\mathcal{S}_2| + o_p(1)$ and $s_2 \ge 3$.

Note that $\hat{\boldsymbol{\beta}}_{\mathcal{S}_1}^{\text{SSR}}(\gamma_n)$ may change the sign of the estimators. We correct this by defining the positive-rule shrinkage estimator.

(c) Positive-rule Stein–Saleh-type R-estimator

$$\begin{aligned}
\hat{\boldsymbol{\beta}}_{\mathcal{S}_1}^{\text{SSR+}}(\gamma_n) &= \hat{\boldsymbol{\beta}}_{\mathcal{S}_1}^{\text{ridgeR}}(\gamma_n) - \left(\hat{\boldsymbol{\beta}}_{\mathcal{S}_1}^{\text{ridgeR}}(\gamma_n) - \hat{\boldsymbol{\beta}}_{1n}^{\text{RR}}\right) d\mathcal{L}_n^{-1} I(\mathcal{L}_n > d) \\
&= \hat{\boldsymbol{\beta}}_{\mathcal{S}_1}^{\text{ridgeR}}(\gamma_n) + \boldsymbol{C}_{n11}^{-1}\boldsymbol{C}_{n12}\hat{\boldsymbol{\beta}}_{\mathcal{S}_2}^{\text{ridgeR}}(\gamma_n) d\mathcal{L}_n^{-1} I(\mathcal{L}_n > d).
\end{aligned}$$

These estimators depend on the tuning parameters and γ_n.

10.6 Asymptotic Properties of the Ridge R-estimators

For the asymptotic properties of the above estimators, and $\hat{\boldsymbol{\beta}}_{n}^{\text{ridgeR}}(\gamma_n)$ in particular, we make the following assumptions:

B(i) The random $\varepsilon_i \sim F(x)$ with p.d.f. $f(x)$ has a finite Fisher information $I(f) < \infty$.

B(ii) $\varrho_{1n}^{-1} = O(n^{-\eta})$ where $\tau < \eta \le 1$ for τ in A(ii).

B(iii) $\log(p_n) = O(n^{\gamma})$ for $0 < \gamma < 1$.

B(iv) There exists a positive definitive matrix $\boldsymbol{C}$ such that $\lim_{n\to\infty} \boldsymbol{C}_n = \boldsymbol{C}$ where the eigenvalues satisfy $0 < \varrho_1 < \varrho_\Sigma < \varrho_2 < \infty$.

Condition B(ii) guarantees the positive eigenvalues of $\boldsymbol{U}^\top \boldsymbol{M}_{\mathcal{S}_1}\boldsymbol{U}$ cannot be too small with a rate associated with the weak signal strengths in $\mathcal{S}_2$. Condition B(iii) permits the ultra-high dimensional case such that the number of variables can grow with the sample size at almost an exponential rate. Condition B(iv) is the regularity condition of $\boldsymbol{X}_{\mathcal{S}_3^c} = (\boldsymbol{z}_1, \dots, \boldsymbol{z}_n)^\top$ to have asymptotic normality.

We now present some linearity results as $n \to \infty$ from Puri and Sen (1985).

(a)

$$n^{-\frac{1}{2}} \left\| \boldsymbol{L}_{n(2)}\left(\hat{\boldsymbol{\beta}}_{1n}^{\mathrm{RR}}, \mathbf{0}\right) - \boldsymbol{L}_{n(2)}\left(\boldsymbol{\beta}_{\mathcal{S}_1}^{*}, \mathbf{0}\right) + \gamma(\phi, f)\boldsymbol{C}_{n21}\left(\hat{\boldsymbol{\beta}}_{1n}^{\mathrm{RR}} - \boldsymbol{\beta}_{\mathcal{S}_1}^{*}\right) \right\| \xrightarrow{P} 0. \tag{10.6.1}$$

(b)

$$n^{-\frac{1}{2}} \left\| \boldsymbol{L}_{n(2)}\left(\hat{\boldsymbol{\beta}}_{1n}^{\mathrm{RR}}, \mathbf{0}\right) - \boldsymbol{L}_{n(2)}\left(\boldsymbol{\beta}_{\mathcal{S}_1}^{*}, \mathbf{0}\right) + \gamma(\phi, f)\boldsymbol{C}_{n11}\left(\hat{\boldsymbol{\beta}}_{1n}^{\mathrm{RR}} - \boldsymbol{\beta}_{\mathcal{S}_1}^{*}\right) \right\| \xrightarrow{P} 0. \tag{10.6.2}$$

(c)

$$n^{-\frac{1}{2}} \left\| \boldsymbol{L}_{n(2)}\left(\hat{\boldsymbol{\beta}}_{1n}^{\mathrm{RR}}, \mathbf{0}\right) - \left\{\boldsymbol{L}_{n(2)}\left(\boldsymbol{\beta}_{\mathcal{S}_1}^{*}, \mathbf{0}\right) - \boldsymbol{C}_{n21}\boldsymbol{C}_{n11}^{-1}\boldsymbol{L}_{n(1)}\left(\boldsymbol{\beta}_{\mathcal{S}_1}^{*}, \mathbf{0}\right)\right\} \right\| \xrightarrow{P} 0. \tag{10.6.3}$$

Then, under $\mathcal{H}_o : \boldsymbol{\beta}_{\mathcal{S}_2}^{*} = \mathbf{0}$, $\left(\boldsymbol{L}_{n(1)}(\mathbf{0},\mathbf{0})^\top, \boldsymbol{L}_{n(2)}(\mathbf{0},\mathbf{0})^\top\right)^\top$ has the same joint distribution as $\left(\boldsymbol{L}_{n(1)}(\mathbf{0},\mathbf{0})^\top, \boldsymbol{L}_{n(2)}(\mathbf{0},\mathbf{0})^\top\right)^\top$ under $\mathcal{H}^* : \boldsymbol{\beta}_{\mathcal{S}_1}^{*} = \mathbf{0}$ and $\boldsymbol{\beta}_{\mathcal{S}_2}^{*} = \mathbf{0}$. Hence,

$$n^{-\frac{1}{2}} \left(\boldsymbol{L}_{n(1)}(\mathbf{0},\mathbf{0})^\top, \boldsymbol{L}_{n(2)}(\mathbf{0},\mathbf{0})^\top\right)^\top \xrightarrow{\mathcal{D}} \mathcal{N}_{p_{1n}+p_{2n}}(0, \eta^2\boldsymbol{C}). \tag{10.6.4}$$

Thus, using Eqs. (10.6.2) and (10.6.3), we can conclude that

$$n^{-\frac{1}{2}} \boldsymbol{L}_{n(2)}\left(\hat{\boldsymbol{\beta}}_{1n}^{\mathrm{RR}}, \mathbf{0}\right) \xrightarrow{\mathcal{D}} \mathcal{N}_{p_{2n}}(0, \eta^2\boldsymbol{C}_{22\cdot 1}). \tag{10.6.5}$$

Under the set of local alternatives $\boldsymbol{\beta}_{\mathcal{S}_2}^{*} = n^{-\frac{1}{2}}\boldsymbol{\delta}_{\mathcal{S}_2}^{*}$ and the contiguity results, one can show that $n^{-\frac{1}{2}}\left(\boldsymbol{L}_{n(1)}(\boldsymbol{\beta}_{\mathcal{S}_1}^{*},\mathbf{0})^\top, \boldsymbol{L}_{n(2)}(\boldsymbol{\beta}_{\mathcal{S}_1}^{*},\mathbf{0})^\top\right)^\top$ has the same joint normal distribution as that of $n^{-\frac{1}{2}}\left(\boldsymbol{L}_{n(1)}(\mathbf{0},\mathbf{0})^\top, \boldsymbol{L}_{n(2)}(\mathbf{0},\mathbf{0})^\top\right)^\top$ under $\boldsymbol{\beta}_{\mathcal{S}_1}^{*} = \mathbf{0}$ and $\boldsymbol{\beta}_{\mathcal{S}_2}^{*} = n^{-1/2}\boldsymbol{\delta}_{\mathcal{S}_2}^{*}$ with mean vector and covariance matrix, respectively, given by

$$\gamma(\phi, f)\begin{pmatrix} \boldsymbol{C}_{11}\boldsymbol{\delta}_{\mathcal{S}_2}^{*} \\ \boldsymbol{C}_{22}\boldsymbol{\delta}_{\mathcal{S}_2}^{*} \end{pmatrix}, \quad \text{and} \quad \eta^2\boldsymbol{C}. \tag{10.6.6}$$

Therefore,

$$\gamma^{-1}(\phi, f)n^{-\frac{1}{2}}\boldsymbol{L}_{n(2)}(\hat{\boldsymbol{\beta}}_{1n}^{\mathrm{RR}}, \mathbf{0}) = \gamma^{-1}(\phi, f)n^{-\frac{1}{2}}\Big[\boldsymbol{L}_{n(2)}\left(\boldsymbol{\beta}_{\mathcal{S}_1}^{*}, \mathbf{0}\right) - \boldsymbol{C}_{n21}\boldsymbol{C}_{n11}^{-1}\boldsymbol{L}_{n(1)}\left(\boldsymbol{\beta}_{\mathcal{S}_1}^{*}, \mathbf{0}\right)\Big]$$

$$\xrightarrow{\mathcal{D}} \mathcal{N}_{p_{2n}}(\boldsymbol{C}_{22\cdot 1}\boldsymbol{\delta}^*_{\mathcal{S}_2}, \eta^2 \boldsymbol{C}_{22\cdot 1}). \tag{10.6.7}$$

We can now state the following theorem.

Theorem 10.1 Suppose the sparse model of Eq. (10.1.1) satisfies the signal strength assumptions A(i)–A(iii) and model assumptions B(i)–B(iii). If we choose

$$\gamma_n = c_1^{-2} c_2 n^{2\alpha} \left(\log(\log(n))\right)^3 \log(n \vee p_n)$$

for some $c_2 > 0$ with $\alpha < (\eta - \gamma - \tau)/3$, then $\hat{\mathcal{S}}_2$ defined by Eq. (10.4.8) satisfies

$$P(\hat{\mathcal{S}}_2 = \mathcal{S}_2 | \hat{\mathcal{S}}_1 = \mathcal{S}_1) \geq 1 - (n \vee p_n)^{-t} \quad \text{for some} \quad t > 0$$

where $0 < \gamma \leq \frac{1}{2}$, τ, η defined in A(ii), B(ii) and B(iii), respectively.

Proof: The proof is due to Shao and Deng (2012) and Gao et al. (2017). Assume n is sufficiently large such that $\log(\log(n)) > 0$. Let $\gamma_n = c_2 a_n^{-2} (\log(\log n))^3 \log(n \vee p_n)$, where $a_n = c_1 n^{-\alpha}$, $0 < \alpha \leq 1/2$, for some $c_1 > 0$. Then, for $u_n = 1 + \log(\log(n))$, we have

$$\frac{|\,\mathrm{ADB}(\hat{\boldsymbol{\beta}}_{jn}^{\mathrm{ridgeR}}(\gamma_n))|}{a_n \log(\log(n))} \leq \frac{\gamma_n n^{\tau - \eta}}{a_n \log(\log(n))} \leq \frac{c_2 (\log(\log(n)))^4}{a_n^3 n^{\eta - \tau - \nu}}$$
$$\leq \frac{c_2 (\log(\log(n)))^4}{c_1^3 n^{\eta - \tau - \nu - 3\alpha}} \to 0 \quad \text{if } \alpha < \frac{1}{3}(\eta - \nu - \tau).$$

From the normal distribution of the rank estimators, we have

$$\mathrm{Cov}(\sqrt{n}\hat{\boldsymbol{\beta}}_{\mathcal{S}_1^c}^{\mathrm{ridgeR}}(\gamma_n)) \leq \eta^2 k_n^{-1} \boldsymbol{I}_{q_n}.$$

Thus,

$$\forall j \notin \mathcal{S}_1, \qquad \mathrm{Var}(\sqrt{n}\hat{\boldsymbol{\beta}}_{jn}^{\mathrm{RR}}(\gamma_n)) = O(\gamma_n^{-1}).$$

Further, note that

$$\sqrt{\gamma_n} a_n (\log(\log(n))) = O(\log(\log(n))) \to \infty.$$

We then find that

$$\frac{a_n \log(\log(n))}{\sqrt{\mathrm{Var}(\hat{\beta}_j^{\mathrm{ridgeR}}(\gamma_n))}} \geq a_n (\log(\log(n))) \sqrt{\gamma_n} \to \infty$$

and for $c_0 > 0$,

$$
\begin{aligned}
&P\left(|\sqrt{n}(\hat{\boldsymbol{\beta}}_{S_1^c}^{\text{ridgeR}}(\gamma_n) - \boldsymbol{\beta}_{S_1^c}^{*})| \geq a_n(\log(\log(n)))\right) \\
&\leq P\left\{ |Z| > \frac{a_n(\log(\log(n)))}{\sqrt{\text{Var}(\hat{\boldsymbol{\beta}}_j^{\text{ridgeR}}(\gamma_n))}} - \frac{|\,\text{ADB}(\hat{\boldsymbol{\beta}}_j^{\text{ridgeR}}(\gamma_n))|}{\sqrt{\text{Var}(\hat{\boldsymbol{\beta}}_j^{\text{ridgeR}}(\gamma_n))}} \right\} \\
&= 2\Phi\left\{ \frac{|\,\text{ADB}(\hat{\boldsymbol{\beta}}_j^{\text{ridgeR}}(\gamma_n))| - a_n(\log(\log(n)))}{\sqrt{\text{Var}(\hat{\boldsymbol{\beta}}_j^{\text{ridgeR}}(\gamma_n))}} \right\} \\
&\leq 2\Phi\left(\frac{-c_0\sqrt{\gamma_n}a_n^2}{\log(\log(n))} \right) \\
&\leq \exp\left(-\frac{c_0^2\gamma_n a_n^2}{\log(\log(n))} \right)
\end{aligned}
$$

is the tail probability of the normal distribution. Thus, we have

$$
\begin{aligned}
&P\left(j \notin \mathcal{S}_1 : |\boldsymbol{\beta}_j^{*}| > a_n(\log(\log(n))) \subset \{ j \notin \mathcal{S}_1 : |\hat{\boldsymbol{\beta}}_j^{\text{ridgeR}}(\gamma_n)| > a_n \} \right) \\
&\geq 1 - P\left(\cup\{ |\hat{\boldsymbol{\beta}}_j^{\text{ridgeR}}(\gamma_n)| \leq a_n \} \right) \\
&\geq 1 - P\left(\cup\left\{ |\hat{\boldsymbol{\beta}}_j^{\text{ridgeR}}(\gamma_n) - \boldsymbol{\beta}_j^{*}| \leq a_n(\log(\log(n))) \right\} \right) \\
&\geq 1 - p_n \exp\left\{ -\frac{c_0^2\gamma_n a_n^2}{(\log(\log(n)))^2} \right\} \\
&\geq 1 - \exp\left\{ -\frac{c_0^2\gamma_n a_n^2}{(\log(\log(n)))^2} - \log(p_n) \right\} \\
&\geq 1 - \exp\{ -(c_0 \log(\log(n-1))) \log(n \vee p_n) \}.
\end{aligned}
$$

For a large enough n, there exists $c_0 \log(\log(n-1)) > t > 0$ for some t. Thus,

$$
\begin{aligned}
\lim_{n\to\infty} P&\left(j \notin \mathcal{S}_1 : |\boldsymbol{\beta}_j^{*}| > a_n(\log(\log(n))) \subset \{ j \notin \mathcal{S}_1 : |\hat{\boldsymbol{\beta}}_j^{\text{ridgeR}}(\gamma_n)| > a_n \} \right) \\
&\geq 1 - (n \vee p_n)^{-t} \to 1.
\end{aligned}
$$

Similarly,

$$\lim_{n\to\infty} P\left(j \notin \mathcal{S}_1 : |\boldsymbol{\beta}_j^*| > a_n(\log(\log(n))) \supset \{j \notin \mathcal{S}_1 : |\hat{\boldsymbol{\beta}}_j^{\text{ridgeR}}(\gamma_n)| > a_n\}\right)$$
$$\geq 1 - (n \vee p_n)^{-t} \to 1.$$

Because of the continuity of the ridge estimators and $\lim_{n\to\infty}((\log(\log(n)))^{-1}+1) = 1$, we have $\lim_{n\to\infty} P(\hat{\mathcal{S}}_2 = \mathcal{S}_2|\hat{\mathcal{S}}_1 = \mathcal{S}_1) = 1$. □

The above Theorem 10.1 is similar to Shao and Deng (2012) and Gao et al. (2017). It says that $\hat{\boldsymbol{\beta}}_{\mathcal{S}_1^c}^{\text{ridgeR}}(\gamma_n)$ is able to single out the sparse set $\mathcal{S}_3$, with high probability, if $\mathcal{S}_1$ is pre-selected in advance such that $P(\hat{\mathcal{S}}_1 = \mathcal{S}_1) = 1$. Note that $\mathcal{S}_1$ can be recovered with a high probability under sparse Riesz condition (SRC) with rank p_1. Here, the matrix $\boldsymbol{X}$ satisfies the SRC with rank q and spectrum bounds $0 < c_* < c^* < \infty$ of

$$c_* < \frac{||\boldsymbol{X}_{\mathcal{S}}\boldsymbol{\nu}||^2}{||\boldsymbol{\nu}||^2} < c^*, \quad \forall \mathcal{S} \quad \text{with} \quad |\mathcal{S}| = q \text{ and } \boldsymbol{\nu} \in \mathbb{R}^p. \tag{10.6.8}$$

The LASSO penalty satisfies this condition under B(i). Suppose A(i) and B(i) are satisfied, and the sparse condition A(ii) holds for some $0 < \eta_1 < O(p_1\sqrt{\frac{\log(p_n)}{n}})$ and $\boldsymbol{X}$ satisfies the SRC with rank p_1. Then $\hat{\mathcal{S}}_1$ generated from the PRD function with a LASSO penalty satisfies

$$\lim_{n\to\infty} P\left(\{\mathcal{S}_1 \subset \hat{\mathcal{S}}_1\} \cap \{\sum_{j\in\mathcal{S}_1} |\beta_j^*| I(\hat{\beta}_{jn}^{\text{PRD}} = 0) = 0\}\right) = \lim_{n\to\infty} P(\hat{\mathcal{S}}_1 = \mathcal{S}_1) = 1. \tag{10.6.9}$$

Corollary 10.1 If Lemma 10.1 and Theorem 10.1 conditions hold, then

$$\lim P\left((\hat{\mathcal{S}}_2 = \mathcal{S}_2) \cap (\hat{\mathcal{S}}_1 = \mathcal{S}_1)\right) = 1. \tag{10.6.10}$$

Corollary 10.1 still requires satisfying the SRC.

Corollary 10.2 If the condition of Theorem 10.1 holds, then

$$\lim_{n\to\infty} P\left(\hat{\mathcal{S}}_2 = \hat{\mathcal{S}}_1^c \cap \mathcal{S}_2\right) = 1. \tag{10.6.11}$$

Theorem 10.1 and its derivatives, Corollaries ?? and 10.2, are important for establishing the efficiency of ridge regression estimators, $\hat{\boldsymbol{\beta}}_{\mathcal{S}_1}^{\text{ridgeR}}(\gamma_n)$ and $\hat{\boldsymbol{\beta}}_{\mathcal{S}_2}^{\text{ridgeR}}(\gamma_n)$.

Theorem 10.2 Consider a sparse model with signal strength A(i)–A(iii) and A(ii) with $0 < \tau < 1/2$. Further, suppose a pre-selected model such as $\mathcal{S}_1 \subset \hat{\mathcal{S}}_1 \subset (\mathcal{S}_1 \cup \mathcal{S}_2)$ is obtained with probability one. If we choose γ_n as in Theorem 10.1 with $\alpha < \{(\eta-\gamma-\tau)/3, 1/4-\tau/2\}$, then we have the following asymptotic normality for the ridge R-estimators:

$$\sqrt{n}\left(\hat{\boldsymbol{\beta}}_{\mathcal{S}_3^c}^{\text{ridgeR}}(\gamma_n) - \boldsymbol{\beta}_{\mathcal{S}_3^c}^*\right) \xrightarrow{\mathcal{D}} \mathcal{N}_{p_{1n}+p_{2n}}(\mathbf{0}, \eta^2\boldsymbol{C}^{-1}). \tag{10.6.12}$$

Proof: In Section 10.4.4, we obtained under $\boldsymbol{\beta}_{\mathcal{S}_3^c}^* = \mathbf{0}$

$$\sqrt{n}(\hat{\boldsymbol{\beta}}_{\mathcal{S}_3^c}^{\text{ridgeR}}(\gamma_n) - \boldsymbol{\beta}_{\mathcal{S}_3^c}^*) = \gamma^{-1}(\phi, f)\boldsymbol{C}_n^{-1}\boldsymbol{L}_{n(2)}(\mathbf{0},\mathbf{0}) + o_p(1)$$

$$\sqrt{n}(\hat{\boldsymbol{\beta}}_{\mathcal{S}_2}^{\text{ridgeR}}(\gamma_n) - \boldsymbol{\beta}_{\mathcal{S}_2}^*) = \gamma^{-1}(\phi, f)\boldsymbol{C}_{n22\cdot1}^{-1}\boldsymbol{L}_{n(2)}(\hat{\boldsymbol{\beta}}_{1n}^{\text{RR}}, \mathbf{0}) + o_p(1).$$

Hence, Eq. (10.6.12) holds but the under local alternatives $\boldsymbol{\beta}_{\mathcal{S}_2}^* = n^{-1/2}\boldsymbol{\delta}_{\mathcal{S}_2}^*$. Note that using contiguity and linearity results, we have

$$\boldsymbol{L}_{n(2)}(\hat{\boldsymbol{\beta}}_{1n}^{\text{RR}}, \mathbf{0}) = \boldsymbol{L}_{n(2)}(\mathbf{0},\mathbf{0}) - \boldsymbol{C}_{n21}^{-1}\boldsymbol{C}_{n11}^{-1}\boldsymbol{L}_{n(1)}(\mathbf{0},\mathbf{0}) + o_p(1) \tag{10.6.13}$$

so that

$$\sqrt{n}\left(\hat{\boldsymbol{\beta}}_{\mathcal{S}_2}^{\text{ridgeR}}(\gamma_n) - n^{-\frac{1}{2}}\boldsymbol{\delta}_{\mathcal{S}_2}^*\right) \xrightarrow{\mathcal{D}} \mathcal{N}_{p_{2n}}(\boldsymbol{C}_{22\cdot1}\boldsymbol{\delta}_{\mathcal{S}_2}^*, \eta^2\boldsymbol{C}_{22\cdot1}^{-1}). \tag{10.6.14}$$

See Puri and Sen (1985) for more details. □

10.7 Asymptotic Distributional L$_2$-Risk Properties

In this section, we provide performance characteristics of the estimators using the asymptotic distributional L$_2$-risk (ADL$_2$-risk).

Let

$$K_{(n)} : \boldsymbol{\beta}_{\mathcal{S}_2}^* = n^{-\frac{1}{2}}\boldsymbol{\delta}_{\mathcal{S}_2}^*, \quad \boldsymbol{\delta}_{\mathcal{S}_2}^* = (\delta_{21}^*, \dots, \delta_{2p_{2n}}^*)^\top \tag{10.7.1}$$

be a sequence of local alternatives. Further, let $\mathcal{H}_\nu(x;\Delta^2)$ be the c.d.f of the non-central chi-squared distribution with ν d.f. and non-centrality parameter $\Delta^2/2$

for $x \geq 0$ and $\mathcal{G}_m(x;\boldsymbol{\mu},\boldsymbol{\Sigma})$ be the c.d.f of the m-variate normal distribution with mean vector $\boldsymbol{\mu}$ and covariance matrix $\boldsymbol{\Sigma}$. Then, we have the following results.

Theorem 10.3 Under $K_{(n)}$ and assumed regularity conditions for the p.d.f. $f(x)$ with the Fisher information $I(f)$

(a)

$$\lim_{n\to\infty} P\left(\mathcal{L}_n \leq x | K_{(n)}\right) = \mathcal{H}_{p_{2n}}(x;\Delta^2), \quad \Delta^2 = \eta^{-2}(\boldsymbol{\delta}_{\mathcal{S}_2}^{*}{}^{\top} \boldsymbol{C}_{22\cdot1}\boldsymbol{\delta}_{\mathcal{S}_2}^{*}).$$

(b)

$$\lim_{n\to\infty} P\left(\sqrt{n}(\hat{\boldsymbol{\beta}}_{1n}^{\mathrm{RR}} - \boldsymbol{\beta}_{\mathcal{S}_2}^{*}) \leq \boldsymbol{x} | K_{(n)}\right) = \mathcal{G}_{p_{1n}}(\boldsymbol{x} + \boldsymbol{C}_{11}^{-1}\boldsymbol{C}_{12}\boldsymbol{\delta}_{\mathcal{S}_2}^{*};\boldsymbol{0},\eta^2\boldsymbol{C}_{11}^{-1}).$$

(c)

$$\begin{aligned}
&\lim_{n\to\infty} P\left(\sqrt{n}(\hat{\boldsymbol{\beta}}_{\mathcal{S}_2}^{\mathrm{PTR}} - \boldsymbol{\beta}_{\mathcal{S}_2}^{*}) \leq \boldsymbol{x} | K_{(n)}\right) \\
&= \mathcal{H}_{p_{2n}}(\lambda^2;\Delta^2)\mathcal{G}_{p_{1n}}(\boldsymbol{x} + \boldsymbol{C}_{11}^{-1}\boldsymbol{C}_{12}\boldsymbol{\delta}_{\mathcal{S}_2}^{*};\boldsymbol{0},\eta^2\boldsymbol{C}_{11}^{-1}) \\
&+ \int_{E(\boldsymbol{\delta}_{\mathcal{S}_2}^{*})} \mathcal{G}_{p_{1n}}(\boldsymbol{x} - \boldsymbol{D}_{11}^{-1}\boldsymbol{D}_{12}\boldsymbol{Z};\boldsymbol{0},\eta^2\boldsymbol{D}_{11\cdot2}^{-1}) \\
&\qquad d\mathcal{G}_{p_{2n}}\left(\boldsymbol{x} - \boldsymbol{D}_{12}\boldsymbol{D}_{22}^{-1}\boldsymbol{Z};\boldsymbol{0},\eta^2\boldsymbol{D}_{22}\right)
\end{aligned}$$

where

$$E(\boldsymbol{\delta}_{\mathcal{S}_2}^{*}) = \{\boldsymbol{Z} : \eta^{-2}(\boldsymbol{Z} + \boldsymbol{\delta}_{\mathcal{S}_2}^{*})^{\top}\boldsymbol{C}_{22\cdot1}(\boldsymbol{Z} + \boldsymbol{\delta}_{\mathcal{S}_2}^{*}) \geq \lambda^2\} \tag{10.7.2}$$

and $\boldsymbol{D} = \boldsymbol{C}^{-1}$.

(d)

$$\sqrt{n}\left(\hat{\boldsymbol{\beta}}_{\mathcal{S}_1}^{\mathrm{ridgeR}}(\gamma_n) - \boldsymbol{\beta}_{\mathcal{S}_1}^{*}\right) \overset{\mathcal{D}}{=} \boldsymbol{D}_1\boldsymbol{U} + \frac{\eta^2\boldsymbol{C}_{11}^{-1}\boldsymbol{C}_{12}(\boldsymbol{D}\boldsymbol{U} + \boldsymbol{\delta}_{\mathcal{S}_2}^{*})}{(\boldsymbol{D}_2\boldsymbol{U} + \boldsymbol{\delta}_{\mathcal{S}_2})^{\top}\boldsymbol{C}_{22\cdot1}(\boldsymbol{D}_2\boldsymbol{U} + \boldsymbol{\delta}_{\mathcal{S}_2})} \tag{10.7.3}$$

where $\boldsymbol{U} \sim \mathcal{N}_{p_{1n}+p_{2n}}(\boldsymbol{0},\eta^2\boldsymbol{C})$ and $\boldsymbol{D} = \begin{pmatrix} \boldsymbol{D}_1 \\ \boldsymbol{D}_2 \end{pmatrix} = \boldsymbol{C}^{-1}$.

Proof: We consider the linearity results under $K_{(n)}$

$$\hat{\boldsymbol{\beta}}_{1n}^{\mathrm{RR}} = \hat{\boldsymbol{\beta}}_{\mathcal{S}_1}^{\mathrm{RR}}(\gamma_n) + \boldsymbol{C}_{n11}^{-1}\boldsymbol{C}_{n12}\hat{\boldsymbol{\beta}}_{\mathcal{S}_2}^{\mathrm{RR}}(\gamma_n) + o_p(1)$$

$$\mathcal{L}_n = \frac{1}{\eta^2}\left[\hat{\boldsymbol{\beta}}_{\mathcal{S}_2}^{\text{ridgeR}}(\gamma_n)\right]^\top \boldsymbol{C}_{n22\cdot 1}\left[\hat{\boldsymbol{\beta}}_{\mathcal{S}_2}^{\text{ridgeR}}(\gamma_n)\right] + o_p(1). \tag{10.7.4}$$

Further, under $K_{(n)}$

$$\left(\sqrt{n}(\hat{\boldsymbol{\beta}}_{\mathcal{S}_1}^{\text{ridgeR}}(\gamma_n) - \boldsymbol{\beta}_{\mathcal{S}_1}^*)^\top, (\hat{\boldsymbol{\beta}}_{\mathcal{S}_2}^{\text{ridgeR}}(\gamma_n) - n^{-\frac{1}{2}}\boldsymbol{\delta}_{\mathcal{S}_2}^*)^\top\right)^\top$$

$$= n^{-\frac{1}{2}}\gamma^{-1}(\phi, f)\boldsymbol{C}_n^{-1}\boldsymbol{L}_n(\boldsymbol{\beta}) + o_p(1) \quad \stackrel{\mathcal{D}}{\to} \quad \mathcal{N}_{p_{1n}+p_{2n}}(\mathbf{0}, \eta^2\boldsymbol{C}^{-1}).$$

Now, by definition, $\boldsymbol{D}_{11} = \boldsymbol{C}_{11}^{-1}\boldsymbol{C}_{12}\boldsymbol{D}_{21}$ and $\boldsymbol{D}_{12} + \boldsymbol{C}_{11}^{-1}\boldsymbol{C}_{12}\boldsymbol{D}_{22} = \mathbf{0}$ and we may write $\hat{\boldsymbol{\beta}}_{1n}^{\text{RR}} = \boldsymbol{L}_1\hat{\boldsymbol{\beta}}_{\mathcal{S}_1}^{\text{ridgeR}}(\gamma_n)$ and $\mathcal{L}_n = \left[\hat{\boldsymbol{\beta}}_{\mathcal{S}_3^c}^{\text{ridgeR}}(\gamma_n)\right]^\top \boldsymbol{L}_2\left[\hat{\boldsymbol{\beta}}_{\mathcal{S}_3^c}^{\text{ridgeR}}(\gamma_n)\right] + o_p(1)$, where $\boldsymbol{L}_1\boldsymbol{C}^{-1}\boldsymbol{L}_2 = \mathbf{0}$. Hence, using Eq. (10.7.4), we conclude $\sqrt{n}(\hat{\boldsymbol{\beta}}_{\mathcal{S}_1}^{\text{ridgeR}}(\gamma_n) - \boldsymbol{\beta}_{\mathcal{S}_1}^*)$ and $\mathcal{L}_n$ are asymptotically independent, while the joint distribution of $(\hat{\boldsymbol{\beta}}_{\mathcal{S}_2}^{\text{ridgeR}}(\gamma_n) - n^{-\frac{1}{2}}\boldsymbol{\delta}_{\mathcal{S}_2}^*)$ and $\mathcal{L}_n$ can be obtained by integrating over the appropriate subspace. □

Following Sen and Saleh (1979), Jurečková and Sen (1984), and Sing and Sen (1985) under the set of local alternatives, $K_{(n)}$, we have the following result.

Theorem 10.4 Under $K_{(n)}$ and assumed regularity conditions

(a) $\text{ADL}_2\text{-risk}(\hat{\boldsymbol{\beta}}_{\mathcal{S}_1}^{\text{ridgeR}}(\gamma_n)) = \eta^2 \operatorname{tr} \boldsymbol{C}_{11\cdot 2}^{-1}$.

(b) $\text{ADL}_2\text{-risk}(\hat{\boldsymbol{\beta}}_{1n}^{\text{RR}}) = \eta^2\left[\operatorname{tr} \boldsymbol{C}_{11}^{-1} + \boldsymbol{\delta}_{\mathcal{S}_2}^{*\top}\boldsymbol{M}\boldsymbol{\delta}_{\mathcal{S}_2}^*\right]$, $\boldsymbol{M} = \boldsymbol{C}_{21}\boldsymbol{C}_{11}^{-2}\boldsymbol{C}_{12}$.

(c)

$$\begin{aligned}\text{ADL}_2\text{-risk}(\hat{\boldsymbol{\beta}}_{\mathcal{S}_1}^{\text{PTR}}(\gamma_n)) = \eta^2\Big[& \operatorname{tr}(\boldsymbol{C}_{11\cdot 2}^{-1})\left(1 - \mathcal{H}_{p_{2n}+2}(\lambda^2; \Delta^2)\right) \\ & + \operatorname{tr}(\boldsymbol{C}_{11}^{-1})\mathcal{H}_{p_{2n}+2}(\lambda^2; \Delta^2) + \boldsymbol{\delta}_{\mathcal{S}_2}^{*\top}\boldsymbol{M}\boldsymbol{\delta}_{\mathcal{S}_2}^* \\ & \times \left(2\mathcal{H}_{p_{2n}+2}(\lambda^2, \Delta^2) - \mathcal{H}_{p_{2n}+4}(\lambda^2, \Delta^2)\right)\Big].\end{aligned}$$

(d)

$$\begin{aligned}\text{ADL}_2\text{-risk}(\hat{\boldsymbol{\beta}}_{\mathcal{S}_1}^{\text{SSR}}(\gamma_n)) = \eta^2\Big[& \operatorname{tr}(\boldsymbol{C}_{11\cdot 2}^{-1}) - d \operatorname{tr}(\boldsymbol{M}\boldsymbol{C}_{22\cdot 1}^{-1}) \\ & \times \left\{2\mathbb{E}\left[\chi_{p_{2n}+2}^{-2}(\Delta^2)\right] - d\mathbb{E}\left[\chi_{p_{2n}+2}^{-4}(\Delta^2)\right]\right\} \\ & + (p_{2n}^2 - 4)\boldsymbol{\delta}_{\mathcal{S}_2}^{*\top}\boldsymbol{M}\boldsymbol{\delta}_{\mathcal{S}_2}^*\mathbb{E}\left[\chi_{p_{2n}+4}^{-4}(\Delta^2)\right]\Big].\end{aligned}$$

(e)

$$\begin{aligned}\text{ADL}_2\text{-risk}(\hat{\boldsymbol{\beta}}_{\mathcal{S}_1}^{\text{SSR+}}(\gamma_n)) &= \text{ADL}_2\text{-risk}(\hat{\boldsymbol{\beta}}_{\mathcal{S}_1}^{\text{SSR}}(\gamma_n)) \\ &-\eta^2\Big[\text{tr}(\boldsymbol{C}_{22\cdot1}^{-1})\Big\{\mathbb{E}\Big[\Big(1-d\chi_{p_{2n}+2}^{-2}(\Delta^2)\Big)^2 I\left(\chi_{p_{2n}+2}^2(\Delta^2)<d\right)\Big]\Big\} \\ &+\boldsymbol{\delta}_{\mathcal{S}_2}^{*\top}\boldsymbol{M}\boldsymbol{\delta}_{\mathcal{S}_2}^{*}\Big\{2\mathbb{E}\Big[\Big(1-d\chi_{p_{2n}+2}^{-2}(\Delta^2)\Big)I\left(\chi_{p_{2n}+2}^2(\Delta^2)<d\right)\Big] \\ &-\mathbb{E}\Big[\Big(1-d\chi_{p_{2n}+4}^{-2}(\Delta^2)\Big)^2 I\left(\chi_{p_{2n}+4}^2(\Delta^2)<d\right)\Big]\Big\}\Big]\end{aligned}$$

where $d = s_2 - 2$ and $\mathbb{E}\left[\chi_q^{-2r}(\Delta^2)\right] = \int_0^\infty x^{-r} d\mathcal{H}_q(x;\Delta^2)$.

It will be shown that for all $\Delta^2 \in \mathbb{R}^+$

(i)

$$\text{ADL}_2\text{-risk}(\hat{\boldsymbol{\beta}}_{\mathcal{S}_1}^{\text{SSR+}}(\gamma_n)) \le \text{ADL}_2\text{-risk}(\hat{\boldsymbol{\beta}}_{\mathcal{S}_1}^{\text{SSR+}}(\gamma_n)) \le \text{ADL}_2\text{-risk}(\hat{\boldsymbol{\beta}}_{\mathcal{S}_1}^{\text{ridgeR}}(\gamma_n)). \tag{10.7.5}$$

(ii) Under $\Delta^2 = 0$,

$$\text{ADL}_2\text{-risk}(\hat{\boldsymbol{\beta}}_{1n}^{\text{RR}}(\gamma_n)) \le \text{ADL}_2\text{-risk}(\hat{\boldsymbol{\beta}}_{\mathcal{S}_1}^{\text{SSR+}}(\gamma_n)) \le \text{ADL}_2\text{-risk}(\hat{\boldsymbol{\beta}}_{\mathcal{S}_1}^{\text{ridgeR}}(\gamma_n)) \tag{10.7.6}$$

where = holds for $s_2 \to \infty$.

10.8 Asymptotic Distributional Risk Efficiency

In this section, we briefly elaborate on the asymptotic distributional risk efficiency (ADRE). Note that if $\boldsymbol{C}_{12} = \boldsymbol{0}$, we have $\boldsymbol{M} = \boldsymbol{0}$ and $\boldsymbol{C}_{11\cdot2} = \boldsymbol{C}_{11}$ and hence, all the ADL$_2$-risk$_2$-risk expressions reduce to the common value $\eta^2 \,\text{tr}\,\boldsymbol{C}_{11}^{-1}$ for all $\Delta^2 \in \mathbb{R}^+$. So, we assume $\boldsymbol{C}_{12} \neq \boldsymbol{0}$. Then, let $\boldsymbol{M}^* = \boldsymbol{C}_{21}\boldsymbol{C}_{11}^{-1}\boldsymbol{C}_{11\cdot2}\boldsymbol{C}_{11}^{-1}\boldsymbol{C}_{12}$ and $\boldsymbol{M}^o = \boldsymbol{C}_{12}\boldsymbol{C}_{22}^{-1}\boldsymbol{C}_{21}\boldsymbol{C}_{11}^{-1}$. Then, we have the following theorem.

Theorem 10.5 Under the conditions of Theorem 10.3, we have

(a)

$$\text{ADL}_2\text{-risk}(\hat{\boldsymbol{\beta}}_{\mathcal{S}_1}^{\text{ridgeR}}(\gamma_n)) \gtreqless \text{ADL}_2\text{-risk}(\hat{\boldsymbol{\beta}}_{1n}^{\text{RR}})$$

accordingly as $\eta^2(\boldsymbol{\delta}_{\mathcal{S}_2}^{*\top}\boldsymbol{M}^*\boldsymbol{\delta}_{\mathcal{S}_2}^{*}) \lesseqgtr \text{tr}(\boldsymbol{M}^o)$.

(b)

$$\text{ADL}_2\text{-risk}(\hat{\boldsymbol{\beta}}_{\mathcal{S}_1}^{\text{ridgeR}}(\gamma_n)) \gtreqless \text{ADL}_2\text{-risk}(\hat{\boldsymbol{\beta}}_{\mathcal{S}_1}^{\text{PTR}}(\gamma_n))$$

accordingly as

$$\boldsymbol{\delta}_{\mathcal{S}_2}^{*\top} \boldsymbol{M}^* \boldsymbol{\delta}_{\mathcal{S}_2}^* \lesseqgtr \frac{\eta^2 \operatorname{tr}(\boldsymbol{M}^o)\mathcal{H}_{p_{2n}+2}(\lambda^2;\Delta^2)}{\left(2\mathcal{H}_{p_{2n}+2}(\lambda^2;\Delta^2) - \mathcal{H}_{p_{2n}+4}(\lambda^2;\Delta^2)\right)}. \tag{10.8.1}$$

(c) If $p_{2n} - 2 \geq 3$, then

$$\text{ADL}_2\text{-risk}(\hat{\boldsymbol{\beta}}_{\mathcal{S}_1}^{\text{SSR}}(\gamma_n)) \leq \text{ADL}_2\text{-risk}(\hat{\boldsymbol{\beta}}_{\mathcal{S}_1}^{\text{ridgeR}}(\gamma_n)), \quad \forall\, \Delta^2 \in \mathbb{R}^+$$

when

$$\frac{\operatorname{tr}(\boldsymbol{M}^*\boldsymbol{C}_{22\cdot1}^{-1})}{\text{Ch}_{\max}(\boldsymbol{M}^*\boldsymbol{C}_{22\cdot1}^{-1})} \geq \frac{p_{2n}+2}{2}.$$

Proof: It is easy to prove (a) and (b), which are left as exercises for the reader. To prove (c), we can write

$$\begin{aligned}
&\text{ADL}_2\text{-risk}(\hat{\boldsymbol{\beta}}_{\mathcal{S}_1}^{\text{SSR}}(\gamma_n)) = \text{ADL}_2\text{-risk}(\hat{\boldsymbol{\beta}}_{\mathcal{S}_1}^{\text{ridgeR}}(\gamma_n)) \\
&- (p_{2n}-2)\eta^2 \operatorname{tr}(\boldsymbol{M}^*\boldsymbol{C}_{22\cdot1}^{-1})\Big\{(p_{2n}-2)\mathbb{E}\left[\chi_{p_{2n}+2}^{-4}(\Delta^2)\right] \\
&\qquad + 2\left[1 - \frac{(p_{2n}+2)\boldsymbol{\delta}_{\mathcal{S}_2}^{*\top}\boldsymbol{M}^*\boldsymbol{\delta}_{\mathcal{S}_2}^*\eta^2}{2\Delta^2 \operatorname{tr}(\boldsymbol{M}^*\boldsymbol{C}_{22\cdot1}^{-1})}\right]\Delta^2\mathbb{E}\left[\chi_{p_{2n}+4}^{-4}(\Delta^2)\right]\Big\}.
\end{aligned}$$

From the above, it follows that

$$\forall \Delta^2 \in \mathbb{R}^+, \qquad \text{ADL}_2\text{-risk}(\hat{\boldsymbol{\beta}}_{\mathcal{S}_1}^{\text{SSR}}(\gamma_n)) \leq \text{ADL}_2\text{-risk}(\hat{\boldsymbol{\beta}}_{\mathcal{S}_1}^{\text{ridgeR}}(\gamma_n))$$

provided

$$\frac{\operatorname{tr}(\boldsymbol{M}^*\boldsymbol{C}_{22\cdot1}^{-1})}{\text{Ch}_{\max}(\boldsymbol{M}^*\boldsymbol{C}_{22\cdot1}^{-1})} \geq \frac{p_{2n}+2}{2},$$

where we used

$$\Delta^2\text{Ch}_{\min}(\boldsymbol{M}^*\boldsymbol{C}_{22\cdot1}^{-1}) \leq \Delta^2 \frac{\boldsymbol{\delta}_{\mathcal{S}_2}^{*\top}\boldsymbol{M}^*\boldsymbol{\delta}_{\mathcal{S}_2}^*}{\boldsymbol{\delta}_{\mathcal{S}_2}^{*\top}\boldsymbol{C}_{22\cdot1}\boldsymbol{\delta}_{\mathcal{S}_2}^*} \leq \Delta^2\text{Ch}_{\max}(\boldsymbol{M}^*\boldsymbol{C}_{22\cdot1}^{-1}).$$

□

The next theorem gives a sufficient condition for $p_{2n} - 2 \geq 3$ for which $\hat{\boldsymbol{\beta}}_{\mathcal{S}_1}^{\text{SSR}}(\gamma_n)$ dominates $\hat{\boldsymbol{\beta}}_{\mathcal{S}_1}^{\text{ridgeR}}(\gamma_n)$.

Theorem 10.6 A sufficient condition for $\hat{\boldsymbol{\beta}}_{\mathcal{S}_1}^{\text{SSR}}(\gamma_n)$ to dominate $\hat{\boldsymbol{\beta}}_{\mathcal{S}_1}^{\text{ridgeR}}(\gamma_n)$, in the ADL$_2$-risk sense, is that $d > 0$, and that the following inequality be satisfied,

$$2\mathbb{E}\left[\chi_{p_{2n}+2}^{-2}(\Delta^2)\right] - d\mathbb{E}\left[\chi_{p_{2n}+2}^{-4}(\Delta^2)\right] + (d+4)h\Delta^2\mathbb{E}\left[\chi_{p_{2n}+4}^{-4}(\Delta^2)\right] \geq 0,$$

for all $\Delta^2 \in \mathbb{R}^+$, which then requires that $d \geq 3$, $h(d+4) \leq 2$, where $h = \text{Ch}_{\max}(\boldsymbol{M}^o)/\operatorname{tr}(\boldsymbol{M}^o)$, so that $0 < h \leq 1$.

For the proof, refer to Sen and Saleh (1987).

Similarly, we have the following theorem.

Theorem 10.7 Under the condition of Theorem 10.3, $\hat{\boldsymbol{\beta}}_{\mathcal{S}_1}^{\text{PTR}}(\gamma_n)$ fails to dominate $\hat{\boldsymbol{\beta}}_{\mathcal{S}_1}^{\text{SSR}}(\gamma_n)$. Also, if for α, the level of significance of the preliminary test ridge R-estimator, we have

$$\mathcal{H}_{p_{2n}+2}(\lambda^2, 0) \geq \frac{q(2(p_{2n}-2)-q)}{p_{2n}(p_{2n}-2)}, \tag{10.8.2}$$

where $q = (p_{2n}-2) \wedge (2/h - 4)$, then $\hat{\boldsymbol{\beta}}_{\mathcal{S}_1}^{\text{ridgeRS}}(\gamma_n)$ fails to dominate $\hat{\boldsymbol{\beta}}_{\mathcal{S}_1}^{\text{PTR}}(\gamma_n)$.

For the proof, refer to Sen and Saleh (1987).

10.9 Summary

To summarize this chapter, we extended the post-selection robust estimation strategy in the high-dimensional sparse regression model using rank-based estimation. Under the separability condition, we defined the preliminary test, Stein–Saleh-type and positive-rule Stein–Saleh-type R-estimators using the post-selection ridge R-estimator, when p_n grows with n.

We derived asymptotic distributions of the post-selection ridge estimators under the contiguity and linearity conditions. Furthermore, we analytically compared the proposed estimators using the asymptotic distributional risk efficiency.

To close this chapter, one may verify that under $\mathcal{H}_o : \boldsymbol{\beta}_{\mathcal{S}_2}^* = \mathbf{0}$

$$\text{ADL}_2\text{-risk}(\hat{\boldsymbol{\beta}}_{1n}^{\text{RR}}) \leq \text{ADL}_2\text{-risk}(\hat{\boldsymbol{\beta}}_{\mathcal{S}_1}^{\text{PTR}}(\gamma_n)) \leq$$

$$\text{ADL}_2\text{-risk}(\hat{\boldsymbol{\beta}}_{\mathcal{S}_1}^{\text{SSR+}}(\gamma_n)) \leq \text{ADL}_2\text{-risk}(\hat{\boldsymbol{\beta}}_{\mathcal{S}_1}^{\text{SSR}}(\gamma_n)) \leq \text{ADL}_2\text{-risk}(\hat{\boldsymbol{\beta}}_{\mathcal{S}_1}^{\text{ridgeR}}(\gamma_n)).$$

Thus, in this chapter, we extended the rank-based theory in Chapter 6 to high-dimensions.

10.10 Problems

10.1. Prove Theorem 10.6.

10.2. Prove Theorem 10.7.

10.3. Prove (a) and (b) in Theorem 10.5.

10.4. Prove for any deterministic design matrix $\boldsymbol{X} : n \times p$, with any n and p, we have

$$(\boldsymbol{X}^\top \boldsymbol{X} + \gamma_n \boldsymbol{I}_p)^{-1} \boldsymbol{X}^\top = \boldsymbol{X}^\top (\boldsymbol{X}\boldsymbol{X}^\top + \gamma_n \boldsymbol{I}_n)^{-1},$$

where $\gamma_n > 0$ is fixed.

10.5. Define the ridge R-estimator of $\boldsymbol{\theta}$ in Eq. (10.2.2) as the minimizer of the following penalized PRD function

$$Q_n(\boldsymbol{\theta}) = D_n(\boldsymbol{\theta}) + \gamma_n \boldsymbol{\theta}^\top \boldsymbol{\theta},$$

where

$$D_n(\boldsymbol{\theta}) = \sum_{i=1}^{n} \left(y_i - \sum_{j=1}^{p} x_{ij}\theta_j \right) a_n(R_{n_i}(\boldsymbol{\theta})). \tag{10.10.1}$$

Then, obtain its asymptotic distributional properties.

10.6. Prove the positive-rule Stein–Saleh-type R-estimator is uniformly superior to the Stein–Saleh-type R-estimator in the asymptotic distributional L_2-risk sense.

11

Rank-based Logistic Regression

11.1 Introduction

In this chapter, we switch gears again and return to the use of rank-based methods in data science and machine learning, especially in binary classification problems. Some readers may not be familiar with data science and machine learning so we will begin with a short description of each in the context of robust methods. This will provide the background for material to be presented later. Then, we examine rank-based logistic regression methods and demonstrate certain untapped capabilities that will be useful in data science and machine learning. Next, we describe pertinent implementation details of rank-based logistic regression and explore its performance characteristics using different data sets from R and Python software environments. Finally, we compare and contrast rank-based logistic regression with standard and best-in-class methods to show the advantages of the new rank-based approach, along with some inherent limitations. Overall, the goal of this chapter is to provide the reader with a broader perspective on the importance of rank-based methods and potential future research directions.

11.2 Data Science and Machine Learning

11.2.1 What is Robust Data Science?

Data science is a field that focuses on creating a coherent data set from a large number of observations, performing a thorough exploratory data analysis and then carefully interpreting results obtained using machine learning and data mining methods. It is a rather expansive field with many layers of depth and a

Rank-Based Methods for Shrinkage and Selection: With Application to Machine Learning.
First Edition. A. K. Md. Ehsanes Saleh, Mohammad Arashi, Resve A. Saleh, and
Mina Norouzirad.

wide variety of applications. We will briefly describe some of these layers and emphasize the key aspects of this discipline.

What may be generally under-emphasized in data science is that there are usually outliers lurking around in the data set that could produce misleading results and, by extension, misleading conclusions about the data. There are potential outliers in every real-world data set that can arise from human error, measurement error, transmission error or it may simply be inherent in the nature of the data being collected. In this and the next chapter, we will show instances of various outliers and how to process them using rank-based methods for robust data science applications. The key to robust data science is to focus on identifying outliers, utilizing data cleaning methods to remove unwanted outliers and applying methods that are largely insensitive to such outliers. We believe that the field will naturally migrate in this direction, given the fundamentals and benefits of the methods to be presented. But before we get there, it is useful to summarize the goals of data science today.

Many of the techniques and analysis methods described in this book assume that the design matrix, $\boldsymbol{X}$, and response vector, $\boldsymbol{y}$, are readily available in some final form. In practice, this is very far from the truth. Before analysis can begin, there is a significant amount of time working with the data itself, perhaps more than 80% of the total effort. Data does not arrive in a complete and coherent form ready to be processed. Someone has to evaluate the data and mine its contents for a long time to build training and test sets suitable for analysis and model building. This is the role of the data scientist.

Data science is the art and science of converting raw data into a form that can be used by machine learning tools to predict outcomes, validate hypotheses and find relationships that were not previously known. Data science, as the name implies, is a data-centric field. Not considering the effort required to decide what raw data to collect and then collecting the data itself, the most time-consuming and intensive task for the data scientist is to build clean data sets from raw data. To begin with, the raw data may not be in one place, in one format, and may be incomplete. The data coming from multiple sources must be combined, unless the data is so massive that it has to be stored and retrieved in an efficient manner (i.e. big data issues). For example, the data may not fit into the computer RAM and must be incrementally read and processed as needed. The details of this issue are beyond the scope of this book. More generally, the first step in data science involves the collection of the needed data from different sources, conversion to the desired format and generation of a coherent and consistent data sets on which to operate.

Data is usually constructed in tabular form, called a *data frame*, and contains either numbers or strings of characters. Rows in a data frame represent observations while columns represent values for each variable (i.e. feature). Strings can be converted to categorical data, if appropriate. For example, the name of a customer may not be useful for analysis, so it may be deleted, but a specific named product purchased by that customer may be extremely important, so it would be declared as categorical data and represented as an integer from 1 to k, where k is the number of different products. The sex of the customer is either male or female, and given its binary nature, this categorical data are usually represented numerically as 0 and 1. Also, if a product purchased for the first time is entered as either yes or no, then this would also qualify as binary categorical data and represented as 0 and 1. In fact, any quantity where there is a finite number of items represented as strings in the table will likely be deemed categorical data and represented as integers.

Numeric data can be variables such as product size and weight, age of the customer, date/time of purchase, or other such quantities, or it could be measurement data such as number of times the product has been purchased in a year or the number of times such products have been searched online by the customer, to name a few. The list of possibilities is rather large. Generally speaking, numeric data is either continuous or discrete and is typically represented as real numbers or integers.

Processing the data may involve removal of errors, scaling/centering certain columns in the table, removing columns of unrelated data, creating new columns by combining existing columns, etc. A common problem encountered by a data scientist is that of missing data entries in one or more columns. These entries are often represented as "NaN" (not a number) or "NA" (not available). The data scientist must decide what to do in those circumstances: replace a piece of missing data, or delete rows with missing entries, or delete columns that have many missing entries. Deleting data may introduce bias in the results. A better approach is to conduct item imputation, perhaps by using distribution sampling to generate "fillin" data for missing entries. For example, the mean, median or mode could be used on different columns. Or the mean and standard deviation of a column could to used as a basis to randomly insert values from the associated normal distribution. The decision as to the best approach to take is based on the experience and expertise of the data scientist. This whole process of removing unneeded data columns and the imputation of missing values is often referred to as data cleaning.

A data scientist spends a lot of time engaged in feature engineering by identifying potential explanatory variables, then aggregating data columns and creating new columns by arithmetic operations on existing columns. The data may be filtered and combined (sometimes called data wrangling) to extract the desired information for analysis and then further validated, through visualizations, plots, cross-tabulation, bounds checking, other sanity checks, and possibly sub-sampling, to ensure that it is trustworthy enough for the intended purposes. Ultimately all useful data must be converted to numerical values since a computer will be used to process the data.

At various stages throughout the process, the data scientist engages in exploratory data analysis whereby histograms, box-plots, 3D plots and other visualization tools are used to understand the nature of the data and inherent relationships that may exist within the data. Essentially the data scientist lives with the data for a long period of time until it is well understood, and the methods needed to build models for the data begin to emerge. All of this is performed with an eye on the types of machine learning methods that would be suitable to analyze the data and to build models to carry out inference. Today, machine learning is typically performed using the R or Python software environments. For example, Python is often executed in Jupyter notebooks using libraries such as numpy, pandas, and matplotlib, to name a few. Machine learning libraries of Scipy and SkLearn contain many packages that may be utilized with little or no effort. Similarly, R contains many packages designed by statisticians to facilitate data science activities. Hence, data scientists must understand machine learning, as well as being good programmers with a strong probability and statistics background, in order to be successful.

A data scientist spends about 80–95% of the time doing tasks just described and the last 5–20% running clean data sets (divided into training and test sets) on machine learning tools and then evaluating the quality of the models until a satisfactory inference engine is built. Unfortunately, when faced with a binary classification problem, not much work was done in the past to address outliers. It is difficult to identify and remove outliers (opposite of missing data problem) since the notion of an outlier is not as easy to define in this context. In this chapter, we will illustrate how outliers may arise in classification problems and explore ways to detect them so that the data scientist can decide what to do about them.

11.2.2 What is Robust Machine Learning?

Machine learning (ML) involves the development of software based on algorithms developed by statisticians, computer scientists, and engineers to analyze the data. Typically, these tools are implemented by programmers in R or Python to process data sets that are produced using the steps described in the previous section. Statistical learning can be viewed as a subset of ML since many ML algorithms are rooted in statistics. However, there are also many ML algorithms developed outside of a statistical framework. The techniques described in this chapter belong to the statistical learning subset – those which, at the core, are based on a regression methodology. Most of the existing techniques use maximum likelihood estimation which is essentially based on the mean of the data. The rank-based methods to be introduced are related to the median of the data. Robust ML is one that uses the median (or other quantiles) rather than the mean which will produce different estimates than the methods using the mean. These two approaches will be compared and contrasted using a variety of data sets. More generally, ML grew out of the computer science field in the 1980s, went dormant for a time in the 1990s, but experienced a reawakening around 2000 and a significant resurgence around 2010, and continues to gain momentum to this day. Of course, in the 1980s, computers were relatively slow, data sets were not readily available, and algorithms for ML were in their infancy. Fast forward a few decades and the landscape changes dramatically. Nowadays, access to high-speed computing is at everyone's fingertips, data is ubiquitous and many advanced ML algorithms have been implemented, tested and made available through open-access online mechanisms.

The algorithms in ML generally fall into four main categories: supervised learning, unsupervised learning, semi-supervised learning and reinforcement learning. However, we focus only on the supervised learning methods in the remainder of this book. To contrast the first two categories, supervised ML requires both $\boldsymbol{X}$ and $\boldsymbol{y}$ while unsupervised ML requires only $\boldsymbol{X}$. For example, linear regression, logistic regression, neural networks, kNN classification, tree-based methods, and so on, fall into the supervised category. On the other hand, k-means clustering, anomaly detection, and principal component analysis are examples of unsupervised learning methods. Again, some of these algorithms are grounded in statistical theory while others are based on heuristic methods. In this chapter, we focus on a supervised ML algorithm for logistic regression using robust rank-based methods.

In particular, we describe the equations required to implement rank-based logistic regression (RLR) and then process various data sets to illustrate how they would be used. We also compare rank-based methods with log-likelihood logistic regression (LLR) and a popular method called support vector machines (SVM). Some details of SVMs are included in this chapter along with results obtained from SVM packages in order to demonstrate the value of the rank-based approach over what may be considered the state-of-the-art in binary classification. The LLR and RLR software used in this chapter were developed from the same code base in order to facilitate a fair comparison. However, we also present results from open Python software libraries, such as Sklearn and Scipy, that contain many machine learning tools and capabilities. Interested readers will find all the necessary documentation online.

11.3 Logistic Regression

Binary classification is a machine learning problem whereby a given set of data falls into one of two possible groups with the goal of finding a suitable boundary, either linear or nonlinear, that acts to separate members of one group from the other. For example, let's say a bank is giving out loans and wants to classify each potential customer as either worthy of getting a loan or not, based on a set of data associated with existing customers. Our target group would be those who received loans and paid them back, and the alternate group would be comprised of those who received loans and did not pay back. This is a classic binary classification problem since we want to identify the characteristics of the customers who tend to pay back loans as the criteria for giving out future loans.

Methods such as the perceptron algorithm, k-nearest neighbours (kNN) classification, random forest, logistic regression and support vector machines, to name the most prominent methods, have been developed to address this problem. Typically one group is labeled with 1 as being the target group while the other is labeled 0. For some methods, +1 and −1 may be more suitable labels. Either way, the objective is to find the separation boundary that allows for accurate prediction of the group to which any new data belongs.

As yet, binary classification problems have not been addressed using rank-based methods, and it is not clear how the approach would perform in this machine learning task. We will examine log-likelihood logistic regression and a new rank-based logistic regression in this chapter. The notion of an outlier in the context of binary classification will be established and methods to detect such outliers will be described using the rank-based approach. There are also some indications that it could establish the maximum accuracy of logistic regression on a training set. As such, it serves as another step towards robust data science and machine learning.

11.3.1 Log-likelihood Setup

Logistic regression is well-suited to binary classification problems and probability prediction (Hosmer and Lemeshow, 1989). We begin with a short description of basic logistic regression. We first define $Y_i \in \{0, 1\}$ as the dichotomous dependent variable and $\boldsymbol{x}_i = (1, x_{i1}, x_{i2}, , x_{ip})^\top$ as a $(p + 1)$-dimensional vector of explanatory variables for the ith observation in a sample set, where $i \in \{1, ..., n\}$. Then, the conditional probability of $Y_i = 1$ given $\boldsymbol{x}_i$ is

$$P(Y_i = 1|\boldsymbol{x}_i) = p(\boldsymbol{x}_i) = \left[\frac{1}{1 + \exp(-\boldsymbol{x}_i^\top \boldsymbol{\beta})}\right], \tag{11.3.1}$$

where $\boldsymbol{\beta} = (\beta_0, \beta_1, \dots, \beta_p)^\top$ is the length p vector of regression parameters of interest, with intercept β_0. Rearranging the above, the logistic regression problem is given by

Figure 11.1 Sigmoid function.

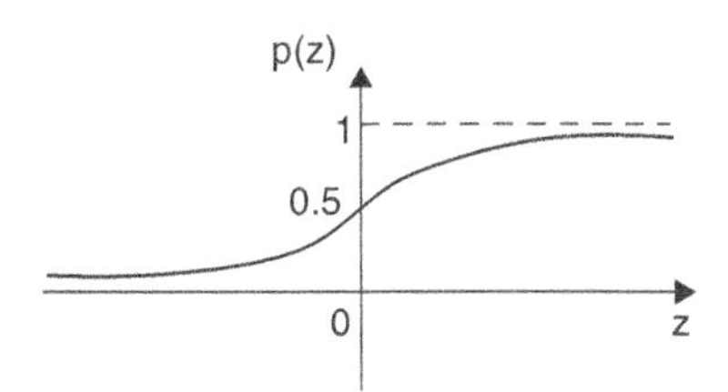

$$\log \frac{p(\boldsymbol{x}_i)}{1 - p(\boldsymbol{x}_i)} = \beta_0 + x_{i1}\beta_1 + \dots + x_{ip}\beta_p. \tag{11.3.2}$$

The left-hand side is referred to as the logit link function, z, also known as the logarithm of the odds,

$$z = \log \frac{p(\boldsymbol{x}_i)}{1 - p(\boldsymbol{x}_i)}. \tag{11.3.3}$$

Suppose we find estimates $\hat{\boldsymbol{\beta}}$ using a suitable estimator and want to predict the probability, $\hat{p}(\boldsymbol{x})$, of some new $\boldsymbol{x}$ that it has a response of $Y = 1$. This can be done in two steps. We first calculate z using

$$\hat{z} = \boldsymbol{x}^\top \hat{\boldsymbol{\beta}} = \hat{\beta}_0 + \hat{\beta}_1 x_1 \dots, \hat{\beta}_p x_p. \tag{11.3.4}$$

Second, we predict the probability from the sigmoid function

$$\hat{p}(\boldsymbol{x}) = \frac{1}{1 + \exp(-\boldsymbol{x}^\top \hat{\boldsymbol{\beta}})} = \frac{1}{1 + \exp(-\hat{z})} = p(\hat{z}). \tag{11.3.5}$$

It is instructive to examine the sigmoid function in more detail. A graph of the function $p(z)$ is shown in Figure 11.1. Note that for large positive values of z, the probability asymptotically approaches 1, whereas for large negative values it asymptotically approaches 0. The ground truth (i.e. true value) of the dependent variable y is either 0 or 1 for any given $\boldsymbol{x}$. However, the predicted value $\hat{p}(\boldsymbol{x})$ lies between 0 and 1 but does not reach either value for reasons that should be clear from Eq. (11.3.5). For binary classification, the predicted value, $\hat{p}(\boldsymbol{x})$, is compared to 0.5. If $\hat{p}(\boldsymbol{x}) > 0.5$, we predict 1, whereas if $\hat{p}(\boldsymbol{x}) < 0.5$, we predict 0.

Alternatively, we could base the decision of 1 or 0 on a sign test using $\hat{z} = \boldsymbol{x}^\top \hat{\boldsymbol{\beta}}$. Conceptually, z represents a line or hyperplane that separates 0s from 1s. We note that $\hat{\beta}_0$ functions as a bias term in this expression. In other words, if $\boldsymbol{x} = 0$ then either $\hat{\beta}_0 > 0$ or $\hat{\beta}_0 < 0$ so there is a built-in bias due to $\hat{\beta}_0$. It effectively shifts the p-dimensional hyperplane in one direction or the other by an amount equal to the bias term. The rest of the coefficients, $\hat{\beta}_1, \dots, \hat{\beta}_p$, establish the actual hyperplane. The linear boundary in p dimensions (i.e. the hyperplane boundary) is positioned where

$$\hat{\beta}_0 + \hat{\beta}_1 x_1 + \cdots + \hat{\beta}_p x_p = 0. \tag{11.3.6}$$

Therefore, we would say we have a predicted response of 1 if

$$\hat{\beta}_0 + \hat{\beta}_1 x_1 + \cdots + \hat{\beta}_p x_p > 0, \tag{11.3.7}$$

and a predicted response of 0 if

$$\hat{\beta}_0 + \hat{\beta}_1 x_1 + \cdots + \hat{\beta}_p x_p \leq 0. \tag{11.3.8}$$

This can also be understood by examining values of z in Figure 11.1. If z is positive, we can declare it as 1 whereas if z is negative, we declare it as 0. To summarize, we can either use $\hat{z}$ or $p(\hat{z})$ to decide the binary response of a given $\boldsymbol{x}$.

In order to determine the estimates, we use maximum likelihood estimation which can be formulated in a number of ways. We will find it convenient to use

$$\mathcal{L}(\boldsymbol{y}, \boldsymbol{p}) = \prod_{i=1}^{n} (p_i)^{y_i} (1 - p_i)^{(1-y_i)}, \tag{11.3.9}$$

where $p_i = p(\boldsymbol{x}_i)$ is the ith element of $\boldsymbol{p} = (p(\boldsymbol{x}_1), \ldots, p(\boldsymbol{x}_n))^\top$, and y_i is the ith element of $\boldsymbol{Y} = (y_1, \ldots, y_n)^\top$. It is more common to minimize the negative log-likelihood (LL) form of the loss function to obtain the estimates, $\hat{\boldsymbol{\beta}}$, as follows

$$J(\boldsymbol{\beta}) = J(\boldsymbol{y}, \boldsymbol{p}) = -\sum_{i=1}^{n} [y_i \log p_i + (1 - y_i) \log(1 - p_i)]. \tag{11.3.10}$$

For one observation, $\boldsymbol{x}$, the loss function is simply

$$J(y, p(\boldsymbol{x})) = -y \log p(\boldsymbol{x}) - (1 - y) \log(1 - p(\boldsymbol{x})). \tag{11.3.11}$$

This function can be interpreted intuitively by noting that y can only be 0 or 1. If $y = 0$, the first term can be ignored and the loss is given by the second term, which simplifies to minimizing $\log(1 - p(\boldsymbol{x}))$. If $y = 1$, then the second term can be ignored, which simplifies to minimizing $\log(p(\boldsymbol{x}))$. Therefore, the function in Eq. (11.3.10) has the desired characteristics for logistic regression. More importantly, $J(\boldsymbol{\beta})$ is a convex function in terms of $\boldsymbol{\beta}$ which facilitates convergence to the global minimum. Hence, the log-likelihood logistic regression (LLR) estimator is given by

$$\hat{\boldsymbol{\beta}}_n^{\mathrm{LLR}} = \underset{\boldsymbol{\beta}}{\operatorname{argmin}} \left\{ -\sum_{i=1}^{n} [y_i \log p_i + (1 - y_i) \log(1 - p_i)] \right\}.$$

However, as we shall see shortly, using this loss function has certain limitations that may affect the prediction accuracy of the estimates.

The numerical optimization technique used in logistic regression is usually some form of gradient descent. We will describe only the simplest form for our purposes here. Basically, we compute the loss function using an initial guess, $\boldsymbol{\beta}^{(0)}$, and then take a step in the gradient direction to reduce the value of $J(\boldsymbol{\beta})$. The process is repeated until the minimum is found. This iterative method requires partial derivatives of $J(\boldsymbol{\beta})$ in Eq. (11.3.10) with respect to $\boldsymbol{\beta}$.

Consider again the case where there is only one observation, $\boldsymbol{x}$, and one known binary response, y. Starting with the loss function, $J(\boldsymbol{\beta})$, we can apply the chain rule to obtain the needed gradients. For some parameter, β_j, we have that

$$\frac{\partial J(\boldsymbol{\beta})}{\partial \beta_j} = \frac{\partial J(y, p(\boldsymbol{x}))}{\partial p(\boldsymbol{x})} \frac{\partial p(\boldsymbol{x})}{\partial \beta_j} = (p(\boldsymbol{x}) - y)x_j. \tag{11.3.12}$$

This comes from the partial derivative in the first term,

$$\frac{\partial J(y, p(\boldsymbol{x}))}{\partial p(x)} = -\frac{y}{p(x)} + \frac{1-y}{1-p(x)}, \tag{11.3.13}$$

and the second term, which involves the partial derivative of the sigmoid function,

$$\frac{\partial p(\boldsymbol{x})}{\partial \beta_j} = \frac{\exp(-\boldsymbol{x}^\top \boldsymbol{\beta})}{(1+\exp(-\boldsymbol{x}^\top \boldsymbol{\beta}))^2} x_j = p(\boldsymbol{x})(1-p(\boldsymbol{x}))x_j. \tag{11.3.14}$$

Next, consider the case with n observations. We have to sum up such quantities in Eq. (11.3.12) associated with each observation for every parameter. Combining the steps above, an update equation for each β_j, $j = 0, \ldots, p$, to compute the next iteration values is given by

$$\beta_j^{(t+1)} = \beta_j^{(t)} - \ell \times \frac{\partial J(\boldsymbol{\beta})}{\partial \beta_j} = \beta_j^{(t)} - \ell \sum_{i=1}^{n} (p(\boldsymbol{x}_i) - y_i)x_{ij}, \tag{11.3.15}$$

where ℓ is a tuning parameter called the learning rate that controls the step size needed in each iteration to produce a smooth yet rapid convergence to the global solution. This is for one observation only.

The equations can be formulated conveniently using matrix notation as follows. We start with $\boldsymbol{X}$, an $n \times (p+1)$ matrix. The n logit values are computed using an initial $\beta_n^{(0)}$ value as

$$\boldsymbol{z} = \boldsymbol{X}\boldsymbol{\beta}_n^{(0)}. \tag{11.3.16}$$

The n predicted responses are computed using the sigmoid function

$$p(\boldsymbol{X}) = \frac{1}{1+\exp(-\boldsymbol{z})}. \tag{11.3.17}$$

Finally, $\boldsymbol{\beta}_n$ is updated at each iteration t of a gradient descent-based optimization routine using

$$\boldsymbol{\beta}_n^{(t+1)} = \boldsymbol{\beta}_n^{(t)} - \ell \times \boldsymbol{X}^\top (p(\boldsymbol{X}) - \boldsymbol{y}). \tag{11.3.18}$$

The needed summations of Eq. (11.3.15) are automatically generated in this elegant form. Essentially, log-likelihood estimation involves finding $\hat{\boldsymbol{\beta}}_n$ that drives the second term of Eq. (11.3.18) to 0 within a given accuracy tolerance, and specifically we desire that at convergence

$$\boldsymbol{X}^\top (p(\boldsymbol{X}) - \boldsymbol{y}) = \mathbf{0}. \tag{11.3.19}$$

The loss is monitored at each step using the log-likelihood function,

$$J(\boldsymbol{\beta}) = -\sum_{i=1}^{n} [y_i \log\, p(\boldsymbol{x}_i) - (1 - y_i) \log(1 - p(\boldsymbol{x}_i))], \tag{11.3.20}$$

until the minimum is reached.

The number of iterations, T, needed to reach an acceptable level of accuracy is controlled by the learning rate, ℓ, which is the size of the step taken in the direction of the gradient. If ℓ is small, the rate of convergence can be slow which implies a large number of iterations. If ℓ is too high, it could lead to non-convergence. A suitable value for ℓ can be obtained by sweeping ℓ over a small range and selecting the appropriate value. Note that the numerical computations will be relatively cheap during each iteration since it only requires calculations of derivatives that have closed-form expressions.

11.3.2 Motivation for Rank-based Logistic Methods

In this section, we explore rank-based logistic regression and compare it against the aforementioned log-likelihood estimation. We motivate the use of a new method based on rank because of some desirable properties in terms of robustness relative to outliers. Rank-based estimation involves the replacement of the log-likelihood loss function with Jaeckel's dispersion function. The implications of this type of replacement are significant. Therefore, it is worthwhile studying this approach in more detail. We note that this is the first detailed description of rank-based logistic regression for machine learning in any publication.

Before describing the method, the notion of an outlier in the context of binary classification should be clarified. Since the response variable is Bernouilli distributed, we can only have a response of 1 or 0 with a given probability of occurrence. The goal of logistic regression is to find the best way to separate the binary data, either linearly or nonlinearly, so that inference can be facilitated. In Figure 11.2, we have a number of binary data points, represented as

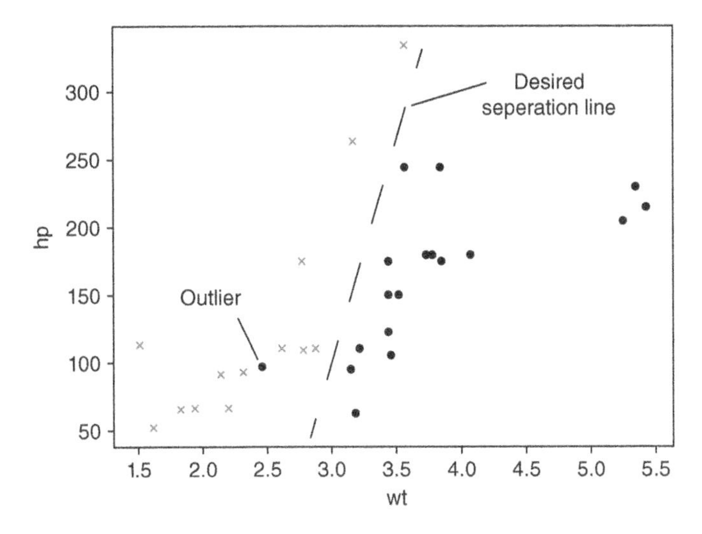

Figure 11.2 Outlier in the context of logistic regression.

× and •, that appear to be linearly separable. However, there is one point that is seemingly out of place. Except for this one data point, a clean line could be drawn to separate the × from the •. This odd point on the left is an outlier in binary classification problems.

An outlier can be produced in a variety of ways but once we establish a decision boundary, it is clear which points are true data and which ones are outliers. Whatever be the cause of outliers, we would like to have a method that is robust enough to produce a good model in the presence of a small percentage of outliers. In this context, robust implies that a certain number of outliers will not greatly affect the estimation of the underlying model. That is, the model building process can withstand a small number of these outliers, and produce results as if they didn't exist, as in Figure 11.2.

Unfortunately, log-likelihood methods include all the data points during estimation. In essence, log-likelihood is based on the mean of the residuals, which is a function of all the data. In the rank-based methods, a quantile near the median is used to build the model. Any quantile in the vicinity of the median effectively ignores outliers. In fact, we could rank the residuals to find potential outliers – they would be the residuals with the extreme order statistics. The breakdown point (Wilcox, 2012) is a measure of the tolerance level of a method to outliers. Using the mean is the least tolerant whereas using a rank-based method may tolerate up to 30% of the data as outliers. This is rather remarkable and provides a strong basis to pursue rank-based methods.

We note here that a precise definition of outliers is difficult to provide. All binary responses are either 0 or 1. Therefore, an outlier is not detectable from the response vector alone. If we pick a specific observation, it is hard to tell if it as outlier simply by examining it on its own, unless more information is provided about the observations surrounding a potential outlier, the so-called nearest neighbors. This can be seen in Figure 11.2 for the outlier on the left. Its nearest neighbors are all of the opposite type. We can use this as a working definition of an outlier – one that is out of place relative to its nearest neighbors. For example, if a "•" exists in the midst of a number of × which surround it in all directions, this observation is declared as an outlier. It also depends on how close it is to the decision boundary. If it is far away, it is an outlier. If it is close, it is not likely to be an outlier but rather an error in classification. Later we will see that the order statistics of residuals will be instrumental in identifying potential outliers.

Clearly, we need to reduce the effect of outliers in some manner when building a robust logistic model. To understand how to do this at a basic level, we review the linear case of Chapter 1 and build on top of it to develop the nonlinear case. Consider a simple linear model as follows

$$\boldsymbol{y} = \beta \boldsymbol{x} + \boldsymbol{\epsilon}, \tag{11.3.21}$$

where β represents the parameter of interest and $\boldsymbol{\epsilon} = (\epsilon_1, \dots, \epsilon_n)^\top$ are i.i.d. residuals with $\boldsymbol{\epsilon} \sim \mathcal{N}_n(0, \sigma^2 \boldsymbol{I}_n)$ and defined as, $\boldsymbol{\epsilon} = \boldsymbol{y} - \hat{\beta}_n \boldsymbol{x}$. We want to estimate β using a robust method. We can do this by defining a loss function based on a so-called pseudo-norm (Hettmansperger and McKean, 2011) as follows,

$$||\boldsymbol{y} - \beta \boldsymbol{x}||_\phi = \sum_{i=1}^{n} (y_i - \beta x_i) a_n (R_{n_i}(y_i - \beta x_i)), \tag{11.3.22}$$

where $a_n(\cdot)$ is a set of scores defined by a scoring function $\phi(u) = \phi(i/(n+1))$ such that $a_n(1/(n+1)) < \dots < a_n(n/(n+1))$. Here, $i = R_{n_i}(y_i - \beta x_i)$, i.e. the rank of the ith residual, $(y_i - \beta x_i)$, among $(y_1 - \beta x_1, \dots, y_n - \beta x_n)$, $1 \leq i \leq n$. For convenience, we use $R_{n_i}(\beta)$ to refer to the rank of the residuals since they are all a function of β. Recall from Chapter 1 that $\phi(u)$, $0 < u < 1$, is a non-decreasing function which is standardized such that $\int_0^1 \phi(u)du = 0$ and $\int_0^1 \phi^2(d)du = 1$. The Wilcoxon pseudo-norm is given by the linear scoring function $\phi(u) = \sqrt{12}(u - 1/2)$.

For R-estimation, we use Jaeckel's dispersion function for linear rank regression. The estimator is based on the pseudo-norm described above,

$$\hat{\beta}_n^{\mathrm{RLR}} = \underset{\beta}{\operatorname{argmin}} \left\{ D_n(\beta) = \sum_{i=1}^{n} (y_i - \beta x_i) a_n (R_{n_i}(\beta)) \right\}. \tag{11.3.23}$$

The linear dispersion function above was described at length in earlier chapters in this book. It can be shown that $D_n(\beta)$ is a non-negative, continuous, and convex function of β and therefore is quite suitable to efficiently find the optimal value of β. This was the essential breakthrough that made linear rank-based methods workable. Extension to multiple linear regression is straightforward as described in detail in earlier chapters. We now consider the nonlinear case.

11.3.3 Nonlinear Dispersion Function

A simple nonlinear model is as follows

$$\boldsymbol{y} = \boldsymbol{f}(\boldsymbol{\beta}; \boldsymbol{x}) + \epsilon, \tag{11.3.24}$$

where $\boldsymbol{f}(\boldsymbol{\beta}; \boldsymbol{x}) = (f(\boldsymbol{\beta}; \boldsymbol{x}_1), \dots, f(\boldsymbol{\beta}; \boldsymbol{x}_n))^{\mathsf{T}}$ is a real-valued nonlinear function. A suitable estimator for β is one that obtains $\hat{\beta}_n$ as the minimizer of the residual norm,

$$\hat{\beta} = \underset{\beta}{\operatorname{argmin}} \,||\boldsymbol{y} - \boldsymbol{f}(\boldsymbol{\beta}; \boldsymbol{x})||. \tag{11.3.25}$$

For the rank-based procedure, we again define a pseudo-norm given by

$$||\boldsymbol{y} - \boldsymbol{f}(\boldsymbol{\beta}; \boldsymbol{x})||_{\phi} = \sum_{i=1}^{n} (y_i - f(\boldsymbol{x}_i^{\mathsf{T}}\boldsymbol{\beta}))a_n(R_{n_i}(y_i - \boldsymbol{x}_i^{\mathsf{T}}\boldsymbol{\beta})). \tag{11.3.26}$$

This is the nonlinear dispersion function associated with the simple nonlinear model of Eq. (11.3.24). Hettmansperger and McKean (2011) established consistency and asymptotic normality of the estimator based on the Wilcoxon score function. We adopt those results and assume that they hold in the extensions to follow.

The general multivariate nonlinear case follows from the discussion above as,

$$D_n(\boldsymbol{\beta}) = \sum_{i=1}^{n} (y_i - f(\boldsymbol{x}_i^{\mathsf{T}}\boldsymbol{\beta}))a_n(R_{n_i}(\boldsymbol{\beta})), \tag{11.3.27}$$

where the nonlinear function, $f(\cdot)$, for the logistic regression case is given by

$$f(\boldsymbol{x}_i^{\mathsf{T}}\boldsymbol{\beta}) = \frac{1}{1 + \exp(-\boldsymbol{x}_i^{\mathsf{T}}\boldsymbol{\beta})}. \tag{11.3.28}$$

The first term of $D_n(\boldsymbol{\beta})$ is simply the residual for each observation. The second term, $a_n(R_{n_i}(\boldsymbol{\beta}))$, is the score for each residual based on the Wilcoxon scoring function, $\phi(u) = \sqrt{12}(u - 1/2)$. The complete scoring function of Eq. (11.3.23) is given by

$$a_n(R_{n_i}(\boldsymbol{\beta})) = \sqrt{12}\left(\frac{R_{n_i}(\boldsymbol{\beta})}{n+1} - 1/2\right) \tag{11.3.29}$$

which essentially applies a higher magnitude scores to the higher and lower ranked residuals and lower magnitude scores to the middle rank residuals. This is how the dispersion function produces a robust result.

The rank-based logistic regression (RLR) estimator is then given by

$$\hat{\boldsymbol{\beta}}_n^{\text{RLR}} = \underset{\boldsymbol{\beta}}{\operatorname{argmin}} \left\{ \sum_{i=1}^{n} (y_i - f(\boldsymbol{x}_i^\top \boldsymbol{\beta}))\sqrt{12}\left(\frac{R_{n_i}(\boldsymbol{\beta})}{n+1} - \frac{1}{2}\right)\right\}. \tag{11.3.30}$$

The optimization technique to solve Eq. (11.3.30) is gradient descent but we will need a set of partial derivatives to utilize it. Following the steps outlined earlier, we obtain

$$\frac{\partial D_n(\boldsymbol{\beta})}{\partial \beta_j} = -\sum_{i=1}^{n} p(\boldsymbol{x}_i)(1 - p(\boldsymbol{x}_i))x_{ij}a_n(R_{n_i}(\boldsymbol{\beta})). \tag{11.3.31}$$

Although it is possible to formulate a matrix equation similar to the log-likelihood case, it is easier to follow the steps by summing the quantities in the above equation. The update equation for each β_j to compute the new parameter values is given by

$$\beta_j^{(t+1)} = \beta_j^{(t)} - \ell \times \frac{\partial D_n(\boldsymbol{\beta})}{\partial \beta_j} \tag{11.3.32}$$

where ℓ is the learning rate that controls the step size needed in each iteration, t, to produce a smooth yet rapid convergence to the optimum solution.

11.4 Application to Machine Learning

This section describes the application of rank-based logistic regression to machine learning on simple binary classification problems to illustrate the benefits of the approach as compared to the log-likelihood method. Recall that the rank-based method simply requires replacing the log-likelihood loss function,

$$J(\boldsymbol{\beta}) = -\sum_{n=1}^{n} y_i \log\, p(\boldsymbol{x}_i) - (1 - y_i)\log(1 - p(\boldsymbol{x}_i)), \tag{11.4.1}$$

with the nonlinear dispersion function,

$$D_n(\boldsymbol{\beta}) = \sum_{i=1}^{n} (y_i - f(\boldsymbol{x}_i^\top \boldsymbol{\beta}))a_n(R_{n_i}(\boldsymbol{\beta})). \tag{11.4.2}$$

Table 11.1 LLR algorithm.

A1:	**Training using log-likelihood logistic regression (LLR).**
1:	Input: training data set $\boldsymbol{X}, \boldsymbol{y}$; number of iterations($T$); learning rate($\ell$)
2:	Set n =number of observations, p =number of parameters
3:	Set $t = 0$. Initialize $\boldsymbol{\beta}^{(0)}$ to small non-zero values.
4:	While $t < T$ do
5:	Update loss function $J(\boldsymbol{\beta}) = -\sum_{i=1}^{n}[y_i \log p(\boldsymbol{x}_i) - (1 - y_i)\log(1 - p(\boldsymbol{x}_i))]$
6:	For j = 1 to p do
7:	Compute gradient of the cost w.r.t to β_j
8:	$\frac{\partial J(\boldsymbol{\beta})}{\partial \beta_j} = \sum_{i=1}^{n}(p(\boldsymbol{x}_i) - y_i)x_{ij}$
9:	Update estimate $\beta_j^{(t+1)} = \beta_j^{(t)} - \ell \times \frac{\partial J(\boldsymbol{\beta})}{\partial \beta_j}$
10:	End for
11:	Update $t = t + 1$
12:	End while

Table 11.2 RLR algorithm.

A2:	**Training using rank-based logistic regression (RLR).**
1:	Input: training data set $\boldsymbol{X}, \boldsymbol{y}$; number of iterations($T$); learning rate($\ell$)
2:	Set n =number of observations, p =number of parameters
3:	Set $t = 0$. Initialize $\boldsymbol{\beta}^{(0)}$ to small non-zero values.
4:	While $t < T$ do
5:	Update loss function $D_n(\boldsymbol{\beta}) = \sum_{i=1}^{n}(y_i - \boldsymbol{x}_i^\top \boldsymbol{\beta})a_n(R_{n_i}(\boldsymbol{\beta}))$
6:	For j = 1 to p do
7:	Compute gradient of the cost w.r.t to β_j
8:	$\frac{\partial D_n(\boldsymbol{\beta})}{\partial \beta_j} = -\sum_{i=1}^{n} p(\boldsymbol{x}_i)(1 - p(\boldsymbol{x}_i))x_{ij}a_n(R_{n_i}(\boldsymbol{\beta}))$
9:	Update estimate $\beta_j^{(t+1)} = \beta_j^{(t)} - \ell \times \frac{\partial D_n(\boldsymbol{\beta})}{\partial \beta_j}$
10:	End for
11:	Update $t = t + 1$
12:	End while

Simplified algorithms in the form of pseudo-code for our software implementations, called LLR and RLR, are provided in Tables 11.1 and 11.2, respectively. Actual implementations can be much more involved. Note that they are almost identical except for lines 5 (loss update) and 8 (gradient calculation).

Therefore, implementation of the rank-approach in existing LLR software is rather straightforward.

The algorithms begin by reading in the $\boldsymbol{X}$ and $\boldsymbol{y}$ data, along with the desired number of iterations and learning rate. The optimal tuning parameter values will be different for the two programs. Then the $\boldsymbol{\beta}$ estimates are initialized to some small values. Next, a while loop is executed until the number of iterations, T, is reached. In the loop, the loss function is computed and monitored for convergence. Then, the derivatives for gradient descent are computed, followed by the parameter update step. The final estimates are produced when the program exits the while loop.

The key to knowing when rank-based methods should be used is that it depends on whether the data set is believed to contain outliers. In particular, binary classification data that are largely separable, but known to contain a number of outliers, are most suitable due to their robustness characteristics. One can show that if there are no outliers, the results of the two programs are very similar, but rank methods often produce better results. This is desirable since we would like rank-based methods to replace log-likelihood methods even when outliers are not present.

Our next step is to explore the advantages and disadvantages of rank-based logistic regression through a number of examples using software we have written in R and Python. The results are based on a single implementation of the two algorithms in one programs with the option of using LLR or RLR. This is due to the fact that only the loss function and its derivative are different in the two algorithms (see lines 5 and 8 in the respective pseudo-codes), implying that any existing log-likelihood program can easily be modified to perform rank-based logistic regression. Of course, additional lines of code to rank the residuals and compute scores for the dispersion function must be written, but this is rather straightforward.

11.4.1 Example: Motor Trend Cars

We begin with a simple illustration of a binary classification problem with a single outlier. The R programming language and analysis environment is a well-known statistical package that contains a number of machine learning programs and built-in data sets. One of data sets in R is a 32-observation sample of car models[1] which we employ here. It contains information about whether or not the car has automatic or manual transmission (am), its horsepower (hp), its weight (wt) and the number of cylinders (cyl). The logistic response is captured in “am”, while the explanatory variables are “hp”,“wt” and “cyl”. The data set is

1 The "mtcars" data set of R is used in this section.

Table 11.3 Car data set.

No.	am	cyl	hp	wt	No.	am	cyl	hp	wt
0	1	6	110	2.620	16	0	8	230	5.345
1	1	6	110	2.875	17	1	4	66	2.200
2	1	4	93	2.320	18	1	4	52	1.615
3	0	6	110	3.215	19	1	4	65	1.835
4	0	8	175	3.440	20	0	4	97	2.465
5	0	6	105	3.460	21	0	8	150	3.520
6	0	8	245	3.570	22	0	8	150	3.435
7	0	4	62	3.190	23	0	8	245	3.840
8	0	4	95	3.150	24	0	8	175	3.845
9	0	6	123	3.440	25	1	4	66	1.935
10	0	6	123	3.440	26	1	4	91	2.140
11	0	8	180	4.070	27	1	4	113	1.513
12	0	8	180	3.730	28	1	8	264	3.170
13	0	8	180	3.780	29	1	6	175	2.770
14	0	8	205	5.250	30	1	8	335	3.570
15	0	8	215	5.424	31	1	4	109	2.780

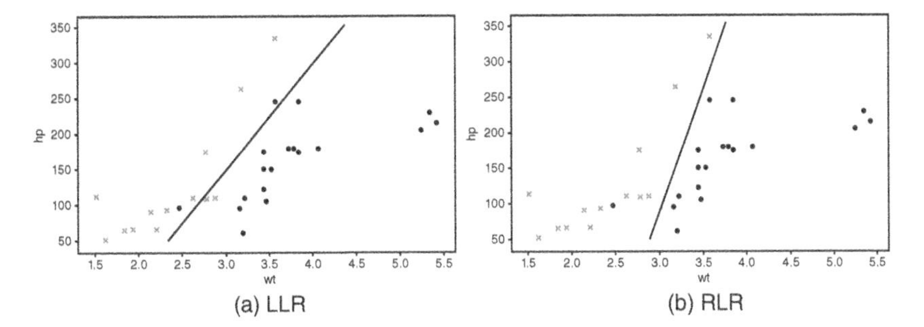

Figure 11.3 LLR vs. RLR with one outlier.

small enough to list in Table 11.3. We eliminate the column "cyl" from further consideration as it does not greatly affect the results but allows us to show 2D plots. The data was shown earlier as a scatter plot in Figure 11.2 along with a desirable separation boundary line.

If we now apply the log-likelihood method using LLR, we obtain the separation boundary shown in Figure 11.3(a). This is a somewhat surprising result at first glance because it is visually clear where the separating line should go and

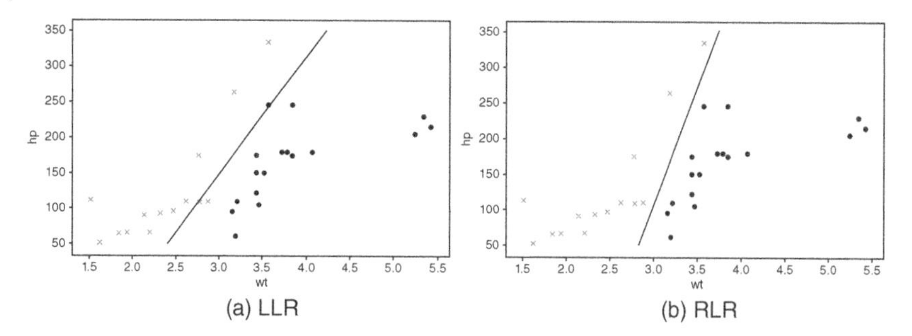

Figure 11.4 LLR vs. RLR with no outliers. LLR is unable to find a proper boundary on linearly separable data. RLR finds the expected decision boundary.

LLR does not produce it because it is influenced by just one outlier. Unfortunately, this is the main issue with LLR – the boundary depends on the mean of the loss terms associated with the data points. As a result, the linear boundary actually incurs four misclassifications in the car example. There are also two points straddling the line that are close enough to be called 1 or 0. Note that the linear boundary denotes the 50% probability line of being 1 or 0. The probability rises to 1 as you move away from the line in one direction and drops to 0 in the other direction moving away from the line (in perpendicular directions). Therefore, even in this ideal setting where a clear delineation exists between 0 and 1 responses, LLR may make a few mistakes if outliers are present.

Now consider the results with RLR shown in Figure 11.3(b). This is exactly what one would expect without any outliers present, and visually this seems like the correct result as discussed previously. The use of rank-based method effectively ignores the outlier in this case. It is also reminiscent of the results from a large margin classifier, such as support vector classifiers to be discussed later. An obvious next experiment is to remove the outlier and compare the two methods. This is shown in Figure 11.4. Note that in Figure 11.4(a), a clean solution on linearly separable data is not obtained by LLR, a known flaw in the basic approach, whereas RLR has hardly changed in Figure 11.4(b) compared to Figure 11.3(b).

Even more startling are the results in Figure 11.5. In this case, LLR is being controlled rather than influenced by the two outliers and produces nine classification errors, whereas RLR does not change much and makes only two misclassifications. This is also exactly what we would expect, given the data set has two outliers. In other words, the maximum number of correct classifications in this 32-observation data set is 30 if a linear boundary is used, and RLR was able to achieve this result. One could argue that LLR failed quite miserably in this case because of two outliers.

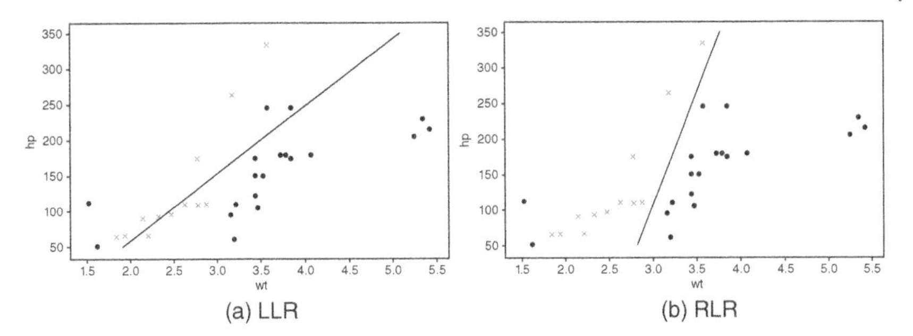

Figure 11.5 LLR vs. RLR with two outliers. The decision boundary for LLR is controlled by the two outliers whereas RLR is unaffected as if there were no outliers.

11.5 Penalized Logistic Regression

Both log-likelihood and rank-based methods for logistic regression can be modified using penalty functions. As with other machine learning methods, they are used for regularization of the parameters so that the results of a training process generalize to unseen test data. This is particularly important for LLR on linearly separable data. In this section, we study the effects of regularization on both methods.

11.5.1 Log-likelihood Expressions

Ridge regression can be applied to log-likelihood logistic regression as follows:

$$J^{\text{LLR-ridge}}(\boldsymbol{\beta}) = -\sum_{i=1}^{n}[y_i \log p_i + (1 - y_i)\log(1 - p_i)] + \lambda_2 \boldsymbol{\beta}^\top \boldsymbol{\beta} \tag{11.5.1}$$

where $\lambda_2 > 0$ is the regularization parameter. The penalized estimates, $\hat{\boldsymbol{\beta}}_n^{\text{LLR-ridge}}$, are obtained by minimizing Eq. (11.5.1). As before, the optimization technique is gradient descent but we will need a modified set of partial derivatives to account for the penalty term. Following the steps shown earlier, we can derive the parameter update equation to be

$$\boldsymbol{\beta}^{(t+1)} = \boldsymbol{\beta}^{(t)} - \ell \times [\boldsymbol{X}^\top(p(\boldsymbol{X}) - \boldsymbol{y}) + 2\lambda_2 \boldsymbol{\beta}^{(t)}] \tag{11.5.2}$$

or equivalently

$$\boldsymbol{\beta}^{(t+1)} = \boldsymbol{\beta}^{(t)}(1 - 2\ell\lambda_2) - \ell \times [\boldsymbol{X}^\top(p(\boldsymbol{X}) - \boldsymbol{y})] \tag{11.5.3}$$

where ℓ is the learning rate that controls the step size needed in each iteration to produce a smooth yet rapid convergence to the solution. In essence, we are shrinking the previous estimates in each iteration by a factor less than 1 given by $(1 - 2\ell\lambda_2)$.

LASSO can be applied to log-likelihood logistic regression as follows:

$$J^{\text{LLR-LASSO}}(\boldsymbol{\beta}) = -\sum_{i=1}^{n}[y_i \log p_i + (1-y_i)\log(1-p_i)] + \lambda_1 \sum_{j=1}^{p} |\beta_j| \quad (11.5.4)$$

where λ_1 is the tuning parameter. The penalized estimates, $\hat{\boldsymbol{\beta}}_n^{\text{LLR-LASSO}}$, are obtained by minimizing Eq. (11.5.4). The optimization technique to minimize Eq. (11.5.4) is gradient descent with a modified set of partial derivatives. The update equation to compute the new parameter values is given by

$$\boldsymbol{\beta}^{(t+1)} = \boldsymbol{\beta}^{(t)} - \ell \times [\mathbf{X}^{\top}(p(\mathbf{X}) - \mathbf{y}) + \lambda_1 \text{sgn}(\boldsymbol{\beta}^{(t)})]. \quad (11.5.5)$$

Similarly, to obtain $\hat{\boldsymbol{\beta}}_n^{\text{LLR-aLASSO}}$, we minimize the objective function

$$J^{\text{LLR-aLASSO}}(\boldsymbol{\beta}) = -\sum_{i=1}^{n}[y_i \log p_i + (1-y_i)\log(1-p_i)] + \lambda_1 \sum_{j=1}^{p} \frac{|\beta_j|}{|\beta_j^{\text{R}}|}. \quad (11.5.6)$$

11.5.2 Rank-based Expressions

Ridge regression can be applied to rank-based logistic regression as follows,

$$J^{\text{RLR-ridge}}(\boldsymbol{\beta}) = D_n(\boldsymbol{\beta}) + \lambda_2 \boldsymbol{\beta}^{\top}\boldsymbol{\beta}, \quad (11.5.7)$$

and in expanded form, we have

$$J^{\text{RLR-ridge}}(\boldsymbol{\beta}) = \sum_{i=1}^{n}(y_i - f(\boldsymbol{x}_i^{\top}\boldsymbol{\beta}))a_n(R_{n_i}(\boldsymbol{\beta})) + \lambda_2 \boldsymbol{\beta}^{\top}\boldsymbol{\beta}. \quad (11.5.8)$$

The penalized rank estimates, $\hat{\boldsymbol{\beta}}_n^{\text{RLR-ridge}}$, are obtained by minimizing Eq. (11.5.8). The optimization technique to minimize Eq. (11.5.8) is gradient descent but we will need a modified set of partial derivatives. For convenience, let us again define $p_i = p(\boldsymbol{x}_i)$. Following the steps outlined earlier, we obtain

$$\frac{\partial J^{\text{RLR-ridge}}(\boldsymbol{\beta})}{\partial \beta_j} = -\sum_{i=1}^{n} p_i(1-p_i)x_{ij}a_n(R_{n_i}(\boldsymbol{\beta})) + 2\lambda_2\beta_j. \quad (11.5.9)$$

The update equation for each β_j to compute the new parameter values is given by

$$\beta_j^{(t+1)} = \beta_j^{(t)} - \ell \times \frac{\partial J^{\text{RLR-ridge}}(\boldsymbol{\beta})}{\partial \beta_j}, \quad (11.5.10)$$

where ℓ is the learning rate that controls the step size needed in each iteration to produce a smooth yet rapid convergence to the solution.

LASSO can be applied to rank-based logistic regression as follows,

$$J^{\text{RLR-LASSO}}(\boldsymbol{\beta}) = D_n(\boldsymbol{\beta}) + \lambda_1 \sum_{j=1}^{p} |\beta_j|, \tag{11.5.11}$$

and again in expanded form, we have

$$J^{\text{RLR-LASSO}}(\boldsymbol{\beta}) = \sum_{i=1}^{n} (y_i - f(\boldsymbol{x}_i^{\top}\boldsymbol{\beta}))a_n(R_{n_i}(\boldsymbol{\beta})) + \lambda_1 \sum_{j=1}^{p} |\beta_j|. \tag{11.5.12}$$

The penalized rank estimates, $\hat{\boldsymbol{\beta}}^{\text{RLR-LASSO}}$, are obtained by minimizing Eq. (11.5.12). The optimization technique is again gradient descent with a modified set of partial derivatives. Here we have

$$\frac{\partial J^{\text{RLR-LASSO}}(\boldsymbol{\beta})}{\partial \beta_j} = -\sum_{i=1}^{n} p_i(1 - p_i)x_{ij}a_n(R_{n_i}(\boldsymbol{\beta})) + \lambda_1 \text{sgn}(\beta_j). \tag{11.5.13}$$

The update equation for each β_j to compute the new parameter values is given by

$$\beta_j^{(t+1)} = \beta_j^{(t)} - \ell \times \frac{\partial J^{\text{RLR-LASSO}}(\boldsymbol{\beta})}{\partial \beta_j}. \tag{11.5.14}$$

Similarly, to obtain $\hat{\boldsymbol{\beta}}_n^{\text{RLR-aLASSO}}$, we minimize the objective function

$$J^{\text{RLR-aLASSO}}(\boldsymbol{\beta}) = \sum_{i=1}^{n} (y_i - f(\boldsymbol{x}_i^{\top}\boldsymbol{\beta}))a_n(R_{n_i}(\boldsymbol{\beta})) + \lambda_1 \sum_{j=1}^{p} \frac{|\beta_j|}{|\beta_j^{\text{R}}|}. \tag{11.5.15}$$

11.5.3 Support Vector Machines

Another approach to binary classification, called support vector machines (SVM) (Vapnik, 1995), is considered to be among the best large-margin classifiers. SVM is not covered in detail in this book. However, there is an excellent description of this approach in chapter 9 of James et al. (2013), which the reader should consult for further information. The method, developed by the computer science community, is not based on fundamental statistics although its methodology can be connected to standard logistic regression.

We first develop the insights that lead to the support vector classifier (SVC). Consider the logistic loss function derived earlier based on log-likelihood as follows,

$$J^{\mathrm{LLR}}(\boldsymbol{\beta}) = \sum_{i=1}^{n}[y_i(-\log p_i) + (1 - y_i)(-\log(1 - p_i))], \tag{11.5.16}$$

where $p_i = (1 + \exp(-\boldsymbol{x}_i^{\mathrm{T}}\boldsymbol{\beta}))^{-1}$ as before. We see that there are two terms, one for the case when $y_i = 1$ (first term) and another for the case when $y_i = 0$ (second term). We also have a penalized version using ridge as follows:

$$J^{\mathrm{LLR\text{-}ridge}}(\boldsymbol{\beta}) = \sum_{i=1}^{n}[y_i(-\log p_i)+(1-y_i)(-\log(1-p_i))]+\lambda_2\boldsymbol{\beta}^T\boldsymbol{\beta}. \tag{11.5.17}$$

The λ_2 parameter can be moved to the first term by defining a C parameter and setting $C = 1/\lambda_2$ to conform to notation used in SVMs. Therefore,

$$J^{\mathrm{LLR\text{-}ridge}}(\boldsymbol{\beta}) = C\sum_{i=1}^{n}[y_i(-\log p_i)+(1-y_i)(-\log(1-p_i))]+\boldsymbol{\beta}^T\boldsymbol{\beta}. \tag{11.5.18}$$

We can now reformulate this objective function by simply replacing the two log terms under the summation with generic functions, $f_1(\boldsymbol{x}_i^{\mathrm{T}}\boldsymbol{\beta})$ and $f_0(\boldsymbol{x}_i^{\mathrm{T}}\boldsymbol{\beta})$, and switching the roles of loss and penalty functions:

$$J^{\mathrm{SVC}}(\boldsymbol{\beta}) = \boldsymbol{\beta}^T\boldsymbol{\beta} + C\sum_{i=1}^{n}[y_i f_1(\boldsymbol{x}_i^{\mathrm{T}}\boldsymbol{\beta}) + (1 - y_i)f_0(\boldsymbol{x}_i^{\mathrm{T}}\boldsymbol{\beta})]. \tag{11.5.19}$$

This produces the basic objective function for a support vector classifier. The reason we may choose to do this is to provide more flexibility than the two functions that arose from the log-likelihood derivation. We can make $f_1(\cdot)$ and $f_0(\cdot)$ anything we want. In particular, we can choose two complementary functions that allow for larger-margin boundaries, and this is exactly what leads to support vector classifiers.

Consider two functions commonly referred to as the *hinge loss* functions:

$$f_1(\boldsymbol{x}_i^{\mathrm{T}}\boldsymbol{\beta}) = \max(0, 1 - \boldsymbol{x}_i^{\mathrm{T}}\boldsymbol{\beta}) \text{ and } f_0(\boldsymbol{x}_i^{\mathrm{T}}\boldsymbol{\beta}) = \max(0, 1 + \boldsymbol{x}_i^{\mathrm{T}}\boldsymbol{\beta}). \tag{11.5.20}$$

Then, plugging into Eq. (11.5.19) and after some rearrangement, we obtain

$$J^{\mathrm{SVC}}(\boldsymbol{\beta}) = \boldsymbol{\beta}^{\mathrm{T}}\boldsymbol{\beta} + C\sum_{i=1}^{n}\max[0, 1 - (2y_i - 1)(\boldsymbol{x}_i^{\mathrm{T}}\boldsymbol{\beta})]. \tag{11.5.21}$$

Re-writing it in the form of a constrained optimization problem, we obtain

$$\min_{\boldsymbol{\beta}} \sum_{j=1}^{p}\beta_j^2 \quad \text{s.t.} \quad (2y_i - 1)(\boldsymbol{x}_i^{\mathrm{T}}\boldsymbol{\beta}) > c, \forall\, i = 1, \ldots, n, \tag{11.5.22}$$

where $c = 1$ for SVC and $c = 0$ for logistic regression. There are many ways to solve this optimization problem, but the details are not important here.

Instead, set $c = 1$ and consider the two cases when $y_i = 1$ (the observation belongs in class 1) and $y_i = 0$ (the observation belongs to class 0). Then, we say that $\boldsymbol{x}_i$ belongs to class 1 if $(\boldsymbol{x}_i^\top \boldsymbol{\beta}) > +1$ and class 0 if $(\boldsymbol{x}_i^\top \boldsymbol{\beta}) < -1$. In the case of logistic regression, we found that $\boldsymbol{x}_i$ belongs to class 1 if $(\boldsymbol{x}_i^\top \boldsymbol{\beta}) > 0$ and class 0 if $(\boldsymbol{x}_i^\top \boldsymbol{\beta}) < 0$ (see Eqs. (11.3.6)–(11.3.8)). By comparison, SVC requires a larger margin to be satisfied than standard logistic regression, so it is called a large-margin classifier. The problem is convex with a unique solution, and therefore amenable to standard optimization techniques (note that the objective function explicitly leaves out β_0 in the minimization problem). This allows us to find a maximal linear boundary to separate one class from the other, assuming they are linearly separable.

Now that the connection to logistic regression has been made, we present a more general approach for SVMs using its standard notation. Let $y_i \in \{-1, 1\}$, and $\boldsymbol{\beta} = (\beta_1, \ldots, \beta_p)^\top)$, i.e. separate out β_0. We can restate Eq. (11.5.22) as follows:

$$\min_{\boldsymbol{\beta}} ||\boldsymbol{\beta}||^2 \quad \text{s.t.} \; y_i(\boldsymbol{x}_i^\top \boldsymbol{\beta} + \beta_0) > M,, \quad \forall i = 1, \ldots, n \tag{11.5.23}$$

where M is related to the classifier margin. Given a new $\boldsymbol{x}$, we can now check whether $(\boldsymbol{x}^\top \hat{\boldsymbol{\beta}}_n^{\text{SVC}} + \hat{\beta}_0) > M$ or $(\boldsymbol{x}^\top \hat{\boldsymbol{\beta}}_n^{\text{SVC}} + \beta_0) < -M$. For the SVM, we use more powerful functions to obtain nonlinear boundaries. Let us re-define

$$\boldsymbol{x}^\top \boldsymbol{\beta} + \beta_0 = \sum_{i=1}^{n} \alpha_i k(\boldsymbol{x}_i, \boldsymbol{x}) + \beta_0, \tag{11.5.24}$$

where $k(\boldsymbol{x}_i, \boldsymbol{x})$ is a kernel function and α_i is a set of new parameters to be estimated. We have effectively replaced the first term with a new function that operates on all the training data $\boldsymbol{x}_i$ and one test data, $\boldsymbol{x}$. Notice that we now have n parameters, essentially one for each observation, rather than one for each column, p. By proper selection, the kernel function allows nonlinear and complex decision boundaries to be created for a given data set, as opposed to the linear boundary in the previous cases. However, in this form, we can no longer directly extract probabilities, except through additional expensive procedures. In contrast, rank-based logistic regression can, in principle, provide probabilities as well as a maximal boundary.

For SVMs, we need only one kernel function, rather than two complementary cost functions. Once selected, the SVM solution has the form given by

$$f(\boldsymbol{x}) = \sum_{i=1}^{n} \hat{\alpha}_i k(\boldsymbol{x}_i, \boldsymbol{x}) + \beta_0. \tag{11.5.25}$$

To obtain different classifiers, we simply choose different kernel functions and compute α_i. Three useful kernel functions are the linear kernel,

$$k(\boldsymbol{x}_i, \boldsymbol{x}) = \boldsymbol{x}_i^\top \boldsymbol{x}, \tag{11.5.26}$$

the polynomial kernel, with expansion up to the dth power, given by

$$k(\boldsymbol{x}_i, \boldsymbol{x}) = (1 + \boldsymbol{x}_i^\top \boldsymbol{x})^d, \tag{11.5.27}$$

and the radial basis function (RBF) kernel or Gaussian kernel,

$$k(\boldsymbol{x}_i, \boldsymbol{x}) = \exp(-\gamma(||\boldsymbol{x}_i - \boldsymbol{x}||^2). \tag{11.5.28}$$

The latter one, the Gaussian kernel, is the most popular and produces a non-linear decision boundary that is quite adept at capturing the true nature of the two classes. As a result, SVM is a powerful alternative to logistic regression. The approach also has certain robustness properties that make a comparison with rank-based methods appropriate. The decision boundary is constructed from only a few data points that belong to the two classes. These key points inside and around the two margins surrounding the boundary are the only ones that have non-zero $\hat{\alpha}_i$ and they are called support vectors[1]. They act to support the decision boundary, and hence their name. All other points are irrelevant to the construction of the boundary, since their $\hat{\alpha}_i = 0$, which makes it robust, albeit in a different sense than rank-based methods.

The RBF kernel has the property of decreasing in value as a function of the distance from a support vector that may produce highly irregular decision boundaries (i.e. it is exponential and depends on the γ parameter). SVM has an added advantage over LLR and RLR in that it does not require specification of all the various polynomial expansions of the variables to create the needed features. It will figure that out for itself from the original variables and data using the RBK function. However, the user must adjust two tuning parameters, γ and C, to obtain satisfactory results.

The above description is meant simply to give the reader a conceptual understanding of SVMs and its connection to logistic regression. Implementation details are quite complex and beyond the scope of this book but many software packages are available that provide this capability. Again, the basic idea is to use so-called kernel functions to increase the feature space of support vector classifiers. These classifiers use a small number of observations, or support vectors, to define the decision boundary. Since SVMs are considered to be a state-of-the-art method, we will use it for comparison purposes and refer the reader to James et al. (2013) for further details.

---

1 The term "vector" is derived from the simple fact that all $\boldsymbol{x}_i$ can be viewed as vectors in p-space.

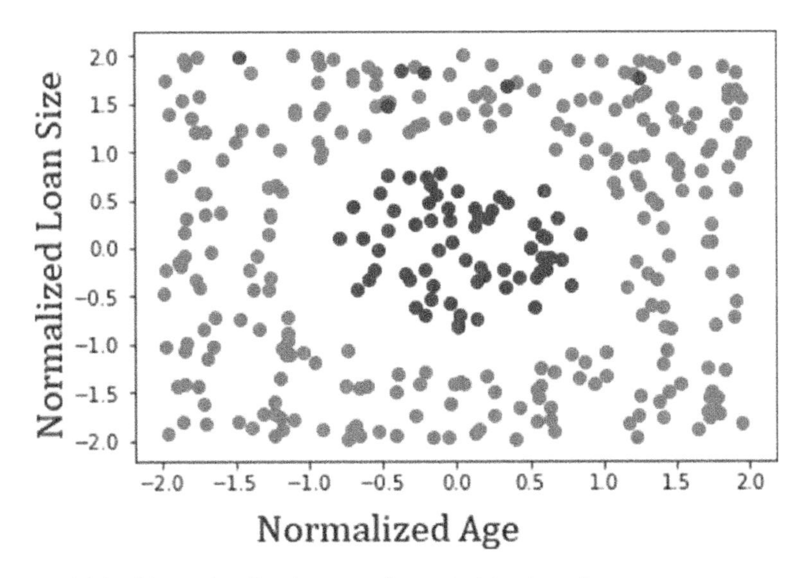

Figure 11.6 Binary classification – nonlinear decision boundary.

11.5.4 Example: Circular Data

The next data set to be considered is a case where the data is separable by a circular or elliptical boundary to illustrate the effect of penalty functions, for both shrinkage (ridge), selection (LASSO) and SVM.[2] The synthesized data set is shown in Figure 11.6. The two dimensional set of points are either gray (0) or black (1) and the goal of binary classification is to find a suitable separation boundary. To connect this example to a toy data science problem, let us assume that Bank ABC has given out 337 loans to customers in its immediate vicinity. After some analysis, it find that 266 loans were not repaid completely (gray points) but 71 loans were repaid (black points). This is not an acceptable ratio so the bank decides to use machine learning to make all future loans. The features to be used are age and loan size of prior customers.

The scatter plot of the data shows that middle-aged customers with mid-size loans have all paid back completely, along with some folks with high value loans. Note that there is a separation region between the two colors and there are six outliers (in black) in the upper region of the figure representing the high-value loans that were paid back. If we ignore these points, the data is clearly separable using logistic regression. We will initially use LLR and RLR to perform binary classification. However, in LLR and RLR, we need x_1, x_2, x_1^2 and

2 We note here that we do not take advantage of circular regression which is beyond the scope of this book.

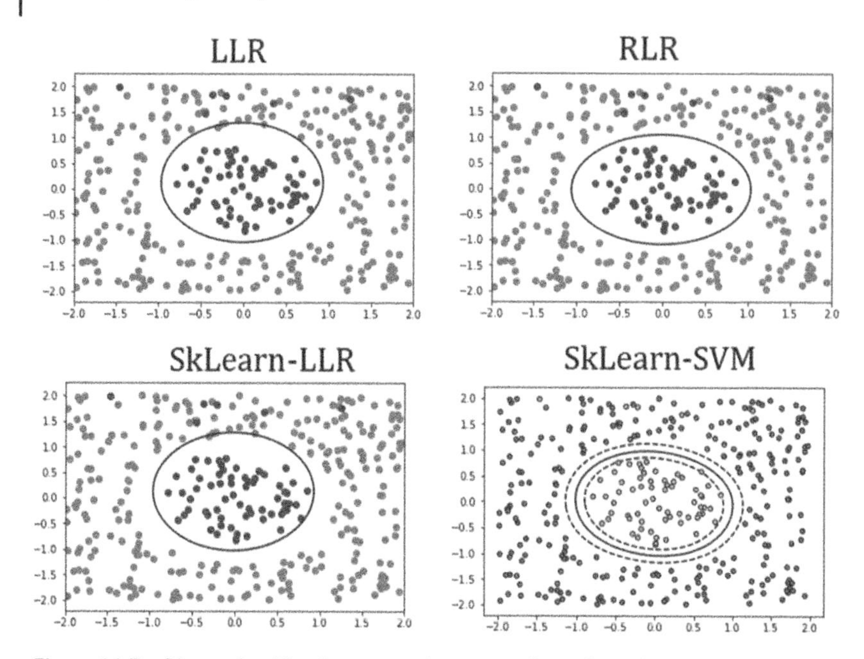

Figure 11.7 Binary classification comparison – nonlinear boundary.

x_2^2 as covariates since the solution has a nonlinear boundary. Therefore, $p = 4$ so that $\boldsymbol{\beta} \in \mathbb{R}^5$ to account for the intercept, β_0, for a total of five parameters.

If we now examine the results for the case of circularly separable data with outliers, the results are very interesting indeed. The log-likelihood method is not able to find a satisfactory solution because of the influence of the outliers, as seen in Figure 11.7 and labeled LLR. In fact, the separation boundary captures black points in the central region (as desired) but inadvertently captures a number of gray points, which implies that some high value loans will be given out in the future even though customers are unlikely to pay it back. The large and somewhat offset boundary is due purely to the outliers in the upper part of the plot. This result is confirmed by a standard package in SkLearn for logistic regression, labeled SkLearn-LLR in Figure 11.7. Neither reaches the target of 331 correct classifications.

The rank-based method, labeled RLR, finds a solution that separates the data in the same manner as it would without outliers. It also apparently obtains a solution similar to a large-margin classifier. This conjecture was validated using SVM from SkLearn using the RBF kernel and tuning parameters, $C = 1, \gamma = 1$. This is labeled SkLearn-SVM[3] in Figure 11.7. The dotted lines show the edges

3 The Python sklearn command used was `clf=svm.SVC(kernel='rbf', C=1, gamma=1)`.

Table 11.4 Ridge accuracy vs. λ_2 with $n = 337$ (six outliers). Learning rate of LLR-ridge = 0.02. Learning rate for RLR-ridge = 0.8. This is plotted in Figure 11.8. Optimal λ_2 at (*).

λ_2	LLR-ridge number of correct results	RLR-ridge number of correct results
0.0	326 (97%)	331 (98%)*
1.0	327 (97%)	331 (98%)
4.0	329 (98%)	331 (98%)
10.0	331(98%)	331 (98%)
20.0	331 (98%)*	314 (94%)
40.0	317 (94%)	266 (79%)
60.0	287 (85%)	266 (79%)
80.0	269 (79%)	266 (79%)
100.0	266 (79%)	266 (79%)

of the inner and outer margins as produced by the SkLearn package, and these dotted lines intersect with support vectors on one side or the other. RLR and SVM were not significantly affected by the outliers and both produce results consistent with a large-margin classifier. In fact, both achieved 331 correct classifications. The two examples thus far, the linearly and nonlinearly separable cases, with and without outliers, demonstrate the inherent value of rank-based methods that makes it worthwhile to investigate further on a larger example as we shall do shortly.

Before that, we need to explore the effect of shrinkage and selection regularization on the results produced by LLR and RLR. To begin with, the ridge penalty function was applied to LLR and RLR. Consider Table 11.4 where the ridge regularization parameter λ_2 is increased for the two methods, with different learning rates as listed in the caption. The number of correct predictions of the training set and their corresponding percentages are listed for each method. We notice that from $\lambda_2 = 10$ to 20, LLR-ridge is able to produce the optimal solution and then the solution degrades. This behavior is due to the bias-variance trade-off. That is, the bias is low initially, but the variance is high. As we increase λ_2, the bias increases while the variance is reduced. At some point in between, the ideal solution is obtained.

However, RLR-ridge does not improve by increasing λ_2. The desired solution is maintained until $\lambda_2 = 10$ and then it begins to degrade. This is due to the fact that the best solution is obtained at $\lambda_2 = 0$ so it is not expected to improve beyond this level. We illustrate the RLR-ridge drop-off of the number of correct decisions in Figure 11.8 which plots the number of correct predictions for the

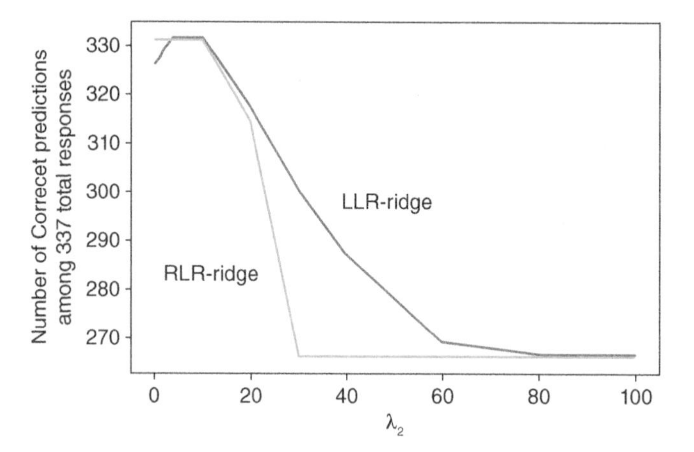

Figure 11.8 Ridge comparison of number of correct solutions with $n = 337$.

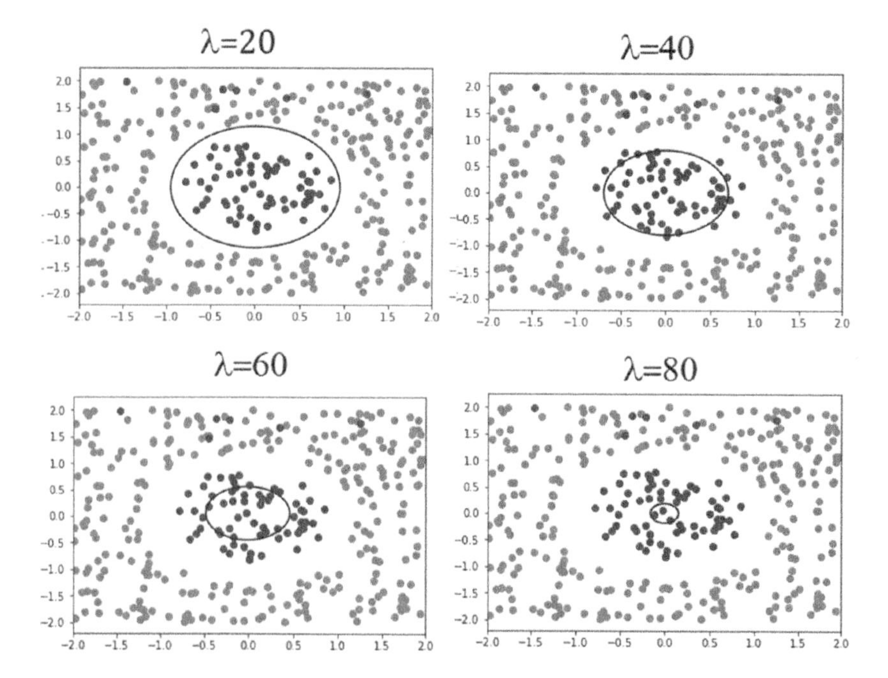

Figure 11.9 LLR-ridge regularization showing the shrinking decision boundary.

training set as a function of λ_2. For the RLR-ridge, there is a rather sharp drop to the worst-case solution. On the other hand, LLR-ridge has a rapid increase to the ideal solution but then a gradual decrease to the worst-case solution. The

Table 11.5 RLR-LASSO estimates vs. λ_1 with number of correct predictions.

λ_1	$\hat{\beta}_0$	$\hat{\beta}_1$	$\hat{\beta}_2$	$\hat{\beta}_3$	$\hat{\beta}_4$	Number correct
0.0	−13.1	0.32	0.41	11.7	11.3	331
1.0	−8.3	−2.6e−04	5.1e−04	7.5	7.2	331
4.0	−5.4	1.1e−03	−1.0e−03	4.9	4.7	331
6.0	4.7	0.006	−0.006	0.002	0.002	266
10.0	5.2	−0.02	−0.012	−0.006	−0.006	266

conclusion is that ridge can act to improve LLR but may not be as useful for RLR in this type of example. The best solution for the RLR-ridge occurs at $\lambda_2 = 0$ so there is no improvement with regularization.

The evolving decision boundaries for several values of λ_2 during LLR-ridge regularization are shown in Figure 11.9. The boundary first shifts downward as λ_2 increases and then begins to shrink in size as λ_2 is increased towards infinity. A similar effect occurs with RLR-ridge (not shown) but the shrinkage occurs much faster than for the LLR-ridge. In fact, the RLR-ridge decision boundary disappears around $\lambda_2 = 30$, whereas this occurs at $\lambda_2 = 80$ for LLR-ridge, as seen in Figure 11.8.

The LASSO penalty function has a similar characteristic as ridge for LLR and RLR. The results improve for LLR-LASSO as λ_1 increases and then eventually the results degrade beyond the optimal value. For RLR-LASSO, the results remain about the same for small λ_1 but degrade beyond a certain large value of λ_1, so it is of no value here. However, subset selection can be carried out using LASSO. In particular, we show the parameter values for a range of values for λ_1 in Table 11.5. From this table, we can see that a value of $\lambda_1 = 1.0$ produces relatively small values of β_1 and β_2, associated with x_1 and x_2, respectively. These can be set to 0 and a rather compact model can be generated using only x_1^2 and x_2^2. Therefore, subset selection using LASSO is still appropriate in this example for LLR and RLR whereas ridge is effective for LLR but not effective for RLR in this case. A key lesson is that LASSO (or aLASSO) should be used for selection and ridge for shrinkage with rank methods.

In the previous examples, the outliers were few in number and only existed in one region of the solution space. In our final example, we show the effect of outliers in the both regions of the circular data which is more realistic. That is, there are gray outliers (loan defaults) in the predominantly black dot region (loan repaying customers) and vice versa. The results from LLR ($\lambda_2 = 0.0$, learning rate = 0.02) and RLR ($\lambda_2 = 0.0$, learning rate = 2.0), LLR-ridge ($\lambda_2 = 10.0$, learning rate = 0.02) and SVM (RBF, $C = 1.0$, $\gamma = 1.0$) are shown in Figure 11.10.

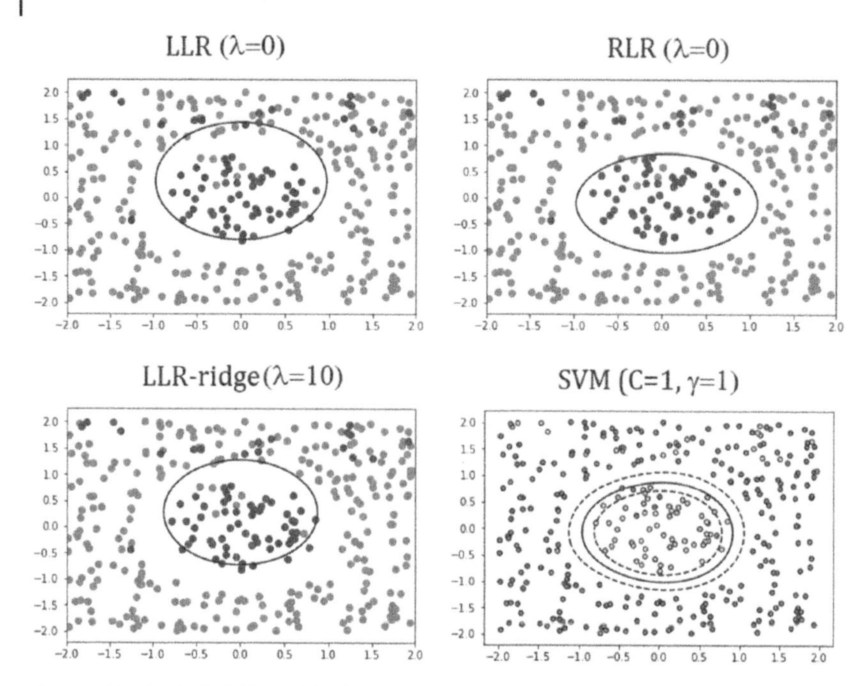

Figure 11.10 LLR, RLR and SVM on the circular data set with mixed outliers.

We see in Figure 11.10 that the LLR ($\lambda_2 = 0$) decision boundary is not satisfactory but is somewhat acceptable with LLR-ridge ($\lambda_2 = 10$). In contrast, both RLR ($\lambda_2 = 0$) and SVM ($C = 1.0, \gamma = 1.0$) carefully capture as many black dots in the center region as possible. This shows that RLR and SVM are superior to LLR-ridge even with optimal regularization. Although not shown, we find that LLR-LASSO regularization ($\lambda_1 = 20.0$) will remove x_1 and x_2 and also produce a similar result to LLR-ridge, but neither can compete with the RLR and SVM decision boundaries.

If we now return to Figure 11.10 with the mixed set of outliers, we note that SVM produces an unnecessarily tight boundary as compared to RLR due in part to the new outliers in the central region. SVMs are, in general, not robust to outliers. RLR is superior with the larger boundary and it can provide probabilities, while SVM does not. However, SVM is robust in that it relies on only a few support vectors. The best option for the loan problem is then RLR and we can use the predicted probabilities to make a more informed decision on each loan. But SVM with the RBF kernel does not require extra features to be defined. RLR and LLR both have this limitation but it will be resolved in the next chapter using neural networks.

11.6 Example: Titanic Data Set

Now that the concepts and value of rank-based logistic regression have been established, it is an appropriate time to use a larger data set with higher dimensions where outliers are not easily observed but their presence is almost certain. The Titanic data set is well-known in data science circles[4]. It will be used to show that RLR is often superior to both LLR and SVM. It contains information about the passengers who survived, and those who did not, on the ill-fated maiden voyage of the Titanic in April of 1912 that struck an iceberg 400 miles off the coast of Canada. Our goal is to use a training set to learn about a random subset of passengers and their survival rates, and use logistic regression to predict the survivors for a given test set.

We first illustrate basic elements of data science and exploratory data analysis. Then the data set is processed to create the needed $\boldsymbol{X}$ and $\boldsymbol{y}$ for both training and testing purposes. Next, the three approaches to binary classification are compared and contrasted. We note up front that the Titanic data set is expected to have many outliers by its very nature, and this leads us to believe that rank-based methods would be well-suited for this application. The goal of this section is to determine if this conjecture is true.

11.6.1 Exploratory Data Analysis

In this section, we carry out rudimentary elements of exploratory data analysis. To begin the analysis, we show the first 10 rows of the Titanic data set in Table 11.6. The first column is the passenger ID for the data set and is only used for bookkeeping. The second column is the binary response indicating whether the passenger survived (1) or not (0). This is used to form the $\boldsymbol{y}$-vector. The rest of the columns are used to derive the $\boldsymbol{X}$ matrix, with one row for each passenger. The definitions associated with these data columns are given in Table 11.7. The original data set contains much more information but this reduced form is sufficient for our purposes here.

The data set is a mixture of strings, categorical data, and real and integer numbers. This is usually the nature of most data sets. They are a mix of all types of data in different formats. Although the data is already in the form of a data frame for our example, the process of getting the data into this form can be a major undertaking in itself. We ignore that aspect here and assume someone else has done that work. Our task at hand is to convert all of this data into numbers so that it can be used for logistic regression.

---

4 The Titanic data set is available on the Kaggle site at www.kaggle.com in the Competitions section.

Table 11.6 Sample of Titanic training data.

ID	Survived	Pclass	Name	Sex	Age	SibSp	Parch	Ticket	Fare	Cabin	Embarked
1	0	3	Braund, Mr. Owen Harris	male	22.0	1	0	A/5 21171	7.2500	NaN	S
2	1	1	Cumings, Mrs. John Bradley (Florence Briggs Th...)	female	38.0	1	0	PC 17599	71.2833	C85	C
3	1	3	Heikkinen, Miss. Laina	female	26.0	0	1	3101282	7.9250	NaN	S
4	1	1	Futrelle, Mrs. Jacques Heath (Lily May Peel)	female	35.0	1	0	113803	53.1000	C123	S
5	0	3	Allen, Mr. William Henry	male	35.0	0	0	373450	8.0500	NaN	S
6	0	3	Moran, Mr. James	male	NaN	0	0	330877	8.4583	NaN	Q
7	0	1	McCarthy, Mr. Timothy J	male	54.0	0	0	17463	51.8625	E46	S
8	0	3	Palsson, Master. Gosta Leonard	male	2.0	3	1	349909	21.0750	NaN	S
9	1	3	Johnson, Mrs. Oscar W (Elisabeth Vilhelmina Berg)	female	27.0	0	2	347742	11.1333	NaN	S
10	1	2	Nasser, Mrs. Nicholas (Adele Achem)	female	14.0	1	0	237736	30.0708	NaN	C

Table 11.7 Specifications for the Titanic data set.

Variables	Description	Data type
Pclass	Passenger class (1, 2, 3)	Categorical
Name	Full name of passenger	String
Sex	Male or female passenger (0, 1)	Categorical
Age	Age of passenger	Real
SibSp	Number of siblings/spouse with passenger	Integer
Parch	number of parent/children with passenger	Integer
Fare	Actual fare paid by passenger	Real
Cabin	Cabin number of passenger	String
Embarked	Starting point for passenger (S, C, Q)	Categorical

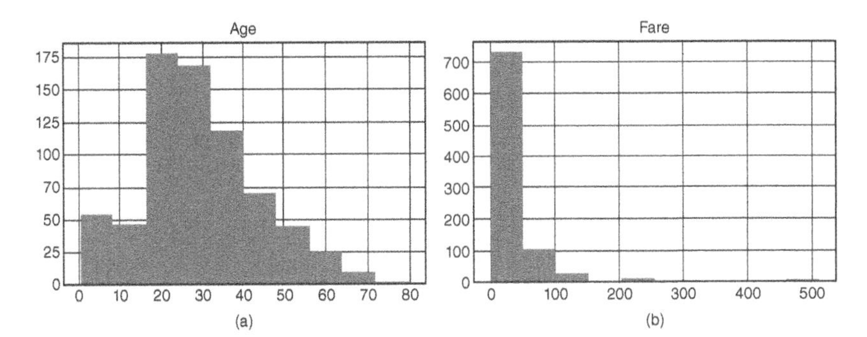

Figure 11.11 Histogram of passengers: (a) age and (b) fare.

We can now begin to carry out an exploratory data analysis. We will only provide a brief glimpse of this task because there are so many ways to represent and visualize a data set[5]. We note that two columns have real numbers: Age and Fare. These are suitable for visualization using histograms as seen in Figure 11.11. For the Age histogram, one can observe a distribution that is neither Gaussian nor uniform. In fact, it seemingly has a step function followed by an exponential decay as Age increases. The Fare histogram also has a sharp exponential decay characteristic. Both are skewed-right. We would typically investigate which categories of Age and Fare have the most survivors to understand if Age and/or Fare are important features to retain. We skip those details here and keep them both because they may turn out to be important. Further,

5 For a thorough analysis, we refer the reader to the Kaggle site and the detailed explanations therein.

Table 11.8 Number of actual data entries in each column.

Data columns	Total number of valid entries
Survived	891
Pclass	891
Name	891
Sex	891
Age	714
SibSp	891
Parch	891
Fare	891
Cabin	204
Embarked	889

we will scale Age by 20 and Fare by 50 to bring them in line with other values in the table. More about scaling will be discussed shortly.

Subsequently, we could examine the total number of missing elements in each column shown in Table 11.8, noting that there are 891 rows in the Titanic training set. The Cabin and Age columns have a significant number missing entries with a label of NaN or NA, whereas Embarked has only two entries missing. Each case would be handled differently. The relevance of the Cabin number seems to be very limited and many entries are missing so this column can be deleted. As an aside, the Name column seems irrelevant too. It is not clear how a person's name would affect their survival, and any family or sibling/spouse relationships are captured in Parch and SibSp columns. Therefore, the Name column can also be deleted. The missing entries in the Age column pose a real problem because there may be some correlation between Age and Survival. The missing values can be simply replaced with the mean of the ages. A better approach is to compute the mean and standard deviation of the histogram in Figure 11.11(a) and sample Age values from the distribution. This approach was taken in producing the final data set.

Next, we examine the categorical data columns: Sex, Embarked and Pclass. These columns can be cross-tabulated with the Survived column to get a snapshot of the relevance of the data. Consider for example the tabulated data in Table 11.9. It indicates that more females survived than males by a significant margin. If we simply guessed that all female passengers survived and all males did not, we would have $233 + 468 = 701$ correct predictions out of 891. That's 78% without using machine learning. Clearly, the Sex of the passenger

Table 11.9 Cross-tabulation of survivors based on sex.

Sex	Survived	Did not survive	Totals
female	233	81	314
male	109	468	577

Table 11.10 Cross-tabulation using Embarked for the Titanic data set.

Embarked	Survived	Did not survive	Totals
S	217	427	644
C	93	75	168
Q	30	47	77
Totals	342	549	889

is a critical feature. Also, we should view 701 out of 891 as a baseline for evaluation of different classification methods for this data set since it is trivial to obtain this value using hand calculations.

We now turn our attention to the Embarked column. Passengers either embarked from Southampton (S), Cherbourg (C), or Queenstown (Q). Table 11.10 lists the number of passengers embarking from each port. The number of survivors from each port seems to be roughly proportional to the total number who embarked from that port. This kind of assessment of the data is very useful because we noted earlier that two data entries are missing. The data scientist performing EDA must now decide how to fill those two missing entries. Since S and C are most frequent, one entry could be set to S and the other to C. This will not change the results significantly so this simple solution is acceptable. The last step is to convert each category into a number. We could choose $S = 3$, $C = 2$ and $Q = 1$ because it roughly correlates to the number of passengers embarking from each port, if for no other reason.

The Pclass column may be examined in the same way as Embarked, with some cross-tabulation work to assess its relative importance. We skip this step here. Since none of the data entries are missing, we may choose to keep this column in case it has relevance. Finally, the Parch and SibSp columns are integer quantities indicating the number of Parent/Children and Sibling/Spouse totals associated with each passenger, so we choose to keep them without modification.

One last operation for this data set before we begin to apply machine learning is to scale certain columns. We note that most of the numbers in the columns range from 0 to 5 except for Age and Fare. Therefore we should scale them to

Table 11.11 Sample of Titanic numerical training data.

Survived	Pclass	Sex	Age/20	SibSp	Parch	Fare/50	Embarked
0	3	1	1.10	1	0	0.145	3
1	1	0	1.90	1	0	1.425	2
1	3	0	1.30	0	1	0.158	3
1	1	0	1.75	1	0	1.062	3
0	3	1	1.75	0	0	0.161	3
0	3	1	1.45	0	0	0.169	1
0	1	1	2.70	0	0	1.037	3
0	3	1	0.1	3	1	0.421	3
1	3	0	1.35	0	2	0.222	3
1	2	0	0.70	1	0	0.601	2

fit into this range, as shown in Table 11.11. More generally, we should calculate the mean and the standard deviation to scale the data in each column. That is, subtract the mean from each value and divide by the standard deviation. It is important for many machine learning methods to have numbers that lie in the same range, especially when interpreting the results, but often to increase the quality of the results obtained.

After applying all the above steps, we would produce the $\boldsymbol{X}$ and $\boldsymbol{y}$ for the training set. Then, we apply all the identical steps to build the test set. We must remove the same columns as in the training set, apply replacements for NaN or NA in the same manner, and scale the data in the same way as the training set. Otherwise, the estimates of logistic regression will be of no value during prediction. There are also some comparative analyses to determine if the distribution of the test set is similar to the training set. We would like it to be so, but if it is not, then prediction may not be as accurate as expected. However, we should not place any part of the test set inside the training set, and we should not generate estimates from machine learning based on the accuracy of prediction on the test set. Instead, we should break up the training set into K subsets and carry out K-fold cross-validation. But under no circumstances should we use elements of the test set to influence the training process.

In the results to follow, we will not use K-fold cross-validation but instead use the entire training set to generate estimates and then perform prediction using the test set. Recall, our goal here is to compare different methods using the same procedure. For this reason, we can skip the normal cross-validation sequence of steps outlined above and simply illustrate the similarities and differences of the

Table 11.12 Number of correct predictions for Titanic training and test sets.

Data set	LLR	SVM	RLR
Train (number correct out of 891)	707	701	725
Train (% correct)	79.3%	78.7%	81.3%
Outliers detected	39	0	163
Test (number correct out of 418)	314	320	324
Test (% correct)	75.1%	76.6%	77.5%

three approaches. Our overall goal is to provide evidence that the rank-based method has some inherent value that should be investigated in future research.

11.6.2 RLR vs. LLR vs. SVM

In this section, we compare the results of running standard packages against the rank-based method on the Titanic data set. The basic steps of EDA in the previous section can be used to create the necessary data, as shown for the first 10 observations in Table 11.11. This data is used to construct the needed $\boldsymbol{X}$ and $\boldsymbol{y}$ quantities. The task of machine learning is to generate $\boldsymbol{\beta}$ estimates for prediction. Model generation is done with the training set ($n = 891, p = 7$); prediction is performed using the test set ($n = 418, p = 7$). We now utilize them to compare three machine learning methods for binary classification.

Since the data set represents a seven-dimensional problem, we are unable to plot or even visualize the hyperplane boundaries produced in each case. However, the $\boldsymbol{\beta}$ estimates allow us to use Eqs. (11.3.7) and (11.3.8) to predict the binary responses for LLR and RLR. Using these values, we can determine the number of correct responses for the training set and test set to compare the three methods against one another. This is shown in Table 11.12. SVM results are from SkLearn. The results indicate that RLR outperforms the other methods in both training and test sets. In fact, SVM does not even exceed the baseline level mentioned earlier of 701 correct predictions out of 891 observations. This may be due to the large number of outliers in the data set. However, SVM does outperform LLR on the more important test set. But the results on the test set only confirm that RLR outperforms the others on both data sets.

Furthermore, RLR is able to identify a large number of potential outliers that may exist in the data set. The methodology to identify outliers is rather straightforward. First, the $\hat{\boldsymbol{\beta}}_n^*$ estimates from any given estimator are used to compute the residuals of logistic regression, r_i, as follows:

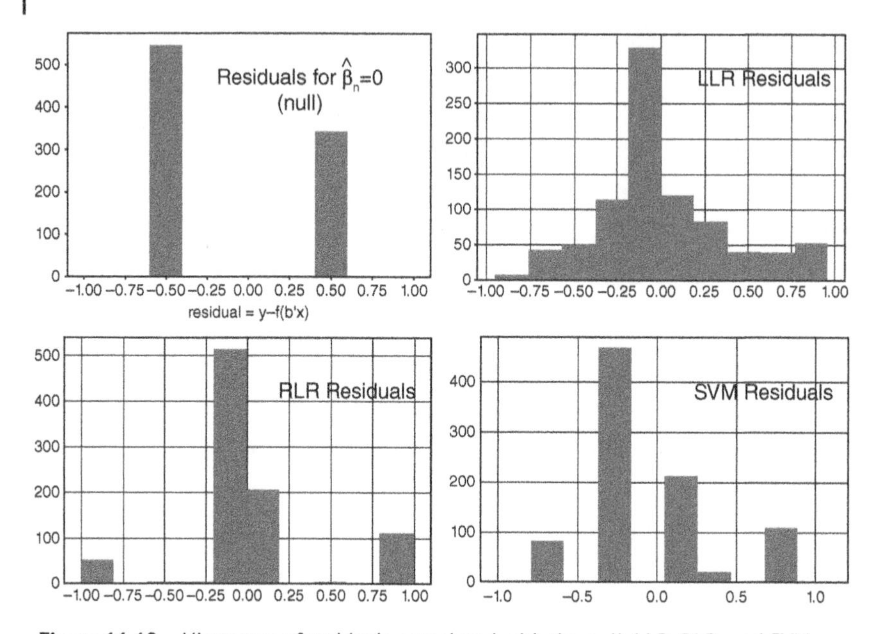

Figure 11.12 Histogram of residuals associated with the null, LLR, RLR, and SVM cases for the Titanic data set. SVM probabilities were extracted from the sklearn.svm package.

$$r_i = y_i - f(\boldsymbol{x}_i^\top \hat{\boldsymbol{\beta}}_n^*) = y_i - \frac{1}{1 + \exp(-\boldsymbol{x}_i^\top \hat{\boldsymbol{\beta}}_n^*)}, \quad i = 1, ..., n. \tag{11.6.1}$$

This was done for LLR, RLR and SVM as shown in the histograms of Figure 11.12. In Figure 11.12(a), $\hat{\boldsymbol{\beta}}_n = 0$ is the null condition. For this case, $r_i = 1/2$ if $y_i = 1$, and $r_i = -1/2$ if $y_i = 0$. The rest of the cases are shown in Figures 11.12(b), (c) and (d). The histogram for LLR appears to be normally distributed between -1 and $+1$, with a large peak around 0. However, RLR has residuals that are either located near 0, 1 or -1. The residuals near -1 and $+1$ are potential outliers.

For SVM, an unexpected histogram of residuals is generated with four peaks. There are clearly some inconsistencies in the way sklearn.svm generates the probabilities. In any case, the outliers are very easy to detect in RLR and virtually impossible to detect in SVM. The RLR and LLR outliers are detected by setting a criteria of $+0.85$ and above for positive outliers, and -0.85 and below for negative outliers. The results of this process are also listed in Table 11.12. LLR reports only 39 outliers. RLR reports that 163 out of 891 observations (i.e. 18.3%) are outliers. This gives us an idea that the maximum achievable accuracy in the Titanic data set is about 82%. Since we know which data points produced these potential outliers, we are able to report and inspect each one. This may serve as an important tool for many data scientists attempting to clean their data set by removing unwanted outliers.

11.6.3 Shrinkage and Selection

Our final step in this section is to explore the effects of penalty functions on the results obtained. We know that RLR produced the best results thus far, at least for the comparison with $\lambda = 0$. The results from ridge regularization are shown in Table 11.13 for LLR-ridge and Table 11.14 for RLR-ridge. We note that better results are obtained by RLR-ridge on the test set. The results from LASSO regularization are shown in Table 11.15 for LLR-LASSO and Table 11.16 for RLR-LASSO. Again better results are obtained by RLR on the test set. Notice

Table 11.13 Train/test set accuracy for LLR-ridge. Optimal value at (*).

Ridge λ_2	Train accuracy (%)	Test accuracy (%)	Test correct
0.0	78	76	318
10.0	80	76	319
20.0	80	76	317
30.0	80	77	320
40.0	79	78	327*
50.0	79	77	323
60.0	78	78	325

Table 11.14 Train/test set accuracy for RLR-ridge. Optimal value at (*).

Ridge λ_2	Train accuracy (%)	Test accuracy (%)	Test correct
0.0	82	77	325
1.0	82	77	325
2.0	81	77	322
5.0	80	78	327
10.0	79	80	333*
15.0	79	78	327
20.0	78	78	327
25.0	79	78	327

Table 11.15 Train/Test set accuracy for LLR-LASSO. Optimal value at (*).

LASSO λ_1	Train accuracy (%)	Test accuracy (%)	Test correct
0.0	78	76	318
10.0	79	76	319
20.0	79	77	320
30.0	80	77	321
40.0	79	77	322*
50.0	79	77	321
60.0	78	78	320

Table 11.16 Train/test set accuracy for RLR-LASSO. Optimal value at (*).

LASSO λ_1	Train accuracy (%)	Test accuracy (%)	Test correct
0.0	82	77	325*
5.0	81	78	325
10.0	79	77	320
15.0	79	77	320
20.0	79	77	320
25.0	79	77	320

that regularization using RLR-LASSO did not improve the results, whereas RLR-ridge regularization produced much better results.

RLR traces for ridge, LASSO and aLASSO are shown in Figures 11.13–11.15. The interpretation of these traces were described in Chapter 2. Of particular interest here is the aLASSO trace. As mentioned in Chapter 2, this trace has the oracle property so we can glean some valuable information about variable importance. Specifically, we find that Sex is a very important parameter in determining whether a passenger survives the sinking of the Titanic since its trace reaches 0 at the largest value of λ_1 in Figure 11.15. Then, following the order of the traces, we see a cluster of (Embarked, Fare, SibSp, Parch) and finally (Age, Pclass). So we can safely conclude that Age and Pclass are the least important variables, Sex is the most important, and the rest lie somewhere in between. The LASSO trace indicates that Sex and Fare are about of equal importance and the rest of the variables are of secondary importance but they are all

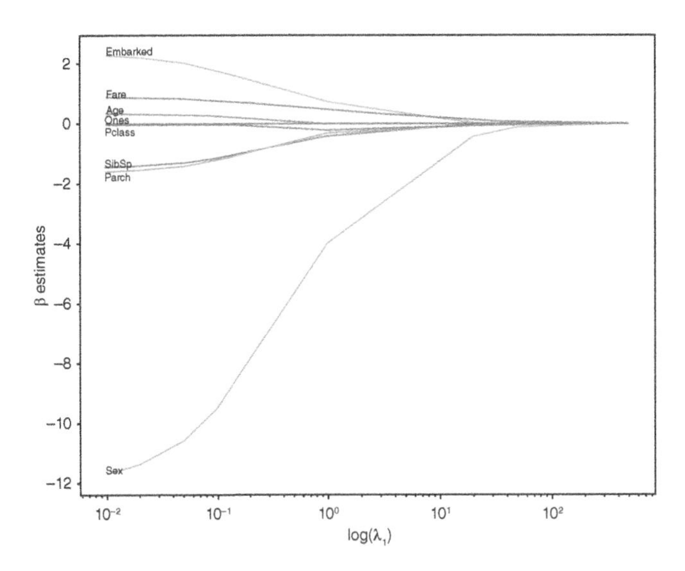

Figure 11.13 RLR-ridge trace for Titanic data set.

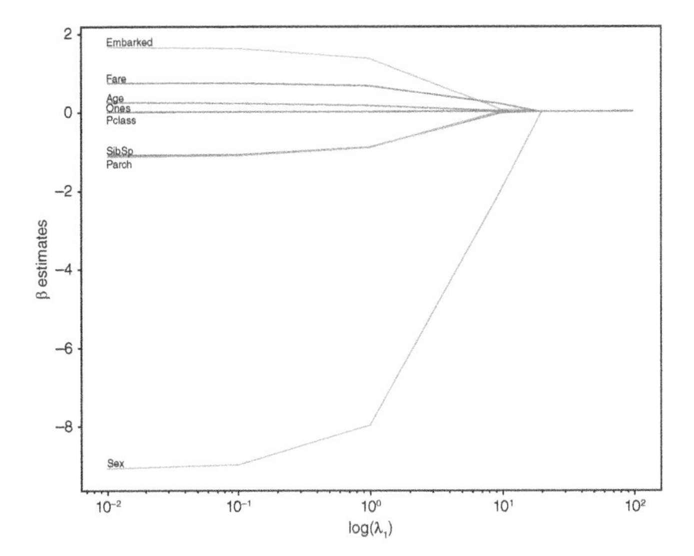

Figure 11.14 RLR-LASSO trace for the Titanic data set.

about the same. Clearly, the hand analysis given earlier showing that Sex is by far the most important feature does not line up with the LASSO result.

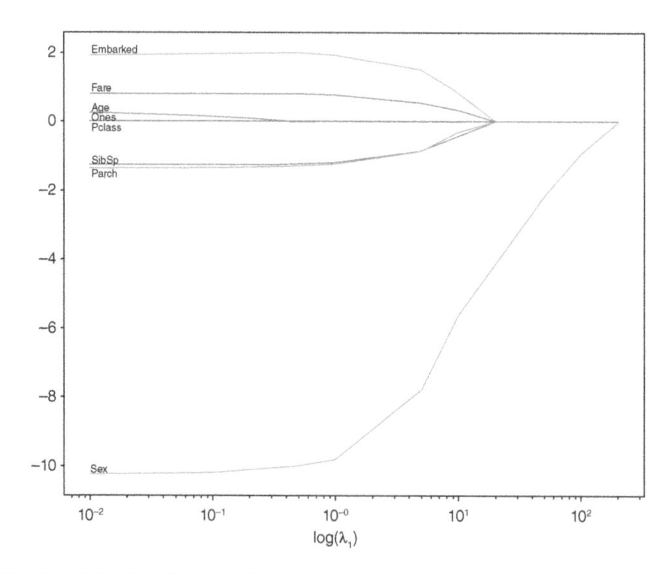

Figure 11.15 RLR-aLASSO trace for the Titanic data set.

As a side note, the correlation matrix can also be used to obtain a first-order assessment of the important explanatory variables. Examining this matrix carefully is another important step used frequently in EDA. Unfortunately, a data set with a large number of outliers will tend to corrupt the correlation matrix, perhaps raising the importance of certain unimportant covariates and lowering the importance of some significant covariates. Rank-based methods with aLASSO traces may be helpful in mitigating this issue. We have now sequenced through an entire data science process with the Titanic data set. Rank outperformed all other approaches and was able to give some idea of the maximum accuracy that can be obtained on the data set, and also provided a list of potential outliers and feature importance.

11.7 Summary

A new rank-based approach to logistic regression has been described in this chapter. It is particularly useful when data sets have a number of outliers. However, if there are no outliers, the rank estimates can still provide better prediction accuracy than the standard log-likelihood methods. Furthermore, it can compete with large-margin classifiers, such as support vector machines, and it was shown to be superior to SVM on a well-known data set for machine learning. The implementation of rank-based methods is rather straightforward since it requires only the modification of a few lines of code in a log-likelihood based program, plus the addition of code for a scoring function. As such, the researcher can compare the two methods on the same code base as done in this chapter for LLR and RLR. The data scientist can use this methodology to identify outliers in the data and for data cleaning purposes, if appropriate. It is also possible to obtain a first-order estimate of the maximum accuracy on a data set set with outliers. RLR-aLASSO was shown to be useful for subset selection, while RLR-ridge was effective for shrinkage to improve model accuracy.

Rank-based methods tend to require more iterations during training and will require different learning rates as compared to log-likelihood methods but this is not a significant price to pay since both methods incur the same cost during prediction. The full story on rank-based logistic regression is not told in this chapter but we are encouraging the reader to pursue this direction further as it will eventually lead to robust data science which is the ultimate goal.

11.8 Problems

11.1. Let the estimates $\hat{\beta}_0 = -1.0, \hat{\beta}_1 = -1.0, \hat{\beta}_2 = -1.0, \hat{\beta}_3 = -1.0$ be derived from a training set for a logistic regression problem where the possible logical response is either 0 or 1.

(a) What is the predicted logical value of the response for $\boldsymbol{x}_1 = (1.0, 2.0, 3.0)^{\mathsf{T}}$?

(b) What about $\boldsymbol{x}_2 = (-1.0, -2.0, -3.0)^{\mathsf{T}}$?

(c) Find a value of $\boldsymbol{x}_3$ where the response may be either 0 or 1 (although it may be categorized as 0)? Note that for this case there is not a unique solution for $\boldsymbol{x}_3$ since it can lie anywhere on the decision boundary.

(d) Compute the probability, $\mathrm{P}(y_i = 1|\boldsymbol{x}_i)$, for each case above using the sigmoid (logistic) function of Eq. (11.3.1).

(Hint: consider using Eqs. (11.3.6)–(11.3.8) for parts (a)–(c).)

11.2. Find the partial derivatives of the nonlinear dispersion function

$$D_n(\boldsymbol{\beta}) = \sum_{i=1}^{n} (y_i - f(\boldsymbol{x}_i^{\mathsf{T}}\boldsymbol{\beta}))a_n(R_{n_i}(\boldsymbol{\beta})),$$

with respect to $\beta_j,\ j = 0, 1$ where the nonlinear function is

$$f(\boldsymbol{x}_i^{\mathsf{T}}\boldsymbol{\beta}) = \frac{1}{1 + \exp(-\boldsymbol{x}_i^{\mathsf{T}}\boldsymbol{\beta})}.$$

(Hint: you need to compute the derivative for the parameter and the intercept, i.e. two cases. See Eq. (11.3.31) for guidance.)

11.3. The linear rank dispersion function can be written as

$$D_n(\beta_0, \boldsymbol{\beta}) = \sum_{i=1}^{n}(y_i - (\beta_0 + \boldsymbol{x}_i^\top \boldsymbol{\beta}))a_n(R_{n_i}(\beta_0, \boldsymbol{\beta})).$$

(a) Show that linear rank is intercept invariant. (Hint: see Chapter 1.)
(b) Next consider the rank dispersion function for the nonlinear case of Problem 11.2. What can be stated about the invariance of the intercept for this case?

11.4. We discussed the support vector classifiers (SVC) in this chapter and mentioned that they are large-margin classifiers. Consider Figures 11.4 and 11.5.
(a) Which of the plots would be more representative of a linear boundary produced by an SVC program? (Hint: SVC has problems with two outliers.)
(b) Identify the likely support vectors in Figure 11.4(b). How many and what type can you find? (Hint: draw two parallel dotted lines around the boundary that touch at least one data point on both sides.)
(c) SVCs are robust in a different sense than rank methods. How are they different? (Hint: consider outliers vs. support vectors.) Also, comment on why SVC would produce Figure 11.4(a) given the data of Figure 11.5.

11.5. (PROJECT) This problem will provide the needed information to run the mtcars example in R using built-in LLR routines and data.
(a) To access the mtcars data set, use the command:
```
mtcars <- datasets::mtcars
```
Next, we want to select the hp and wt features and use am as the response. We can invoke glm() to obtain the estimates using:
```
logitMod <- glm(am~hp+wt,
   family="binomial", data = mtcars)
```
Finally, we can display the estimates using:
```
summary(logitMod)
```
(b) Next, we want to show a scatter plot of the data that we have just processed.
```
input <- mtcars[,c("wt","hp")]
# Plot the chart for cars with wt and hp.
plot(x = input$wt,y = input$hp,
     xlab = "WT", ylab = "HP", main = "HPvsWT",
     xlim = c(1.5,5.5), ylim = c(50,350))
```

(Adjust the scatter plot to show 1 and 0 with different colors, if desired.)

(c) Using the coefficients from the summary(), plot a line representing the decision boundary on the previous plot. You will only need two points to construct the line. Note that the estimates define the boundary as $\beta_0 + \beta_1 wt + \beta_2 hp = 0$. This can be used to create two points to draw a line. Here is an example of a possible pair of coordinates.

```
xwt <- c(2.5,4.0)
yhp <- c(37, 370)
lines(xwt, yhp, col="black", lty=3)
```

Compare your results with the results shown in Figure 11.3(a).

11.6. (PROJECT) In this problem, you will write a simple program in R for rank-based logistic regression (RLR) and test it using the data from the last problem (assumes you ran the code shown above) to reproduce the results shown in Figure 11.3(b). Note that to compute $D_n(\boldsymbol{\beta})$, you will need to compute $a_n(R(\boldsymbol{\beta}))$ which requires ranking the residuals and applying the scoring function as described in this chapter. In particular, the equation below will prove to be very useful,

$$\hat{\boldsymbol{\beta}}_n^{RLR} = \underset{\boldsymbol{\beta}}{\operatorname{argmin}} \left\{ \sum_{i=1}^{n} (y_i - f(\boldsymbol{\beta}^T \boldsymbol{x}_i))\sqrt{12} \left(\frac{R_{n_i}(\boldsymbol{\beta})}{n+1} - \frac{1}{2} \right) \right\}.$$

(a) The first step is to write a routine for the sigmoid function and then test it to ensure that it operates properly. Enter and plot this function in R.

```
sigmoid <- function(z){
    p = 1/(1+exp(-z))
    return(p)
}
```

(b) Implement, test and plot the scoring function for $0 < u < 1$ as follows:

```
phi <- function(u){
    result <- sqrt(12.)*(u-0.5)
    return(result)
}
```

(c) Next, insert a column of 1s into the data for the intercept term β_0, and then define the needed matrix and vector quantities.

```
mtcars$ones <- rep(1,32)
```

```
x <- model.matrix(am~hp+wt+ones, mtcars)[,-1]
y <- mtcars$am
n <- length(y)
an <- rep(0,n)
Dn <- rep(0,n)
```

(d) Finally, construct the function to minimize the dispersion function.

```
minimize.RLR <- function(par, X, Y) {
    diff <- Y-sigmoid(X %*% par)
    Rn <- rank(diff,ties.method = "first")
    for ( i in 1:n) {
        an[i] <- phi(Rn[i]/(n+1))
        Dn[i] <- diff[i]*an[i]
    }
    Dnsum <- sum(Dn)
    return(Dnsum)
}
```

Test this function with an initial guess as the first parameter. What is your resulting value?

```
minimize.RLR(c( 0.05,-10.,20.),x,y)
```

(e) Compare the RLR boundary line with that of the previous problem. Use

```
op.result <- optim(c( 0.05,-10.,20.),
        fn = minimize.RLR, method = "BFGS",
        X = x, Y = y)
betaR <- op.result$par
```

What are the estimates when the R optimizer is used on minimize.RLR? Plot the boundary line using the estimates. Here is an example of a possible pair of coordinates.

```
xwt <- c(2.5,4.0)
yhp <- c(-50, 420)
lines(xwt, yhp, col="red", lty=3)
```

11.7. (PROJECT) Modify the minimize.RLR routine to include ridge regression.

(a) Update your R program to minimize $D_n(\boldsymbol{\beta})+\lambda_2\sum_{j=1}^{p}\beta_j^2$ and ensure that β_0 is not included as follows:

```
lambda2 <- 0.0
```

```
Dnsum <- sum(Dn)+
  lambda2*(par[1]*par[1]+par[2]*par[2])
```

(b) Plot a ridge trace on a $\log(\lambda_2)$ scale by choosing λ_2 values from 0 to 1.0 such as $0.001, 0.01, 0.1, 1.0$.

11.8. (PROJECT) Update your program to use LASSO regularization and generate the plot for LASSO trace on a $\log(\lambda_1)$ scale.

(a) Minimize $D_n(\boldsymbol{\beta}) + \lambda_1 \sum_{j=1}^{p} |\beta_j|$. Here, you can use:

```
lambda1 <- 0.0
Dnsum <- sum(Dn)+
  lambda1*(abs(par[1])+abs(par[2]))
```

(b) Choose $\lambda_1 = 0$ to 10, with values such as $0.01, 0.1, 1.0$, and plot the results.

(c) What is the most important feature based on your RLR-LASSO plot, hp or wt?

11.9. (PROJECT) Update your program to use aLASSO regularization and generate the plot for the aLASSO trace.

(a) First the unpenalized estimates must be captured. Execute the following two lines once:

```
bR1 <- betaR[1]
bR2 <- betaR[2]
```

Note that bR1= $\hat{\beta}_1^{\text{RLR}}$ and bR2 $= \hat{\beta}_2^{\text{RLR}}$. These are the rank estimates for $\lambda_1 = 0$.

(b) Next, to minimize $D_n(\boldsymbol{\beta}) + \lambda_1 \sum_{j=1}^{p} |\beta_j|/|\beta_j^{\text{RLR}}|$, the `minimize.RLR` routine must be modified as follows:

```
lambda1 <- 0.0
Dnsum <- sum(Dn)+
lambda1*(abs(par[1]/bR1)+abs(par[2]/bR2))
```

(c) Re-run the optimization multiple times with different `lambda1` values. Use the range given by $\lambda_1 = 0$ to 100, using values such as $0.01, 0.1, 1.0, 10.0$. Plot the results on a $\log(\lambda_1)$ scale.

(d) What is the most important feature based on your RLR-aLASSO plot: hp or wt?

11.10. Examine the ridge, LASSO and aLASSO traces of the Titanic data set of Figures 11.13–11.15.

(a) Qualitatively speaking, which provides the smoothest shrinkage characteristics.

(b) Which one provides the most accurate variable ordering?

(c) Which one appears to be a combination of the two?

11.11. For logistic regression, it is not possible to use the adjusted R^2 for subset selection as presented in Chapter 2. Describe how a 5-fold or 10-fold cross-validation scheme can be used to select the "optimal" size of the model for the Titanic data set. In particular, consider the test set MSE as a criterion for model selection.

12

Rank-based Neural Networks

12.1 Introduction

Our objective in this chapter is to build upon Chapter 11, which covered rank-based logistic regression, and investigate the feasibility of applying rank-based methods to neural networks. The question here becomes, can rank-based methods be effectively used in conjunction with neural networks? Since we already know that rank-based methods are applicable to logistic regression, it follows naturally that it should be applicable to neural networks. Then the issue at hand is to determine what are the advantages of rank-based neural networks over log-likelihood neural networks for machine learning. This is the main question we will explore in this chapter. The advanced nature of the this topic requires that the reader have some degree of familiarity with neural networks and how they operate. An excellent book to consult for details on the subject is Haykin (2009). We strongly urge the novice reader to consult with this reference, or perhaps any other suitable book, as needed.

Work on neural networks (NNs) began in earnest by computer scientists in the mid-to-late 1980s and early 1990s in the pursuit of artificial intelligence systems. Neural networks were initially touted as a way to mimic the behavior and operation of the brain, although that connection is not completely embraced today. There was much excitement at the time regarding the potential of NNs to solve many unsolvable problems due to breakthroughs such as the use of the sigmoid function for activation in conjunction with development of efficient methods for back propagation. But there were technological barriers to its success at that time. In particular, computers were slow and the era of big data had not arrived and so NNs were relegated to being a niche technology and lay dormant for many years. In the intervening years, computer power increased significantly, new and efficient algorithms for neural networks were developed,

Rank-Based Methods for Shrinkage and Selection: With Application to Machine Learning.
First Edition. A. K. Md. Ehsanes Saleh, Mohammad Arashi, Resve A. Saleh, and
Mina Norouzirad.

and the Internet emerged to provide easy access to large data sets for users around the world.

Around 2010, the neural network emerged from its dormant state and made an impact in many important applications, most notably speech recognition, image processing and natural language processing. In medical applications, image recognition was able to quickly and accurately identify cancerous tissue and, as a result, NNs gained a strong foothold in radiology. Other advancements in supervised and unsupervised learning (James et al., 2013) were developed in the same period. Collectively, these methods were re-branded as "machine learning". This term caught the attention of the media and is used to describe almost all applied statistical methods in use today.

Our purpose in this chapter is rather limited to a comparison of new rank-based NNs (RNNs)[1] against traditional log-likelihood neural networks (LNNs) for the application of image recognition. Here, we view NNs as simply multiple layers of logistic regression. If we have only one layer, it is standard logistic regression. If we have two or three layers it is a shallow NN. If there are many layers of logistic regression, it is called a deep NN, which is used for deep learning. Thc key difference between the early work on NNs and its ubiquitous presence today is the availability of big data. Images, video files, audio files, etc., are well-known forms of data for deep learning, but there are an uncountable number of other data sets that are also being processed using deep NNs.

Today, when the term "deep learning" is used, it refers to, for example, advanced convolutional NNs (CNNs) where the first few layers involve convolutional and maxpool layers, and the last layers are fully connected. The architecture of CNNs can be quite complex. These and other aspects of neural networks are well beyond the scope of this book. Fortunately, there are an unlimited number of resources for readers who seek an in-depth study of neural networks. Here, we focus on fully connected layers in order to compare RNNs with LNNs. It is tempting to use the term "multi-level logistic regression" to describe the fully connected architecture. However, the terms neural networks and deep learning are much more provocative and captivating than multi-level logistic regression, so they are now commonly used to describe work in this area. The meaning of layers and networks will become clear later in this chapter but the basic idea is that complex relationships between the independent variables can be realized by increasing the feature space of the problem.

Like logistic regression, neural networks are well-suited to binary classification problems and probability prediction, i.e. the likelihood that an event will

1 The acronym RNN is commonly used for recurrent neural networks. However, here we use it exclusively for rank-based neural networks.

occur. It is also used for multi-class classification. We will look at both types of classification in the context of image recognition. Both LNNs and RNNs have some advantages over support vector machines (SVMs), as described in Chapter 11, in that they both provide probability prediction and handle multi-class problems directly. Further, we will demonstrate that RNNs will outperform LNNs in image classification with outliers present and will provide equivalent performance when they are not present. We will present the loss functions for both methods that are used to estimate parameters of the network. Optimization of these types of networks is carried out using gradient descent but they are usually accelerated using "momentum", which will also be described. The method of creating the gradient terms through back propagation will also be included. In addition, accuracy metrics suitable for binary and multi-class classification will also be presented. Finally, three different applications will be used to compare and contrast RNNs and LNNs.

12.2 Set-up for Neural Networks

We begin with what appears to be a short review of logistic regression but actually forms the basis of neural networks. Recall that $Y_i \in \{0, 1\}$ is the dichotomous dependent variable and $\boldsymbol{x}_i = (1, x_{i1}, \ldots, x_{ip})^T$ is the $(p + 1)$-dimensional vector of independent variables for the ith observation. Then, the conditional probability of $Y_i = 1$ given $\boldsymbol{x}_i$ is

$$P(Y_i = 1|\boldsymbol{x}_i) = (1 + \exp(-\boldsymbol{x}_i^{\top}\boldsymbol{\beta}))^{-1}, \tag{12.2.1}$$

where $\boldsymbol{\beta} = (\beta_0, \beta_1, \ldots, \beta_p)^T$ is the (p+1)-dimensional vector of regression parameters. The corresponding logistic regression model is as follows

$$\log\frac{p(\boldsymbol{x}_i)}{1 - p(\boldsymbol{x}_i)} = \beta_0 + x_{i1}\beta_1 + \ldots + x_{ip}\beta_p, \tag{12.2.2}$$

where the sigmoid function is

$$p(\boldsymbol{x}_i) = \frac{1}{1 + \exp(-\boldsymbol{x}_i^{\top}\boldsymbol{\beta})}. \tag{12.2.3}$$

We can develop an estimator for $\boldsymbol{\beta}$ using maximum likelihood principles starting with

$$\mathcal{L}(\boldsymbol{\beta}) = \prod_{i=1}^{n}(p_i)^{y_i}(1 - p_i)^{(1-y_i)}, \tag{12.2.4}$$

where

$$p_i = p(\boldsymbol{x}_i) = \frac{1}{1 + \exp(-z_i)}, \tag{12.2.5}$$

and

$$z_i = \beta_0 + x_{i1}\beta_1 + \dots + x_{ip}\beta_p. \tag{12.2.6}$$

The (negative) log-likelihood loss function is

$$J(\boldsymbol{\beta}) = -\sum_{i=1}^{n}[y_i \log p_i + (1 - y_i)\log(1 - p_i)]. \tag{12.2.7}$$

Then, the resulting estimator of log-likelihood neural network (LNN) is

$$\hat{\boldsymbol{\beta}}_n^{\text{LNN}} = \underset{\boldsymbol{\beta}}{\text{argmin}}\left\{-\sum_{i=1}^{n}[y_i \log p_i + (1 - y_i)\log(1 - p_i)]\right\}.$$

We minimize this convex function using some form of gradient descent to obtain the $\boldsymbol{\beta}$ estimates.

In neural networks, parameter shrinkage is typically conducted using ridge so we will focus on that penalty function. LASSO, aLASSO and Enet penalties will not be covered in his chapter, although extensions of the equations to include these penalties are rather straightforward. We should mention that a method referred to as *dropout*[2] is widely used for regularization and the reader should investigate this method and other penalties further, if interested. We will focus on the ridge form of the penalized log-likelihood estimator as follows,

$$\hat{\boldsymbol{\beta}}_n^{\text{LNN-ridge}} = \underset{\boldsymbol{\beta}}{\text{argmin}}\left\{J(\boldsymbol{\beta}) + \lambda_2 \sum_{j=1}^{p} \beta_j^2\right\}.$$

Note that, typically, we only employ the penalty on $(\beta_1, \dots, \beta_p)^\top$ and not on the intercept parameter, β_0.

For rank-based methods, the same steps are applied except that the log-likelihood loss function above is replaced by the nonlinear dispersion function,

2 Dropout is a method whereby hidden units randomly selected in the neural network are literally dropped out of the network in each iteration to improve the generality of the model. Anywhere from 1% to 10% may be removed in a given iteration. This is probably the most well-known form of regularization in neural networks. Ridge is also used in conjunction with dropout for shrinkage and to control parameter magnitudes.

$$D_n(\boldsymbol{\beta}) = \sum_{i=1}^{n}(y_i - f(\boldsymbol{x}_i^{\top}\boldsymbol{\beta}))a_n(R_{n_i}(\boldsymbol{\beta})), \tag{12.2.8}$$

where the nonlinear function is

$$f(\boldsymbol{x}_i^{\top}\boldsymbol{\beta}) = \frac{1}{1+\exp(-\boldsymbol{x}_i^{\top}\boldsymbol{\beta})}. \tag{12.2.9}$$

Then, the resulting estimator for the rank neural network (RNN) is

$$\hat{\boldsymbol{\beta}}_n^{\text{RNN}} = \underset{\boldsymbol{\beta}}{\text{argmin}}\left\{\sum_{i=1}^{n}(y_i - f(\boldsymbol{x}_i^{\top}\boldsymbol{\beta}))a_n(R_{n_i}(\boldsymbol{\beta}))\right\}.$$

The penalized R-estimator for neural networks with ridge is

$$\hat{\boldsymbol{\beta}}_n^{\text{RNN-ridge}} = \underset{\boldsymbol{\beta}}{\text{argmin}}\left\{D_n(\boldsymbol{\beta}) + \lambda_2\sum_{j=1}^{p}\beta_j^2\right\}.$$

Thus far, the neural network appears to be the same as logistic regression, with the attendant consistency and asymptotic normality conditions satisfied. However this is because we are considering a neural network with an input layer and an output layer. There are no hidden layers or hidden units as yet. For this case, the neural network and logistic regression are identical, considering the sigmoid activation function. When we begin to add layers between the inputs and outputs, and then furnish each layer with a specified number of hidden units, it becomes a proper neural network. The number of layers, L, and the number of units in each hidden layer, U_k, where $k = 1, \ldots, L$, define the architecture of a fully connected neural network. The details of the architecture of neural networks and how to implement them are described in the next section.

12.3 Implementing Neural Networks

In this section, we describe practical aspects of the implementation of neural networks in the traditional log-likelihood form and the new rank-based approach. Any description of statistical learning algorithms in software involves the use of several tuning parameters, called hyper-parameters (to distinguish them from regression parameters), that are adjusted when applied to specific use cases. For example, the learning rate (ℓ), penalty factor (λ), number of layers (L), number of units in each layer (U_k, $k = 1, .., L$), momentum factor (γ) and ROC threshold (p_{T}) are among the many possible hyper-parameters.

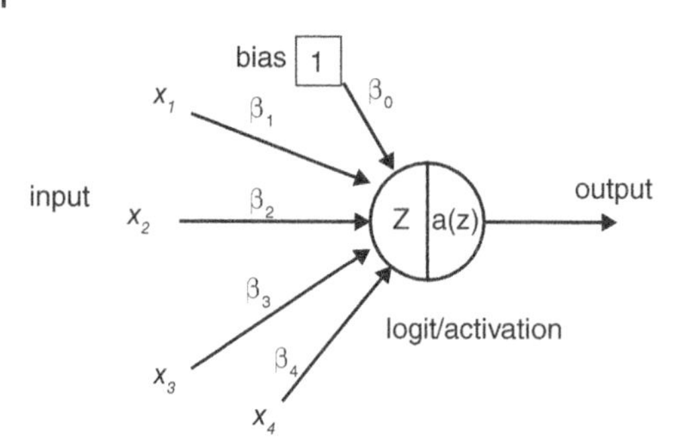

Figure 12.1 Computational unit (neuron) for neural networks. It is comprised of a set of inputs, a logit/activation function and an output. Neural networks are constructed by replicating this basic unit to form hidden layers to increase the feature space.

These hyper-parameters will all be defined in the sections to follow along with their purpose and typical ranges.

12.3.1 Basic Computational Unit

The basic building block of a neural network is comprised of a set of inputs, a logit/activation unit, and an output. The computation involves a linear combination of the inputs and the application of an activation function. The activation function is some type of nonlinear function that operates on the linear combination of inputs to produce the output. We can represent this two-step sequence of operations in terms of a data flow graph, as shown in Figure 12.1. The inputs are combined to form z and the output as determined by $a(z)$, the activation function.

Figure 12.1 illustrates a specific example: the case of four independent variables (plus the bias term[3], β_0) and one dependent variable. Each input, x_i, is multiplied by its respective edge weights, β_i, and summed together to form z,

$$z = \beta_0 + \beta_1 x_1 + \beta_2 x_2 + \beta_3 x_3 + \beta_4 x_4. \tag{12.3.1}$$

Then z is applied to the activation function, $a(z)$, to produce the output of the computational unit. The reason for the data flow representation will become clear shortly when it is used in a neural network diagram.

12.3.2 Activation Functions

The nonlinear part of the computational unit is referred to as the activation function. There are a number of possible choices for this function: sigmoid(z), relu(z), leaky-relu(z), elu(z), maxout(z), tanh(z), and so on. In Chapter 11, we

3 The bias term as used here can be viewed as an intercept term and is unrelated to the bias-variance tradeoff.

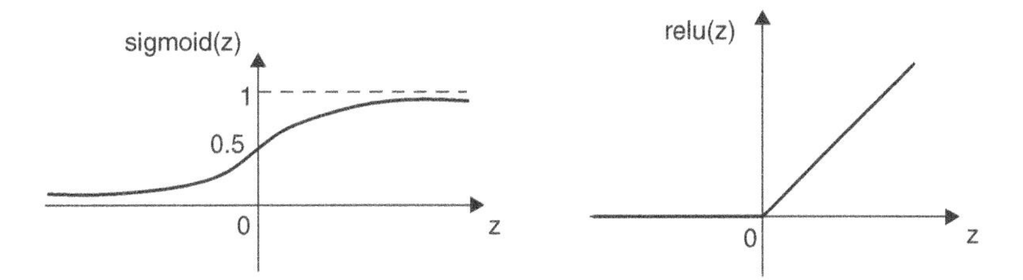

Figure 12.2 Sigmoid and relu activation functions. Typically, relu is applied to hidden units to speed convergence whereas sigmoid is applied to the output layer to provide probabilities.

described the sigmoid function in detail, shown as sigmoid(z) in Figure 12.2. In the computational unit, we would set

$$a(z) = \text{sigmoid}(z) = \frac{1}{1 + \exp(-z)}. \quad (12.3.2)$$

In expanded form, it becomes

$$a(z) = \frac{1}{1 + \exp(-(\beta_0 + \beta_1 x_1 + \beta_2 x_2 + \beta_3 x_3 + \beta_4 x_4))}. \quad (12.3.3)$$

For the moment, we should consider (12.3.3) and Figure 12.1 to be identical to one another. One is a mathematical equation while the other is a data flow representation of the equation – two sides of the same coin. The figure represents one neuron that we will replicate to build a four-layer neural network shortly.

We will need the derivatives of the activation function for use in gradient descent during optimization. The derivative of the sigmoid function is given by

$$\frac{\partial p(z)}{\partial z} = -\frac{\exp(-z)}{(1 + \exp(-z))^2} = -p(z)(1 - p(z)), \quad (12.3.4)$$

where

$$p(z) = \frac{1}{1 + \exp(-z)}.$$

We could use the sigmoid function as the activation throughout the neural network but unfortunately convergence is very slow during optimization. Therefore, alternatives were sought to speed up the rate of convergence. The most popular activation function in neural networks is called the rectified linear unit or relu function[4]. It has a number variants such as the leaky-relu and the elu, but we will not discuss them here. The relu is widely used in

4 Often, the acronym ReLU is used rather than relu. We will use relu in this chapter.

the hidden layers of neural networks. This function can be viewed as a "poor man's" sigmoid function, as can be seen from a quick glance at Figure 12.2. It is comprised of two line segments that can be represented as follows

$$\text{relu}(z) = zI(z) = \begin{cases} 0 & \text{if } z \leq 0 \\ z & \text{if } z > 0 \end{cases}.$$

The problem with the sigmoid function is that its derivative in Eq. (12.3.4) approaches zero for large positive or negative values of z. These near-zero gradients present major convergence problems in cases when large values of z (positive or negative) arise in the network. Convergence will be very slow since the gradient direction is essentially zero in these two cases. Instead, the relu function has gradients given by

$$\frac{\partial \text{relu}(z)}{\partial z} = I(z) = \begin{cases} 0 & \text{if } z \leq 0 \\ 1 & \text{if } z > 0 \end{cases}.$$

Now there is a slope of 1 in cases where $z > 0$. So the optimization method does not get trapped in local ravines as readily as in the case of the sigmoid function. There may be a concern that this relu function is too simple to be useful. In contrast, relu works extremely well in practice and is straightforward to implement. The relu function is used instead of the sigmoid function in all neurons except the output layer. There, we want to use the sigmoid function since the output represents a probability that must lie in (0,1). The problem with the relu function at the output is that it can take on any value greater than 0. Once the a relu output exceeds 1, it can no longer be viewed as a probability so we insist that it not be used in the output layer.

12.3.3 Four-layer Neural Network

We are now ready to take the next step with neural networks by increasing the number of layers from 1 to some number greater than 1. Consider Figure 12.1 once again. There are four inputs, x_1, x_2, x_3, x_4, to which we could attach four computational units, one at each input. That would create a new layer and 16 new inputs. If we work backwards and add another 4 units and then connect those inputs to the outputs of the new layer, we have two layers. If we repeat this one more time, we obtain three layers, which we will call *hidden layers* because they are not part of the input or the output layers. In doing so, we create a shallow neural network. The resulting fully connected four-layer neural network is illustrated in Figure 12.3 with the bias terms removed to reduce clutter in the diagram. It has three hidden layers and one output layer for a total of

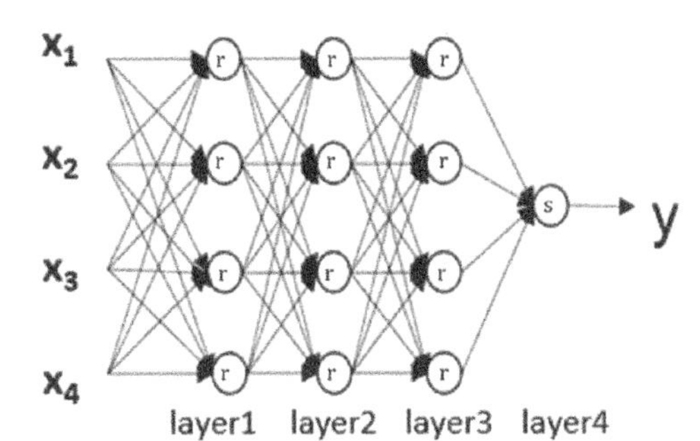

Figure 12.3 Four-layer neural network with relu (r) activation for hidden layers and sigmoid (s) activation for output layer (bias terms not shown). This is a fully connected architecture.

four layers [5]. This type of data flow diagram from input to output is often used when describing the architecture and implementation of neural networks. The hidden layers all use the relu activation function (annotated as "r") while the output layer uses the sigmoid activation function (annotated as "s"). This will be the case in all the examples to follow in this chapter.

The architecture of this neural network is symmetric but actual networks can be highly asymmetric. That is, the number of units in each layer may vary considerably, and the number of layers used depends on the characteristics of the application and the associated data set. We call this a fully connected neural network because every output of a prior layer connects to every input of a subsequent layer. In some sense, it seems like a waste of resources. But when the initial values are set (usually to very small random non-zero values), each one produces a different result. We could remove some of them if we wanted to, but it is not clear which ones to keep and which ones to kill. Dropout is a heuristic approach that randomly keeps and kills neurons in each cycle for regularization purposes.

Note that each arrow in the diagram represents a different parameter or feature, as shown in Figure 12.1. Notice what happens when we increase the depth of the neural network. The number of parameters in this network has increased considerably in this process. We can easily count the number of parameters associated with the four-layer neural network. There are three layers with 20 parameters per layer (16 arrows plus 4 more for the missing bias terms), and the output layer has 4+1 parameters, for a total of 65 parameters. We have greatly increased the feature space. As an aside, logistic regression would have used only 5 parameters which may lead to under-fitting the data set.

5 Note that we are not counting the input layer since it is fixed, or alternatively it could be called layer 0. There is a minor debate in the NN community of how to count layers. For our purposes, the number of layers L is determined by the number of activation layers. That is, the input does not count as a layer.

Depending on the number of observations, this architecture with 65 parameters could lead to over-fitting. Therefore, regularization using the ridge penalty function or dropout would be needed to balance the two effects. In any case, ridge is needed to keep the parameters from exploding due to the exponential in the sigmoid function. The next step is to figure out how to estimate the parameters.

12.4 Gradient Descent with Momentum

12.4.1 Gradient Descent

Neural networks inherently have a large number of variables, which are collectively referred to as the feature space. This implies that a large number of parameters must be estimated using an optimization method. In order to optimize multi-layered neural networks, a gradient descent approach is used. This is a numerical method that involves finding the minimum of a convex objective function using an iterative approach. Essentially, given an initial guess of the solution, the slope of the objective function is used to decide which direction to take in the solution space for the next step. So we will need to take partial derivatives of the objective function with respect to every parameter in the network. The size of the step taken is controlled using a factor called the learning rate, ℓ. If the step is too small, many iterations will be required. If too large, the process may not converge at all. After a proper selection of ℓ, the objective function is repeatedly evaluated, gradients are recomputed and parameters are updated, until the iterations smoothly converge to the minimum.

While this approach is theoretically appealing, the characteristics of the solution space will dictate the number of iterations needed to converge. The true nature of the solution space is hard to grasp since the number of dimensions is very high. Step size adjustment can only go so far in navigating a complex configuration space. Many different learning rates must be tried before a suitable one is found. But even with a suitable learning rate, the optimization procedure for neural networks may get stuck in ravines with sharp edges on each side. It may take hundreds of thousands of iterations to escape the ravine and move faster towards the minimum value.

To address this type of problem, a well-known method called "gradient descent with momentum" has been developed to speed up convergence in difficult solution spaces. The Adam optimizer contains this method as well as other heuristic techniques to improve convergence of neural networks. We will not discuss the Adam optimizer here but the interested reader should consult the literature online.

Gradient descent operates much like the name implies. The derivative of the objective function is used to compute the gradient which directs the optimization method towards the minimum of the convex function. We start with some initial value, $\boldsymbol{\beta}^{(0)}$, then take a step towards the minimum, and then repeat this in each iteration until we reach the minimum. The parameters are updated at each iteration using some fraction of the gradient direction controlled by the learning rate, ℓ, shown symbolically as follows, using t is the iteration count:

$$\boldsymbol{\beta}^{(t+1)} = \boldsymbol{\beta}^{(t)} - \ell \times \left(\frac{\partial J(\boldsymbol{\beta})}{\partial \boldsymbol{\beta}}\right)^{(t)}. \tag{12.4.1}$$

Here, the log-likelihood function, $J(\boldsymbol{\beta})$, is being minimized and we seek $\hat{\boldsymbol{\beta}}_n^{\text{LNN}}$. This iterative method requires partial derivatives of $J(\boldsymbol{\beta})$ in Eq. (12.2.7) with respect to all the parameters represented by $\boldsymbol{\beta}$. We will consider logistic regression first and ignore the penalty term for the purposes of this discussion. Later we will show how to do this for neural networks.

We must produce the partial derivatives of the log-likelihood function over all the n observations in the summation in Eq. (12.2.7). For convenience, let us consider the ith observation and let $p_i = p(\text{x}_i)$. We can apply the chain rule to obtain the needed gradients for the ith observation using

$$\frac{\partial J(\boldsymbol{\beta})}{\partial \beta_j} = \frac{\partial J(p_i)}{\partial p_i}\frac{\partial p_i}{\partial \beta_j}. \tag{12.4.2}$$

Then, computing the derivative of the first term with respect to p_i yields

$$\frac{\partial J(p_i)}{\partial p_i} = -\frac{y_i}{p_i} + \frac{1-y_i}{1-p_i}, \tag{12.4.3}$$

and for the second term with respect to each parameter, β_j, $j = 1 \dots p$, which involves the derivative of the sigmoid function given earlier, we obtain

$$\frac{\partial p_i}{\partial \beta_j} = p_i(1-p_i)x_j. \tag{12.4.4}$$

Combining the steps above, an update equation for each β_j to compute the next iteration value is given by

$$\beta_j^{(t+1)} = \beta_j^{(t)} - \ell\frac{\partial J(p_i)}{\partial p_i}\frac{\partial p_i}{\partial \beta_j} = \beta_j^{(t)} - \ell\sum_{i=1}^{n}(p_i - y_i)x_j. \tag{12.4.5}$$

Note here that we have to sum the quanities over the entire n observations in the data set, and this is done for each of the parameters, $j = 1, \dots, p$.

For a data set, X, which is an $n \times (p+1)$ matrix, the equations can be more conveniently formulated for use in programming languages with optimized matrix-vector capabilities, especially in R and Python. The logit vector is computed using an initialized set of $\boldsymbol{\beta}$ values as

$$z = X\beta. \tag{12.4.6}$$

The predicted values are computed using the sigmoid function, which would be implemented in R or Python using the expression:

$$p(X) = \frac{1}{1+\exp(-X\beta)} = \frac{1}{1+\exp(-z)}. \tag{12.4.7}$$

Here, the vector, $p(X)$, is a produced via a term-by-term application of the sigmoid function to vector, z. Next the loss is determined using the log-likelihood function

$$J(\beta) = -\frac{1}{n}\sum_{i=1}^{n}[y_i \log p(\mathbf{x}_i) + (1-y_i)\log(1-p(\mathbf{x}_i))]. \tag{12.4.8}$$

Finally, the β values are updated at each iteration t of a gradient descent-based optimization routine using

$$\beta^{(t+1)} = \beta^{(t)} - \frac{\ell}{n}X^T(p(X) - Y). \tag{12.4.9}$$

Here we seek to drive the second term to zero within a given accuracy tolerance using numerical techniques, and specifically we desire that at convergence

$$\mathrm{X}^T(p(\mathrm{X}) - \mathrm{Y}) = \mathbf{0}. \tag{12.4.10}$$

For the equivalent steps in the rank-based approach, define $p_i = p(x_i)$. Starting with Eq. (12.2.8), we follow the same steps outlined above to obtain

$$\frac{\partial D_n(\beta)}{\partial \beta_j} = -x_j \sum_{i=1}^{n} p_i(1-p_i)a_n(R_{n_i}(\beta)). \tag{12.4.11}$$

The update equation for each β_i to compute the new parameter values is given by

$$\beta_j^{(t+1)} = \beta_j^{(t)} - \ell \times \frac{\partial D_n(\beta)}{\partial \beta_j}. \tag{12.4.12}$$

Again, we note that the quantities are obtained by summing over the entire data set rather than a single observation, and that it is applied to all parameters, $j = 1, \ldots, p$.

12.4.2 Momentum

We now have established the procedure to perform gradient descent to obtain the estimates. However, one problem still remains. The number of iterations needed to reach an acceptable level of accuracy is controlled by the learning rate, ℓ. If ℓ is small, the rate of convergence can be slow which implies a large number of iterations. Slow convergence may be due to oscillations in the gradients in one or more dimensions in the optimization space. If we damp out the oscillations by averaging the gradients over several past iterations, it would allow for a more rapid convergence. A method called gradient descent with momentum was developed for this purpose. But how many past iterations should be used and how should they be combined? Momentum works by taking a combination of the previous gradients and the new gradient to produce an exponentially weighted moving average using the following equation for log-likelihood,

$$\boldsymbol{v}_{\mathrm{dB}}^{(t+1)} = \gamma \boldsymbol{v}_{\mathrm{dB}}^{(t)} + (1-\gamma)\left(\frac{\partial J(\boldsymbol{\beta})}{\partial \boldsymbol{\beta}}\right)^{(t+1)}, \tag{12.4.13}$$

and for the rank-based case,

$$\boldsymbol{v}_{\mathrm{dB}}^{(t+1)} = \gamma \boldsymbol{v}_{\mathrm{dB}}^{(t)} + (1-\gamma)\left(\frac{\partial D_n(\boldsymbol{\beta})}{\partial \boldsymbol{\beta}}\right)^{(t+1)}. \tag{12.4.14}$$

Initially, $\boldsymbol{v}_{\mathrm{dB}}^{(0)} = \mathbf{0}$. The smoothing parameter, γ, controls the number of gradients from previous iterations used in the averaging process and is typically set to 0.9. The rolling exponentially weighted average is used to update the parameters. For example, with $\gamma = 0.9$, the last nine gradients are averaged and added to the latest gradient. The weights for the trailing gradients decay exponentially. The resulting smoothed out composite gradient is then used to establish the direction of the next step. Then, the update equation for $\boldsymbol{\beta}$ becomes

$$\boldsymbol{\beta}^{(t+1)} = \boldsymbol{\beta}^{(t)} - \ell \boldsymbol{v}_{\mathrm{dB}}^{(t+1)}, \tag{12.4.15}$$

The use of momentum in conjunction with gradient descent has been found to be highly effective in neural networks. As mentioned above, momentum has been incorporated into an advanced optimization package referred to as the Adam optimizer. Although we will not describe further details of this optimizer, the interested reader should investigate this method in the literature.

12.5 Back Propagation Example

Thus far we have only discussed the use of gradients associated with one layer of the neural network – the output layer. Suppose there are two layers, a hidden layer and an output layer, as in Figure 12.4. In order to compute the full set of gradients, we must produce a gradient for every parameter with respect to the objective function. More specifically, we must produce a vector of five gradients between layers 1 and 2, i.e. $\beta_1, ...\beta_4, \beta_0$. These values are usually organized as a vector for the output layer. Between the input and layer 1, a matrix of 12 gradients must be computed, i.e. $\beta_{11}, \beta_{12}, \beta_{13}, ..., \beta_{23}, ..., \beta_{33}, \beta_{34}$. They are organized as a matrix for ease of storage and access. A total of 17 gradient terms are needed and the question is, how do we do it for a large neural network in an efficient manner?

Back propagation is a formalized algorithm to compute the vectors and matrices of gradients by starting at the output of the neural network and propagating backwards to the input, hence, its nickname: backprop. The basic idea of backprop is to use the chain rule to systematically compute the entries. This was already shown in Eqs. (12.4.2)–(12.4.4) for one layer. We simply need to apply it recursively as we traverse the entire network backwards. The easiest way to explain back propagation is to examine Figure 12.4 and walk through the steps for three different parameters in this two-layer network. The rest of the parameters are determined in a similar fashion. Although the example will use direct derivative calculations, the backprop algorithm stores data during a forward propagation step, and then uses that stored data during backward propagation to compute the gradients. In other words, the backprop algorithm is very efficient in the way it computes the gradients and this made neural networks computationally feasible.

12.5.1 Forward Propagation

The entire sequence of operations will now be described for one iteration using Figure 12.4. Each light back-arrow in this figure indicates the back-traces taken for the three parameters to be computed. There are a total of $3 \times 4 + 5 = 17$ parameters to compute in this example. The first step is to initialize all parameters to some small values. Let us assume that the parameters in the first-layer matrix are β_{ik}, $i = 1, 2, 3$, $k = 1, 2, 3, 4$ and the parameters in the second-layer vector are β_j, $j = 0, 1, 2, 3, 4$. Then, we can apply one input vector to the network, say $\boldsymbol{x} = (x_1, x_2, 1)^{\mathrm{T}}$, and follow the steps for forward and backward propagation. Figure 12.5 illustrates the steps for forward propagation for the network in Figure 12.4.

First, the input values and bias (i.e. 1) are multiplied by the $\boldsymbol{\beta}_{\text{matrix}}$ to produce the $\boldsymbol{z}$-vector in step 1 as follows

$$\boldsymbol{z} = \boldsymbol{\beta}_{\text{matrix}} \boldsymbol{x}, \tag{12.5.1}$$

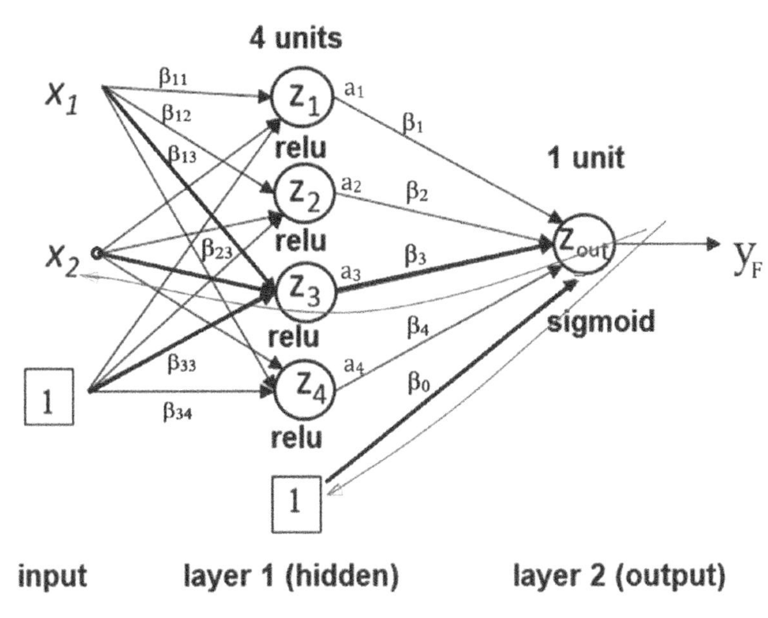

Figure 12.4 Neural network example of back propagation. In this case, β_3, β_{23} (upper light back-arrow) and β_0 (lower light back-arrow) are to be computed using backprop.

where

$$\boldsymbol{\beta}_{\text{matrix}} = \begin{bmatrix} \beta_{11} & \beta_{12} & \beta_{13} \\ \beta_{21} & \beta_{22} & \beta_{23} \\ \beta_{31} & \beta_{32} & \beta_{33} \\ \beta_{41} & \beta_{42} & \beta_{43} \end{bmatrix}.$$

These $\boldsymbol{z}$ values are "activated" in step 2 with relu to produce the $\boldsymbol{a}$-vector,

$$\boldsymbol{a} = relu(\boldsymbol{z}). \tag{12.5.2}$$

The $\boldsymbol{a}$-vector is augmented with 1s for the bias term to form the $\boldsymbol{a}1$-vector which is multiplied by the $\boldsymbol{\beta}_{\text{vector}} = (\beta_1, \beta_2, \beta_3, \beta_4, \beta_0)^T$ to produce z_{out} and passed through the sigmoid activation function to produce y_{F}. This is step 3 given by

$$z_{\text{out}} = \boldsymbol{a}1^{\top}\boldsymbol{\beta}_{\text{vector}}, \tag{12.5.3}$$

and step 4 as

$$y_{\text{F}} = p(z_{\text{out}}) = \frac{1}{1 + \exp(-z_{\text{out}})}. \tag{12.5.4}$$

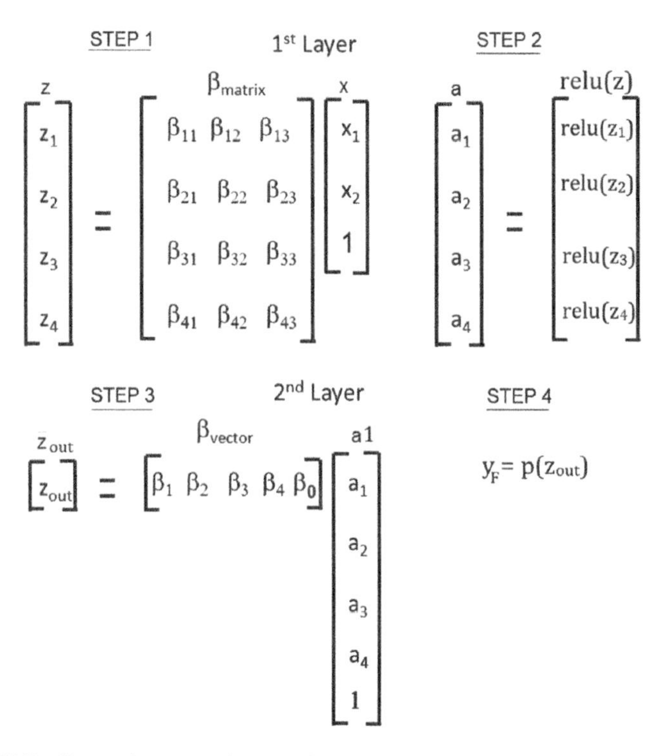

Figure 12.5 Forward propagation matrix and vector operations.

Therefore, the result of forward propagation through the second layer is

$$y_F = \frac{1}{1 + \exp(-(\beta_0 + \beta_1 a_1 + \beta_2 a_2 + \beta_3 a_3 + \beta_4 a_4))}. \tag{12.5.5}$$

Assume that the true value of the output is known to be y. The output error after completing the forward path is given by $y - y_F$. We need to drive this error to 0 by changing $\boldsymbol{\beta}$ in the network. The log-likelihood loss is computed using

$$J(\boldsymbol{\beta}) = -y\log y_F - (1-y)\log(1-y_F). \tag{12.5.6}$$

12.5.2 Back Propagation

Initially, y_F produced by forward propagation will not be correct. In order to improve its value on the next iteration, we must compute the gradients of the loss function to update all 17 parameters of the matrix and vector of Figure 12.5 and then apply one step of gradient descent with momentum. These 17 gradients are computed using back propagation which requires using the chain rule backwards through the network. This is a rather straightforward process mathematically but difficult to implement efficiently as an algorithm. The name back propagation is given to the set of algorithms that efficiently codify the use of the chain rule to obtain the needed gradients. We show the steps for β_3, β_0 and β_{23}.

We begin backprop by going backwards through the network, i.e. finding the partial derivatives of the objective function with respect to each of the parameters. The objective function is the log-likelihood loss, and the activation function is the sigmoid function. The loss is given by

$$J(y, y_{\mathrm{F}}) = -y \log y_{\mathrm{F}} - (1-y)\log(1-y_{\mathrm{F}}). \tag{12.5.7}$$

To obtain the partial derivative term associated with β_3, we use the chain rule:

$$\frac{\partial J(y, y_{\mathrm{F}})}{\partial \beta_3} = \frac{\partial J(y, y_{\mathrm{F}})}{\partial y_{\mathrm{F}}} \frac{\partial y_{\mathrm{F}}}{\partial \beta_3}. \tag{12.5.8}$$

Using the loss and sigmoid functions, the needed derivatives are evaluated to obtain

$$\frac{\partial J(y, y_{\mathrm{F}})}{\partial \beta_3} = \left(-\frac{y}{y_{\mathrm{F}}} + \frac{1-y}{1-y_{\mathrm{F}}}\right) y_{\mathrm{F}}(1-y_{\mathrm{F}})a_3 = (y_{\mathrm{F}} - y)a_3. \tag{12.5.9}$$

This is exactly what we would expect since the second layer is essentially logistic regression. To obtain the five parameters in the second layer, we use

$$\beta_j^{(t+1)} = \beta_j^{(t)} - \ell \times \frac{\partial J(\boldsymbol{\beta})}{\partial \beta_j} = \beta_j^{(t)} - \ell \times (y_{\mathrm{F}} - y)a_j \tag{12.5.10}$$

for $j = 0, 1, 2, 3, 4$. The update for the bias term can be directly obtained from this equation. For example, the arrow in Figure 12.4 associated with β_0 has an input of $a_0 = 1$. For this parameter, the update equation is given by

$$\beta_0^{(t+1)} = \beta_0^{(t)} - \ell \times \frac{\partial J(\boldsymbol{\beta})}{\partial \beta_0} = \beta_0^{(t)} - \ell \times (y_{\mathrm{F}} - y). \tag{12.5.11}$$

Next, the goal is to compute the update equation for β_{23}. We need to back propagate through two activation functions, as indicated by the long back arrow of Figure 12.4. Noting that $y_{\mathrm{F}} = (1 + \exp(\beta_0 + \beta_1 a_1 + \beta_2 a_2 + \beta_3 a_3 + \beta_4 a_4))^{-1}$, we take the path back through unit z_3 back to the input, x_2. The chain rule is as follows

$$\frac{\partial J(y, y_{\mathrm{F}})}{\partial \beta_{23}} = \frac{\partial J(y, y_{\mathrm{F}})}{\partial y_{\mathrm{F}}} \frac{\partial y_{\mathrm{F}}}{\partial a_3} \frac{\partial a_3}{\partial z_3} \frac{\partial z_3}{\partial \beta_{23}} = (y_{\mathrm{F}} - y)\beta_3 I(z_3)x_2. \tag{12.5.12}$$

The first two partial derivatives are straightforward. The last two partial derivatives depend on the type of activation function, which in this case is a relu function. The relu derivative is either 1 or 0, since it is a "keep or kill" term that depends on the value of z_3. From Figure 12.4, the bold arrows pointing into z_3 imply that

$$z_3 = \beta_{33} + \beta_{13}x_1 + \beta_{23}x_2 \tag{12.5.13}$$

and therefore

$$a_3 = z_3 I(z_3), \tag{12.5.14}$$

where $I(\cdot)$ is the indicator function. Hence, to obtain Eq. (12.5.12), we use

$$\frac{\partial J(y, y_{\mathrm{F}})}{\partial y_{\mathrm{F}}} \frac{\partial y_{\mathrm{F}}}{\partial a_3} = (y_{\mathrm{F}} - y)\beta_3 \quad \text{and} \quad \frac{\partial a_3}{\partial z_3} \frac{\partial z_3}{\partial \beta_{23}} = I(z_3)x_2. \tag{12.5.15}$$

Clearly, the term $I(z_3)$ is either 1 or 0. This aspect of the relu function may seem harsh at first but it is intended to improve the convergence of deep neural networks for $z > 0$. On the other hand, if the partial derivative is zero, the coefficient does not change which could slow convergence. Thus, variants of the relu, such as the leaky-relu and the elu that have non-zero derivatives in the $z \leq 0$ region, were proposed to address this issue. For our purposes, the standard relu function is sufficient.

All of the 12 partial derivatives of the gradient matrix would be computed in the same way for the first layer using the general equations for one observation,

$$\frac{\partial J(\boldsymbol{\beta})}{\partial \beta_{jk}} = (y_{\mathrm{F}} - y)\beta_k x_j I(z_k) \quad \text{and} \quad \beta_{jk}^{(t+1)} = \beta_{jk}^{(t)} - \ell \times \frac{\partial J(\boldsymbol{\beta})}{\partial \beta_{jk}}, \tag{12.5.16}$$

where $j = 1, 2, 3, 4$ and $k = 1, 2, 3$. Using the steps outlined above, all 17 gradients can be calculated to update the $\boldsymbol{\beta}$ matrix and vector in Figure 12.5. For n observations, we must sum all n gradients to produce the updated parameters.

12.5.3 Dispersion Function Gradients

For the rank-based approach, a similar set of operations are needed for back propagation. To obtain β_0, we would use

$$\frac{\partial D_n(\boldsymbol{\beta})}{\partial \beta_0} = -\sum_{i=1}^{n} y_{\mathrm{F}_i}(1 - y_{\mathrm{F}_i})a_n(R_{n_i}(\boldsymbol{\beta})), \tag{12.5.17}$$

where y_{F_i} is the ith value of the output associated with input $\boldsymbol{x}_i$. The needed quantities can be obtained after performing forward propagation.

To obtain β_3, we would use

$$\frac{\partial D_n(\boldsymbol{\beta})}{\partial \beta_3} = -\sum_{i=1}^{n} a_{3,i}\, y_{F_i}(1 - y_{F_i}) a_n(R_{n_i}(\boldsymbol{\beta})), \tag{12.5.18}$$

and if momentum is applied, we have that

$$v_{dB}^{(t+1)} = \gamma v_{dB}^{(t)} + (1 - \gamma)\frac{\partial D_n(\boldsymbol{\beta})}{\partial \beta_3}, \tag{12.5.19}$$

and, therefore,

$$\beta_3^{(t+1)} = \beta_3^{(t)} + \ell v_{dB}^{(t+1)}. \tag{12.5.20}$$

For β_{23}, the same type of formulation is used as for log-likelihood except for the use of the dispersion function. The needed partial derivatives are

$$\frac{\partial D_n(\boldsymbol{\beta})}{\partial \beta_{23}} = \frac{\partial D_n(y, y_F)}{\partial y_F}\frac{\partial y_F}{\partial a_3}\frac{\partial a_3}{\partial z_3}\frac{\partial z_3}{\partial \beta_{23}}. \tag{12.5.21}$$

Of course, these steps would have to be followed for all 17 parameters. In the general case, it can quickly become unwieldy, especially if there are many outputs, many hidden layers and many units per layer. The number of paths through the network can grow exponentially as the size of the network grows making deep learning intractable. However, the back propagation algorithm developed in the 1980s has rendered this a solved problem and led to the widespread use of neural networks and deep learning that exists today. We note that aside from the objective function itself, as well as its derivative being different from the log-likelihood case, the rest of the neural network derivative calculations are identical. This makes it very easy to replace LNNs with RNNs in existing machine learning programs.

12.5.4 RNN Algorithm

We can now combine all the elements discussed thus far into a simple algorithm for the rank-based neural network (RNN). The algorithm is provided in Table 12.1. The algorithm shown is for RNN-ridge. The algorithm for LNN proceeds in a similar manner, except for the use of the log-likelihood loss function in lines 12 and 15, so it is not shown.

In the setup phase, the algorithm begins by reading in the input data, $\boldsymbol{X}$ and $\boldsymbol{y}$. In the data set, $\mathbf{X}$, which is an $n \times (p + 1)$ matrix, we find that n is often very large, especially in the case of deep learning problems. As such, the whole data set is not usually processed all at once, which would be batch gradient descent. Instead, we create mini-batches of equal-sized subsets of rows of the data and process them one mini-batch at a time. This is called mini-batch gradient descent and each pass through all the data is called an epoch. Once completed, we return to the first mini-batch of rows and repeat the process until convergence, which may require hundreds of thousands of epochs. If we process one row of data at a time, it is referred to as stochastic gradient descent. In our case, we perform batch gradient descent (all the rows at once) such that one iteration is always equal to one epoch.

Next, the neural network architecture is defined by the number of hidden layers, and the number of units in each layer, U_k, $k = 1, ..., L$. The activation of the output is by the sigmoid function while all hidden layers use the relu function. Then, the remaining hyper-parameters are set: number of epochs (maxIter), ridge parameter (λ_2), learning rate (ℓ), and momentum factor (γ). Finally, matrices and vectors of parameters are initialized to random small values. These are referenced as $\boldsymbol{\beta}_k$, $k = 1, ..., L$, as there is one such entity for each layer.

The algorithm enters a while loop and executes a series of forward and backward propagation steps to determine the parameter values. The output is determined by a forward propagation step. The objective function is computed using $\boldsymbol{y}_{\mathrm{F}}$ and the sum of squares of the parameters. The backprop operation involves computing the gradients for each matrix or vector of parameters in the neural network and then taking one step of gradient descent with momentum. Each pass requires starting at the output layer L and traversing the network backwards to layer 1 to compute the gradients, $\nabla\boldsymbol{\beta}_k$, $k = 1, ..., L$.

After the while loop is completed, the objective function should be plotted graphically for each iteration to ensure that the proper value of ℓ and maxIter have been used. The convergence plot should be smooth and should reach a consistent final value at the latter stages of the iteration process. If not, the process must be repeated with different values of ℓ and/or maxIter. The final model is stored in $\boldsymbol{\beta}_k$, $k = 1, ..., L$.

12.6 Accuracy Metrics

To assess the accuracy of a binary image classification model, we could simply take the number of correct predictions and divide it by the total number of observations to determine the success rate. However, this type of analysis, although useful as a first-order assessment, may be misleading if most of the responses are 0 and only a few are 1. Then by guessing 0 in all cases, a high accuracy can be obtained but it would not be a true indication of the accuracy of the model. Instead, it is common practice to compute the F_1 score as follows. First, we categorize each image based on whether or not the model predicted the correct label of 1 or 0. There are four possible scenarios as listed in Table 12.2.

Table 12.1 RNN-ridge algorithm.

	Training procedure for rank-based neural network (RNN)
1:	Input: training data set $\boldsymbol{X}, \boldsymbol{y}$;
2:	Set n =number of observations, p =number of variables
3:	Network: number of layers (L); number of units in each layer (U_k), $k = 1, \ldots, L$
4:	Number of inputs = $p + 1$, , number of outputs= m ($m \leq 2$, binary; $m > 2$, multi-class)
5:	Activation: layer L = sigmoid, layers $1, \ldots, L-1$ = relu
6:	Set number of epochs (maxIter), learning rate (ℓ), ridge parameter (λ_2)
7:	Momentum (γ).
8:	Initialize matrices and vectors $\boldsymbol{\beta}_k^{(0)}$, $k = 1, \ldots, L$ to small non-zero values
9:	Set $t = 0$
10:	While $t <$ maxIter do
11:	Complete forward propagation through network using $\boldsymbol{X}$ and $\boldsymbol{\beta}_k^{(t)}$, $k = 1, \ldots, L$
12:	Update objective $D_n(\boldsymbol{\beta}) = \sum_{i=1}^{n}(y_i - y_{F_i})a_n(R_{n_i}(\boldsymbol{\beta})) + \lambda_2 \sum_{k=1}^{L} \boldsymbol{\beta}_k^{\top}\boldsymbol{\beta}_k$
13:	Use backprop to compute gradients, as follows:
14:	For $k = L$ downto 1 do
15:	Compute gradients $\nabla\boldsymbol{\beta}_k^{(t+1)}$ of $D_n(\boldsymbol{\beta})$
16:	Gradient descent w/momentum $v_{\mathrm{dB},k}^{(t+1)} = \gamma v_{\mathrm{dB},k}^{(t)} + (1-\gamma) \times \nabla\boldsymbol{\beta}_k^{(t+1)}$
17:	Update estimates $\boldsymbol{\beta}_k^{(t+1)} = \boldsymbol{\beta}_k^{(t)} - \ell \times v_{\mathrm{dB},k}^{(t+1)}$
18:	End for
19:	Update t = t + 1
20:	End while
21:	Final model: $\boldsymbol{\beta}_k$, $k = 1, \ldots, L$

For example, if the true value is 1 and the quantized model prediction is 1, it is referred to as a true positive (tp). Similarly, a true value of 0 and a predicted value of 0 is called a true negative (tn). On the other hand, incorrect predictions are either false positives (fp) or false negatives (fn), i.e. the model is confused. The combination of true values along the diagonal and false values on the off-diagonals forms a confusion matrix. This matrix is very useful in telling us what type of error we are making using a particular classifier. To obtain it, we can simply quantize each predicted probability and count up the numbers in each matrix entry. Then using this information, we compute the recall and precision, which are both standard terms in binary classification (cf. James et al., 2013). Finally, the F_1 score is computed using these two quantities to assess the overall effectiveness of the model.

Table 12.2 Interpretation of the confusion matrix. The four possible outcomes of tn, fn, tp or tp are accumulated in this table format from the ground truth and predicted binary values.

	True-0	True-1	Interpretation of matrix entries
Predicted-0	tn	fn	tn = true negative, fn = false negative
Predicted-1	fp	tp	fp = false positive, tp = true positive

Table 12.3 Confusion matrix for Titanic data sets using RLR (see Chapter 11).

Training set	True-0	True-1
Predict-0	494	114
Predict-1	55	228
Test set	**True-0**	**True-1**
Predict-0	221	51
Predict-1	43	105

In particular, the recall value is given by

$$\text{recall} = \frac{\text{tp}}{\text{tp} + \text{fn}}, \tag{12.6.1}$$

and the precision value is given by

$$\text{precision} = \frac{\text{tp}}{\text{tp} + \text{fp}}. \tag{12.6.2}$$

Then the F_1 score is computed using

$$F_1 = \frac{2 \times \text{precision} \times \text{recall}}{\text{precision} + \text{recall}}. \tag{12.6.3}$$

Two examples of the confusion matrix are given in Table 12.3 using the Titanic data set described extensively in Chapter 11. The data is shown using RLR for the training and test sets. From the training set matrix, the recall $= 228/(228 + 114) = 0.67$, the precision $= 228/(228 + 55) = 0.81$, and the F_1 score $= 0.73$. The maximum value of an F_1 score is 1 so this would be a relatively good performance for the classifier. Furthermore, the test set matrix produces a recall $= 0.67$, a precision $= 0.71$ and an F_1 score $= 0.69$. This indicates that the model produces similar results on the training set and the test set, which is a desirable characteristic.

The next issue is to decide how to quantize a predicted probability value. We have not discussed the issues around this process in detail for reasons that will be clear shortly. The actual (observed) value of the dependent variable y_i for any given input $\boldsymbol{x}_i$ is either 0 or 1. However, the predicted value $\tilde{p}$ from an LNN or RNN model lies between 0 and 1 but rarely reaches either value for reasons that should be clear from Eq. (12.3.2). The meaning of $\tilde{p}$ is that it represents the probability of a particular event occurring. For binary classification, the predicted output is compared to a threshold, typically $p_T = 0.5$. If $\tilde{p} > p_T$, it is taken as 1 whereas if $\tilde{p} \leq p_T$, it is taken as 0. This is how a probability is quantized into a binary value and then used to build a confusion matrix, from which F_1 scores can be calculated.

Using $p_T = 0.5$ as the threshold is not necessarily the optimal choice. Some other threshold may be more appropriate to improve accuracy of the classifier. Rather than trying to figure out the ideal p_T, a different metric is used. The receiver operating characteristics (ROC) can be an effective measure of the performance of a classifier for different thresholds when comparing two classifiers. The ROC is a plot produced using the true positive rate vs. the false positive rate for different p_T as follows. For each value of p_T, the true positive rate (TPR) is given by

$$\text{TPR} = \frac{\text{tp}}{\text{tp} + \text{fn}}, \tag{12.6.4}$$

and the true negative rate is given by

$$\text{TNR} = \frac{\text{tn}}{\text{tn} + \text{fp}}. \tag{12.6.5}$$

Then, the false positive rate (FPR) is given by

$$\text{FPR} = 1 - \frac{\text{tn}}{\text{tn} + \text{fp}}. \tag{12.6.6}$$

In Figure 12.6, we plot a representative TPR against FPR from the Titanic data set used in Chapter 11. The trajectory of the plot starts from (0,0) and traces a path to (1,1). Two such paths are shown. First, the diagonal line represents the case of a random value classifier. That is, a classifier that simply guesses the outcomes rather than computing the outcomes would produce a diagonal line which represents a worst-case classifier. Second, the line labeled ROC is the plot for an actual classifier that is clearly better than the random guess classifier since its trajectory is above the diagonal line. This particular curve is produced by selecting different values of p_T starting at 0 and ending at 1 in steps of 0.01 – a total of 101 points. The quality of a classifier can be established by finding the area under the ROC curve (AUROC). Specifically, the area under this ROC curve is 0.74 whereas the area under the random guess classifier is 0.5. Therefore, the former is better than the latter. More importantly, we now have one

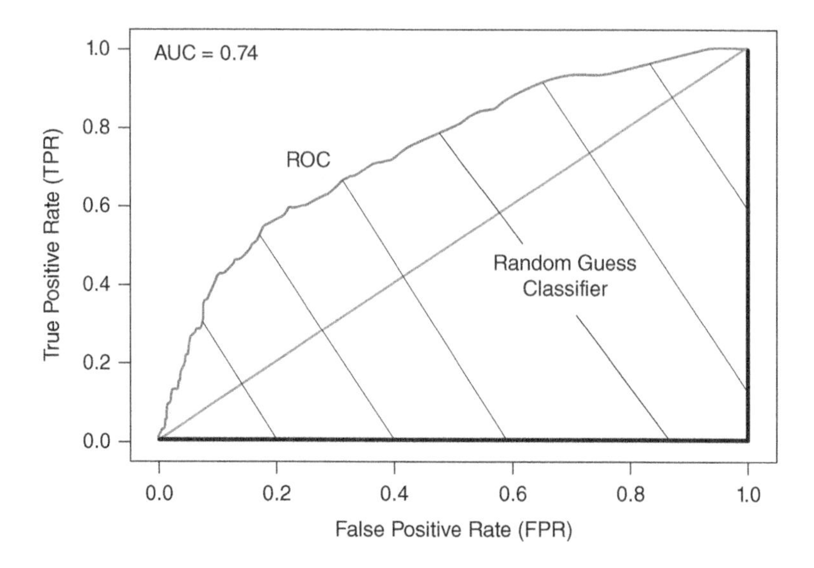

Figure 12.6 ROC curve and random guess classifier line based on the RLR classifier on the Titanic data set of Chapter 11.

metric, the AUC for short, that quantitatively compares two or more classifiers. The ideal value of course is AUROC = 1.0.

12.7 Example: Circular Data Set

Having established the two methods to be compared, along with the details of their implementation and the accuracy metrics, we can proceed to the next step of using standard data sets to understand the advantages and disadvantages of each approach. The first data set is intended to build up a basic understanding of the methods on a simple nonlinear classification problem from Chapter 11 whereby the data is largely separable by a circular or elliptical boundary, except for a few outliers. We compared log-likelihood logistic regression (LLR) and rank-based logistic regression (RLR). However, in both LLR and RLR, we needed to specify x_1, x_2, x_1^2 and x_2^2 as input variables since the solution has a nonlinear boundary. In other words, in logistic regression, a suitable polynomial expansion of the variables must be provided to obtain satisfactory results. Recall that, for support vector machines (SVM), only x_1 and x_2 are necessary because SVM will find a complex boundary using radially expanded regions around the data points controlled by a parameter called γ. Therefore, SVM has an advantage over LLR and RLR in that it does not require specification of the needed variables. But it has a shortcoming since it does not directly report probabilities that may be needed in many applications.

A similar capability exists in neural networks because there are usually many parameters in each model. There may be many different combinations of x_1 and x_2 required to create a highly nonlinear boundary for the classification problem but, with the right number of layers and hidden units, neural networks will find such a boundary. The trade-off here is that it will not be easy to determine *a priori* the best architecture for the application, and *a posteriori* what relationships exist between the variables and the estimates. If the neural network used x_1^2 and x_2^2 for example, we will not be able to tell from the final estimates. The parameters values are related to the variables but it will be difficult to make any connections between them as there are typically a large number of parameters in a neural network.

Furthermore, while regularization is very important in neural networks, the traces for ridge and LASSO are not meaningful here because of the large number of parameters and their unknown connection to the variables. Although penalty functions are useful, it is also difficult to glean any useful information about their associated traces as we did for logistic regression. As a practical matter, regularization techniques such as "dropout" (where random hidden units are simply dropped from the network during the iterative process) are more effective than ridge or LASSO. But we will need to control the size of the parameters and for that reason the shrinkage property of ridge, and to some extent LASSO, will be very useful.

Our objective at this stage is to compare fully connected neural networks based on log-likelihood and rank-based methods on a data set with a known circular/elliptical boundary. For this purpose, we will only examine the performance on a training set in this section, as we did in the previous chapter. We would also like to assess the relative importance of a penalty function in each case. For this purpose, we will apply only ridge regression to the circular data set since we know that there are only two variables and both x_1 and x_2 are important. Hence, subset selection is not needed.

The circular data set can be handled using the NN architecture of Figure 12.7, which should be quite familiar by now. We feed x_1, x_2, and a constant of 1 (representing the bias term) to a hidden layer with 4 units, which then feeds the output layer. The 4 hidden units use the relu activation function while the output unit uses the sigmoid function. The choice of a hidden layer with 4 units is based on knowledge about the problem itself. In general, it is not easy to determine the number of hidden units needed, or the number of layers

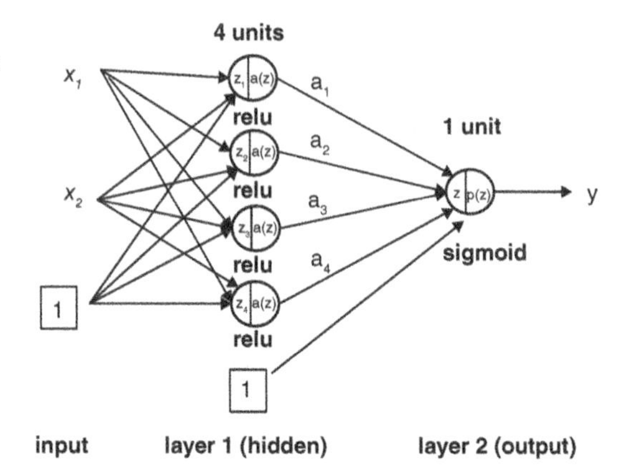

Figure 12.7 Neural network architecture for the circular data set. Hidden layer activation a(z) is relu, output layer activation p(z) is sigmoid.

for that matter.[6] In this case, we believe that it is likely we will need x_1, x_2, x_1^2 and x_2^2 which implies 4 hidden units at a minimum. In practice, a number of different NN architectures should be attempted with different numbers of layers and hidden units, and then the best architecture selected among all the combinations tried. This can be done using trial-and-error or a grid search.

The original data set is shown in Figure 12.8(a) which illustrates the two sets of points in black and gray. Note the black outliers in the sea of gray points at the top of the figure. There are 337 points, of which 6 are outliers. So, at best, the classier should get 331 correct out of 337. The other three plots show the color-coded predicted values on a synthesized test set, each producing a different decision boundary. We note that true boundaries in neural networks are not easy to specify with a contour line so we predicted thousands of points in the region to obtain the plots.

We see that RNN finds a solution that separates the data in a manner as if there were no outliers, as shown in Figure 12.8(b). However, the same cannot be said for LNN as they inadvertently capture incorrect data in their solutions and miss some correct data, shown in Figure 12.8(c). In other words, for this example, the LNN was influenced by the outliers. At first glance, it may appear that the LNN may not have converged to a final solution because of the irregularity of the boundary. However, if we check the convergence plots for RNN and LNN (loss vs. iteration) shown in Figure 12.9, we see that both have properly converged to their solutions smoothly and monotonically.

The characteristics of the convergence plots are rather interesting. In the case of LNN, there is a steep slope initially, followed by a plateau, and then finally another step to the final value. Many convergence plots for LNN exhibit

6 The website playground.tensorflow.org provides the user with some facilities to develop intuition about the number of layers and hidden units needed for the circular data set.

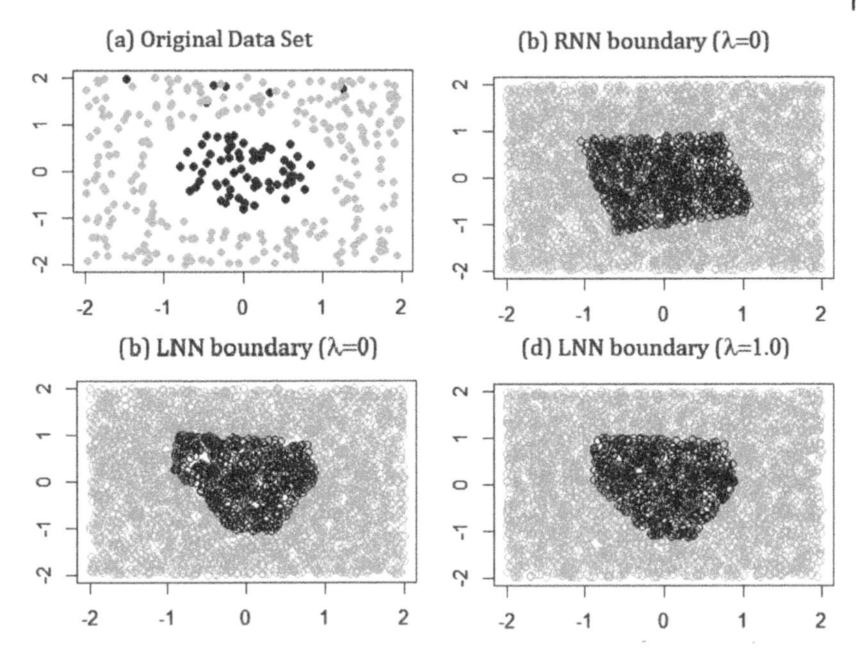

Figure 12.8 LNN and RNN on the circular data set (n = 337) with nonlinear decision boundaries. Part (a) shows the original data set. Parts (b), (c), and (d) show the irregular boundaries of each method. Number of correct predictions were (b) = 331, (c) = 327 and (d) = 331.

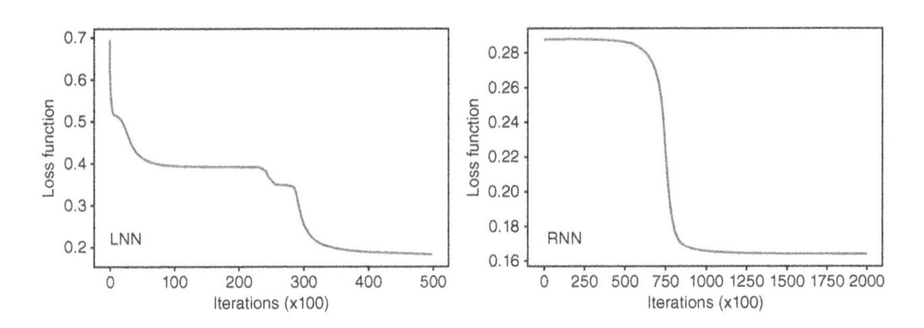

Figure 12.9 Convergence plots for LNN and RNN for the circular data set.

this type of behavior whereby the gradient descent method will get stuck in a narrow ravine and then eventually begin the drop to the minimum. A total of 50,000 iterations were needed to reach convergence.

The RNN convergence plot is also typical of rank-based methods. Initially we are walking along a plateau until we reach the edge and drop to the next level down and continue in this manner until we reach the solution. In

Table 12.4 Number of correct predictions (percentages) and AUROC of LNN-ridge and RNN-ridge on circular data set vs. λ_2 ($n = 337$ with 6 outliers).

λ_2 parameter	LNN-ridge correct	LNN-ridge AUROC	RNN-ridge correct	RNN-ridge AUROC
0.0	227 (97%)	0.974	331 (98%)	0.958
0.2	328 (97%)	0.978	331 (98%)	0.955
0.4	331 (98%)	0.977	331 (98%)	0.953
1.0	331 (98%)	0.974	331 (98%)	0.951
2.0	226 (97%)	0.803	331 (98%)	0.950

this case, the initial roll down the hill produces the solution. However, RNN required about 150,000 iterations to reach convergence. From the plots, we know that both LNN and RNN did indeed reach their minimum values, that RNN required three times as many iterations and that LNN was not able to produce a satisfactory boundary.

The last step is to determine if a ridge penalty would improve the results for either a LNN or RNN. It is hard to imagine how it would improve RNNs since it already has the desired solution, but we will compare the two in any case. If ridge is applied to LNN, we find that with $\lambda_2 = 0.4$ to 1.0, the results are beginning to resemble RNN. This is shown in Figure 12.8(d) for $\lambda_2 = 1.0$. Beyond this point, the solution gets worse. Hence, for LNN, ridge regression can be used to find a solution that is equivalent to RNN in this example. If we apply ridge to RNN, the results remain roughly the same. The optimal solution for RNN occurs immediately at $\lambda_2 = 0$. It appears that ridge when used with LNN has the effect of improving the solution, but not in the case of RNN since it already produces the proper boundary.

The results of the analysis for LNN and RNN with different values of λ_2 are provided in Table 12.4. We list the λ_2 values along with the number of correct classifications by LNN and RNN, as well as the area under the ROC curve (AUROC). Clearly, the results for LNN improve and then degrade as λ_2 is increased, while the RNN results remain the same (although the AUROC value does degrade slightly).

It is interesting to examine the ROC curves for the two methods when $\lambda_2 = 0$. This is shown in Figure 12.10. We see the two ROC curves for LNN and RNN are very similar, providing a near perfect classifier in both cases for this idealized example. The area under the ROC curve (AUROC) indicates that LNN is better than RNN but this may be misleading. Because of the outliers, is not able to produce the optimal boundary but the AUROC performance is higher. Hence,

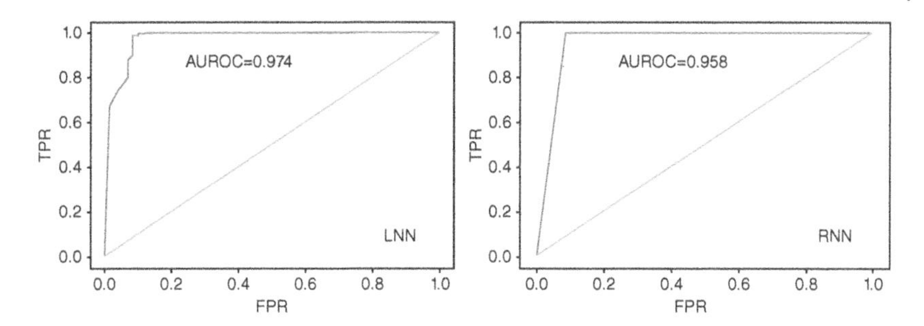

Figure 12.10 ROC plots for LNN and RNN for the circular data set.

AUROC by itself may not be sufficient if there are outliers in the data set as it does not tell the full story about the quality of the classifier. RNN is clearly better but AUROC would select LNN in this particular example.

12.8 Image Recognition: Cats vs. Dogs

In this section, we describe a well-known application of supervised learning for binary classification using neural networks, that of recognizing cats vs. dogs. This is a classic image recognition problem that is often used to demonstrate the power of neural networks. Here, we build neural network models using a small amount of training data and then validate the accuracy using a small number of test images and illustrate the effect of penalty functions. Then, we use a very large set of test data to capture a sense of the overall performance of the two methods, LNN and RNN.

The standard procedure in machine learning is shown in Figure 12.11. First, the supervised training data $(X_{\text{train}}, Y_{\text{train}})$ is applied to the selected architecture

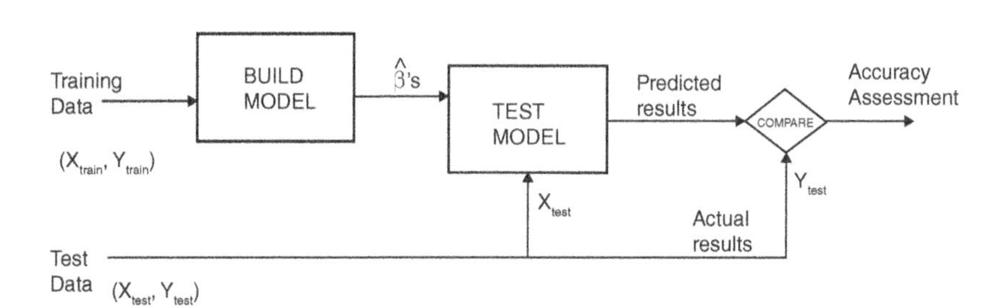

Figure 12.11 Typical setup for supervised learning methods. The training set is used to build the model. The test set is used to evaluate the model (i.e. estimates). The predicted and actual results are used to assess overall accuracy.

for neural networks in order to compute the $\boldsymbol{\beta}$ parameters. The accuracy of the model generated using the training data is usually quite high when measured on the same training data. Typically, it is 80–100% depending on the application. This can be misleading since the goal of the optimization is to have high accuracy on unseen data that is not in the training set. Therefore, it must also be evaluated on an independent set of test data ($X_{\text{test}}, Y_{\text{test}}$) to properly assess its accuracy. This is carried out by applying X_{test} to the model derived from neural network training and then comparing the predicted result $\tilde{P}$ against Y_{test}.

When given an initial data set, it is important to divide it into two sets, one for training and another for testing. For small data sets, 70% is used for training and 30% is used for testing. For larger data sets, it is possible to use 90% for training, with a 5-fold or 10-fold cross-validation scheme, and then 10% on the test set for a final model evaluation. Here we will consider a small training set (without cross-validation) and a small test set to simplify the presentation, and then use a very large unseen test set to assess the overall performance of LNN against RNN.

12.8.1 Binary Image Classification

In this section, we will compare LNN and RNN in the cat vs. dog problem. Two representative members of the data set are shown in Figure 12.12. The goal is to demonstrate the viability of RNN on a realistic application, so we will not strictly adhere to the guidelines of the size requirements for proper training and testing. Specifically, the training set has 100 images, with 50 cats and 50 dogs, and we label each one appropriately and use this to train a neural network. No outliers are inserted into the training set. After training, we test the model and provide detailed analysis on a set of 35 test images, with 19 cats and 16 dogs, to evaluate the accuracy of the model. Then we run the model on a larger 700 image test set to determine if RNN is competitive with LNN in situations without any outliers in the training set.

12.8.2 Image Preparation

In order to carry out binary classification, images must be converted into numerical values to build a design matrix $\boldsymbol{X}$ and a binary response vector $\boldsymbol{y}$. The basic idea is to partition an image into picture elements (pixels) and assign a number to each pixel. Color images have pixels that can be either red, green or blue (RGB) and the intensity level of each is specified by a numerical value. Therefore, each pixel is represented by three integer values in the range 0–255, and the entire image is then represented by the number of pixels vertically multiplied by the number of pixels horizontally.

Figure 12.12 Examples from test data set with cat = 1, dog = 0.

X_{33} X_{34} X_{35} X_{36} X_{40} X_{44} X_{48} B

X_{17} X_{18} X_{19} X_{20} X_{24} X_{28} X_{32} G

X_1	X_2	X_3	X_4
X_5	X_6	X_7	X_8
X_9	X_{10}	X_{11}	X_{12}
X_{12}	X_{14}	X_{15}	X_{16}

R

R	G	B
X_1 X_2 X_3 ... X_{16}	X_{17} X_{18} X_{19} ... X_{32}	X_{33} X_{34} X_{35} ... X_{48}

Figure 12.13 Unrolling of an RGB image into a single vector. Each image has three channels of data. In this example, each channel is a 4×4 image of either R, B or G. They are unrolled into one vector of length $4 \times 4 \times 3 = 48$.

Each image is pre-processed as shown in Figure 12.13. For simplicity in our illustration, each image is divided into 4×4 pixels as shown in the figure to produce a total of 16 pixels. However, there are three possible color combinations (RGB) for each pixel resulting in a total of $4 \times 4 \times 3 = 48$ pixels in this example. Next, the values are "unrolled"to create one long $\boldsymbol{x}$ vector that represents one image[7]. The same operation is carried out for all images

7 A CNN is most appropriate for image recognition purposes. However, here we are using a fully connected architecture so we require that the input be a vector. Hence, "unrolling" of the image is necessary.

to produce the $\boldsymbol{X}$ matrix, with one row for each image, and the $\boldsymbol{y}$ vector with observed value of 1 (= cat) or 0 (= dog).

In the actual case to be presented in the experimental results, images are divided into $64 \times 64 \times 3 = 12288$ pixels to obtain higher resolution for image recognition purposes. A total of 100 training images are used to build models, and 35 test images to evaluate the model. Therefore, the training set $\boldsymbol{X}$ matrix is 100×12288 with $\boldsymbol{y}$ as 100×1, while the test set is 35×12288. As a side note, we should state that typically 50,000 images may be used in the training set in a five-fold validation scheme to produce the parameter estimates. In our case, we are trying to demonstrate the feasibility of rank-based neural networks, and then comparing LNN to RNN so we have used smaller data sets.

Table 12.5 shows representative results from this type of pre-processing of images. Column 1 indicates the training set observation number, where we have listed only 10 samples of the training set. The next p columns are the input values $x_{i1}, x_{i2}, \ldots, x_{i12288}$ associated with each training image. Only a few of the 12288 data columns are shown here. The next column in the table is y_i which is the true label (0 or 1) for the image. And finally, $\tilde{p}(\boldsymbol{x}_i)$ is the value predicted by the model which will range between 0 and 1 since it is the output of a sigmoid

Table 12.5 Input (x_{ij}), output (y_i) and predicted values $\tilde{p}(\boldsymbol{x}_i)$ for the image classification problem. $p = 12288$ for unrolled images. Input pixels are converted to intensity values and scaled to lie in [0,1]. Output labels lie in the set {0,1}. Predicted values are in the range (0,1) and then quantized to be in the set {0,1}. Final column indicates confusion matrix category.

i	x_{i1}	x_{i2}	$\cdots$	x_{i12288}	y_i	$\tilde{p}(\boldsymbol{x}_i)$	
1	0.306	0.898	$\cdots$	0.902	1	$0.602 = 1$	tp
2	0.294	0.863	$\cdots$	0.871	1	$0.507 = 1$	tp
3	0.361	0.835	$\cdots$	0.714	1	$0.715 = 1$	tp
4	0.310	0.902	$\cdots$	0.507	0	$0.250 = 0$	tn
5	0.298	0.871	$\cdots$	0.222	1	$0.644 = 1$	tp
6	0.365	0.863	$\cdots$	0.625	1	$0.369 = 0$	fn
7	0.314	0.741	$\cdots$	0.306	1	$0.628 = 1$	tp
8	0.302	0.733	$\cdots$	0.173	0	$0.602 = 1$	fp
9	0.369	0.745	$\cdots$	0.122	0	$0.456 = 0$	tn
10	0.318	0.757	$\cdots$	0.174	0	$0.343 = 0$	tn

function. This value is compared to 0.5 to quantize it to a value of either 0 or 1, which is also provided in the table alongside its confusion matrix category.

12.8.3 Over-fitting and Under-fitting

Binary classification problems also face the issue of bias-variance trade-offs, more commonly known as over-fitting vs. under-fitting. The method of using a training set to build a model has a potential problem of over-fitting. When there are a large number of variables, such as 12288 in this case, there is a good chance that the parameters will be tailored to fit the training set but may not generalize to an unseen test set. In other words, we may produce parameters that predict the training set with 100% accuracy but fail to reach that level of accuracy on the test set. The root cause of this problem is due to the bias-variance trade-off described in Chapter 2. The parameter values will have a high variance and low bias, which implies over-fitting. We want to reduce the variance using ridge regularization. However, if we apply ridge and use a large tuning value of λ_2, we may produce parameter values with a low variance but a rather high bias, which implies under-fitting. This result will not generalize either. Therefore, we need to find the "goldilocks" value of λ_2 to provide a set of parameters from the training set that generalizes to the test set.

In Figure 12.14, we illustrate conceptually the effect of regularization in a two-dimensional space. The 1s represent cats while 0s represent dogs. The model will define the decision boundary between the two classes. In Figure 12.14(a), when ridge is not used, the decision boundary is some complex path through the data as shown in bold. There are so many parameters, 12288 to be specific, that we can over fit the training set. This over-fitted situation produces 100% accuracy on the training data. However, if we use ridge with a very large value of λ_2, it would eventually lead to under-fitting as depicted by the straight line in the same figure.

With a proper choice of λ_2 for ridge, a more reasonable boundary is produced, as in Figure 12.14(b). In this case, the training data accuracy is below 100% because some cats have been misclassified as dogs and vice versa. On the other hand, it will likely be more accurate on test data since the decision boundary is more generalized. The inherent errors are due to the fact that some cat images may appear to look like dogs, while dog pictures may appear to look like cats, and those occur at or near the boundary. But in general, the smoother curve is preferred as opposed to the over-fitted or under-fitted decision boundary. In addition, the use of a penalty function tends to constrain the parameter values to smaller values and prevents numerical overflow during the optimization process. In fact, this is one key reason why the ridge penalty function is widely used in neural networks.

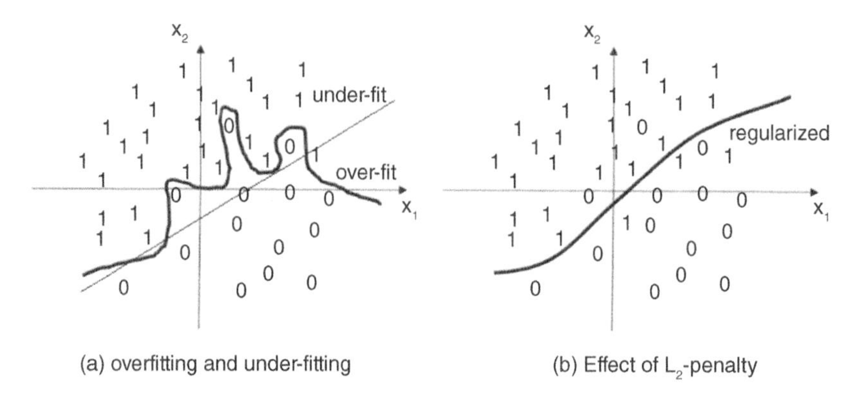

Figure 12.14 Effect of over-fitting, under-fitting and regularization.

12.8.4 Comparison of LNN vs. RNN

A set of $n = 100$ images[8] of cats and dogs (50 each) were used as training data for a series of neural network regression models. The test set consists of 35 images (19 cats, 16 dogs). As mentioned above, there are $p = 12288$ independent variables (the "unrolled" pixels) and one dependent variable (1 = cat, 0 = dog). The architecture of the neural network was the same for both LNN and RNN, specifically, a two-layer NN with one hidden layer containing 5 units. Numerical methods were used to optimize the $\boldsymbol{\beta}$ parameters for each regression model by minimizing the loss function using gradient descent, with a step size of $\ell = 0.002$ for the LNN and $\ell = 0.01$ for the RNN. The solution space was assumed to be convex and so a simple gradient descent was sufficient to obtain the minimum loss value and the corresponding $\boldsymbol{\beta}$ values. Momentum was not needed in this case. The number of iterations of gradient descent was adjusted until convergence was achieved. Typically, LNN required 8000 iterations while RNN required 18000 iterations. However, each method required more iterations as the ridge tuning parameter λ_2 was increased.

The convergence behavior of the loss, training accuracy and test accuracy is plotted in Figure 12.15. The loss function is monotonically decreasing indicating that the proper learning rate has been selected. Note the steep-slope characteristic of the LNN loss function compared to the plateau-hill characteristic for RNN. Both are typical of their respective methods. On the other hand, the training and testing accuracy plots are somewhat erratic in their behavior,

8 Most of the images were obtained from kaggle.com in the Competition section.

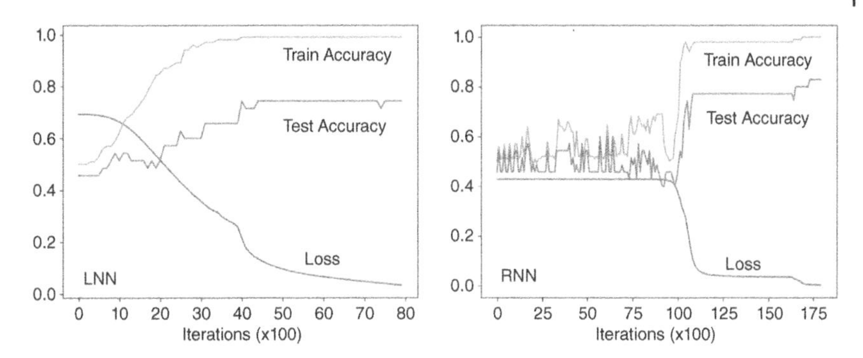

Figure 12.15 Convergence plots for LLN and RNN (test size = 35).

more so for RNN. However, as convergence is achieved, both LNN and RNN reach their final values. Note that we have plotted both the training and testing characteristics for illustrative purposes. In practice, only the training set accuracy can be plotted since the test set is not available. The test set should only be used at the completion of the training process rather than during the training process.

The results of the training and testing processes are shown Tables 12.6 and 12.7. The confusion matrix for RNN indicates two fewer errors compared to LNN. For this small test case, the LNN has six false positives compared to RNN which has only four. The accuracy, precision, recall and F_1 scores provided in Table 12.7 indicate that RNN performs slightly better than LNN for this combination of training and test sets. This is confirmed by the area under the ROC curve of Figure 12.16. The next question is whether or not we can improve the results using a penalty function.

If we consider whether to use ridge or LASSO for regularization, it is useful to examine Figure 12.12. Every pixel is a data column and therefore an input variable. We can understand that none of these variables can or should be eliminated, as LASSO would attempt to do. In fact, it was mentioned in (Saleh et al., 2019) that ridge outperforms LASSO for these types of images because variable sparsity is not present in the data. Hence, shrinkage estimation using ridge is the sensible option for this application.

Table 12.8 contains the results for LNN with ridge: as λ is increased from 0.0 to 10.0, the effect of the penalty function increases. This can be understood by examining the columns labeled train accuracy, test accuracy and F_1 score. The train accuracy remains roughly the same although, in general, it tends to degrade as λ is increased. Initially, the test accuracy increases as λ is increased until an optimal value of the test accuracy is achieved. The optimal result is

Table 12.6 Confusion matrices for RNN and LNN (test size = 35).

RNN ($\lambda = 0$)	True-0	True-1
Predict-0	12	2
Predict-1	4	17
LNN ($\lambda = 0$)	**True-0**	**True-1**
Predict-0	10	2
Predict-1	6	17

Table 12.7 Accuracy metrics for RNN vs. LNN (test size = 35).

	Train accuracy	Test accuracy	Recall	Precision	F_1 score	Iterations
RNN	100%	82%	0.89	0.81	0.85	18000
LNN	99%	77%	0.89	0.73	0.80	8000

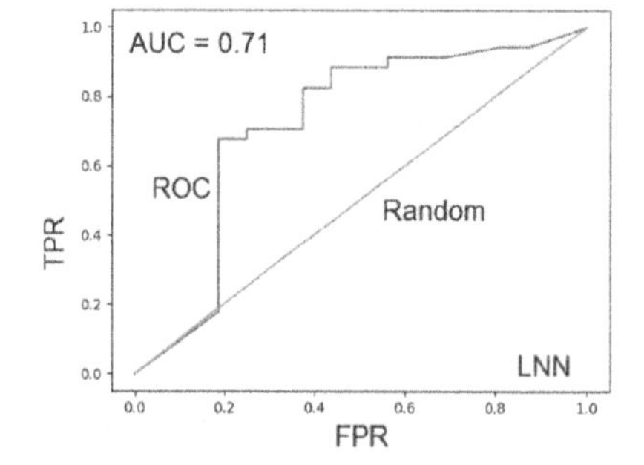

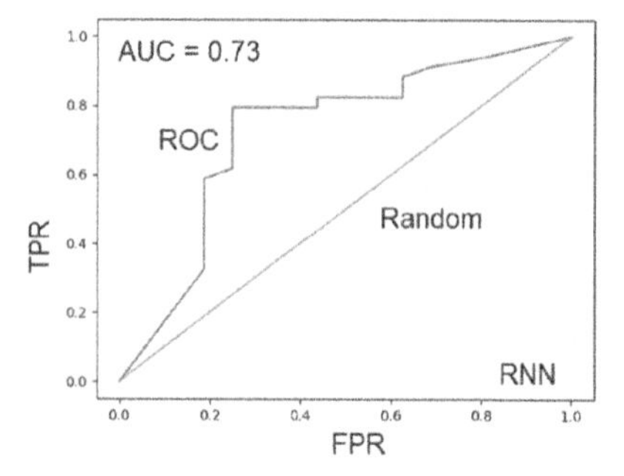

Figure 12.16 ROC plots for LLN and RNN (test size = 35).

obtained for $\lambda = 2.0$ and this case delivers a test accuracy of 82.0% along with the highest F_1 score of 0.85. This implies that 82% of the test images were correctly identified. However, increasing the value of λ further does not improve the test accuracy and, in fact, it begins to degrade at very high values.

A rather interesting result is obtained for RNN as shown in Table 12.9. As λ is increased from 0.0 to 1.0, the train accuracy remains at 100%, while the test accuracy and the F_1 score decreases. The optimal result is obtained for $\lambda = 0.0$ and this case delivers a test accuracy of 82.0% along with the highest F_1 score of 0.85. Therefore, it appears that ridge regression on RNN is not effective in the

Table 12.8 Train/test set accuracy for LNN. F_1 score is associated with the test set.

Ridge λ	Train accuracy (%)	Test accuracy (%)	F_1 score	Runtime (s)
0.0	99	77	0.80	130
1.0	99	80	0.82	154
2.0	99	82	0.85	251
4.0	99	77	0.79	311
10.0	99	77	0.70	332

Table 12.9 Train/test set accuracy for RNN. F_1 score is associated with the test set.

Ridge λ	Train accuracy (%)	Test accuracy (%)	F_1 score	Runtime (s)
0.0	100	82	0.85	330
0.1	99	80	0.82	324
0.2	99	77	0.79	351
0.4	99	77	0.79	311
1.0	99	77	0.79	332

case of image recognition. This is not surprising because the rank method is able to find the best solution without the assistance of a penalty function. This is also consistent with the findings with the circular data set shown earlier. The important point here is that we may not need to do an expensive search for the optimal λ for RNN whereas it is imperative for LNN. This may lead to an overall lower compute cost for RNN since the search for the optimal λ can be avoided.

We state again that this table is shown for illustrative purposes. It is not practical to use the test set to determine λ for LNN. Instead, we need access to thousands of cat and dog pictures to run a five-fold cross-validation scheme, and then check if the results generalize to an unseen test set. The steps used above are intended to show that LNN may require the use of an L_2 penalty to improve results compared to RNN. However, both require some degree of regularization to keep the parameters from getting too large.

The training set used here is exceedingly small for an image recognition problem but the results on the small test set were quite good. To demonstrate that a much larger training set should be used, the LNN and RNN models were applied to a test set with 700 images. The results are shown in Tables 12.10 and

Table 12.10 Confusion matrices for RNN and LNN (test size = 700).

RNN	True-0	True-1
Predict-0	107	127
Predict-1	158	308
LNN	**True-0**	**True-1**
Predict-0	119	130
Predict-1	146	305

Table 12.11 Accuracy metrics for RNN vs. LNN (test size = 700).

	Train accuracy	Test accuracy	Recall	Precision	F_1 score	Iterations
RNN	100%	59%	0.71	0.66	0.68	18000
LNN	99%	60%	0.70	0.67	0.69	8000

12.11. We note the dramatic reduction in accuracy across all metrics. However, both LNN and RNN produce similar results. This is expected because there are no outliers in the data set, but also reassuring since RNN can be used even when outliers are not present.

12.9 Image Recognition: MNIST Data Set

In this section, we take on the more challenging problem of multi-class classification. We compare RNN with LNN using the MNIST data set[9]. Note that we consider only fully connected networks again rather than the more advanced convolutional neural network (CNN) in order to establish a proper comparison baseline. A CNN is more appropriate for image recognition but our goal here is to compare LNN and RNN on a level playing field. As such, we will use the same NN architecture for both methods as well as the same learning rate. The objective is to tackle a multi-class problem, which gives SVM some difficulty, and determine if RNN is suitable for this purpose. No regularization is applied here for reasons described earlier.

The MNIST data set consists of 42,000 hand-written numbers from 0 to 9 (total of 10 digits) in the form of 28 × 28 gray-scale images with the goal of

9 This data set is available on kaggle.com in the Competitions section.

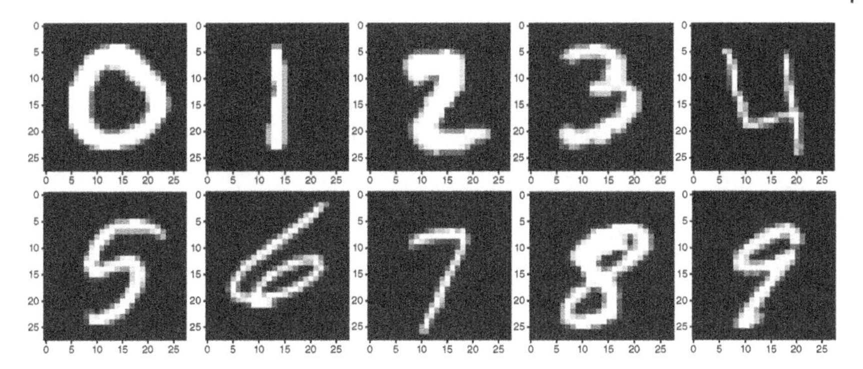

Figure 12.17 Ten representative images from the MNIST data set.

recognizing the correct number using neural networks. Each image must be unrolled to 784 variables with a corresponding label of 0–9, which is the digit being represented by the image. A sampling of 10 such digits are provided in Figure 12.17. They are hand-drawn digits such as would be used in a US zip code or street address on letters to be mailed. Recognizing digits is challenging since every person writes each digit in a slightly different way. The NN is given a subset of these images for training purposes, while the rest are used for testing purposes.

The MNIST data set falls into the category of multi-class classification. Rather than one output of the network, we need 10 outputs – one for each digit to be recognized. The standard training approach is to apply a given image at the input and then set the corresponding output to 1 while setting all other outputs to 0. For example, an image of the number 3 would be encoded at the output as 0001000000, the number 7 would be encoded as 0000000100, and so on. This type of encoding is referred to as one-hot encoding. Hence, the response is now a matrix $\boldsymbol{Y}$ whereby each element $y_{ij} \in [0, 1]$. The neural network is then trained on a data set of images, each with 784 input values and a 10-bit binary code representing the one-hot encoded output value. The activation function at the output is typically the softmax function for log-likelihood methods. For a given predicted value $\tilde{p}_j$, $j = 0, \ldots, 9$, the softmax activation is

$$\text{softmax}(\tilde{p}_j) = \frac{e^{\tilde{p}_j}}{||e^{\tilde{p}}||}, \tag{12.9.1}$$

where $||e^{\tilde{p}}||$ is the norm of the output vector. The corresponding loss function for softmax is given by

$$\mathbb{L}(y, \tilde{p}) = -\sum_{i=1}^{n} \sum_{j=0}^{9} y_{ij} \log \tilde{p}_{ij}, \tag{12.9.2}$$

where y_{ij} are the true values and $\tilde{p}_{ij}$ are the predicted values.

The NN architecture to be used is as follows: 784 inputs, 25 units in the hidden layer and 10 units in the output layer. Each image is unrolled and normalized into 784 pixel values used as input. The number of hidden units was determined by experimentation with different values and eventually settling on 25 units as optimal for this case. Roughly speaking, this amounts to $784 \times 25 + 25 \approx 20000$ parameters. Each image is labeled from 0 to 9 which is one-hot encoded to form a binary output for the NN. Each encoded value is assigned to the corresponding output of the NN. Again, these are all standard steps in mapping an image and label for training purposes onto a fully connected neural network.

With this setup, a data set of 42,000 images of hand-written characters were used to compare RNN and LNN. To make the problem tractable for demonstration purposes, the training set was comprised of 1000 properly labeled images plus some additional images that were purposely mislabeled so that they would be outliers. The outliers were inserted manually and varied between 0 to 300 images. Then, a very large test set of the remaining 40,000 images were used to compare the prediction accuracy of LNN against RNN. We chose this approach because the goal is to carry out a detailed comparison of the two methods rather than to demonstrate how to perform machine learning on digital hand-written images.

In the results to follow, we modify a percentage of the labels so that the incorrect labels are explicitly used for 0, 90, 180 and 270 images used for training. These outliers are purposely introduced to determine if either RNN or LNN can detect them. The overall objective is to determine what happens to prediction accuracy as a function of the percentage of outliers. Before presenting the results, it is important to ensure that both methods converge to a solution using gradient descent with momentum.

The convergence plots are shown for LNN and RNN, respectively, in Figure 12.18. We can notice the steep-slope type convergence characteristic of LNN and the plateau-rolling hill type convergence of RNN, as seen previously in this chapter. RNN generally requires 2–5 times as many iterations to converge. Specifically, LNN required 70,000 iterations while RNN required 200,000 iterations. Next, consider the first set of results in Table 12.12 with no outliers inserted. We find that LNN was able to fit to 100% of the training data while RNN was able to fit only 95% of the data. At first glance, one may believe that LNN is doing a better job and, in some respects, that is true. However, there are so many parameters (i.e. approx. 20,000) and only 1000 images such that overfitting is likely to occur in LNN and RNN. Both methods reached convergence as shown above so there must be some other reason for this discrepancy.

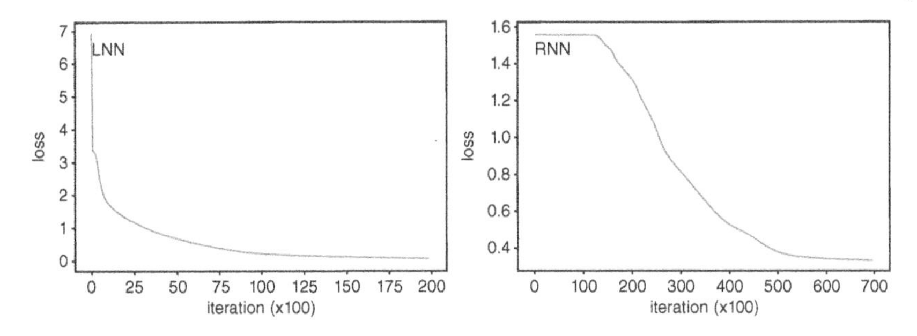

Figure 12.18 LNN and RNN convergence traces – loss vs. iterations (×100).

Table 12.12 MNIST training set size is $n = 1000$. In this case, 0 outliers were artificially added; however, 50 outliers were detected after training by RNN. 0 detected by LNN. The resulting models were used to predict on a test set with $n = 40000$. We see that the training set accuracy of RNN is closer to the test set accuracy than is the case for LNN.

Train	Accuracy	Matched	Outliers
LNN	100%	1000/1000	0
RNN	95%	950/1000	50
Test	**Accuracy**	**Matched**	**Outliers**
LNN	88.6%	35441/40000	—
RNN	87.3%	34820/40000	—

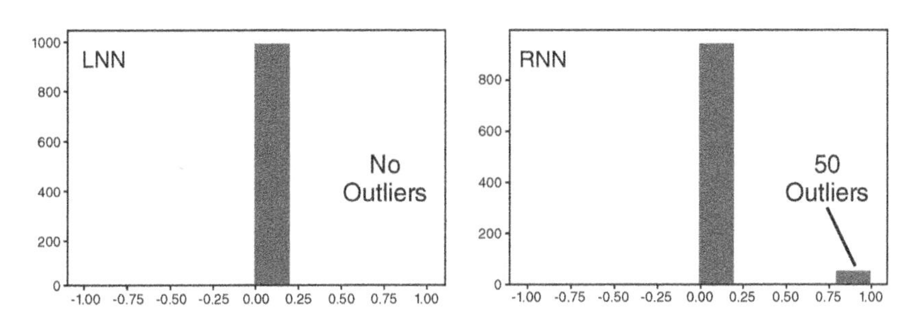

Figure 12.19 Residue histograms for LNN (0 outliers) and RNN (50 outliers).

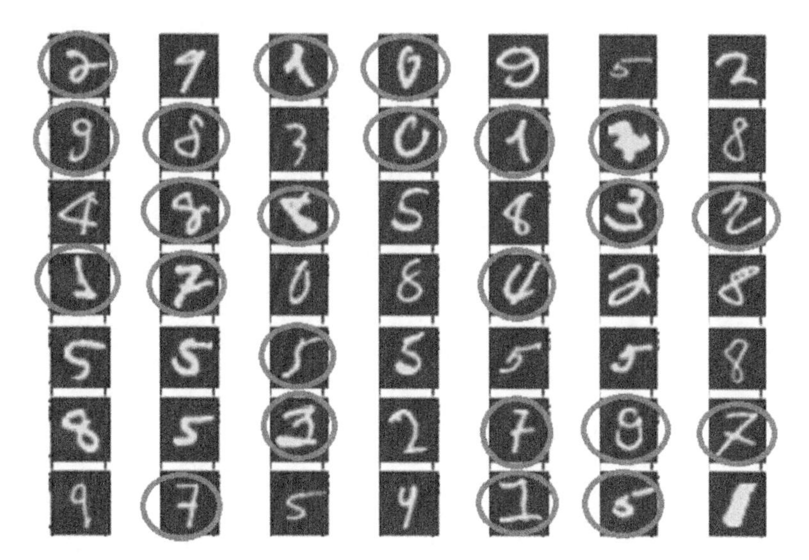

Figure 12.20 These are 49 potential outlier images reported by RNN. Of these, the circled digits are not easily recognized by humans as a number. The data scientist would decide if they are to be kept or removed.

The reason turns out to be outliers detected in the original unaltered data by RNN. Even though no outliers were added artificially, RNN still detected 50 outliers. The histogram of residuals in Figure 12.19 indicates that all of the data were fitted in the case of LNN, while in the case of RNN, only 95% of the digits were recognized and fitted. The remaining 50 images were designated as potential outliers by RNN. We show 49 such images in Figure 12.20 and note that it is difficult to decide which digit is being represented in each circled case. Clearly, RNN is able to identify many images in the MNIST data set that are not easily recognizable as a known digit. At this point, the digits identified as outliers can be kept or removed from the data set. In our case, we will continue to keep them to see their effect as we introduce more and more artificial outliers.

Returning to Table 12.12, the models from LNN and RNN are used to predict 40,000 images of digits in the test set. We find that LNN is able to correctly predict 88.6% of test images correctly (i.e. 35,441 out of 40,000) while RNN predicted 87.3% correctly (34,820 out of 40,000). They are roughly the same (1.3% difference), although it could easily be argued that LNN is more accurate. However, given the small size of the training set, a different combination of 1000 images would produce different results. The issue with running a training set with 40000 images is simply runtime to convergence. It could take many long hours to reach convergence. Since our goal here is to introduce rank-based neural networks on realistic data sets and compare to log-likelihood neural networks, the use of 1000 images is sufficient for this purpose. Advanced

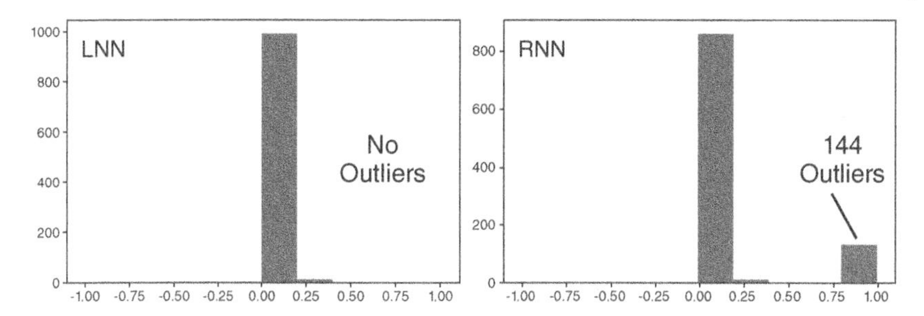

Figure 12.21 LNN (0 outliers) and RNN (144 outliers) residue histograms.

methods such as mini-batch gradient descent would be required to perform full-scale analysis, along with the Adam optimizer and a convolutional neural network architecture. All of this is beyond the scope of this book, but worth pursuing as future research.

Now consider the results for the cases when artificial outliers are inserted, as shown in Tables 12.13–12.15. The training set accuracy remains 100% for LNN in all cases since there are 784 variables and over-fitting the data of 1000 images is likely to occur. However, as the number of outliers increases, RNN identifies most of the outliers and fits only those images that are not considered to be outliers. This is exactly what we expect from RNN. When predicting on the test set, the results indicate that LNN accuracy drops to 81.4% (90 outliers), 76.0% (180 outliers) and 70.7% (270 outliers). For RNN, the corresponding accuracy values on the test set are 83.7% (90 outliers), 82.0% (180 outliers) and 79.1% (270 outliers). This is substantially better than LNN for obvious reasons.

We show the histograms of residuals for the case with 90 outliers manually inserted in Figure 12.21 and observe the significant increase in the number of outliers in the RNN case (total of 144 out of 1000), whereas none are detected in the LNN case. Again, we find that LNN has 100% accuracy on the training set, indicating that it over-fits across outliers. Recall that there are approximately 20000 free parameters in the neural network and only 1000 images. So over-fitting is a realistic possibility regardless of whether there are outliers or not. This is true in all instances where outliers were added.

The accuracy of multi-class classifiers can not be directly assessed using the methods used in binary classification. However, further insight when comparing LNN and RNN can be obtained by examining the confusion matrices shown in Figure 12.22. This is now a 10×10 matrix of correct and incorrect predictions for the case with 180 artificial outliers that were manually inserted. The number of correctly identified digits are along the diagonal and the specific incorrect predictions are on the off-diagonals for each case, starting with 0 and ending with 9. We note that the LNN is making substantial mistakes due

Table 12.13 MNIST training set size is $n = 1000$. In this case, 90 outliers have been manually added; After training, the model was used to predict on a test set with $n = 40000$. We see that the training set accuracy of RNN is close to the test set accuracy. This is not the case for LNN.

Train	**Accuracy**	**Matched/total**	**Outliers**
LNN	100%	1100/1100	0
RNN	87.6%	967/1100	133
Test	**Accuracy**	**Matched/total**	**Outliers**
LNN	81.4%	32591/40000	—
RNN	83.7%	33464/40000	—

Table 12.14 MNIST training set size is $n = 1200$. Here, 180 outliers have been manually added; After training, the model was used to predict on a test set with $n = 40000$. We see that the training set accuracy of RNN is close to the test set accuracy. This is not the case for LNN.

Train	**Accuracy**	**Matched/total**	**Outliers**
LNN	100%	1200/1200	0
RNN	82.0%	992/1200	208
Test	**Accuracy**	**Matched/total**	**Outliers**
LNN	76.0%	30485/40000	—
RNN	82.0%	32795/40000	—

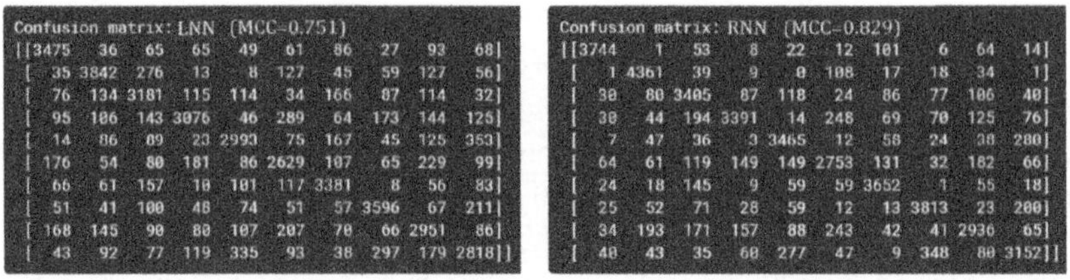

```
Confusion matrix: LNN (MCC=0.751)
[[3475   36   65   65   49   61   86   27   93   68]
 [  35 3842  276   13    8  127   45   59  127   56]
 [  76  134 3181  115  114   34  166   87  114   32]
 [  95  106  143 3076   46  289   64  173  144  125]
 [  14   86   89   23 2993   75  167   45  125  353]
 [ 176   54   80  181   86 2629  107   65  229   99]
 [  66   61  157   10  101  117 3381    8   56   83]
 [  51   41  100   48   74   51   57 3596   67  211]
 [ 168  145   90   80  107  207   70   66 2951   86]
 [  43   92   77  119  335   93   38  297  179 2818]]
```

```
Confusion matrix: RNN (MCC=0.829)
[[3744    1   53    8   22   12  101    6   64   14]
 [   1 4361   39    9    0  108   17   18   34    1]
 [  30   80 3405   87  118   24   86   77  106   40]
 [  30   44  194 3391   14  248   69   70  125   76]
 [   7   47   36    3 3465   12   58   24   38  280]
 [  64   61  119  149  149 2753  131   32  182   66]
 [  24   18  145    9   59   59 3652    1   55   18]
 [  25   52   71   28   59   12   13 3813   23  200]
 [  34  193  171  157   88  243   42   41 2936   65]
 [  40   43   35   60  277   47    9  348   80 3152]]
```

Figure 12.22 LNN and RNN confusion matrices and MCC scores.

Table 12.15 MNIST training set size is $n = 1300$. Here, 270 outliers added have been manually added; After training, the model was used to predict on a test set with $n = 40000$. We see that the training set accuracy of RNN is close to the test set accuracy. This is not the case for LNN.

Train	Accuracy	Matched/total	Outliers
LNN	100%	1300/1300	0
RNN	78.6%	1021/1300	279
Test	**Accuracy**	**Matched/total**	**Outliers**
LNN	70.7%	28268/40000	—
RNN	79.1%	31625/40000	—

to over-fitting. Since it is difficult to assess 100 numbers in each table at a glance, the Matthews correlation coefficient (MCC) (Matthews, 1975), which is similar to the F_1 score, is also provided in the figure. For RNN, we find that MCC = 0.829 whereas for LNN we have MCC = 0.751. Clearly, RNN outperforms LNN when outliers are present.

12.10 Summary

In this chapter, we have opened the door to rank-based neural networks and illustrated some of the novel capabilities available with this approach. We have demonstrated that it is superior in many respects to log-likelihood based methods and support vector machines. It was used in conjunction with well-known data sets, such as cats vs. dogs and MNIST, to demonstrate its viability and multi-class capability. However, we have only skimmed the surface of the possible research topics in this area. We see a large potential in this direction and expect there to be fruitful outcomes for those who pursue this topic. Our hope is that this book and all of the new concepts and ideas will resonate with the readers and provide motivation to pursue some of these ideas in more detail in the near future.

12.11 Problems

12.1. How many partial derivative terms are computed for the neural network in Figure 12.3? Hint: consider the sizes of the matrices between layers and add up the total number of entries in all matrices. Make sure you include the bias terms.

12.2. In this chapter, we discussed the sigmoid and relu activation functions. The function, tanh(z), is another often-used activation function and is the subject of this problem.
   (a) Write the equation for this function and construct one plot that includes tanh(z) with the sigmoid and relu functions of Figure 12.2.
   (b) Specify the domain and range of z and tanh(z), respectively.
   (c) Determine the partial derivative of tanh(z) with respect to z.
   (d) Explain why this is a valid activation function and its advantages or disadvantages relative to the sigmoid and relu activation functions.

12.3. Find the expression for $\partial D_n(\boldsymbol{\beta})/\partial \beta_0$ of Figure 12.4 for the rank objective function of Eq. (12.2.8). (Hint: derive Eq. (12.5.17) noting that the bias term is fixed at 1.)

12.4. Apply the chain rule as in Eq. (12.5.21) to find the complete expression for β_3 of Figure 12.4 for the rank dispersion loss function with a ridge penalty applied. (Hint: derive Eq. (12.5.18) and include the ridge gradient term)

12.5. Apply the chain rule as in Eq. (12.5.21) to find the complete expression for β_{23} of Figure 12.4 for the rank objective function of Eq. (12.2.8). (Hint: you will need to use the indicator function since the relu function is involved.)

12.6. Consider Table 12.16 of true responses, predicted probabilities and quantized probabilities from a binary classifier. Compute the accuracy, recall, precision and F_1 score based on this data. Set undefined values to 0. Why is the accuracy so high relative to the other metrics? What is your assessment of this classfier?

Table 12.16 Table of responses and probability outputs.

i	y_i	$\tilde{p}_i$	quantized $\tilde{p}_i$
1	0	0.37	0
2	0	0.37	0
3	1	0.37	0
4	0	0.37	0
5	0	0.37	0
6	0	0.37	0
7	1	0.37	0
8	0	0.37	0
9	0	0.37	0
10	0	0.37	0

12.7. Gradient descent with momentum is important for RNN. It uses an exponential weighting of previous derivative terms using a momentum factor which is typically set to $\gamma = 0.9$. The update equation is given by

$$\boldsymbol{v}_{\text{dB}}^{(t+1)} = \gamma \boldsymbol{v}_{\text{dB}}^{(t)} + (1-\gamma)\left(\frac{\partial D_n(\boldsymbol{\beta})}{\partial \boldsymbol{\beta}}\right)^{(t+1)} \tag{12.11.1}$$

Initially, $\boldsymbol{v}_{\text{dB}}^{(0)} = 0$. Compute this term for ten iterations and plot the effective weights on the previous 10 derivatives as a function of iteration number. First, set $\left(\frac{\partial D_n(\boldsymbol{\beta})}{\partial \boldsymbol{\beta}}\right)^{(t+1)} = \Delta^{(t+1)}$ to make it easier to work out the recursion formula. (Hint: the weight on each new derivative is $1-\gamma = 0.1$. Then, our first iteration would be $\boldsymbol{v}_{\text{dB}}^{(1)} = 0.9 \times 0 + 0.1\Delta^{(1)} = 0.1\Delta^{(1)}$. Now let $t = 1$ and continue the recursion from there. You will end up with $\Delta^{(1)}$ to $\Delta^{(10)}$ in the final iteration.)

12.8. This problem concerns some of the applications used in this chapter.
   (a) Assume that you have 42,000 images of cats and dogs. How would you set up the training, cross-validation and test sets for this problem to obtain results that generalize to the test set. What actual numbers of images would you use in each data set? Explain why.
   (b) The MNIST data set is typically addressed using mini-batch gradient descent but could also be done using stochastic gradient descent where only one image is processed at a time. What potential problem/solution do you foresee using rank-based neural networks with stochastic gradient descent?
   (c) How many parameters are actually used in the digit recognizer neural network if there is one hidden layer with 25 units as used in this chapter? What if the number of units in the hidden layer were increased to 35?

12.9. FINAL PROJECT: Part 1 – LNN program
   (a) Build a circular data set similar to Figure 12.8(a) using R. Assign six data points to be outliers as shown in the upper region of the figure. The R code below is suitable for this purpose. Enter this R code and check the results.

```
set.seed(123)
#Pick 400 random points
n <- 400
x1 <- runif(n, min = -2, max=2)
x2 <- runif(n, min = -2, max=2)
plot(x1,x2)
#Cut out a circular region and
#  apply y labels. Add 6 outliers
```

```
y <- rep(0,n)
cnt <- 0
for (i in 1:n) {
  if ((x1[i]^2 + x2[i]^2) < 0.8) {
    y[i] = 0
  } else if ((x1[i]^2 + x2[i]^2) > 1.5) {
    if ((x2[i]>1.8) & (cnt<6)) {
      y[i]=0
      cnt <- cnt +1
    } else {
      y[i] = 1
    }
  } else {
    y[i] = -1
  }
}
#Make a circle dataframe
ones <- c(rep(1,n))
circle <-  data.frame("x1"= x1, "x2"=x2,
                       "ones"=1,"y"=y)
circle <- circle[circle$y>-1,]
#Plot the data with color-coding
set1 <-circle[circle$y == 1,]
plot(set1$x1,set1$x2,col="red")
set2 <-circle[circle$y == 0,]
points(set2$x1,set2$x2,ylim=c(-2,2),
       xlim=c(-2,2),col="blue")
```

(b) Next, define the sigmoid function in R as done in the chapter problems in Chapter 11 using the follow R code.

```
sigmoid <- function(z ) {
  p <- 1.0/(1.0+exp(-z))
  return(p)
}
```

(c) Write R code for a neural network code that performs LLN using the information in this chapter. The architecture of the NN is given in Figure 12.7. The forward propagation step should implement what is shown in Figure 12.5. The log-likelihood loss function should be computed using the results. The backward propagation step should use the partial derivatives to compute the matrix and vector gradients. In each step, the gradient matrix and vector for one observation should be computed and added to the full gradient matrix and vector. Then the learning rate should be applied to obtain the new estimates. The R code below should be useful for this purpose if you have trouble writing your own code. If are are well-versed in R, you could improve the code by writing it in a form that is easier to follow. Enter the code and run it on the circular data set just created. It will take a few minutes to produce the estimates.

```
# 2-layer neural network (NN) using
# log-likelihood loss
# Build X and y data set from circle data
y <- circle$y
X <- model.matrix(y~x1+x2,circle)
n <- length(y)

# NN Architecture L=2, U1=4, U2=1

#Initialize parameters, vector, matrix
set.seed(123)  #Start ALL run from here
lambda2 = 0.0 #ridge regularization factor
gamma = 0.0   #momentum factor
learning_rate = 0.02
Bvector <- rnorm(5)*0.01
Bmatrix <- matrix(rnorm(12),nrow=4,ncol=3)*0.01
maxIter <- 20000
iterations <- rep(0,maxIter)

#MAIN LOOP
for (iter in 1:maxIter){

  #FWD PROP
  z <- Bmatrix %*% t(X)  #Step 1
  a <- z * (z>0)         #Step 2
  ones <- matrix(1,nrow=1,ncol=n)
  a1 <- rbind(a, ones)
  zout <- Bvector %*% a1 #Step 3
  yF <- sigmoid(zout)    #Step 4

  #Back prop
  delBmatrix <- matrix(0,nrow=4,ncol=3)
  bvec <- Bvector[1:4]
  for (i in 1:n) {   #Create gradient matrix
```

```
      d <- z[,i]>0
      d2 <- rbind(d,d)
      Indicator <- rbind(d2,d)  # Indicator matrix
      fullmatrix <- ((bvec) %*% t(X[i,])) *
        (yF[i]-y[i])
      delBmatrix <- delBmatrix +
        fullmatrix * t(Indicator)
    }
    delBvector <- rnorm(5)*0.0
    for (i in 1:n) {    #Create gradient vector
      delBvector <- delBvector +
        a1[,i]*(yF[i]-y[i])
    }

    #Calculate loss function
    loss <- -(1/n)*sum(y*log(yF)+(1-y)*log(1-yF))
    if (iter %% 1000 == 1) {
      print(iter)
      print(loss)
    }
    iterations[iter] <- loss

    #Parameter update using learning rate
    Bvector <- Bvector -
      learning_rate * delBvector/n
    Bmatrix <- Bmatrix -
      learning_rate * delBmatrix/n
  }
```

(d) Plot the values of the loss function as a function of the iteration count to ensure proper convergence. Then plot the predicted probabilities y_F vs. y to get a sense of the quality of the solution. Explain this plot and categorize regions as tp, tn, fp, or fn using a threshold of 0.5.

```
plot(1:maxIter,iterations,type = "l",
    ylim=c(0,0.7), ylab = "Loss function",
    xlab = "iteration count")
plot(yF,y)
```

(e) In order to obtain the nonlinear decision boundary, one option is to create thousands of points in the space (either a grid or random) and then run the resulting $\boldsymbol{X}$ matrix through the forward propagation steps to get the predicted values. These can be color-coded to obtain the resulting boundary. Use the codes given in the above to construct a plot of the nonlinear decision boundary similar to Figure 12.8. (Hint: R code is shown in Problem 12.10(f)).

(f) Add code to the program to implement LNN-ridge. This involves changing the code for the loss function and modifying the derivatives to add the penalty term. First, comment out the loss statement in the original R code. Then, for the loss function and the ridge penalty for the gradients, insert the following R code above the commented loss statement and re-run from the beginning. Try different values of λ_2 such as 0, 1.0 and 10.0 to see what values produce acceptable results.

```
# Add the penalty term to Bmatrix
#   after unpenalized gradients are computed
ridgeBmatrix <- Bmatrix
ridgeBmatrix[,1] <- 0
ridgeBmatrix <- 2.*lambda2*ridgeBmatrix/n
delBmatrix <- delBmatrix + ridgeBmatrix

#Add the penalty term to Bvector
#  after unpenalized gradients are computed
ridgeBvector <- Bvector
ridgeBvector[5] <- 0
ridgeBvector <- 2.*lambda2*ridgeBvector/n
delBvector <- delBvector + ridgeBvector

#Calculate penalized loss function
#loss <- -(1/n)*sum(y*log(yF)+(1-y)*log(1-yF))
ridge <-sum(ridgeBvector*ridgeBvector)+
  sum(ridgeBmatrix*ridgeBmatrix)
loss <- -(1/n)*sum(y*log(yF)+
  (1-y)*log(1-yF))+lambda2*ridge/n
```

(g) The last (optional) step is to add momentum to the gradient descent process. This capability should not be needed in this small example, but it is useful to see how it is implemented. Prior to the update of the parameters, the following code should be added. Note that some previous lines must be removed as well as shown below. Try different values of gamma to see if it has any effect (mainly used in deep NNs).

```
#Initialize momentum matrix and vector
# PRIOR to entering main loop
vdBV <- rnorm(5)*0.0
vdBM <- matrix(rnorm(12),nrow=4,ncol=3)*0.0
```

```
#At the end of the main loop
# REPLACElines as follows
vdBV <- gamma*vdBV + (1.0-gamma)*delBvector
vdBM <- gamma*vdBM + (1.0-gamma)*delBmatrix

Bvector <- Bvector - learning_rate * vdBV/n
Bmatrix <- Bmatrix - learning_rate * vdBM/n
#Remove these lines
#Bvector <- Bvector -
  #learning_rate * delBvector/n
#Bmatrix <- Bmatrix -
  #learning_rate * delBmatrix/n
```

12.10. FINAL PROJECT: Part 2 – RNN program

In this problem, you will replace parts of the LNN program to produce an RNN program, optionally with ridge and momentum if you have completed those portions of the final project part 1. Make sure you study the LNN code in detail before proceeding. To begin the conversion to RNN, it is best to increase the learning rate 0.1 and the number of iteration to 25000 to produce a solution quickly.

```
learning_rate = 0.1
maxIter <- 25000
```

(a) In this part, your task is to replace the loss function of the LNN code with the dispersion function. First, we need to define a scoring function at the beginning of the program (after the sigmoid function is defined). The R code below will be useful for this purpose.

```
phi <- function(u) {
  result <- sqrt(12.)*(u-0.5)
  return(result)
}
an <- rep(0,n)
Dn <- rep(0,n)
```

(b) Next you need to INSERT the terms of the dispersion function AFTER step 4 of the forward propagation step.

```
diff <- y-yF
Rn <- rank(diff,ties.method = "first")
for (i in 1:n) {
```

```
    an[i] <- phi(Rn[i]/(n+1.))
    Dn[i] <- diff[i]*an[i]
  }
```

(c) The loss function is computed by simply adding up the Dn quantities. This should REPLACE the log-likelihood line of code.

```
#loss <- -(1/n)*sum(y*log(yF)+(1-y)*log(1-yF))+
     #lambda2*ridge/n
loss <- (1/n)*sum(Dn)+lambda2*ridge/n
```

(d) Next, you will need to carefully REPLACE the gradient matrix and vector codes for LNN with the following R code. It is based on the gradients derived earlier in this problem set. You should end up with something similar to the following R code.

```
#REPLACE
#fullmatrix <- ((bvec) %*% t(X[i,])) *
   #(yF[i]-y[i])
#WITH
fullmatrix <- -((bvec) %*% t(X[i,])) *
    (yF[i])*(1.0-yF[i])*an[i]
```

Note the sign change in the gradient matrix and the replacement of one term with the other. Next, carefully REPLACE the lines as follows.

```
#REPLACE
#delBvector <- delBvector + a1[,i]*
   #(yF[i]-y[i])
#WITH
delBvector <- delBvector - a1[,i]*
   (yF[i])*(1.0-yF[i])*an[i]
```

Again, there is a the sign change in the gradient vector and the replacement of one term with the other. From these changes, we see that only a few lines of code separate RNN from LNN. Therefore, existing programs can easily be modified to include rank-based methods.

(e) If you have placed the given code in the appropriate places in the original code, you should be able to run the RNN program at this point. Remember to start at the beginning and also reset lambda2 and gamma to 0.0. Once completed, you can check the convergence plot, and the logic plot. To examine the results of RNN, the color-coded predicted values can be checked using the following code.

```
yP <- rep(0,n)
for (i in 1:n) {
  if (yF[i] >0.5) {
```

```
      yP[i] <- 1
    } else {
      yP[i] <- 0
    }
  }

  circle$yP <- yP
  set1 <-circle[circle$yP == 1,]
  plot(set1$x1,set1$x2,col="red")
  set2 <-circle[circle$yP == 0,]
  points(set2$x1,set2$x2,ylim=c(-2,2),
    xlim=c(-2,2),col="blue")
```

(f) Finally, if you want to observe the nonlinear decision boundary, this can be accomplished using the following code fragment, which is also useful for the LNN decision boundary.

```
xt1 <- runif(4000, min = -2, max=2)
xt2 <- runif(4000, min = -2, max=2)
ones <- c(rep(1,4000))
yt <- c(rep(1,4000))
circle3 <-  data.frame("x1"= xt1, "x2"=xt2,
    "ones"=ones,"yt"=yt)
Xtest <- model.matrix(yt~xt1+xt2,circle3)
z <- Bmatrix %*% t(Xtest)  #Step 1
a <- z * (z>0)          #Step 2
ones <- matrix(1,nrow=1,ncol=4000)
a1 <- rbind(a, ones)
zout <- Bvector %*% a1 #Step 3
ytest <- sigmoid(zout)    #Step 4
yP <- rnorm(4000)*0.0
for (i in 1:4000) {
  if (ytest[i] >0.5) {
    yP[i] <- 1
  } else {
    yP[i] <- 0
  }
}
circle3$yP <- yP
sett1 <-circle3[circle3$yP == 1,]
plot(sett1$x1,sett1$x2,ylim=c(-2,2),
  xlim=c(-2,2),col="red")
sett2 <-circle3[circle3$yP == 0,]
```

```
points(sett2$x1,sett2$x2,ylim=c(-2,2),
  xlim=c(-2,2),col="blue")
```

Assuming the code ran successfully, you have now completed the final project and you can now begin to run your own experiments to compare LNN and RNN. You can also run any two-dimensional binary classification problem to further your understanding and start your research in this area. Greatness awaits.

Bibliography

Adichie, J. N. (1967). Estimates of regression parameters based on rank tests, *The Annals of Mathematical Statistics*, **38**(3), 897–904.

Akdeniz, F., and Kaciranlar, S. (1995). On the almost unbiased generalized Liu estimator and unbiased estimation of the bias and mse, *Communications in Statistics-Theory and Methods*, 24(7), 1789–1797.

Akdeniz, F., and Kaciranlar, S. (2001). More on the new biased estimator in linear regression, *Sankhya*, **63**, 321–325.

Akdeniz, F., and Ozturk, F. (2005). The distribution of stochastic shrinkage biasing parameters of the Liu type estimator, *Applied Mathematics and Computation*, **163**(1): 29–38.

Akdeniz, F., Akdeniz Duran, E., Roozbeh, M., and Arashi, M. (2015). Efficiency of the generalized difference-based Liu estimators in semiparametric regression models with correlated errors, *Journal of Statistical Computation and Simulation*, **85**(1), 147–165.

Akritas, M. G., and Johnson, R. A. (1982). Efficiencies of tests and estimators for p-order autoregressive processes when the error distribution is nonnormal, *Annals of the Institute of Statistical Mathematics*, **34**(1), 579–589.

Anderson, T. W. (2003). *An Introduction to Multivariate Statistical Analysis* (Third Edition), John Wiley & Sons.

Anderson, T. W., and Gill, R. D. (1982). Cox's regression model for counting processes: a large sample study, *The Annals of Statistics*, **10**(4), 1100–1120.

Arashi, M., Kibria, B. M. G., Norouzirad, M., and Nadarajah, S., (2014). Improved preliminary test and Stein-rule Liu estimators for the ill-conditioned elliptical linear regression model, *Journal of Multivariate Analysis*, **124**, 53–74.

Arashi, M., Asar, Y., and Yuzbasi, B. (2017). LLASSO: A linear unified LASSO for multicollinear situations, *arXiv:1710.04795 [stat.ME]*.

Rank-Based Methods for Shrinkage and Selection: With Application to Machine Learning. First Edition. A. K. Md. Ehsanes Saleh, Mohammad Arashi, Resve A. Saleh, and Mina Norouzirad.

Arashi, M., Nadarajah, S., and Akdeniz, F. (2017). The distribution of the Liu type estimator of the biasing parameter in elliptically contoured models, *Communication in Statistics–Theory and Methods*, **46**(8), 3829–3837.

Arashi, M., Norouzirad, M., Ahmed, S. E., and Yuzbasi, B. (2018). Rank-based Liu regression, *Computational Statistics*, **33**(3), 1525–1561.

Arashi, M., Asar, Y., and Yuzbasi, B. (2021). SLASSO: A scaled LASSO for multicollinear situations, *Journal of Statistical Computation and Simulation*, **91**(15), 3170–3183.

Asar, Y., Erisoglu, M., and Arashi, M. (2017). Developing a restricted two-parameter Liu-type estimator: A comparison of restricted estimators in the binary logistic regression model, *Communications in Statistics-Theory and Methods*, **46**(14), 6864–6873.

Asar, Y. and Genc, A. (2016). New shrinkage parameters for the Liu-type logistic estimators, *Communications in Statistics-Simulation and Computation*, **45**(3), 1094–1103.

Bancroft, T. A. (1944). On biases in estimating due to use of preliminary tests of significance, *The Annals of Mathematical Statistics*, **15**, 195–204.

Bancroft, T. A. (1964). Analysis and inference for incompletely specified models involving the use of preliminary test(s) of significance, *Biometrics*, **20**, 427–442.

Bancroft, T. A. (1965). Inference for incompletely specified models in the physical sciences (with discussion), *Bull. ISI, Proc. 35th Session*, **41**(1), 497–515.

Belloni, A., and Chernozhukov, V. (2013). Least squares after model selection in high-dimensional sparse models, *Bernoulli*, **19**, 521–547.

Breiman, L. (1995). Better subset regression using the nonnegative garrote, *Technometrics*, **37**, 373–384

Breiman, L. (1996). Heuristics of instability and stabilization in model selection, *The Annals of Statistics*, **24**, 2350–2383.

Casella, G., and Hwang, J. T. (1987). Employing vague prior information in the construction of confidence sets, *Journal of Multivariate Analysis*, **21**, 79–104.

Chiou, P., and Saleh, A. K. Md. E. (2002). Preliminary test confidence sets for the mean of a multivariate normal distribution, *Journal of Probability and Statistics*, **2**, 177–189.

Corder, G. W., and Foreman, D.I. (2014). *Nonparametric Statistics:A Step-by-Step Approach* (Second Edition), John Wiley & Sons, New York.

Donoho, D. L., Johnstone, I. M. (1994). Minimax estimation via wavelet shrinkage, *The Annals of Statistics*, **26**, 879–921.

Draper, N. R., and Van Nostrand, R. C. (1979). Ridge regression and James-Stein estimation: review and comments, *Technometrics*, **21**, 451–466.

Efron, B. and Morris, C. (1973). Stein's estimator rule and its competitors - An empirical Bayes approach, *Journal of the American Statistical Association*, **68**(341), 117–130.

Efron, B., and Morris, C. (1975). Data analysis using Stein's estimator and its generalizations, *Journal of the American Statistical Association*, **70**(350), 311–319.

Efron, B. and Morris, C. (1975). Stein's paradox in statistics, *Scientific American*, **236**(5), 119–127.

Fan, J., and Li, R. (2001). Variable selection via nonconcave penalized likelihood and its oracle properties, *Journal of the American Statistical Association*, **96**, 1348–1360.

Feng, L., Zou, C., Wang, Z., and Chen, B. (2013). Rank-based score tests for high-dimensional regression coefficients, *Electronic Journal of Statistics*, **7**, 2131–2149.

Filzmoser, P., and Kurnaz, F. S. (2018). A robust Liu regression estimator, *Communications in Statistics-Simulation and Computation*, **47**(2), 432–443.

Frank, L. E.,and Friedman, J. H. (1993). A statistical view of some chemometrics regression tools, *Technometrics*, **35**, 109–135.

Gao, J. (1995). Asymptotic theory for partially linear models, *Communication in Statistics-Theory and Methods*, **22**, 3327–3354.

Gao, J. (1997). Adaptive parametric test in a semi-parametric regression model, *Communication in Statistics-Theory and Methods*, **26**, 787–800.

Gao, Z., Ahmed, S. E., Feng, Y. (2017). Post selection shrinkage estimation for high-dimensional data analysis, *Applied Stochastic Models in Business and Industry*, **33**(2), 97–120.

Geyer, C. J. (1994). On the convergence of Monte-Carlo maximum likelihood calculations, *Journal of the Royal Statistical Society, Series B (Methodological)*, **56**(1), 261–274.

Geyer, C. J. (1996). On the asymptotic of convex stochastic optimization, Unpublished manuscript.

Greene, W. H. (2012). *Econometric Analysis* (Seventh Edition), Prentice-Hall.

Gruber, M. H. J. (2012). Liu and Ridge estimators-A comparison, *Communications in Statistics-Theory and Methods*, **41**(20), 3739–3749.

Genc, M., and Ozkale, M. R. (2020). Usage of the GO estimator in high dimensional linear models, *Computational Statistics*, **36**, 217–239.

Hajek, J. (1969). *A Course in Nonparametric Statistics*, Holden-Day.

Hajek, J., and Sidák, Z. (1967). *Theory of Rank Tests*, New York: Academic.

Hajek, J., Sidák, Z., and Sen, P. K. (1999). *Theory of Rank Tests* (Second Edition), New York: Academic.

Hollander, M., and Wolfe, D. A. (1999). *Nonparametric Statistical Methods* (Second Edition), John Wiley & Sons.

Hall, P., and Heyde, C. C. (1981). Rates of convergence in the matringle central limit theorem, *The Annals of Probability*, **9**(3), 395–404.

Hansen, B. E. (2016). The risk of James-Stein and LASSO shrinkage, *Econometric Reviews*, **35**(8–10), 1456–1470.

Hardel, W., Liang, H., Gao, J. (2000). *Partially Linear Models*, Springer.

Haykin, S. (2009). *Neural Networks and Machine Learning* (Third Edition), Prentice-Hall.

Hettmansperger, T. P. and McKean, J. W. (2011). *Robust Nonparametric Statistical Methods*, 2nd Ed. Chapman Hall, New York.

Hodges, J. L., and Lehmann, E. L. (1963). Estimates of location based on rank tests, *The Annals of Mathematical Statistics*, **34**, 598–611.

Hoeffding, W. (1948). A class of statistics with asymptotically normal distribution, *The Annals of Mathematical Statistic*, **19**(3), 293–325.

Hoerl, E., and Kennard, R. W. (1970). Ridge regression: Biased estimation for nonorthogonal problems, *Technometrics*, **12**, 55–67.

Hosmer, D. W., and Lemeshow, S. (1989). *Applied Logistic Regression*, John Wiley & Sons, Inc., New York.

Huang, J., Breheny, P., Lee, S., Ma, S., and Zhang, C. H. (2016). The Mnet method for variable selection, *Statistica Sinica*, **26**(3), 903–923.

Hwang, J. T. and Casella, G. (1982). Minimax confidence sets for the mean of a multivariate normal distribution, *The Annals of Mathematical Statistics*, **10**, 868–881.

Hwang, J. T., and Casella, G. (1984). Improved set estimators for a multivariate normal mean, *Statistics & Decisions*, **1**, 3–16.

James, W. and Stein, C. (1961). Estimation with Quadratic Loss, *Proceedings of the Fourth Berkeley Symposium on Mathematical Statistics and Probability, Volume 1: Contributions to the Theory of Statistics*, 361–379.

James, G., Witten, D., Hastie, T., and Tibshirani, R. (2013). *An introduction to statistical learning with application to R*, Springer.

Jaeckel, L. A. (1972). Estimating regression coefficients by minimizing the dispersion of the residuals, *The Annals of Mathematical Statistics*, **43**(5), 1449–1458.

Johnson, B. A., and Peng, L. (2008). Rank-based variable selection, *Journal of Nonparametric Statistics*, **20**(3), 241–252.

Joshi, V. M. (1967). Inadmissibility of the usual confidence sets for the mean of a multivariate normal population, *The Annals of Mathematical Statistic*, **38**, 1868–1875.

Jurečková, J. (1969). Asymptotic linearity of a rank statistic in regression parameter, *The Annals of Mathematical Statistics*, **40**(6), 1889–1900.

Jurečková, J. (1971). Nonparametric estimate of regression coefficients, *The Annals of Mathematical Statistics*, **42**, 1328–1338.

Jurečková, J., and Sen, P. K. (1984). On adaptive scale-equivariant M-estimators in linear models, *Statist. Decisions*, **2**(1), 31–46.

Jurečková, J., and Sen. P. K. (1996). *Robust Statistical Procedures: Asymptotics and Interrelations*. John Wiley & Sons, New York.

Kaciranlar, S., Sakallioglu, S., Akdeniz, F., Styan, G. P. H., and Werner, H. J. (1999). A new biased estimator in linear regression and detailed analysis of the widely analysed dataset on Portland cement, *Sankhya* Series B, **61**, 443–459.

Karbalaee, M. H., Arashi, M., and Tabatabaey, S. M. M. (2018). Performance analysis of the preliminary test estimator with series of stochastic restrictions, *Communication in Statistics-Theory and Methods*, **47**(1), 1–17.

Karbalaee, M. H., Tabatabaey, S. M. M., and Arashi, M. (2019). On the preliminary test generalized Liu estimator with series of stochastic restrictions, *Journal of Iranian Statistical Society (JIRSS)*, **18**(1), 113–131.

Koenker, R. and Bassett, G. (1978). Regression quantiles, *Econometrica*, **46**, 33–50.

Kibria, B. M. G. (2012). On some Liu and ridge type estimators and their properties under the ill-conditioned Gaussian linear regression model, *Journal of Statistical Computation and Simulation*, **82**(1), 1–17.

Kloke, J. D., and McKean, J. W. (2012). Rfit: Rank-based estimation for linear models, *The R Journal*, **4**(2), 57–64.

Knight, K., and Fu, W. (2000). Asymptotics for LASSO-type estimators, *The Annals of Statistics*, **28**(5), 1356–1378.

Koenker, R. (2005). *Quantile Regression*, Cambridge University Press.

Koul, H. L., and Saleh, A. K. Md. E. (1993). R-estimation of the parameters of autoregressive [Ar(p)] models, *The Annals of Statistics*, **21**(1), 534–551.

Kurnaz, F. S., and Akay, K. U. (2015). A new Liu-type estimator, *Statistical Papers*, **56**(2), 495–517.

Kurtoglu, F., and Ozkale, M. R. (2016). Liu estimation in generalized linear models: application on gamma distributed response variable, *Statistical Papers*, **57**, 911–928.

Liang, H., and Hardle, W. (1999). Large sample theory of the estimation of the error distribution for semi-parametric models, *Communications in Statistics-Theory and Methods*, **28**(9): 2025–2036

Liu, K. (1993). A new class of biased estimate in linear regression, *Communications in Statistics-Theory and Methods*, **22**, 393–402.

Liu, K. (2003). Using Liu-type estimator to combat collinearity, *Communications in Statistics-Theory and Methods*, **32**(5), 1009–1020.

Mann, H. B., and Whitney, D. R. (1947). On a test of whether one of two random variables is stochastically larger than the other, *The Annals of Mathematical Statistics*, **18**(1), 50–60.

Matthews, B. W. (1975). Comparison of the predicted and observed secondary structure of T4 phage lysozyme, *Biochimica et Biophysica Acta (BBA) - Protein Structure*, **405 (2)**, 443–451.

Mayer, L. S., and Willke, T. A. (1973). On biased estimation in linear models, *Technometrics*, **15**, 497–508.

McDonald, G. C., and Schwing, R. C. (1973). Instabilities of regression estimates relating air pollution to mortality, *Technometrics*, **15**, 463–481.

Montgomery, D. C., Peck, E. A., and Vining G. G. (2012). *Introduction to Linear Regression Analysis* (Fourth Edition), Wiley.

Muller, M., and Ronz, B. (2000). Credit scoring using semiparametric models, *Lecture Notes in Statistics*, **147**, Springer.

Ozkale, M. R., Kaciranlar, S. (2007). The restricted and unrestricted two-parameter estimators, *Communications in Statistics-Theory and Methods*, **36**, 2707–2725

Pollard, D. (1991). Asymptotic for least absolute deviation regression estimators, *Econometric Theory*, **7**(2), 186–199.

Puri, M. L., and Sen, P. K. (1985). *Nonparametric methods in general linear models*, Wiley, New York.

Randles, R. H., and Wolfe, D. A. (1979). *Introduction to the Theory of Nonparametric Statistics*, Wiley, New York.

Robert, C., and Saleh, A. K. Md. E. (2001). Recentered Confidence Sets: A Review, In *Data Analysis from Statistical Foundations - A Festschrift in Honor of the 75th birthday of D.A.S. Fraser*, A. K. Md. E. Saleh, ed. Nova Science Publishers, New York, 295–308.

Rohatgi, V. K., and Saleh, A. K. Md. E. (2000). *An Introduction to Probability and Statistics*, Wiley, New York.

Roozbeh, M., and Arashi, M. (2016). Shrinkage ridge regression in partial linear models, *Communications in Statistics-Theory and Methods*, **45**(20), 6022–6044.

Roozbeh, M., Arashi, M., and Hamzah, N. A. (2020). Generalized cross-validation for simultaneous optimization of tuning parameters in ridge regression, *Iranian Journal of Science and Technology, Transactions A: Science*, DOI: 10.1007/s40995-020-00851-1.

Rousseeuw, P. J., and Leroy, A. M. (1987). *Robust Regression and Outlier Detection*, Wiley Series in Probability and Mathematical Statistics: Applied Probability and Statistics, John Wiley & Sons, New York.

Safariyan, A., Arashi, M., and Arabi Belaghi, R. (2019). Improved estimators for stress-strength reliability using record ranked set sampling scheme, *Communication in Statistics - Simulation and Computation*, **48**(9), 2708–2726.

Saleh, A. K. Md. E., Arashi, M., and Tabatabaey, S. M. M. (2014). *Statistical Inference for Models with Multivariate t-Distributed Errors*, Wiley, New Jersey.

Saleh, A. K. Md. E., Arashi, M., and Kibria, B. M. G. (2019). *Theory of Ridge Regression Estimation with Applications*, Wiley.

Saleh, A. K. Md. E., Navrátil, R., and Norouzirad, M. (2018). Rank theory approach to ridge, LASSO, preliminary test and Stein-type estimators: A comparative study: rank theory approach to improved estimators, *Canadian Journal of Statistics*, **46**(4), 690–704.

Saleh, A. K. Md. E., and Sen, P. K. (1978). Nonparametric estimation of location parameter after a preliminary test on regression, *Annals of Statistics*, **6**, 154–168.

Saleh, A. K. Md. E., and Sen, P. K. (1983). Nonparametric tests of location after a preliminary test on regression in the multivariate case, *Communications in Statistics-Theory and Methods*, **12**(16), 1855–1872.

Saleh, A. K. Md. E., and Sen, P. K. (1984). Least squares and rank order preliminary test estimation in general multivariate linear models, *Proceedings ISI Golden Jubilee Conference on Statistics: Application and New Directions (Dec. 16–19, 1981)*, 237–253.

Saleh, A. K. Md. E., and Sen, P. K. (1984). *Nonparametric preliminary test inference. Handbook of Statistics 4* (eds. P. R. Krishnaiah and P. K. Sen) North-Holland, Amsterdam, 275–297.

Saleh, A. K. Md. E., and Sen, P. K. (1985a). On shrinkage M-estimators of location parameters, *Communications in Statistics-Theory and Methods*, **14**(10), 2313–2329.

Saleh, A. K. Md. E., and Sen, P. K. (1985b). Nonparametric shrinkage estimation in a parallelism problem, *Sankhya*, **47**A, 156–165.

Saleh, A. K. Md. E., and Sen, P. K. (1985c). Preliminary test predicted in general multivariate linear models, *Proceedings Pacific Area Statistical Conference (ed. M. Matusita) (Dec. 15–19, 1982)*, North-Holland, Amsterdam, 619–638.

Saleh, A. K. Md. E., and Sen, P. K. (1985d). Shrinkage least squares estimation in a general multivariate linear model, *Proceedings 5th Pannonian Symposium on Mathematical Statistics (eds. J. Mogyorodi, I. Vincze, and W. Wertz)*, 307–325.

Saleh, A. K. Md. E., and Sen, P. K. (1985e). On shrinkage least squares estimation in a parallelism problem, *Communications in Statistics-Theory and Methods*, **15**, 1451–1466.

Saleh, A. K. Md. E., and Sen, P. K. (1986). On shrinkage R-estimation in a multiple regression model, *Communications in Statistics-Theory and Methods*, **15**(7), 2229–2244.

Saleh, A. K. Md. E., and Sen, P. K. (1986). Relative performance of Stein-rule and preliminary test estimators in linear models: least squares theory, *Communications in Statistics-Theory and Methods*, **16**, 461–476.

Saleh A. K. Md. E. (2006). *Theory of Preliminary Test and Stein-Type Estimators with Applications*, John Wiley & Sons, New York.

Saleh A. K. Md. E., Arashi, M., and Kibria, B. M. G. (2019). *Theory of Ridge Regression Estimation with Applications*, John Wiley & Sons, New York.

Sen, P. K. (1963). *On the estimation of relative potency in dilution(-direct) assays by distribution-free methods, Biometrics*, **19**(4), 532–552.

Sen, P. K. (1981). *Sequential Nonparametrics: Invariance Principle and Statistical Inference*, John Wiley, New York.

Sen, P. K. (1984). A James-Stein type detour of U-statistics, *Communications in Statistics-Theory and Methods*, **13**, 2725–2747.

Sen, P. K., and Saleh, A. K. Md. E. (1979). Nonparametric estimation of location parameter after a preliminary test on regression in the multivariate case, *Journal of multivariate analysis*, **15**(4), 1580–1592.

Sen, P. K., and Saleh, A. K. Md. E. (1985). On some shrinkage estimators of multivariate location, *Annals of Statistics*, **13**, 272–281.

Sen, P. K., and Saleh, A. K. Md. E. (1987). On preliminary test and shrinkage M-estimation in linear models, *Annals of Statistics*, **15**(4), 1580–1592.

Sengupta, D., and Jammalamadaka, S. R. (2003). *Linear models: an integrated approach*, World Scientific Publishing Company, Singapore.

Shao, J., and Deng, X. (2012). Estimation in high-dimensional linear models with deterministic design matrices, *The Annals of Statistics*, **40**(2), 812–831.

Shi, J., and Lau, T. S. (2000). Empirical likelihood for partially linear models, *Journal of Multivariate Analysis*, **72**(1), 132–148

Sing, J. M., and Sen, P. K. (1985). M-methods in multivariate linear models, *Journal of Multivariate Analysis*, **17**(2), 168–184.

Speckman, P. (1988). Kernel smoothing in partial linear models, *Journal of the Royal Statistical Society. Series B (Methodological)*, **50**(3), 413–436.

Stein, C. (1956). Inadmissibility of the usual estimator for the mean of a multivariate normal distribution, *Proceedings of the Third Berkeley Symposium on Mathematical Statistics and Probability*, 1954–1955.

Stein, C. (1962). Confidence sets for the mean of a multivariate normal distribution, *Journal of the Royal Statistical Society. Series B*, **24**, 265–296.

Theil, H. (1950). A rank-invariant method of linear and polynomial regression analysis. I, II, III, *Nederl. Akad. Wetensch., Proc.*, 53: 386–392, 521–525, 1397–1412.

Tibshirani, R. (1996). Regression shrinkage and selection via the lasso, *Journal of the Royal Statistical Society. Series B (Methodological)*, **58**(1), 267–288.

Tikhonov, A. N. (1963). Solution of incorrectly formulated problems and the regularization method, *Soviet Mathematics Doklady*, 4(4), 1035–1038.

Turkmen, A., and Ozturk, O. (2014). Rank-based ridge estimation in multiple linear regression, *Journal of Nonparametric Statistics*, **26**(4), 737–754.

Vapnik, V. N. (1995). *The Nature of Statistical Learning Theory*, Springer, New York.

Wilcox, R. (2012). *Introduction to robust estimation and hypothesis testing, Statistical Modeling and Decision Science* (Third Edition), Elsevier/Academic Press, Amsterdam.

Wilcoxon, F. (1945). Individual comparisons by ranking methods, *Biometrics*, **1**, 80–83.

Wu, J., Asar, Y., and Arashi, M. (2018). On the restricted almost unbiased Liu estimator in the logistic regression model, *Communications in Statistics-Theory and Methods*, **47**(18), 4389–4401.

Xu, J., and Yang, H. (2012). On the Stein-type Liu estimator and positive-rule Stein-type Liu estimator in multiple linear regression models, *Communications in Statistics-Theory and Methods*, **41**(5), 791–808.

Yang, H., and Chang, X. (2010). A new two-parameter estimator in linear regression, *Communications in Statistics-Theory and Methods*, **39**(6), 923–934.

Yuzbasi, B., Arashi, M., and Ahmed, S. E. (2020). Shrinkage estimation strategies in generalised ridge regression models: Low/High-dimension regime, *International Statistics Review*, **88**(1), 229–251.

Zhang, C. H. (2010). Nearly unbiased variable selection under minimax concave penalty, *Annals of Statistics*, **38**, 894–942

Zou, H. (2006). The adaptive lasso and its oracle properties, *Journal of the American statistical association*, **101**, 1418–1429.

Zou, H., and Hastie, T. (2005). Regularization and variable selection via the elastic net, *Journal of the Royal Statistical Society. Series B (Methodological)*, **67**, 301–320.

Author Index

a
Akay, K. U. 265
Akdeniz, F. 264, 267, 280
Akritas, M. G. 296
Anderson, T. W. 220, 234, 249
Arashi, M. 241, 263–266, 277, 280
Asar, Y. 264

b
Bancroft, T. A. 118
Breiman, L. xxix

c
Chang, X. 265
Corder, G. W. 10

d
Deng, X. 264, 309, 318, 320
Donoho, D. L. 169, 171, 227, 229, 253

f
Fan, J. xxix, 233, 283
Filzmoser, P. 264
Foreman, D. I. 10
Frank, L. E. 282
Friedman, J. H. 282
Fu, W. 175, 235, 236

g
Gao, J. 243
Gao, Z. 318, 320
Genc, M. 264, 266
Geyer, C. J. 175, 235, 236
Gill, R. D. 234
Greene, W. H. 192
Gruber, M. H. J. 264

h
Hajek, J. 161, 162, 196
Hall, P. 294
Hardle, W. 241, 243
Hastie, T. xxix, 31, 39, 49, 82, 139, 266, 283, 284
Haykin, S. 375
Hettmansperger, T. P. 17, 18, 22, 30, 51, 152, 340, 341
Heyde, C. C. 294
Hoerl, E. xxviii, 34, 48, 53, 263, 264
Hollander, M. 7
Hosmer, D. W. 334
Huang, J. 284

j
James, G. xxvii, 49, 61, 79, 81, 349, 352, 378, 397
James, W. xxviii, 89
Jaeckel, L. A. xxviii, 20, 104, 159, 172, 196, 233, 266, 292
Johnson, B. A. xxx, 3, 47, 73, 85, 236
Johnson, R. A. 296
Johnstone, I. M. 169, 171, 227, 229, 253
Jurečková, J. 20, 106, 133, 175, 196, 215, 217, 219, 293, 323

Rank-Based Methods for Shrinkage and Selection: With Application to Machine Learning.
First Edition. A. K. Md. Ehsanes Saleh, Mohammad Arashi, Resve A. Saleh, and Mina Norouzirad.

k

Kaciranlar, S. 264, 267
Karbalaee, M. H. 122, 264
Kennard, R. W. xxviii, 34, 48, 53, 263, 264
Kibria, B. M. G. 264
Kloke, J. D. xxx, 7, 20, 28, 42, 151
Knight, K. 175, 235, 236
Koul, H. L. 293
Kurnaz, F. S. 264, 265
Kurtoglu, F. 264

l

Lau. T. S. 243
Lemeshow, S. 332
Leroy, A. M. 4
Li, R. xxix, 233, 283
Liang, H. 243
Liu, K. 263, 264

m

Matthews, B. W. 420
Mayer, L. S. 264, 267
McKean, J. W. xxx, 7, 17, 18, 20, 22, 28, 30, 42, 51, 151, 152, 340, 341
Montgomery, D. C. 52, 65
Muller, M. 242

o

Ozkale, M. R. 264, 266
Ozturk, F. 280
Ozturk, O. 47, 266, 267

p

Peng, L. xxx, 3, 47, 73, 85, 236
Pollard, D. 234
Puri, M. L. 19, 316, 321

r

Ronz, B. 242
Roozbeh, M. 241, 265
Rousseeuw, P. J. 4

s

Safariyan, A. 122
Saleh, A. K. Md. E. xxviii, 31, 47, 48, 52, 64, 72, 110, 120, 122, 123, 137, 158, 165, 169, 241, 247, 252, 257–260, 263, 264, 267, 293, 296, 297, 307, 315, 323, 326, 411
Sen, P. K. xxviii, 19, 133, 297, 315, 316, 321, 323, 326
Shao, J. 264, 309, 318, 320
Shi, J. 243
Sidák, Z. 161, 162
Sing, J. M. 323
Speckman, P. 242
Stein, C. xxviii, 89, 264

t

Theil, H. 7
Tibshirani, R. xxviii, 31, 36, 48, 169, 229, 232, 253, 294
Tikhonov, A. N. xxviii
Turkmen, A. 47, 266, 267

v

Vapnik, V. N. 349

w

Wilcox, F. 4
Wilcox, R. 339
Willke, T. A. 264, 267
Wolfe, D. A. 7
Wu, J. 264

x

Xu, J. 264

y

Yang, H. 264, 265
Yuzbasi, B. 287, 288

z

Zhang, C. H. xxix, 283
Zou, H. xxix, 31, 38, 39, 48, 49, 73, 74, 82, 139, 266, 282–284

Subject Index

F_1 score 398
L_1 loss 16
L_2 loss 20, 50
z-score 10

a
Activation function 382
  relu, 382, 383, 385, 391
  sigmoid, 381–383, 385, 386, 391
Adam optimizer 386, 389, 419
Adaptive elastic net 84
Adaptive LASSO 38, 73, 173, 233, 282
Adjusted-R^2 statistic 81
aLASSO trace 74, 78, 86, 370
aLASSO-type R-estimator 138
ANOVA 149
  one-way, 151
  robust, 151, 152
Asymptotic 233, 249, 271, 316, 321
Asymptotic distributional
  bias, 102
  efficiency, 103, 299
  risk, 102
  variance, 102
Asymptotic linearity results 196, 314
Asymptotic quadratic approximation 196
Asymptotic uniform linearity 217, 293
AUROC 399, 404, 405
Auto-regressive 291

b
Back propagation 390, 393
Bias-variance trade-off 36, 65, 85, 355, 409
Binary classification 333, 409
Bootstrap method 60
Boxplot 61, 151
Breakdown point 4, 7, 19, 339

c
Cauchy-Schwartz 314
Collinearity 51
Computational complexity 15
Consistent 234, 309
Constrained estimator 52
Contiguity 321
Convergence plots 403, 411, 416
Courant theorem 220, 249
Cramér-Wold 293

d
Data sets
  Belgium telephone, 7
  cats and dogs, 406
  circular, 353, 400
  diabetes, 85, 280
  MNIST, 414
  Motor Trend cars, 344
  Swiss fertility, 56
  Titanic, 359
Data-adaptive 310
Deterministic design matrix 309
Diagonal ridge-type estimator 55

Rank-Based Methods for Shrinkage and Selection: With Application to Machine Learning.
First Edition. A. K. Md. Ehsanes Saleh, Mohammad Arashi, Resve A. Saleh, and Mina Norouzirad.

Dispersion function 19, 21, 32, 51
Dropout 380, 385, 386, 401

e
Effective coefficients 308
Efficiency 200
Eigenvalues 316
Elastic net 39, 82, 283
Elastic net-type R-estimator 127, 169, 205, 296
Enet trace 93
Epi-convergence 175, 236

f
F-statistic 152
Finite Fisher information 102
Forward propagation 390

g
Gaussian kernel 352
Generalized inverse 311
Generalized Liu estimator 264
Gradient descent 266, 337, 386
Gradient test 22, 23

h
Hard threshold R-estimator 109, 136, 157, 169, 227, 294
High-dimensional 265, 307, 309
High-leverage point 14
Hodges-Lehmann estimator 16

i
Identifiability 309
Ill-conditioned 51, 277
Inadmissibility 277
Index set 308
Invariant 21

j
Jaeckel's rank dispersion 158, 244

k
K-fold Cross-validation 68, 70, 90, 284
Kernel 241

l
LASSO 36, 71, 232, 282
LASSO trace 78, 155, 369
LASSO-type R-estimator 126, 180, 204, 212, 218
Linear kernel 352
Linear space 309
Linear Unified (Liu) 263
Lipschitz condition 244
Liu-type R-estimator 268, 277
Location model 16, 102
Logistic distribution 103
Lower bound 199, 208
LSE 6, 31, 50

m
MAD 3, 118, 156
MAE 89
Marginal theory 267
Matthews correlation coefficient 421
MCP 283
Mean 1
Median 2
MNET 284
Model complexity 33
Moderately large 308
Momentum 389
Moore-Penrose 311
MPE 285
MSE 65, 88
Multi-class classification 379, 414, 415, 420
Multicollinearity 31, 51, 52, 265
Multiple linear regression 30, 49, 215

n
Naive elastic Net 82
Naive elastic-net-type R-estimator 139, 231
Non-orthonormal 257
Nonparametric 10, 15, 149, 242, 243

o
OLS 50, 192, 263, 265
Optimum 280
Oracle property 73, 173, 235
Ordered Statistics 5
Orthogonal ridge-type estimator 54
Orthonormal 257
Orthonormal-LS estimator 54
Outlier 2–5, 7–9, 11–15, 60, 85, 344, 353, 359, 400, 416, 417, 419
Over-fitting 33, 85, 409

p

Parsimonious model 309
Partial-residuals 7, 9, 15, 21, 23
Partially linear model 241
Penalized Jeackel's rank dispersion 309
Penalty 154, 246, 281
Penalty estimator 52
Penalty function 32
Pixel unrolling 407
Polynomial kernel 352
Positive-rule Saleh-type R-estimator 110, 111, 126
Positive-rule shrinkage Liu-type R-estimator 269
Positive-rule Stein-Saleh-type R-estimator 158, 165, 182, 202, 211, 225, 229, 251, 296, 316
Post-selection 307, 310, 313, 315
Prediction errors 285
Preliminary test Liu-type R-estimator 269
Preliminary test R-estimator 118, 163, 169, 200, 209, 222, 226, 250, 300, 312
Preliminary test subset selector R estimator 109, 169, 232, 313
Projection 172, 178, 309

q

Q-Q plot 60, 61, 153
Quadratic loss 3, 20, 102
Quadratic risk 65

r

Rank dispersion function 17
Rank estimator 17
Rank Statistics 5
Rank-based methods 5, 20
RBF kernel 352, 354
Receiver operating characteristics 399
Regularization 32
Residuals 7
Restricted R-estimator 107, 134, 162, 250
Ridge definition 34, 53, 282
Ridge R-estimator 154, 296, 312
Ridge regression 34, 53, 158, 231
Ridge trace 55, 62, 70, 91, 369
Ridge-type R-estimator 117, 135, 180, 199, 208
Robust data science 329
Robust estimator 47
Robust machine learning 332
Robust regression 267
Robust Statistics 3
Robustness 4
RSE 9
RSS 6

s

Saleh-type R-estimator 110, 123, 137
SCAD 283
Scipy 332
Seemingly unrelated 191
Separation 312
Shrinkage 156
Shrinkage R-estimator 114, 135
Sigmoid function 335, 337, 371, 377, 379, 382
Signed-ranked statistics 104
Simple linear regression 6
  t-statistic, 10
  intercept, 6
  residuals, 12
  slope, 6, 19
Singular 51, 311
Skew-symmetric 193
SkLearn 332, 354
Slutsky's theorem 174, 175, 236
Small effect 308
Smoothing 243
Soft threshold 171, 229, 294
Sparse Riesz condition 320
Sparsity 154, 157, 169, 218, 221, 242, 285, 307, 411
Spectral decomposition 271
SSE 20, 32
Standard deviation 3
Standardization 90
Stein-Saleh-type R-estimator 164, 201, 210, 224, 251, 296, 300, 312
Stein-type shrinkage Liu-type R-estimator 269
Subset selection 33, 36, 71, 73
Subsets 198, 285, 308
Support vector classifier 349
Support vector machines 349

t

Test set 68
Theil's Estimator 7

Threshold 313
Time-series 291
Training set 68
Tuning parameter 34, 310

u
Ultra-high dimensional 316
Uncentered data 60
Uncertain prior information 118
Uncorrelated trajectory 64
Under-fitting 33, 85, 409
Unrestricted R-estimator 105, 158

v
VIF 31, 52, 281

w
Walsh average 17
Wilcoxon 293
Wilcoxon score function 24
Wilcoxon test 10

z
Zero-crossing 73
Zero-crossing order 74

Printed and bound by CPI Group (UK) Ltd, Croydon, CR0 4YY